Soda Lakes of East Africa

Michael Schagerl
Editor

Soda Lakes of East Africa

Springer

Editor
Michael Schagerl
Department of Limnology and Bio-Oceanography
University of Vienna
Vienna
Austria

ISBN 978-3-319-28620-4 ISBN 978-3-319-28622-8 (eBook)
DOI 10.1007/978-3-319-28622-8

Library of Congress Control Number: 2016949068

Printed on acid-free paper

This Springer imprint is published by Springer Nature
The registered company is Springer International Publishing AG Switzerland

Preface

It was around 16 years ago, when I got the opportunity to take my first small steps in Kenya. I remember so well the flurry of excitement for this "unknown" continent and my first visit to Baringo/Bogoria. After fortifying myself with the traditional foods Njama Joma and Samaki, I put on my sandals for night sampling because of the hot temperatures, but was fortunately stopped just in time by a local resident who carefully lifted up stones and showed me—the "mzungu" (Swahili word for white people)—some scorpions and dangerous centipedes. On the next day, we travelled to Bogoria, and an astounding new world opened itself up to me: heat, geysers, hundreds of thousands of flamingos, the characteristic pungent smell of the soda lake, the soapy water, and an incredibly vast amount of the cyanoprokaryote *Arthrospira*. This switched a button in me and I decided on the spot to focus on these astonishing systems. I am quite sure that I share this experience with the authors of this book.

It is a privilege to work in systems where tourists spend their holidays, and we scientists are often envied for our workplaces. Of course, unforgettable moments are inseparably linked to such sites, but we also face a wide range of challenges and hardships. Very basic accommodations, heat, wild and dangerous animals, bad roads with terrible traffic, and frequent power cuts impede the exhausting field work. Not to mention getting our equipment and instruments from our home universities to their deployment sites. These hurdles are part of the game, but they are far outweighed by the highly interesting systems and their organisms, by the incredible scenery, and by new friendships. I hope that the contributions of this book prompt the spark to leap over to you readers.

Without my colleagues, all of them renowned authorities in their fields, such a book would have never been possible. I am deeply indebted to them for their patience, for sharing knowledge and investing their valuable time in this book project, and for finding a way to deal with my many inquiries and special requests. I am greatly obliged to many reviewers, who helped further shape the contributions. They are (in alphabetical order): Janine Adams, Bertram Boehrer, Jens Boenigk, Margaret Brock, Paul Antony Chakkiath, Gudrun De Boeck, Pawan K Dadheech, Jenny A Day, Genoveva F. Esteban, Bland J. Finlay, Franz Jirsa, Thomas C Johnson, Elizabeth Kebede Westhead, Noha M. Mesbah, Paul Oberholster, Eric Odada, Aharon Oren,

Judit Padisák, Friedrich Schiemer, Manfred Strecker, Petra Sumasgutner, Dirk Verschuren Johann Waringer, and 4 anonymous reviewers. Ulrike Stricker-Komba and Stephan Klapp (both Springer publishing house) took very personal care of the book project, my daughter Hannah provided some beautiful graphs, Barbara Mähnert checked for typing errors, edited hundreds of references, and compiled the indices, and Michael Stachowitsch completed the "production chain" with professional copy-editing.

As common for scientific publications, the authors and reviewers provided their knowledge and time for free. The royalties of the editor are being fully donated for Africa projects to MSF Medecins Sans Frontieres (Ärzte ohne Grenzen), who daily risk their lives helping others. Although travels to remote areas are always linked to risks, our families allowed us to pursue our vocations. They enabled us to live our dreams and, therefore, *I dedicate this book to the families of all authors*. May this book, like all wonderful ecosystems, be much more than the sum of its parts.

December 2015 Michael Schagerl

Challenges. (**a**) corruption; (**b**) long discussions with ethnic groups for sampling permission; (**c**) electrical connections and power failures; (**d**) sometimes local food; (**e**, **f**) dangerous animals; (**g**) bad roads and quicksand; (**h**) very basic accommodations

Unforgettable moments. (**a**, **b**) stunning landscapes; (**c**, **d**, **e**) dipping into other cultures; (**f**, **g**) hospitality; (**h**) friendship; (**i**) Buna (Ethiopian coffee) in a street stall during a sampling trip and (**j**) Tusker baridi (cold beer) after a long and hard field trip

Contents

List of Contributors

Bock, Christina (1), Department of Biodiversity, University of Duisburg-Essen, Universitätsstraße 5, 45141 Essen, Germany, E-Mail: christina.bock@uni-due.de

Bell, Olivia (2), Department of Biology, University of Leicester, Leicester LE1 7RH, UK, E-Mail: olivia-b@hotmail.co.uk

Burian, Alfred (3), Department of Ecology, Environment and Plant Sciences, Stockholm University, 106 91 Stockholm, Sweden, E-Mail: flinserl@hotmail.com

Dadheech, Pawan K. (4), Department of Microbiology, Central University of Rajasthan, NH 8, Bandar Sindri, 305817 Ajmer, Rajasthan, India, E-Mail: pdadheech@curaj.ac.in

Deocampo, Dan M. (5), Department of Geosciences, Georgia State University, 33 Gilmer Street SE, Atlanta, Georgia, USA, E-Mail: deocampo@gsu.edu

Dindi, Edwin (6), Department of Geology, University of Nairobi, Chiromo Campus, Riverside Drive, PO Box 30197-00100, Nairobi, Kenya, Africa, E-Mail: eedindi@yahoo.com

Grant, William D. (7), Department of Infection, Immunity and Inflammation, University of Leicester, Leicester LE1 9HN, UK, E-Mail: wdg1@leicester.ac.uk

Harper, David M. (8), Emeritus Professor in Limnology, University of Leicester, LE1 7RH, UK; Director Aquatic Ecosystem Services Ltd (UK); Affiliated Researcher, Department of Zoology, National Museums of Kenya, Nairobi, Kenya, E-Mail: dmh@le.ac.uk

Jones, Brian E. (9), formerly: DuPont Industrial Biosciences/Genencor International BV, Leiden 2333CN, The Netherlands, E-Mail: bejones1948@gmail.com

Kavembe, Geraldine D. (10), Department of Biology, South Eastern Kenya University, PO Box 170-90200 Kitui, Kenya, Africa, E-Mail: dkavembe@seku.ac.ke

Kipkemboi, Julius (11), Department of Biological Sciences, Egerton University, PO Box 536-20115, Egerton, Kenya, Africa, E-Mail: jkipkemboi@egerton.ac.ke

Kotut, Kiplagat (12), Department of Biological Sciences, Embu University College, PO Box 6-60100, Embu, Kenya, Africa, E-Mail: kkotut@gmail.com

Krienitz, Lothar (13), Department of Experimental Limnology, Leibniz-Institute of Freshwater Ecology and Inland Fisheries, D-16775 Stechlin, Germany, E-Mail: krie@igb-berlin.de

Luo, Wei (14), SOA Key Laboratory for Polar Science, Polar Research Institute of China, Shanghai 200136, China, E-Mail: luowei@pric.org.cn

MacIntyre, Sally (15), Department of Ecology, Evolution, and Marine Biology, University of California at Santa Barbara, Santa Barbara, CA 93106, USA, E-Mail: sally@eri.ucsb.edu

Mähnert, Barbara (16), Department of Limnology and Bio-Oceanography, University of Vienna, Althanstrasse 14, A-1090 Vienna, Austria, E-Mail: barbara1990@gmx.at

Melack, John M. (17), Department of Ecology, Evolution and Marine Biology, University of California at Santa Barbara, Santa Barbara, CA 93106, USA

Bren School of Environmental Science and Management, University of California at Santa Barbara, Santa Barbara, CA 93106, USA, E-Mail: john.melack@lifesci.ucsb.edu

Mengistou, Seyoum (18), Department of Zoological Sciences, Fisheries and Aquatic Sciences Stream, College of Natural Sciences, Addis Ababa University, King George VI St, Addis Ababa 1000, Ethiopia, Africa, E-Mail: seyoumeng@gmail.com

Meyer, Axel (19), Department of Biology, University of Konstanz, Universitätsstrasse 10, 78457 Konstanz, Germany, E-Mail: axel.meyer@uni-konstanz.de

Oduor, Steve Omondi (20), Department of Biological Sciences, Egerton University, PO Box 536-20115, Egerton, Kenya, Africa, E-Mail: oduorsomondi@yahoo.co.uk

Olago, Daniel (21), Institute for Climate Change and Adaptation and Department of Geology, University of Nairobi, Chiromo Campus, Riverside Drive, PO Box 30197-00100, Nairobi, Kenya, Africa, E-Mail: dolago@uonbi.ac.ke

Owor, Michael (22), Department of Geology and Petroleum Studies, Makerere University, PO Box 7062, Kampala, Uganda, Africa, E-Mail: mmowor@yahoo.com

Renaut, Robin W. (23), Department of Geological Sciences, University of Saskatchewan, 114 Science Place, Saskatoon, SK S7N 5E2, Canada, E-Mail: robin.renaut@usask.ca

Robinson, Victoria J. (24), Department of Biology, University of Leicester, Leicester LE1 7RH, UK, E-Mail: vjr7@le.ac.uk

Schagerl, Michael (25), Department of Limnology and Bio-Oceanography, University of Vienna, Althanstrasse 14, A-1090 Vienna, Austria, E-Mail: michael.schagerl@univie.ac.at

Taylor, William D. (26), Department of Biology and Department of Earth and Environmental Sciences, University of Waterloo, Waterloo, Ontario, N2L 3G1, Canada, E-Mail: wdtaylor@uwaterloo.ca

Tebbs, Emma (27), Department of Geography, King's College London, Strand Campus, London WC2R 2LS, UK, E-Mail: emma.tebbs@gmail.com

WoldeGabriel, Giday (28), EES-14/MS D462, Earth Environmental Sciences Division, Los Alamos National Laboratory, Los Alamos, NM 87545, USA, E-Mail: wgiday@lanl.gov

Wood, Chris M. (29), Department of Zoology, University of British Columbia, Vancouver, BC, CanadaV6T 1Z4, E-Mail: woodcm@zoology.ubc.ca

Yasindi, Andrew W. (30), Department of Biological Science, Egerton University, PO Box 536-20115, Kenya, Africa, E-Mail: ayasindi@yahoo.com

1
2
3
4
5
6
7
8
9
10
11
12
13
14
15
16
17
18
19
20
21
22
23
24
25
26
27
28
29
30

Part I

Genesis, Physics and Chemistry

How are saline lakes formed and how can they be delimited from freshwater lakes? The following chapters provide general information about the special features of salt lakes. Although the term 'saline lake' suggests a high and narrow salinity range and uniform ion composition, salt lakes are diverse in both chemistry and biology. We provide classification systems of saline lakes with a special focus on saline alkaline or 'soda' lakes. Their genesis and chemistry are highly linked to tectonic forces and volcanic bedrocks. The East African soda lakes are young systems, some of them showing major fluctuations in water level. These so-called amplifier lakes act as sentinels of climate change. The topography and lake morphology determine the mixing regime and influence the turbidity, which in turn help determine water chemistry and set the conditions for both the fauna and flora. Soda lake chemistry is complex, with sodium and carbonate species being prevalent, making them interesting for soda-ash mining and other exploitations.

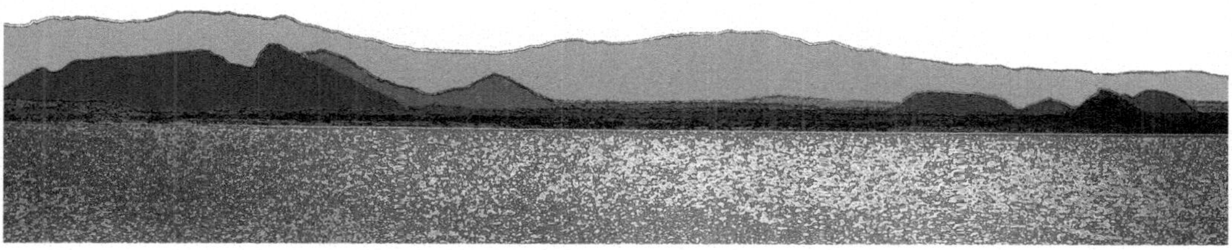

Dipping into the Soda Lakes of East Africa 1

Michael Schagerl and Robin W. Renaut

Abstract
Saline alkaline lakes are the focus of this book, but how do they form and how do they differ from freshwater lakes? We provide a brief introduction to soda lakes, their origin and their inhabitants, many of which are extremophiles. These closed lakes are highly sensitive ecosystems, and many experience the impacts of climate change. Some saline alkaline lakes provide economic resources, for both tourism ('flamingo lakes') and industry. A few lakes have long been commercially exploited but are ecologically damaged; others are threatened because of their potential mineral resources and changing land use in their drainage basins, including siltation and pollution from both agricultural and industrial sources. These unusual ecosystems must be protected (1) because they are 'hot spots' of haloalkaliphiles and rich genetic pools that might soon be tapped for biotechnological applications, and (2) for their exceptional natural environments, threatened increasingly by anthropogenic intervention. These environments must be preserved for future generations.

1.1 Genesis, Occurrence and Definition of Saline Lakes

Most inland saline lakes lie in semiarid regions in hydrologically closed, or almost closed, basins that lack lake outflow through rivers or significant water loss by subsurface groundwater discharge. In such endorheic basins, where annual water loss by evaporation and evapotranspiration commonly exceeds the annual inflow from precipitation, rivers and groundwater, the lake waters become saline (Fig. 1.1). Such closed lakes vary frequently in level, salinity and

M. Schagerl (✉)
Department of Limnology and Bio-Oceanography, University of Vienna, Althanstrasse 14, 1090 Vienna, Austria
e-mail: michael.schagerl@univie.ac.at

R.W. Renaut
Department of Geological Sciences, University of Saskatchewan, 114 Science Place, Saskatoon, SK S7N 5E2, Canada
e-mail: robin.renaut@usask.ca

M. Schagerl (ed.), *Soda Lakes of East Africa*, DOI 10.1007/978-3-319-28622-8_1

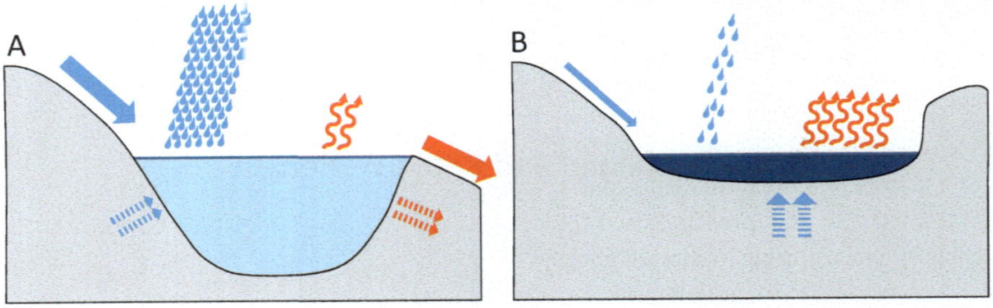

Fig. 1.1 Open (reservoir, flow-through) lakes versus closed (endorheic) lakes. (**a**) Open lakes receive water from rivers, underground springs and precipitation (*blue*). The water exits via seepage, evaporation and outflows (*red*). (**b**) Closed lakes receive water from rivers; some of these are enriched in salts due to the underground and springs (salty). Closed lakes also have less rain. The main way for water to exit is via evaporation—'distilled water' is evaporating; the salts remain trapped in the lake

hydrochemistry in response to climate changes over timescales ranging from years to decades to millennia and longer (Deocampo and Jones 2014; Eugster and Hardie 1978; Renaut and Gierlowski-Kordesch 2010; Rosen 1994). Tectonics, including fault movements and basin-floor subsidence, and related volcanic and geothermal activity also contribute to the evolution of many saline lakes, especially those in East Africa and the western interior regions of North America.

Saline lakes are common and are found worldwide. The estimated total volume of saline lakes (104×10^3 km^3) is comparable to that of freshwater lakes (124×10^3 km^3; Dodds and Whiles 2010). The salt content of oceans is nowadays expressed as 'practical salinity units' (PSU) (Fofonoff 1985); formerly, salinity also was used. The PSU parameter was introduced because accurate measuring devices for conductivity became available. PSU is a unit to determine the salt content from conductivity measurements. Salinity was originally 'conceived as a measure of the mass of dissolved salts in a given mass of solution' (APHA 1995, pp. 2–48), which implies that salinity is a unitless parameter. In limnology, salinity is usually given in parts per thousand (ppt) or ‰, and in older literature in g L^{-1} (gram of salt per litre of water). For better readability, we have retained ‰ in this volume. Early measurements until the nineteenth century were done gravimetrically, but this method has disadvantages (loss of volatile compounds, additional dissolved compounds other than salts). Later on, titrimetry and densitometry were applied; today, conductivity meters are common to determine salinity because they are robust, sensitive and easy to handle. Conductivity is a non-specific measure of the total ion content of solutions and is usually given in μS cm^{-1} for freshwaters and mS cm^{-1} for saline waters. The earlier units of conductivity were μmhos cm^{-1} or mmhos cm^{-1} and are found in older literature ('mho' is the reverse writing of the resistance unit 'ohm', because conductivity $=$ resistance^{-1}). A conversion factor based on field data to convert conductivity values to salinity for the East African saline alkaline lakes (EASL) is given in the Appendix.

How are freshwaters distinguished from saline waters? Of the many different classification systems, we follow that of Hammer (1986), who defined fresh lake waters as having a salinity < 0.5‰ and saline lakes as those with salinities ≥ 3.0‰ (Fig. 1.2). The range from 0.5 to 3.0‰, termed subsaline, is a transitional zone in which both freshwater organisms and halophiles can coexist. The term 'saline' may give the wrong impression of a narrow salinity range and ion composition comparable to that of sea water (the average global salinity of oceans is about 35.5‰), but it covers a wide range of salt concentrations and also various compositional types (Figs. 1.2 and 1.3; Eugster and Jones 1979; Smoot and Lowenstein 1991). Sodium (Na^+) is the dominant cation in most

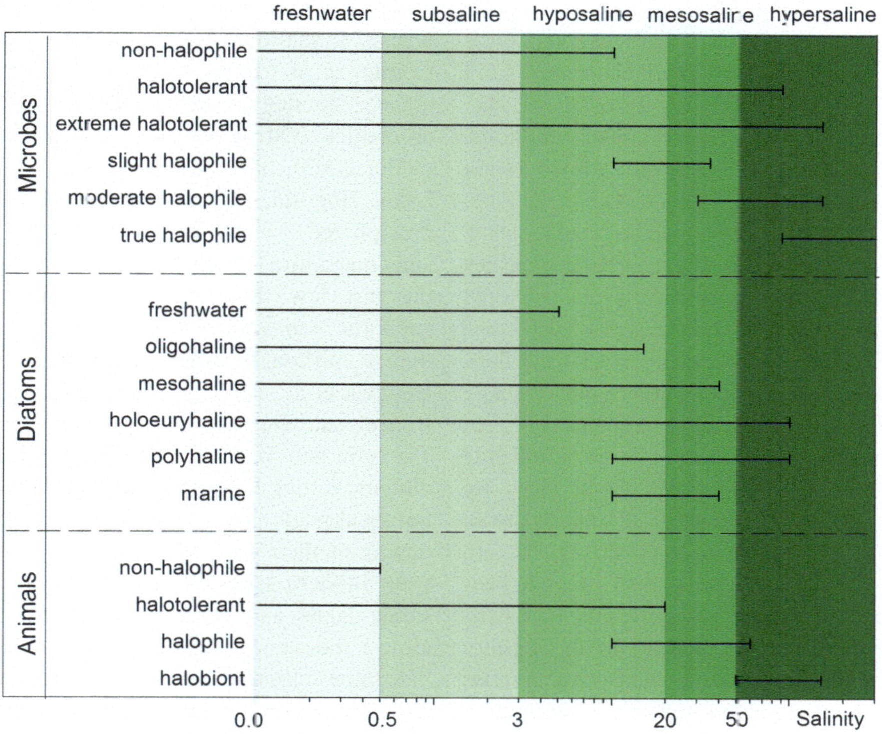

Fig. 1.2 Classification of lakes according to their salinity ranges (Hammer 1986) versus classification of organisms inhabiting saline lakes. Microbes—data combined from Banciu and Sorokin (2013), Madigan et al. (2011), Mesbah and Wiegel (2011), and Oren (2006). Diatoms according to Carpelan (1978). Animals according to Williams (1978). Conversion factor 1 M NaCl to salinity = 58.5‰

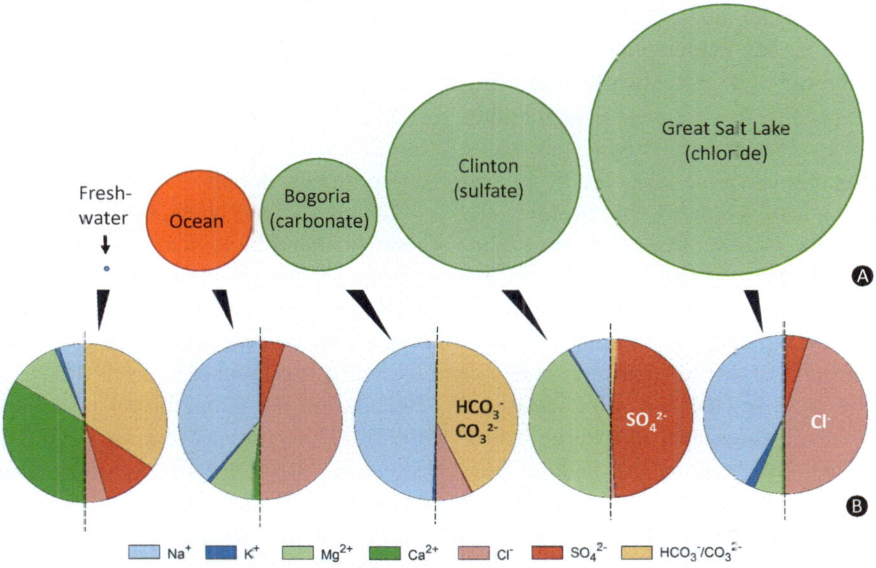

Fig. 1.3 Lake comparisons. (**a**) Salinity ranges of freshwater, oceans and saline lakes; the circle areas correspond to salinity levels on a relative basis. (**b**) Lakes classified according to their chemical composition, basis meq L^{-1}

saline lakes, but for the anion composition, three main types are distinguished: carbonate (HCO_3^-/ CO_3^{2-}), sulfate (SO_4^{2-}) and chloride (Cl^-) systems (Fig. 1.3b). The combination of Na^+ and carbonates results in alkaline conditions, and such systems are therefore termed 'soda lakes'. The focus of this book is on this lake type.

Saline alkaline lakes—'soda lakes'—have saline waters in which Na^+ and carbonate species (HCO_3^-, CO_3^{2-}) are the dominant ions and a pH that is typically > 9. Soda lake waters also commonly have high concentrations of Cl^- and fluoride (F^-) ions and aqueous silica, and variable concentrations of SO_4^{2-} and potassium (K^+), but they have very low concentrations of alkaline earths (Ca^{2+}, Mg^{2+}; Fig. 1.3b). Soda lakes are present worldwide (Garrett 1992; Grant 2004; Hammer 1986; Pecoraino et al. 2015), but are especially common in regions with volcanic bedrock, including the eastern branch of the East African Rift and several lake basins of the western interior of North America (e.g. Mono Lake, Searles Lake and Owens Lake in California, Goodenough Lake and Last Chance Lake in British Columbia). Soda lakes are also present in Europe, Australia, South America, China and elsewhere.

The EASL, which include many lakes in Ethiopia, Kenya and Tanzania (see Appendix), are the most spectacular soda lakes in the world, but most of them lack adequate protection (see Chap. 14). The EASL include playa lakes fed by ephemeral run-off and groundwater that occasionally evaporate to become dry mudflats encrusted with efflorescent trona and halite (Fig. 1.4a); perennial saline lakes fed by run-off and springs, some of which become stratified and meromictic (Figs. 1.4b, c); perennial saline lakes that rely on hydrothermal recharge to retain surface water in most years (Fig. 1.4d); sheltered crater lakes that typically lie in small drainage basins in dormant or extinct volcanic cones or calderas (Fig. 1.4e, g, h); and saline pans fed mainly by hot springs that discharge saline alkaline waters either along permeable lake-marginal faults or directly from faults or fractures along their lake floors (Fig. 1.4f).

In contrast, some lakes in the same geographic regions have fresh or only mildly saline and alkaline waters. For example, Lake Baringo in Kenya, a shallow dilute lake with freshwaters (Tarits et al. 2006), supports Cyanobacteria, algae, sponges, fish, hippopotamus and crocodiles (Odada et al. 2006; Schagerl and Oduor 2003), but it lies only ~25 km north of Lake Bogoria, a saline alkaline lake that precipitated sodium carbonate salts and magadiite [$NaSi_7O_{13}(OH)_3{\cdot}4H_2O$] during the last few thousand years, and frequently supports *Arthrospira* blooms that feed more than a million flamingos (Harper et al. 2003; Kaggwa et al. 2013; Oduor and Schagerl 2007; Renaut and Tiercelin 1993; Tebbs et al. 2013; Tiercelin and Vincens 1987; Figs. 1.5a, c). The climatic setting for both lakes is almost identical, but local geological and hydrological conditions determine their very different ecology. Evidence from modern lakes in East Africa shows that saline lakes and fresh lakes often coexist in almost the same place at the same time under the same climatic regime. In the geological record, therefore, the appearance of sedimentary, mineralogical and paleontological indicators of soda lakes need not always imply an increase in regional aridity.

Paleaolimnologial records obtained from sediment cores and local rock outcrops from many modern EASL confirm that most had deeper fresher precursors (Figs. 1.6b, e, f), especially during the end Pleistocene–early Holocene 'African Humid Period' (Bard 2013; de Menocal et al. 2000). In contrast, some dilute lakes in the East African Rift had more saline precursors during the Quaternary.

Saline alkaline lakes are dynamic from both geological and biological perspectives. This was clearly shown between 2011 and 2015, when many soda lakes in Kenya attained their highest levels in more than 50 years (Figs. 1.7b, d), some exceeding historical high levels in 1961 (Conway 2002; Lamb 1966) and the early 1900s (Johnston 1902). This timing implies a possible, but untested, 50–60-year cycle of unusually wet intervals in East Africa. For example, hippos and crocodiles temporarily moved into Lake Bogoria in 2013–2014 (W. Kimosop, personal communication), and the several hundred

Fig. 1.4 The different types of soda lakes in East Africa. (**a**) Southern end of Lake Elmentaita. Kenya, July 2008. Volcanic ash cones in distance. Lake Elmentaita is a 'playa lake' that periodically evaporates to leave a thin (<2 cm) crust of efflorescent sodium-carbonate salts (trona, thermonatrite and minor halite) upon lake clays (mainly smectite) that also contain detrital feldspar and calcite. Lake Elmentaita is partly fed by warm springs. (**b**) Lake Bogoria, Kenya. Photograph taken in July 2013 along the SE shoreline, with the Bogoria Escarpment (a major fault-scarp) to the east. This saline alkaline lake is perennial with contributions from runoff, ~200 shoreline hot springs and sublacustrine hot-spring discharge. The lake is normally stratified, but may not be fully meromictic (DeCort et al. 2013). (**c**) Alkaline hot springs discharging into perennial Lake Shala in the Main Ethiopian Rift (January 2014). Lake Shala occupies a volcanic crater up to 260 m deep. (**d**) Southern end of Nasikie Engida, Kenya, during July 2011. This alkaline lake, fed mainly by hot springs near its northern shoreline, precipitates trona and nahcolite at its southern end, but fish (*Alcolapia*) inhabit the northern lake water. (**e**) Lesser Flamingos in 'Lake Magadi' in the steep-sided volcanic Ngorongoro caldera, northern Tanzania. (**f**) The saline pan of Lake Magadi, southern Kenya Rift, July 2009. The trona salt crust, precipitated from hypersaline alkaline brines, is broken by metre-scale polygons linked to lateral pressures from salt crystallization. Saturated sodium-carbonate brines lie a few decimetres below the salt crust surface. (**g**) The small crater lake Chitu is one of the few spots, where flocks of flamingos can be viewed in Ethiopia. (**h**) Lake Arenguade (Ethiopia) on the rim of the Ethiopian Rift Valley experiences substancial changes in salinity

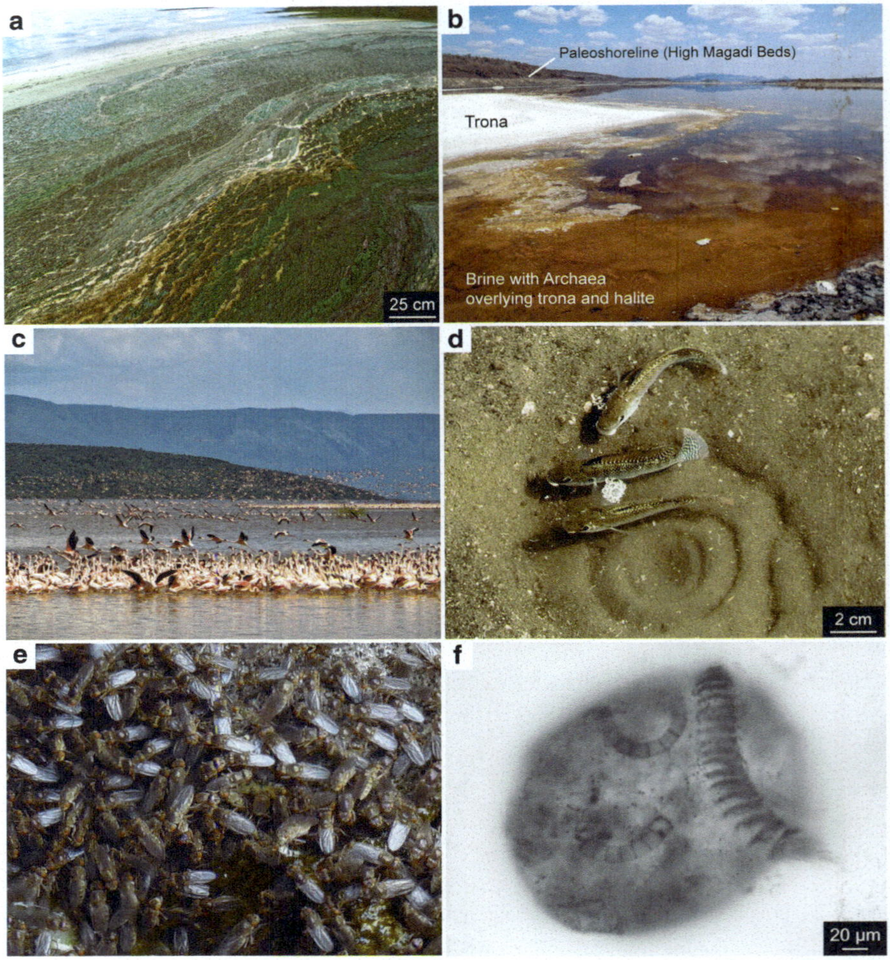

Fig. 1.5 The biology of East African soda lakes. (**a**) *Arthrospira fusiformis* forming littoral scums along the southern shoreline of Lake Bogoria (northeastern Ng'wasis), Kenya, July 2009. In places the green surficial scums of Cyanobacteria are more than 5 cm thick. (**b**) Archaea in the East Lagoon of Lake Magadi, Kenya, July 2009. Photograph also shows recent trona crusts and a paleoshoreline (+13 m = ~613 m elevation) of the late Pleistocene–early Holocene High Magadi Beds. The High Magadi paleoshoreline there is thinly coated with calcite stromatolites. During that period, fish lived in the higher, fresher paleolake. (**c**) Lesser Flamingos (*Phoenicopterus minor*) along the northern shoreline of Lake Bogoria in July 2013 after the lake had been diluted by heavy rains and lake level had risen by more than a metre since 2011; the enhanced microbial productivity was sufficient to support several hundred thousand flamingos. (**d**) *Alcolapia* sp. with a concentric fish nest in feldspathic sands and silts, East Lagoon, Lake Magadi, July 2009. (**e**) Mass occurrence of Ephydridae at Lake Bogoria. (**f**) The large ciliate *Linostoma* (Heterotrichida) ingested some *Arthrospira* filaments

thousand flamingos present in July 2014 (Fig. 1.5c) had left by July 2015, when only a few thousand were present. Some EASL also dry out from time to time; examples include Nakuru and Amboseli in Kenya and Eyasi in Tanzania (Figs. 1.7a, c). This can sometimes be inferred from their local names: Nakuru, for example, is derived from the Maasai word 'Nakurro', which means 'a dusty place' (S. Oduor, personal communication).

Many saline lakes respond to minor changes in climate and have large fluctuations in their levels. Some have been termed 'amplifier lakes' (Street 1980). Examples of amplifier lakes in the East African Rift include Lake Nakuru in Kenya and Lake Abbe in Ethiopia (Trauth et al. 2010).

Fig. 1.6 Witnesses of precursor lakes. (**a**) Early Holocene diatomaceous silts (Galana Boi Formation) near Koobi Fora, Lake Turkana. (**b**) The High Magadi Beds (late Pleistocene–early Holocene) in the dry lagoon, 1 km south of Lake Magadi. Green lake muds with fish bones are overlain by waterlain volcanic ash. (**c**) Holocene calcite stromatolites coating bedrock at Ng'wasis, Lake Bogoria, that record a mid-Holocene phase of fresh, Ca-bearing lake water. (**d**) Former late Pleistocene high-level (985 m) shoreline (arrow) of Lake Baringo composed of sand and mollusc shells (inset photograph). Paleoshoreline is located near Logumukum, southeast of the lake. (**e**) Diatomite is mined at Kariandusi near Gilgil in the Elmentaita area (Kenya). (**f**) Microphoto of diatomite showing freshwater species

Similar to glaciers, such lakes can sometimes be viewed as climate sentinels.

1.2 Geological Setting of the Soda Lakes in East Africa

Most EASL lie in the eastern arm of the two branches that form the East African Rift (Fig. 1.8). That arm of the Afro-Arabian Rift System has its northeastern focus in the Afar Depression, a plate triple junction in Ethiopia where three separate rifts diverge (Morley 1999; Wolfenden et al. 2004). The first branch extends northwards to form the Red Sea, where new oceanic crust is forming by seafloor spreading along its axis. A second branch, also with active extension, forms the Gulf of Aden Ridge. The third rift branch, the East African Rift system, extends southwards more than 4000 km

Fig. 1.7 Lake level changes. (**a**) Dried Lake Amboseli (Kenya; 2004). (**b**) Increased lake levels of Nakuru flooding buildings at the entrance of the Nakuru National Park (2014). (**c**) Dried Lake Eyasi (Tanzania; 2010) and (**d**) flooding of Lake Elmentaita (2010)

from Afar towards Mozambique and Botswana but has two separate limbs—an eastern limb in Djibouti, Eritrea, Ethiopia, Kenya and Tanzania, where it splays outwards, and a western limb that extends from Uganda into western Tanzania, Rwanda, Burundi, eastern Zambia, Malawi, Mozambique and Botswana (Fig. 1.8).

At the largest scale, the East African Rift is associated with two large (~1000 km diameter) lithospheric domes or elevated plateaus, linked to underlying mantle plumes, termed the Afar Plateau (mean elevation, ~1500 m) and the East African Plateau (mean elevation, ~1200 m). The East African Plateau includes smaller domal uplifts termed the Kivu Dome and Kenya Dome, respectively (Morley et al. 1999). The orientation and geological evolution of both rift branches have been strongly influenced by rigid cold Archaean cratonic basement rocks, particularly the Tanzanian and Zambian Cratons, and tectonic lineaments of several ages in older rocks. Both rift branches have developed in the younger, less rigid, orogenic belts that surround the ancient Tanzanian Craton (Daly et al. 1989).

On a smaller scale, the overall architecture of the East African Rift that contains the modern soda lakes is one of a linear series of faulted depressions, each typically 40–70 km wide. Most fault troughs are half-grabens that have a high-angle boundary (border) fault with a large (km-scale) displacement on one side and an opposing faulted flexural margin without a single dominant fault scarp. The orientations of these half-grabens (i.e. which side has the border fault) commonly alternate, *en echelon*, along the predominant N–S rift axes that veer eastwards in Ethiopia. Full grabens, where two opposing faults have similar vertical displacements and morphology, are rare. The axial rift troughs are compartmentalised into many sectors, 60–100 km long, some separated by transverse structural highs termed transfer faults or accommodation zones (Morley et al. 1990; Rosendahl 1987). In Kenya and Tanzania, many of these transverse structures have NW–SE orientations

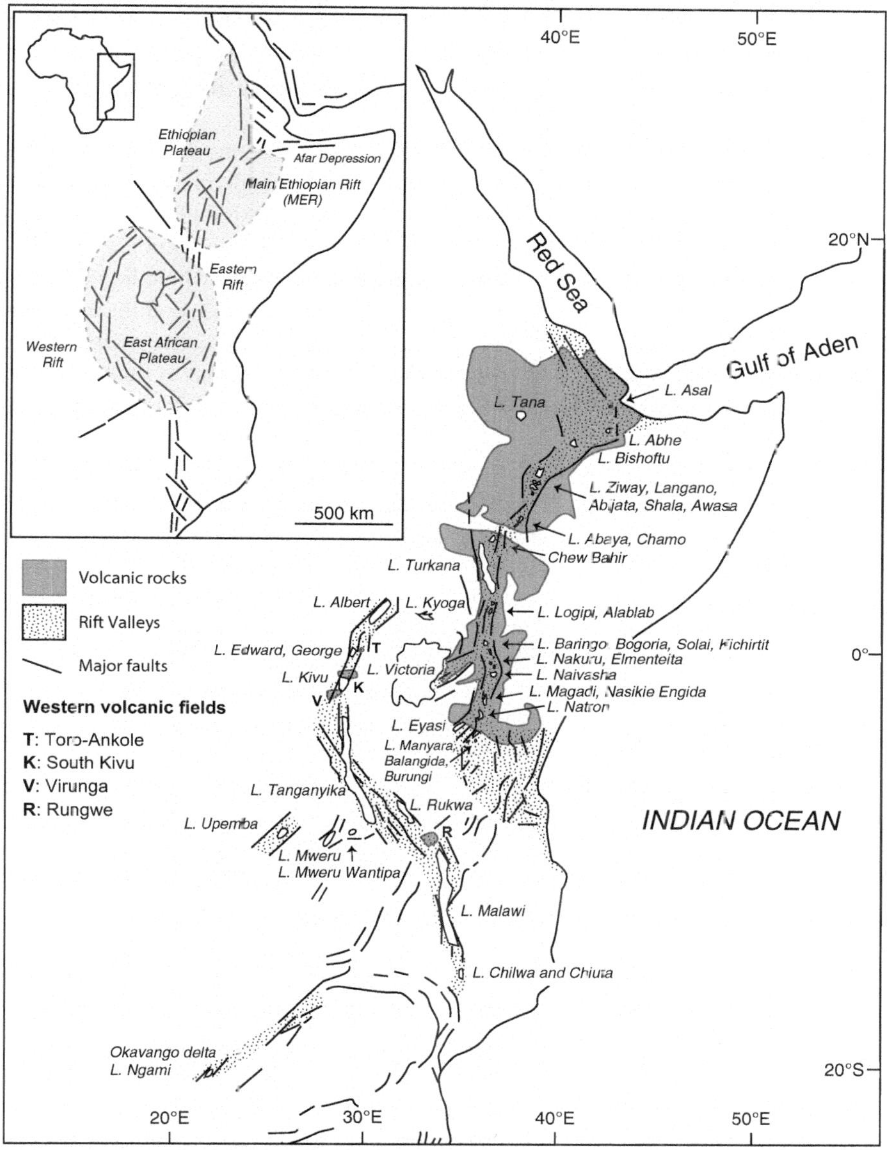

Fig. 1.8 Geological setting of the East African Rift, adapted from Tiercelin and Lezzar (2002)

that reflect the inherited influence of older, deep tectonic lineaments in the shallow crust (e.g. at Lake Baringo). The overall extension of the divergent East African Rift and morphological details of many lake basins have been controlled by these older tectonic structures and lineaments (Chorowicz 2005; Le Turdu et al. 1995; Tiercelin and Lezzar 2002).

The eastern and western branches of the rift are very different in character (Ring 2014), which is reflected in contrasting types of lakes. Volcanic rocks of Oligocene to Recent age, some extending beyond the modern rift margins, shallow lake basins (mainly soda lakes) and a semiarid climate characterise the eastern branch. Local evidence for older Palaeogene rift activity and volcanism is present in northern Kenya and Ethiopia (Hendrie et al. 1994; McGregor 2015). Many Quaternary to recent central volcanoes lie along the floor of the eastern branch of the rift (Abbate et al. 2015; Dawson 2008; Dunkley et al. 1993; MacDonald 2003). These volcanic centres, with dykes intruded along the rift axes, provide heat for most of the high-temperature geothermal systems along the eastern branch of the East African Rift. Volcanoes and lava flows, many of which have been faulted, also provide topographic closure for many rift lakes bordered by deep-seated marginal faults.

The western branch of the rift system, though seismically active, has experienced Neogene volcanism in only four regions: Rungwe, southern Kivu, Virunga and Toro-Ankole. The bedrock in the western branch consists mainly of Precambrian basement rocks composed of Archaean cratonic nuclei welded by Paleoproterozoic to early Palaeozoic orogenic belts. Large, very deep, fresh lakes (e.g. Lake Tanganyika > 1400 m and Lake Malawi > 700 m) occupy much of the western branch, reflecting the higher rainfall and the lack of infilling by volcanic rocks and sediments, both during and after rifting (Yuretich 1982).

Older Permo-Triassic (Karoo age), Jurassic–Cretaceous and Palaeogene rift systems (paleorifts) have also influenced the location and orientation of the Tertiary East African Rift and locations of accommodation zones. Some Mesozoic rift structures were reactivated during Neogene rifting. The location of several of the Quaternary volcanic centres and modern hydrothermal activity have been strongly influenced by the intersection of the generally N–S Tertiary rifts (except Ethiopia) with some of these paleorifts and older, deeply rooted Precambrian basement structures (including major NW–SE shear zones) (Chorowicz 2005). The locations and development of many modern soda lakes in East Africa are related to long, sometimes complex, geological histories involving both tectonics and volcanism.

1.3 Importance of Volcanism, Tectonics and Geothermal Activity in East Africa

The faulted volcanic terrains that surround most EASL play a vital role in their hydrochemistry and evolution. Although saline alkaline lakes exist in nonvolcanic terrains, most EASL and many soda lakes elsewhere lie in drainage basins where volcanic rocks dominate the local landscape. Others are present where volcanic rocks underlie the lake basins at shallow depth (e.g. the Cariboo Plateau in British Columbia: Cummings 1940; Renaut 1990).

Neogene and Quaternary basalt, trachyte, phonolite and related volcanic rocks and their tephra (ash, tuffs) are widespread along most of the eastern branch of the East African Rift (Fig. 1.9a, b). Chemical weathering of these volcanic rocks provides most of the solutes (ions and dissolved species) that flow into closed, endorheic lake basins on the Rift Valley floors. The dominant weathering process is silicate hydrolysis, during which feldspars (K-feldspar and plagioclase), volcanic glass, pyroxenes, amphiboles and other silicate minerals react with water charged with CO_2 as carbonic acid (H_2CO_3). Rains falling across the region are neutral to mildly acidic, but much of the acidity that chemically corrodes the silicate minerals in the bedrock originates in soils where organic acids released from organic matter decay intensify chemical weathering processes (Deocampo and Jones 2014; Chap. 4; Eugster and Hardie 1978).

Silicate hydrolysis reactions commonly produce new minerals on site, especially clay minerals or zeolites, with compositions that vary with weathering intensity. Iron-bearing minerals in parent rocks are oxidised, producing ferric oxyhydroxides that give soils in many lake catchments a reddish colour, especially in the higher wetter parts of many drainage basins. Ions produced during silicate weathering, including Na^+, K^+, Ca^{2+} and Mg^{2+} and SiO_2 (as monosilicic acid: H_4SiO_4), dissolve in dilute surface waters (rivers and unchannelled flow),

Fig. 1.9 Impacts of volcanism on soda lake development. (**a**) Faulted trachyte flows (Pleistocene Magadi trachyte) at Nasikie Engida, a soda lake north of Lake Magadi, Kenya. (**b**) Bedded volcanic tephra, Hell's Gate National Park, Kenya. (**c**) Lake Logipi, northern Suguta Valley, Kenya, ponded against the Barrier volcanic complex. (**d**) Geyser eruption, southern Loburu, Lake Bogoria, Kenya, in July 2006 when lake level was relatively low

and some percolate downwards to the water table and groundwater through permeable faulted rocks and sediments. The carbonic acid (H_2CO_3) that is consumed during silicate weathering results in more alkaline solutions because bicarbonate ions (HCO_3^-) are released. The Cl^- ions in the soda lakes originate mainly from rainfall and dissolved older salts, and much of the SO_4^{2-} derives from oxidation of sulfide minerals (e.g. pyrite) in the parent rocks or derives from volcanogenic sources. Local geology commonly controls the anion compositions of the soda lakes.

Solutes acquired from rainfall and chemical weathering move lakewards as run-off and groundwater, but many chemical and biochemical processes can modify the composition of the waters that flow into the lakes. These include the loss of most calcium and magnesium, which precipitate as minerals (calcite, aragonite, dolomite) in the drainage basin or in the lakes themselves when they are relatively dilute, by several chemical and biochemical processes (Deocampo and Jones 2014; Chap. 4; Eugster and Hardie 1978). The inflow waters, depleted in Ca^{2+} and Mg^{2+}, then progressively evaporate to produce saline alkaline lake waters rich in Na^+ and bicarbonate– carbonate (HCO_3^-, CO_3^{2-}), commonly with high Cl^- concentrations, while SO_4^{2-} is often lost by microbial reduction. The salinity of those lake waters depends on the local environmental conditions (e.g. annual recharge vs. loss by evaporation and evapotranspiration) that can change through time; some lakes remain mildly to moderately saline but alkaline; other lakes evaporate and become hypersaline and strongly alkaline with a pH >10. A few hypersaline lakes, most of which are groundwater fed and lie in regional sumps, precipitate highly soluble sodium carbonate salts such as trona [$Na_3(CO_3)(HCO_3){\cdot}2(H_2O)$], nahcolite ($NaHCO_3$) and halite ($NaCl$) (Fig. 1.4f).

Another role played by volcanism is physical. Central volcanoes and lava flows, some erupted from axial fissures, commonly provide barriers that contain lakes bounded by rift faults (Fig. 1.10). The Barrier volcanic complex in northern Kenya (Fig. 1.9c), for example, ponds both Lake Turkana to the north and Lake Logipi to the south (Dunkley et al. 1993). Lakes Natron and Magadi are separated by extinct central volcanoes, but both lakes were united several times during the Pleistocene when the climate was wetter (Hillaire-Marcel and Casanova 1987). Volcanic ponding of lakes, however, is not always effective—Lake Baringo and Lake Naivasha each occupy faulted basins that are topographically closed by central volcanoes which form physical barriers, but they are not hydrologically closed and maintain fresh to mildly saline waters. Both lakes have long been thought to leak through underlying permeable sediments and faulted bedrock, flowing northwards and southwards, respectively (Becht et al. 2006). Saline alkaline lakes also commonly lie in volcanic craters and calderas in Ethiopia (e.g. lakes Chitu and Arenguade, Figs. 1.4g, h), Kenya and Tanzania, including seasonal Lake Magadi (Tanzania) in Ngorongoro Crater (Fig. 1.4e, not to be confused with Lake Magadi in Kenya), fed by springs, groundwater discharge and local run-off. Lake Katwe, a crater lake in Uganda in the western rift branch, is also saline alkaline and precipitates sodium carbonate evaporites (Kasedde et al. 2014).

The third role of volcanism is its contribution to geothermal activity along the rift systems. Many EASL are fed partly by geothermal fluids, discharged as hot springs or, rarely, as geysers (Fig. 1.9d). In some lakes, such as Nasikie Engida and Lake Magadi in southern Kenya, hot spring fluids provide most of the annual recharge; in other lake basins, thermal springs are minor but sometimes important contributors. In Lake Beseka, Ethiopia, a recent rise in lake level has been attributed to increased thermal discharge (Goerner et al. 2009, but see also below). The very hot (>70 °C) springs have fluids of mainly Na–HCO_3 composition with variable Cl^-, but some warm springs (<40 °C) have Na–Ca–HCO_3 waters. Most of these spring fluids originate from surface waters that have permeated to different depths along faults and fractures (Dunkley et al. 1993; Renaut et al. 2013). Those that have penetrated to great depths tend to be hotter and richer in Cl^- and SiO_2. Those with shallower circulation and shorter residence time are warm (<45 °C) and may retain some alkaline earths. The proportion of thermal fluids flowing into some soda lakes varies with climatic change, often being highest during periods of drought when surface run-off is reduced. Thermal contributions to soda lake recharge, however, are not independent of climate change.

1.4 Precursor Lakes

Most of the modern EASL have had more dilute precursors (Figs. 1.6 and 1.10). During the late Pleistocene and early-to-mid-Holocene, the climate was much wetter across East Africa (de Menocal et al. 2000). Many lakes that today are saline and lie in hydrologically closed basins were then open lakes with much fresher waters that supported different ecosystems (e.g. Dühnforth et al. 2006; Junginger and Trauth 2013; Olago et al. 2009). About 10,000 years ago, Lake Magadi, the most hypersaline of the larger EASL, was an open lake with fish (*Tilapia* sp.) and freshwater diatoms and was often united with an expanded Lake Natron in northern Tanzania (Eugster 1980; Fig. 1.6b). Lake Nakuru and Lake Elmentaita in Kenya were then joined as a single lake with water levels ~ 180 m higher than today (Olaka et al. 2010). The Ethiopian lakes too were often interlinked (Benvenuti et al. 2002; Foerster et al. 2012; Gasse and Street 1978; Grove et al. 1975). The Ethiopian Ziway–Shala area was covered by a large lake ~9000 years ago, and Lake Turkana and paleo-lake Suguta probably had outlets (Garcin et al. 2009; Junginger and Trauth 2013).

This geologically recent history is important because of its implications for the biology and

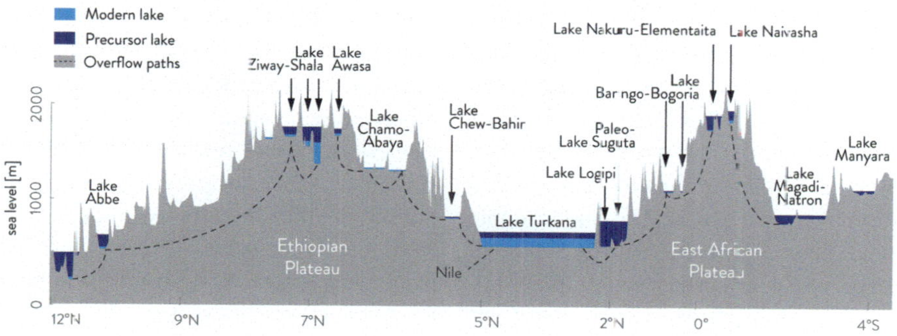

Fig. 1.10 North–south cross-section of the eastern branch of the East African rift system, with both modern and precursor lakes in the Holocene (by courtesy of Junginger and Trauth 2013, with modifications)

ecology of the modern soda lakes. Many of the EASL have not been isolated closed systems for more than ten thousand years. Each lake basin has had a unique geological and palaeoclimatic history; local tectonic, geomorphological and hydrological controls have occurred under an overriding climatic control. This should always be considered when evaluating differences between individual lake basins and their biology. This pattern is not a recent trend. Ephemeral soda lakes were present more than 15 million years ago as the East African Rift began to develop. Those earliest soda lakes also responded to local and regional variations (Renaut et al. 1999).

1.5 Biology of Soda Lakes

Soda lakes are remarkable for their biology and ecology. Lesser Flamingos (*Phoeniconaias minor* Geoffroy Saint-Hilaire; Fig. 1.5c) and many other organisms including cichlid fish (Fig. 1.5d), brine flies (*Ephydra* spp.; Fig. 1.5e) and large ciliates (Fig. 1.5f) depend on these unusual environmental conditions. Although often considered 'extreme' aquatic environments because of their exceptionally high alkalinity and salinity, soda lakes are amongst the most productive ecosystems on earth (Oduor and Schagerl 2007), with abundant microbial life dominated by Prokaryotes such as *Arthrospira fusiformis* (Voronikhin) Komárek and Lund (Fig. 1.5a) and Archaea (Fig. 1.5b). Some groups have productivity that ranks amongst the highest known. The high pH and salinity of soda lakes exclude many organisms, but those adapted to, or tolerant of, such conditions thrive in an otherwise hostile environment.

Organisms inhabiting extreme environments are commonly termed extremophiles and, depending on their specific environmental conditions, classification systems have been developed for organismic groups (Fig. 1.2). These schemes do not always match because different requirements are needed according to the taxonomic groups. Two criteria are met for organisms thriving in soda lakes that make them polyextremophiles: because of the alkaline and saline environment, such organisms are referred to as haloalkaliphiles. Some haloalkaliphiles that depend on high sodium concentrations to maintain pH homoeostasis are termed natronophiles. The cyanoprokaryote *Arthrospira fusiformis* is a classic example (Schlesinger et al. 1996). Both stenohaline (restricted to a narrow salinity range) and euryhaline taxa live in the EASL and cover a wide salinity range, the latter often occurring at higher salinity ranges (see Chap. 12).

Organisms thriving in such biotopes have special adaptations at the molecular level. For osmoregulation, synthesis of osmoprotectants (compatible solutes, osmolytes) is often observed. These are of low molecular weight and accumulate in high concentrations without impeding metabolism. Common substances are glycine betaine, proline and glutamine (Erdmann and Hagemann 2004; Roberts 2005). A fascinating example of morphological adaptation is the Soda Tilapia *Alcolapia grahami* Boulenger. This fish survives extremely high salinities, which implies constant drinking to maintain homoeostasis. The high amounts of ingested alkaline water, however, create two major challenges, which are a full stomach and neutralisation of stomach acid. The Soda Tilapia circumvented these problems with a so-called pyloric bypass of the gut (Bergman et al. 2003). This morphological diversion ensures that water passes around the stomach (for other examples, see Chap. 12).

Microbial diversity is exceptionally high in subsaline to mesosaline systems (see Chap. 5, Lanzén et al. 2013). Saline alkaline and well-buffered systems provide ideal conditions for the haloalkaliphilic microbial community (Sorokin et al. 2014); very high concentrations of dissolved organic carbon (Jirsa et al. 2013) also boost microbial life. In contrast, the diversity of Eukaryotes in EASL has long been assumed to be extremely low compared to freshwaters and oceans (Vareschi 1987). A recent study, however, has provided an entirely different picture with an unexpectedly high hidden diversity (Luo et al. 2013).

The EASL and adjacent areas attract many species, some of which are only indirectly linked to the ecosystems such as scavengers and insectivores. These lakes are home to an avifauna that is rich both in individual numbers and biodiversity (Harper et al. 2003). Some waterbodies, informally known as 'flamingo lakes', attract hundreds of thousands of tourists annually from all over the world to gaze at the pink spectacle and in turn generate income for the region. In the EASL, two flamingo species are known, which are the Greater Flamingo (*Phoenicopterus roseus* Pallas) and the Lesser Flamingo, the latter taxon accounting for around 90 % of the large flamingo flocks and a flagship species of EASL. The largest population of Lesser Flamingos worldwide inhabits the chain of EASL; three smaller populations exist in southern and West Africa and India (Childress et al. 2008). A population exchange between the EASL and southern Africa was recently confirmed (Zaccara et al. 2011), but it remains unclear how such long distances are achieved. Flamingos move from one lake to another. This may be related to fluctuating food availability (Kaggwa et al. 2013), but also to other causes yet to be studied (Brown 1975). The two flamingo species have different feeding strategies. The Lesser Flamingo is a primary consumer that prefers to ingest microphytes from the lake water surface; the Greater Flamingo is a secondary consumer and feeds mainly on small invertebrates from the lake sediments (see Chap. 10).

The temperature variations of EASL, which lie in a tropical region, are much less pronounced than those of temperate regions. This affects the organisms that inhabit the lakes. A typical succession of communities throughout the year, comparable to those of temperate systems (Sommer et al. 2012), is not discernable. Community shifts are erratic and still not fully understood. Two sets of controlling factors can be distinguished. Allogenic factors, which are mainly climate driven, provide an overall framework. These are linked to changing lake water levels that, in turn, are typically associated with changes in lake hydrochemistry through either dilution (high lake levels) or concentration (low lake levels and evaporation) of solutes. Within this context, autogenic factors such as competition, viral infection and grazing further shape the communities. During periods of low salinity, freshwater taxa are frequently observed. Under mesosaline conditions, *Arthrospira fusiformis* often dominates and is responsible for the greenish, turbid appearance of the lakes. These blooms, however, are sometimes replaced by benthic diatoms or nanoplankton. At high salinity, *Picocystis salinarum* Lewin and other picoplankton become common (Schagerl and Oduor 2008; Schagerl et al. 2015; for further details see Chap. 12).

How are the organisms and trophic levels of the food web linked to each other? Vareschi (1987) and other scientists suggested that saline lakes are ideal systems for modelling because of their low degree of complexity. About 30 years ago, Vareschi and Jacobs (1985) presented energy flows between food web compartments of Lake Nakuru, Kenya. This model did not consider heterotrophic Bacteria, Archaea and viruses, simply because the major role of these groups was not yet understood. At about the same time, the significance of heterotrophic Prokaryotes for carbon cycling in the marine environments was recognised by Azam et al. (1983), who introduced the term 'microbial loop' for recirculation of dissolved organic carbon to higher trophic levels of the food web via its incorporation into Bacteria. A few years later, the importance of viruses in the oceans was discovered by Fuhrman (1999) and Wilhelm and Suttle (1999). This so-called 'viral shunt' is a viral-mediated transfer of substances from organisms to dissolved and nonliving particulate organic matter. Although we must admit that we know little about the importance of Bacteria and Archaea for energy flows in the EASL, we do know that the microbial community is highly involved in element cycling in saline lakes (Sorokin et al. 2014). Moreover, virus-like particles and Bacteria numbers of EASL are amongst the highest worldwide (Gruber-Dorninger 2015), and there is strong evidence that cyanophage infections of *Arthrospira fusiformis* are a major cause for erratic breakdowns of *Arthrospira* blooms and that these trigger flamingo movements (Peduzzi et al. 2014). If we consider the hidden diversity of these ecosystems, the food webs of soda lakes are clearly much more complex than assumed 30 years ago.

1.6 Soda Lakes as Economic Resources and Resulting Threats to the Lakes

Evaporation of sodium carbonate brines in East Africa leads to crystallisation of salts in hypersaline lakes. The dominant salt is trona, but nahcolite ($NaHCO_3$) forms in waters with high PCO_2 (e.g. Nasikie Engida, Kenya), and thermonatrite ($Na_2CO_3{\cdot}H_2O$) crystallises when temperatures are unusually high (Eugster 1980). Geothermal CO_2 may also contribute to trona and nahcolite crystallisation (Earman et al. 2005; Renaut and Tiercelin 1994) and might explain why most economic soda deposits are present in volcanic regions. Trona has been mined at Lake Magadi in southern Kenya since 1911 and more recently at Lake Abijata in Ethiopia (see below). The trona, extracted as a partly fluid crystal mush, is processed to produce soda ash (Na_2CO_3), which has many commercial applications including glass manufacture, chemical industries, dyes, fertilisers, water treatment, soaps and detergents, metallurgy and others (Tiercelin 1991; Fig. 1.11a). At Lake Magadi, halite is also produced in ponds where the brine has been evaporated beyond trona saturation (Fig. 1.11b) for local distribution.

Mining trona for soda ash from ponded evaporated brine was recently contemplated at Lake Natron in northern Tanzania (Baker 2011), which is the primary breeding site of flamingos in East Africa. As for many other EASL, Lake Natron is not protected by law, but international protests and a cost-benefit analysis which showed that trona mining is commercially not viable (Kadigi et al. 2014) have prevented the lake from this worst-case scenario. An instructive example of ecosystem destruction caused mainly by soda ash mining is Lake Abijata in Ethiopia (Getaneh et al. 2015). Water abstraction had already led to a fall in lake water level; the lake shrunk dramatically and the former lakeshore became dusty dry grassland (Alemayehu et al. 2006). Moreover, the present company operators (formerly a state-run facility, but now privately owned) are searching for ways to eliminate Cyanobacteria because their pigments stain the soda ash, which results in a lower quality. A channel to Lake Shala, which would provide more water for Lake Abijata and enable future soda ash mining, has been discussed, but these plans have met with strong disapproval by experts.

Diatomite (fossil diatoms, diatomaceous earth) is found near Lake Elmentaita, Kenya. The diatomite layers are witnesses of a former large freshwater lake from the early-to-mid-Pleistocene (Bergner et al. 2009; Durkee and Brown 2014; McCall 1967). The sediments comprise layers of freshwater diatoms, which have been mined since the early 1940s (Figs. 1.6e, f).

Fig. 1.11 Economic resources and contemporary issues. (**a**) The soda ash plant at Lake Magadi (currently Tata Chemicals Magadi). (**b**) Halite and trona in evaporating ponds, Tata Chemicals Magadi. The halite crystallises from brines concentrated beyond trona saturation by solar evaporation. (**c**) Oil shales that formed in a middle Miocene (~12 Ma) soda lake at Kapkiamu in the Tugen Hills west of Lake Baringo (Member C, Ngorora Formation). Geological hammer (arrow) for scale. In stratified saline alkaline lakes, microbial organic matter is preserved in the anoxic monimolimnion. That organic matter is 'oil prone' and, after burial and heating (50–120 °C), can generate oil. (**d**) Visitors at Loburu hot springs, Lake Bogoria, Kenya. These unusual travertine deposits have been badly damaged by trampling. (**e**) Concrete diversion channels at Sandai village northeast of Lake Bogoria, Kenya. Water diverted for drinking and irrigation has reduced the natural inflow to the lake. (**f**) Silts deposited by channels and sheetfloods north of Lake Magadi, Kenya. Increased siltation, some related to deforestation and changing land use to the north, has had an impact on soda ash production at Lake Magadi

Diatomite is used as filtering material for beverages.

Other minerals that form in the EASL include zeolites and sodium silicates and chert, but these have not been developed commercially beyond local use. Sepiolite ($Mg_4Si_6O_{15}(OH)_2{\cdot}6H_2O$) has been mined at Amboseli, a Kenyan ephemeral playa lake near Kilimanjaro, from old alkaline lake sediments (Williams 1972). Some trace elements in hyperalkaline brines such as zirconium (Zr) have unusually high concentrations (Kerrich et al. 2002) but have not been evaluated.

When saline alkaline lakes become chemically stratified or meromictic, microbial organic matter from plankton has been preserved in dense anoxic water (monimolimnion) below a chemocline. That lipid-rich organic matter, where preserved in fine lake centre muds, can become a potential oil-prone source rock (type I and II Kerogen) after burial and thermal maturation, which is rapid in subsiding continental rifts. Ancient soda lake muds in the East African Rift might have sourced oil (Fig. 1.11c), providing an ecological dilemma: soda lakes and their related geothermal inflow fluids are a potential energy source, but they are also a global ecological reservoir of biodiversity that cannot be replaced.

Hot spring waters flow into many EASL. The geothermal energy resource in Ethiopia and Kenya is immense, so the pressure to replace imported fuels by developing national geothermal sources is very easy to understand. In the Kenya Rift, geothermal energy is already used at Olkaria near Hell's Gate National Park just south of Lake Naivasha. Tapping deep geothermal fluids for energy, however, can sometimes have a major impact at the land surface. In New Zealand, once pristine geothermal areas such as Wairakei and Rotorua were damaged by geothermal power development, which lowered the local groundwater tables (Barrick 2007). For some soda lakes, this issue is more critical because, unlike New Zealand, the EASL lie in a semiarid environment where rainfall is often unreliable and the groundwater table is locally very shallow.

Pollution has already affected several lakes, and some wildlife has suffered. Natural and anthropogenic toxins from industrial sources have been reported, for example, at lakes Nakuru, Bogoria and Natron (e.g. Ballot et al. 2004; Bettinetti et al. 2011). Additional pressures have come from recent changes in land use in the drainage basins of some of the lakes (Raini 2009). A good example is Lake Beseka (Ethiopia), which experienced an increase in water level of 7.6 m and an increase in surface area from 2.7 km^2 to around 50 km^2 during the past 50 years (Dinka 2012). Comparisons with meteorological data imply that changes in groundwater flow are responsible for this dramatic increase (Dinka et al. 2014), which can be linked to tectonic activity (Goerner et al. 2005), increased discharge of hot springs (Goerner et al. 2009) and anthropogenic interferences such as irrigation (Alemayehu et al. 2006) and the Koka Dam construction. Moreover, this lake experienced underwater detonations for seismic surveys (Klemperer and Cash 2007). The lake water level rise is not climate driven in this case (Goerner et al. 2009), so Lake Beseka is not categorised as an amplifier lake. Underwater explosions for seismic experiments are also discussed as one major factor for the strong decrease of *Arthrospira* biomass in the Ethiopian Lake Arenguade ('Arenguade' means green in Amharic; the name is a hint of the former high phytoplankton density; Girma et al. 2012; Fig. 1.4h).

Diversion of inflow waters (from streams and dilute springs) for drinking water, hotel 'spa' pools and irrigation can alter the delicate balance of some lakes, and such pressures will increase (Fig. 1.11e). Regional deforestation and changes in traditional land use practices have often increased erosion, leading to a higher suspended sediment load (silt and clay) in streams seasonally flowing into lakes (Mwaura and Moore 1991). This, in turn, has affected water clarity, as has run-off from agricultural land where fertiliser usage has increased, making eutrophication more likely. Siltation, resulting from soil erosion near Narok, for example, has recently affected soda processing at Lake Magadi by lowering the purity of extracted salts (Fig. 1.11f).

Some organisms living in the lakes have high economic potential. Besides the scenery and wildlife that annually attract hundreds of thousands of tourists to the EASL, some microorganisms that inhabit the lakes are of special interest for biotechnological purposes (see Chap. 13). *Arthrospira*, for example, is already mass cultivated on other continents and marketed under the name 'spirulina' (e.g. Cyanotech Corp., Hawaii; Earthrise Nutritionals, California). Also microbial enzymes such as cellulases, proteinases and lipases are the focus of other enterprises (Chakkiath et al. 2013;

Sorokin et al. 2014). The major advantage is the high stability of enzymes in alkaline and saline solutions, sometimes combined with thermostability. Two enzymes originating from haloalkaliphiles of Kenyan soda lakes are already used in textile industries (Genencor International; Grant and Heaphy 2010).

1.7 Concluding Remarks

The EASL, which are unique systems, are famous for their stunning environments and wildlife, but are also valuable in many other respects. Based on kinetic, mass-balance and thermodynamic considerations, Kempe and Degens (1985), for example, postulated an ancient Precambrian ocean with a much higher alkalinity and pH that was more similar to soda lakes than the modern oceans. They proposed that a transfer from a Precambrian soda ocean to the modern halite sea took place around one billion years (Ga) ago, which according to Kempe and Kazmierczak (1994) was responsible for the onset of biocalcification and a major trigger for evolution. Modern soda lakes are essential for testing the 'soda ocean hypothesis', which may provide insights into the geological history and the diversity of life. This hypothesis, nonetheless, remains controversial.

EASL have long been recognised as global treasures that should be preserved, but they are also important sources of revenue because of the tourists that they attract. Balancing conservation and pressures from tourism is difficult (Fig. 1.11d), and only a few areas have met the challenge. Some National Parks and reserves have been established in Ethiopia (Abijata–Shala, Awash), Tanzania (Lake Manyara) and Kenya (Nakuru, Amboseli, Lake Bogoria and others), but this does not guarantee protection of the lakes, as has been seen at Lake Abijata. A balanced approach, but based on scientific data rather than ideology, is urgently needed to preserve the integrity of the EASL. They share an ecology unmatched by saline alkaline lakes elsewhere. Nonetheless, East African nations must develop their resources for the benefit of all their citizens first, then visitors. This also includes clear regulations for scientists. In the past, there have been unfortunate cases of export and exploitation of genetic resources (Sheridan 2004), clearly demonstrating the urgent need for simple, effective and fast processes for granting research permission for bona fide scientists. If such permissions become difficult, both local and international scientists might focus their research efforts elsewhere.

The soda lakes are there partly because of the present climate—too wet, they become fresh; too dry, they will desiccate, leaving trona and halite deposits. Both extremes have occurred in the past few millennia, but the lakes have always recovered. Over the longer term, many soda lakes will survive and regenerate, whereas others will dry up. New soda lakes will form when and where tectonics, volcanism and hydrology combine to create favourable conditions. They have existed in East Africa for at least 15 million years. Their biota will change but the soda lakes should survive.

References

Abbate E, Bruni P, Sagri M (2015) Geology of Ethiopia: a review and geomorphological perspectives. In: Billi P (ed) Landscapes and landforms of Ethiopia. Springer, Dordrecht, pp 33–64

Alemayehu T, Ayenew T, Kebede S (2006) Hydrogeochemical and lake level changes in the Ethiopian Rift. J Hydrol 316:290–300

APHA (1995) Standard methods for the examination of water and wastewater, 19th edn. American Public Health Association, Washington, DC

Azam F, Fenchel T, Field JG, Gray JS, Meyer-Reil LA, Thingstad F (1983) The ecological role of water-column microbes in the sea. Mar Ecol Prog Ser 10:257–263

Baker M (2011) Lake Natron, flamingos and the proposed soda ash factory. http://www.tnrf.org/en/natronbriefing. Accessed Dec 2015

Ballot A, Krienitz L, Kotut K, Wiegand C, Metcalfe JS, Codd GA, Pflugmacher S (2004) Cyanobacteria and cyanobacterial toxins in three alkaline Rift Valley lakes of Kenya-Lakes Bogoria, Nakuru and Elmenteita. J Plankton Res 26:925–935

Banciu HL, Sorokin DY (2013) Adaptation in haloalkaliphiles and natronophilic bacteria. In: Seckbach J, Oren AE, Stan-Lotter H (eds) Polyextremophiles, series cellular origin, life in extreme habitats and astrobiology, vol 27. Springer, Dordrecht, pp 121–178

Bard E (2013) Out of the Africa humid period. Science 342:808–809

Barrick KA (2007) Geyser decline and extinction in New Zealand—energy development impacts and implications for environmental management. Environ Manage 39:783–805

Becht R, Mwango F, Muno FA, (2006) Groundwater links between Kenyan Rift Valley lakes. In: Proceedings of 11th world lake congress, Nairobi, Vol 2, pp 7–14

Benvenuti M, Carnicelli S, Belluomini G, Dainelli N, Di Grazia S, Ferrari GA, Iasio C, Sagri M, Ventra D, Atnafu B, Kebede S (2002) The Ziway–Shala lake basin (main Ethiopian Rift, Ethiopia): a revision of basin evolution with special reference to the Late Quaternary. J Afr Earth Sci 35:247–269

Bergman AN, Laurent P, Otiang'a-Owiti G, Bergman HL, Walsh PJ, Wilson P, Wood CM (2003) Physiological adaptations of the gut in the Lake Magadi tilapia, *Alcolapia grahami*, an alkaline- and saline-adapted teleost fish. Comp Biochem Physiol A Mol Integr Physiol 136:701–715

Bergner AGN, Strecker MR, Trauth MH, Deino A, Gasse F, Blisniuk P, Dühnforth M (2009) Tectonic and climatic control on evolution of rift lakes in the Central Kenya Rift, East Africa. Quat Sci Rev 28:2804–2816

Bettinetti R, Quadroni S, Crosa G, Harper D, Dickie J, Kyalo M, Mavuti M, Galassi S (2011) A preliminary evaluation of the DDT contamination of sediments in Lakes Natron and Bogoria (Eastern Rift Valley, Africa). Ambio 40:341–350

Brown LH (1975) 6 East Africa. In: Kear J, Duplaix-Hall N (eds) Flamingos. T. & A.D. Poyser, Berkhamsted, p 248

Carpelan LH (1978) Revision of Kolbe's System der Halobien based on diatoms of California lagoons. Oikos 31:112–122

Chakkiath PA, Kumaresan D, Hunger S, Drake HL, Murrell JC, Shouche YS (2013) Microbiology of Lonar Lake and other soda lakes. ISME J 7:1252

Childress B, Nagy S, Hughes B (2008) International single species action plan for the conservation of the lesser Flamingo *Phoeniconaias minor*. CMS Technology Series 18, AEWA Technology Series 34. UNEP/CMS Secretariat, Bonn, p 59

Chorowicz J (2005) The East African Rift system. J Afr Earth Sci 43:379–410

Conway D (2002) Extreme rainfall events and lake level changes in East Africa: recent events and historical precedents. In: Odada EO, Olago DO (eds) The East African Great Lakes: limnology, palaeolimnology and biodiversity. Kluwer, Amsterdam, pp 62–92

Cummings JM (1940) Saline and hydromagnesite deposits of British Columbia. Br Col Dept Mines Bull 4:160

Daly MC, Chorowicz J, Fairhead JD (1989) Rift basin evolution in Africa: the influence of reactivated step basement shear zones. In: Cooper MA, Williams CD (eds) Inversion tectonics, vol 44, Geological Society of London Special Publications. Geological Society, London, pp 309–334

Dawson JB (2008) The Gregory Rift Valley and Neogene-Recent volcanoes of Northern Tanzania. Mem Geol Soc Lond 33:102

De Cort G, Bessems I, Keppens E, Mees F, Cumming B, Verschuren D (2013) Late-Holocene and recent hydroclimatic variability in the central Kenya Rift Valley: the sediment record of hypersaline lakes Bogoria, Nakuru and Elmenteita. Palaeogeogr Palaeoclimatol Palaeoecol 388:69–80

de Menocal PB, Ortiz J, Guilderson T, Adkins J, Sarnthein M, Baker L, Yarusinsky M (2000) Abrupt onset and termination of the Africa Humid Period: rapid climate responses to gradual insolation forcing. Quat Sci Rev 19:347–361

Deocampo DM, Jones BF (2014) Geochemistry of saline lakes. In: Holland HD, Turekian KK (eds) Treatise on geochemistry, vol 7, 2nd edn. Elsevier, Oxford, pp 437–469

Dinka MO (2012) Analysing the extents (size and shape) of Lake Basaka expansion (Main Ethiopian Rift Valley) using remote sensing and GIS. Lake Reserv Res Manag 17:131–141

Dinka MO, Loiskandl W, Ndambuki JM (2014) Hydrologic modelling for Lake Basaka: development and application of a conceptual water budget model. Environ Monit Assess 186:5363–5379

Dodds WK, Whiles MR (2010) Freshwater ecology: concepts and environmental applications of limnology, 2nd edn. Academic Press, San Diego, CA

Dühnforth M, Bergner AGN, Trauth MH (2006) Early Holocene water budget of the Nakuru-Elmenteita basin, Central Kenya Rift. J Palaeolimnol 36:281–294

Dunkley PN, Smith M, Allen DJ, Darling WG (1993) The geothermal activity and geology of the northern sector of the Kenya Rift Valley. British Geological Survey Research Report SC/93/1, p 185

Durkee H, Brown FH (2014) Correlation of volcanic ash layers between the Early Pleistocene Acheulean sites of Isinya, Kariandusi, and Olorgesailie, Kenya. J Arch Sci 49:510–517

Earman S, Philips FM, McPherson JOL (2005) The role of "excess" CO_2 in the formation of trona deposits. Appl Geochem 20:2217–2232

Erdmann N, Hagemann M (2004) Salt acclimation of algae and cyanobacteria: a comparison. In: Rai LC, Gaur JP (eds) Algal adaptation to environmental stresses. Springer, Berlin, pp 323–361

Eugster HP (1980) Lake Magadi, Kenya, and its precursors. In: Nissenbaum A (ed) Hypersaline brines and evaporitic environments. Elsevier, Amsterdam, pp 195–232

Eugster HP, Hardie LA (1978) Saline lakes. In: Lerman A (ed) Lakes: chemistry, geology, physics. Springer, New York, pp 237–293

Eugster HP, Jones B (1979) Behavior of major solutes during closed-basin brine evolution. Am J Sci 279:609–631

Foerster V, Junginger A, Langkamp O, Gebru T, Asrat A, Umer M, Lamb H, Wennrich V, Rethemeyer J,

Nowaczyk N, Trauth MH, Schäbitz F (2012) Climatic change recorded in the sediments of the Chew Bahir basin, southern Ethiopia, during the last 45,000 years. Quat Int 274:25–37

Fofonoff NP (1985) Physical properties of seawater: a new salinity scale and equation of state of seawater. J Geophys Res 90:3332–3342

Fuhrman JA (1999) Marine viruses and their biogeochemical and ecological effects. Nature 399:541–548

Garcin Y, Junginger A, Melnick D, Olago DO, Strecker MR, Trauth MH (2009) Late Pleistocene–Holocene rise and collapse of Lake Suguta, northern Kenya Rift. Quat Sci Rev 28:911–925

Garrett DE (1992) Natural soda ash: occurrences, process and use. Springer, Berlin

Gasse F, Street A (1978) Late Quaternary lake-level fluctuations and environments of the northern Rift Valley and Afar region (Ethiopia and Djibouti). Palaeogeogr Palaeoclimatol Palaeoecol 24:279–325

Getaneh G, Temesgen G, Damena E (2015) Environmental degradation and its effect on terrestrial and aquatic diversity in the Abijata-Shala Lakes National Park, Ethiopia. Point J Agric Biotechnol Res 1:1–12

Girma MB, Kifleb D, Jebessab H (2012) Deep underwater seismic explosion experiments and their possible ecological impact – the case of Lake Arenguade – Central Ethiopian highlands. Limnologica 42:212–219

Goerner A, Jolie E, Gloaguen R (2005) Tectonic induced growth of Lake Beseka (Main Ethiopian Rift)? Geophys Res Abst 7:957

Goerner A, Jolie E, Gloaguen R (2009) Non-climatic growth of the saline Lake Beseka, Main Ethiopian Rift. J Arid Environ 63:287–295

Grant WD (2004) Half a lifetime in soda lakes. In: Ventosa A (ed) Halophilic microorganisms. Springer, Berlin, pp 17–31

Grant WD, Heaphy S (2010) Metagenomics and recovery of enzyme genes from alkaline saline environments. Environ Technol 31:1135–1143

Grove AT, Street FA, Goudie AS (1975) Former lake levels and climatic change in the Rift Valley of southern Ethiopia. Geogr J 141:171–194

Gruber-Dorninger M (2015) East African Soda Lakes – biogeochemistry and interactions between organismic compartments. Dissertation, University of Vienna, Vienna

Hammer UT (1986) Saline lake ecosystems of the world. Junk, Dordrecht

Harper DM, Childress RB, Harper MM, Boar RR, Hickley P, Mills SC, Otieno N, Drane A, Vareschi E, Nasirwa O, Mwatha WE, Darlington JPEC, Escuté-Gasulla X (2003) Aquatic biodiversity and saline lakes: Lake Bogoria, National Reserve, Kenya. Hydrobiologia 500:259–276

Hendrie DB, Kusznir NJ, Morley CK, Ebinger CJ (1994) Cenozoic extension in northern Kenya: a quantitative model of rift basin development in the Turkana region. Tectonophysics 236:409–438

Hillaire-Marcel C, Casanova J (1987) Isotopic hydrology and paleohydrology of the Magadi (Kenya)-Natron (Tanzania) Basin during the Late Quaternary. Palaeogeogr Palaeoclimatol Palaeoecol 58:155–181

Jirsa F, Gruber M, Stojanovic A, Omondi SO, Mader D, Körner W, Schagerl M (2013) Major and trace element geochemistry of Lake Bogoria and Lake Nakuru, Kenya, during extreme draught. Chem Erde Geochem 73:275–282

Johnston HH (1902) The Uganda Protectorate, vol 1. Hutchinson, London

Junginger A, Trauth MH (2013) Hydrological constraints of paleo-Lake Suguta in the Northern Kenya Rift during the African Humid Period (15–5 ka BP). Glob Planet Change 111:174–188

Kadigi RMJ, Mwathe K, Dutton A, Kashaigili J, Kilima F (2014) Soda ash mining in Lake Natron: a reap or ruin for Tanzania? J Environ Conserv Res 2:37–49

Kaggwa MN, Gruber M, Oduor SO, Schagerl M (2013) A detailed time series assessment of the diet of Lesser Flamingos: further explanation for their itinerant behaviour. Hydrobiologia 710:83–93

Kasedde H, Kirabira JB, Bäbler MU, Tilliander A, Jonsson S (2014) Characterization of brines and evaporites of Lake Katwe, Uganda. J Afr Earth Sci 91:55–65

Kempe S, Degens ET (1985) An early soda ocean? Chem Geol 53:95–108

Kempe S, Kazmierczak J (1994) The role of alkalinity in the evolution of ocean chemistry, organization of living systems and biocalcification processes. In: Doumenge F (ed) Past and present biomineralization processes. Considerations about the carbonate cycle. Bull Inst Oceanogr Monaco 13:61–117

Kerrich R, Renaut RW, Bonli T (2002) Trace-element composition of cherts from alkaline lakes in the East African Rift: a probe for ancient counterparts. In: Renaut RW, Ashley GM (eds) Sedimentation in continental rifts. SEPM Spec Publ 73:275–294

Klemperer SL, Cash MD (2007) Temporal geochemical variation in Ethiopian Lakes Shala, Arenguade, Awasa, and Beseka: possible environmental impacts from underwater and borehole detonations. J Afr Earth Sci 48:174–198

Lamb HH (1966) Climate in the 1960s. Geogr J 13:183–212

Lanzén A, Simachew A, Gessesse A, Chmolowska D, Jonassen I, Øvreås L (2013) Surprising prokaryotic and eukaryotic diversity, community structure and biogeography of Ethiopian Soda Lakes. PLoS One 8: e72577

Le Turdu C, Coussement C, Tiercelin JJ, Renaut RW, Rolet J, Richert JP, Xavier JP, Coquelet D (1995) Rift basin structure and depositional patterns interpreted using a 3D remote sensing approach: the Baringo and Bogoria basins, central Kenya Rift, East Africa. Bull Centres Rech Explor-Prod Elf-Aquitaine 19:1–37

Luo W, Kotut K, Krienitz L (2013) Hidden diversity of eukaryotic plankton in the soda lake Nakuru, Kenya, during a phase of low salinity revealed by a SSU rRNA gene clone library. Hydrobiologia 702:95–103

MacDonald R (2003) Magmatism of the Kenya Rift Valley: a review. Trans R Soc Edinburgh Earth Sci 93:239–253

Madigan MT, Martinko JM, Stahl DA, Clark DP (2011) Brock biology of microorganisms, 13th edn. Pearson Education, San Francisco, CA

McCall GJH (1967) Geology of the Nakuru-Thomson's Falls-Lake Hannington Area, Report 78. Geological Survey of Kenya, Republic of Kenya, Nairobi, p 122

McGregor D (2015) History of the development of the East African Rift System: a series of interpreted maps through time. J Afr Earth Sci 101:232–252

Mesbah N, Wiegel J (2011) Halophiles exposed concomitantly to multiple stressors: adaptive mechanisms of halophilic alkalithermophiles. In: Ventosa A, Oren A, Ma Y (eds) Halophiles and hypersaline environments. Springer, Berlin, pp 249–273

Morley CK (1999) Geoscience of rift systems–evolution of East Africa. AAPG Stud Geol 44:242 + appendices

Morley CK, Nelson RA, Patton TL (1990) Transfer zones in the East African Rift System and their relevance to hydrocarbon exploration in rifts. Am Assoc Petrol Geol Bull 74:1234–1253

Morley CK, Ngenoh DK, Ego JK (1999) Introduction to the East African Rift System. In: Morley CK (ed) Geoscience of rift systems–evolution of East Africa. AAPG Stud Geol 44:1–18

Mwaura F, Moore T (1991) Forest and woodland depletion in the Lake Elementeita Basin, Kenya. Geoforum 22:17–26

Odada EO, Onyando JO, Obudho PA (2006) Lake Baringo: addressing threatened biodiversity and livelihoods. Lake Reserv Res Manag 11:287–299

Oduor SO, Schagerl M (2007) Phytoplankton photosynthetic characteristics in three Kenyan Rift Valley saline-alkaline lakes. J Plankton Res 29:1041–1050

Olago D, Opere A, Barongo J (2009) Holocene palaeohydrology, groundwater and climate change in the lake basins of the Central Kenya Rift. Hydrol Sci J 54:765–780

Olaka LA, Odada EO, Trauth MH, Olago DO (2010) The sensitivity of East African rift lakes to climate fluctuations. J Paleolimnol 44:629–644

Oren A (2006) Life at high salt concentrations. In: Dworkin M, Falkow S, Rosenberg E, Schleifer K-H, Stackebrandt E (eds) The prokaryotes. Springer, New York, pp 263–282

Pecoraino G, D'Alessandro W, Inguaggiato S (2015) The other side of the coin: geochemistry of alkaline lakes in volcanic areas. In: Rouwet D et al (eds) Volcanic lakes, Advances in volcanology. Springer, Berlin, pp 219–237

Peduzzi M, Gruber M, Gruber M, Schagerl M (2014) The virus's tooth: cyanophages affect an African flamingo population in a bottom-up cascade. ISME J 8:1346–1351

Raini J (2009) Impact of land use changes on water resources and biodiversity of Lake Nakuru catchment basin, Kenya. Afr J Ecol 47:39–45

Renaut RW (1990) Recent carbonate sedimentation and brine evolution in the saline lake basins of the Cariboo Plateau, British Columbia, Canada. Hydrobiologia 197:67–81

Renaut RW, Gierlowski-Kordesch EG (2010) Lakes. In: James NP, Dalrymple R (eds) Facies models 4. Geological Association of Canada, St. Johns, pp 541–575

Renaut RW, Tiercelin JJ (1993) Lake Bogoria, Kenya: soda, hot springs and about a million flamingoes. Geol Today 9:56–61

Renaut RW, Tiercelin JJ (1994) Lake Bogoria, Kenya Rift Valley: a sedimentological overview. In: Renaut, RW, Last WM (eds) Sedimentology and geochemistry of modern and ancient saline lakes. SEPM Spec Publ 50:101–123

Renaut RW, Ego J, Tiercelin JJ, Le Turdu C, Owen RB (1999) Saline, alkaline palaeolakes of the Tugen Hills–Kerio Valley region, Kenya Rift Valley. In: Andrews P, Banham P (eds) Late Cenozoic environments and hominid evolution: a tribute to Bill Bishop. Geological Society, London, pp 41–58

Renaut RW, Owen RB, Jones B, Tiercelin JJ, Tarits C, Ego JK, Konhauser KK (2013) Impact of lake level changes on the formation of thermogene travertine in continental rifts: evidence from Lake Bogoria, Kenya Rift Valley. Sedimentology 60:428–468

Ring U (2014) The East African Rift system. Aust J Earth Sci 107:132–146

Roberts MF (2005) Organic compatible solutes of halotolerant and halophilic microorganisms. Saline Syst 1:5

Rosen MR (1994) The importance of groundwater in playas: a review of playa classifications and the sedimentology and hydrogeology of playas. In: Rosen MR (ed) Paleoclimate and basin evolution of playa systems. Geol Soc Am Spec Pap 289:1–18

Rosendahl BR (1987) Architecture of continental rifts with special reference to East Africa. Annu Rev Earth Planet Sci 15:445–504

Schagerl M, Oduor SO (2003) On the limnology of Lake Baringo (Kenya): II. Pelagic primary production and algal composition of Lake Baringo, Kenya. Hydrobiologia 506–509:297–303

Schagerl M, Oduor SO (2008) Phytoplankton community relationship to environmental variables in three Kenyan Rift Valley saline-alkaline lakes. Mar Freshw Res 59:125–136

Schagerl M, Burian A, Gruber-Dorninger M, Oduor SO, Kaggwa MN (2015) Algal communities of Kenyan soda lakes with a special focus on *Arthrospira fusiformis*. Fottea 15:245–257

Schlesinger P, Belkin S, Boussiba S (1996) Sodium deprivation under alkaline conditions causes rapid death

of the filamentous cyanobacterium *Spirulina platensis*. J Phycol 32:608–613
Sheridan C (2004) Kenyan dispute illuminates bioprospecting difficulties. Nat Biotechnol 22:1337
Smoot JP, Lowenstein TK (1991) Depositional environments of non-marine evaporites. In: Melvin JN (ed) Evaporites, petroleum and mineral resources. Elsevier, Amsterdam, pp 189–347
Sommer U, Adrian R, De Senerpont-Domis L, Elser JJ, Gaedke U, Ibelings B, Jeppesen E, Lürling M, Molinero JC, Mooij WM, van Donk E, Winder M (2012) Beyond the Plankton Ecology Group (PEG) model: mechanisms driving plankton succession. Annu Rev Ecol Evol Syst 43:429–448
Sorokin DY, Berben T, Melton ED, Overmars L, Vavourakis CD, Muyzer G (2014) Microbial diversity and biogeochemical cycling in soda lakes. Extremophiles 18:791–809
Street FA (1980) The relative importance of climate and local hydrogeological factors in influencing lake level fluctuations. Palaeoecol Afr 12:137–158
Tarits C, Renaut RW, Tiercelin JJ, Le Hérissé A, Cotten J, Cabon JY (2006) Geochemical evidence of hydrothermal recharge in Lake Baringo, central Kenya Rift Valley. Hydrol Proc 20:2027–2055
Tebbs EJ, Remedios JJ, Harper DM (2013) Remote sensing of chlorophyll-a as a measure of cyanobacterial biomass in Lake Bogoria, a hypertrophic, saline–alkaline, flamingo lake, using Landsat ETM+. Remote Sens Environ 135:92–106
Tiercelin JJ (1991) Natural resources in the lacustrine facies of the Cenozoic rift-basins of East Africa. In: Anadón P, Cabrera Ll, Kelts K (eds) Lacustrine facies analysis. Spec Publ Int Assoc Sediment 13:3–37
Tiercelin JJ, Lezzar KA (2002) 300 million years history of rift lakes in Central and East Africa: an updated broad review. In: Odada EO, Olago DO (eds) The East African Great lakes: limnology, palaeolimnology and biodiversity. Kluwer, Amsterdam, pp 3–60
Tiercelin JJ, Vincens A (eds) (1987) Le demi-graben de Baringo-Bogoria, Rift Gregory, Kenya. Bull Centres Rech Explor-Prod Elf Aquitaine 11:249–540
Trauth MH, Maslin MA, Deino AL, Junginger A, Lesoloyia M, Odada EO, Olago DO, Olaka LA, Strecker MR, Tiedemann R (2010) Human evolution in a variable environment: the amplifier lakes of Eastern Africa. Quat Sci Rev 29:2981–2988
Vareschi E (1987) Saline lake ecosystems. In: Schulze E-D, Zwölfer H (eds) Potentials and limitations of ecosystem analysis, vol 61. Springer, Berlin, pp 347–364
Vareschi E, Jacobs J (1985) The ecology of Lake Nakuru. 6. Synopsis of production and energy-flow. Oecologia 65:412–424
Wilhelm SW, Suttle CA (1999) Viruses and nutrient cycles in the sea—viruses play critical roles in the structure and function of aquatic food webs. Bioscience 49:781–788
Williams LAJ (1972) Geology of the Amboseli area. Rep Geol Surv Kenya 90:86
Williams WD (1978) Limnology of Victorian salt lakes, Australia. Verh Int Ver Limnol 20:1165–1174
Wolfenden E, Ebinger C, Yirgub G, Deino A, Ayalew D (2004) Evolution of the northern Main Ethiopian Rift: birth of a triple junction. Earth Planet Sci Lett 224:213–228
Yuretich RF (1982) Possible influences upon lake development in the East African rift valleys. J Geol 90:329–337
Zaccara S, Crosa G, Vanetti I, Binelli G, Childress B, McCulloch G, Harper DM (2011) Lesser Flamingo *Phoeniconaias minor* as a nomadic species in African shallow alkaline lakes and pans: genetic structure and future perspectives. Ostrich 82:95–100

Genesis of the East African Rift System 2

Giday WoldeGabriel, Daniel Olago, Edwin Dindi, and Mike Owor

Abstract

The East African Rift System (EARS) started in Late Oligocene to Early Miocene time and gradually propagated southwards from the Afar Depression, beginning in the Middle Miocene. The hot, low-density mantle material of the Afar Plume heated the overlying lithosphere, causing thinning, regional doming, and the earliest basaltic volcanism in southern Ethiopia. In Ethiopia, the Afar Depression, the Main Ethiopian Rift, and the broadly rifted zone of southwestern Ethiopia represent the northern segment of the EARS.

In the Kenyan sector of the EARS, uplift and doming also gave rise to the Kenya Dome. The radial flow patterns of the initial phonolites provide evidence for doming. Another important observation is that the rift geometry was greatly influenced by pre-existing structures of the underlying Mozambique Mobile Belt. Rifting proceeded through alternating episodes of volcanism and tectonics. Crossing into Tanzania, the influence of the neighbouring Tanzania Craton becomes evident. Here, the rift is expressed only in the northern part, splaying out in diverging half-graben valleys that are outside the Kenya Dome.

G. WoldeGabriel (✉)
EES-14/MS D462, Earth Environmental Sciences Division, Los Alamos National Laboratory, Los Alamos, NM 87545, USA
e-mail: wgiday@lanl.gov

D. Olago
Institute for Climate Change and Adaptation and Department of Geology, University of Nairobi, Chiromo Campus, Riverside Drive, PO Box 30197-00100, Nairobi, Kenya
e-mail: dolago@uonbi.ac.ke

E. Dindi
Department of Geology, University of Nairobi, Chiromo Campus, Riverside Drive, PO Box 30197-00100, Nairobi, Kenya
e-mail: eedindi@yahoo.com

M. Owor
Department of Geology and Petroleum Studies, Makerere University, PO Box 7062, Kampala, Uganda
e-mail: mmowor@yahoo.com

M. Schagerl (ed.), *Soda Lakes of East Africa*, DOI 10.1007/978-3-319-28622-8_2

Large boundary faults and opposing flexural margins, producing mobile asymmetrical full and half-graben basins that are individually linked along the rift axis, mark the Western Rift Valley. These basins are frequently occupied by elongate and narrow lakes (largely freshwater) separated by accommodation zones and containing significant hydrocarbon resources especially in the Albertine Graben. Small to large lakes existed in the EARS during the Plio–Pleistocene. Lakes in the Western Rift are large and deep, whereas those in the Kenya, Main Ethiopian, and Afar Rifts are generally small and shallow. Geological records indicate that the lakes sensitively responded to orbital forcing as well as to local, regional, and global climatic, environmental, and tectonic changes, resulting in fluctuating lake sizes and even desiccation.

2.1 Introduction

The East Africa Rift System (EARS) is one of the most extensive (>3000 km), densely faulted, and generally north- to south-oriented continental extensional volcano-tectonic systems of basins and horsts. It extends from the Red Sea in the northeastern part of the continent and continues southwestwards across two distinct structural and topographic swells defined by the Ethiopian and Kenya–Tanzania Domes, before it bifurcates and diffuses into a broader tectonic zone in southeastern Africa (Chorowicz 2005 and references therein). From north to south, four major rift zones constitute the EARS: the Afar, Main Ethiopian Rift (MER), Kenya–Tanzania, and the Western Rifts (Fig. 2.1, Index map). These basins vary in width, length, depth, and orientation and generally exhibit along-axis segmentation and occasionally en-échelon displacements that are connected by kinematically complex transfer zones. Currently, the rift basins exhibit distinct tectonic, volcanic, sedimentological, hydrological, and geomorphic characteristics and features of variable ages attributed to differences in volcanic and tectonic processes along the axis of the EARS and related climatic conditions. The physiographic features, pre-rift tectonics, and basement anisotropies and the Cenozoic tectonic framework of the major rift basin from the Afar Depression to Lake Malawi are described. This is preceded by an introductory summary on the geodynamic setting of Africa that led to the genesis of the EARS. We highlight the complex volcanic and tectonic processes that triggered rifting and the formation of rift basins of the EARS, the influence of pre-rift tectonics and structures on the location, and orientation of the EARS and related sedimentological, geomorphic, hydrological, and paleoenvironmental records, starting from the north with the Ethiopian Rift System, the Kenya–Tanzania Rift, and the Western Rift.

2.2 Geodynamic Characteristics of the EARS

When and why the EARS developed along the eastern continental sector of the African Plate appears to be closely related to Africa's plate tectonic history and to the two major mantle plumes (i.e. Réunion and Afar Plumes) that impacted it, beginning in the late Cretaceous and in the early Tertiary (Burke and Cannon 2014; Cande and Stegman 2011 and references therein). Africa's plate tectonic motion has been relatively slow since the middle Mesozoic even though the plate was partly underlain by mantle with above-average temperature and subjected to contemporaneous rifting and basin subsidence in the continental interior (Gaina et al. 2013 and

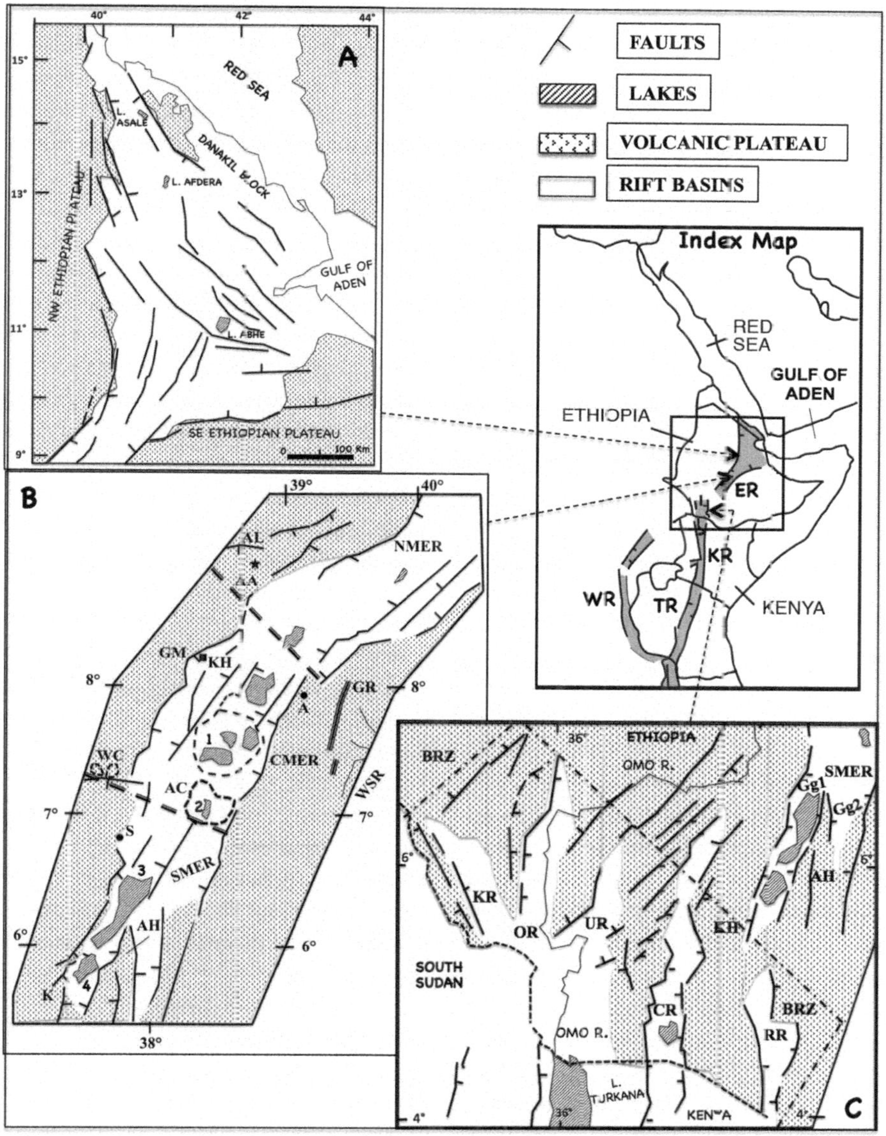

Fig. 2.1 The index map shows the location of the Eastern Africa Rift System, including the Ethiopian (ER), Kenyan (KR), Tanzanian (TR), Western (WR), and Malawi (MR) Rifts. The Ethiopian Rift System includes from north to south the Afar Rift (**a**), the Main Ethiopian Rift (**b**), and the broad rift zone of SW Ethiopia (**c**). The abbreviations in **b** represent the Ambo Lineament (AL), Addis Ababa (AA), the northern Main Ethiopian Rift (NMER), the central MER (CMER), the Galama Range (GR), Asela (A), Wabi Shebele River (WSR), Kella Horst (KH), Guraghe Mountains (GM), Awasa Caldera (AC), Wagebeta Caldera (WC), Soddo (S), southern MER (SMER), Amaro Horst (AH), and Konso (K). The numbers represent Lake Shala and nearby lakes (1), Lake Awasa (2), Lake Abaya (3), and Lake Chamo (4). In **c**, the Ganjuli (Gg1) and Galana (Gg2) Grabens, the Amaro Horst (AH), the Konso Highlands (KH), the broadly rifted zone (BRZ), the Ririba (RR), the Chew Bahir (CR), Usno (UR), Omo (OR), and Kibish (KR) Rifts are marked

references therein). Furthermore, the geographic distribution of the EARS along the eastern continental edge of the African Plate and its proximity to Mesozoic basins and spreading zones led to the break-up and separation of East Gondwana from the African Plate. This appears to be a Tertiary analogue to the older geodynamic processes that affected the continental plate. Successive tectonic processes intermittently continued to transform the plate after the break-up and separation of East Gondwana. These processes were responsible for the split of India from East Gondwana in the early Cretaceous, Australia from Antarctica in the late Cretaceous, and rifting and drifting of the continental blocks of India, Seychelles, and Madagascar also since the late Cretaceous (Cande and Stegman 2011; Eagles and König 2008; Gaina et al. 2007; Gibbons et al. 2013; Yatheesh et al. 2013). Cande and Stegman (2011) proposed that a force generated from the Réunion Plume head drove the coupled tectonic activities of the African and Indian Plates during the late Mesozoic and early Tertiary.

By the early Tertiary, hot, low-density mantle material related to the rising Afar Plume beneath the region now occupied by the proto-oceanic Afar Depression and the oceanic Red Sea and Gulf of Aden Rifts started to heat the lithosphere. This caused lithospheric thinning, regional uplift, and the initial eruptions of Eocene flood basalts in what is now the broadly rifted zone of SW Ethiopia and the southern sector of the MER (Davidson and Rex 1980; Ebinger et al. 1993, 2000; Ebinger and Sleep 1998; WoldeGabriel et al. 1991; Yemane et al. 1999). As the Afar Plume head continued to expand and become shallower, continued thermal thinning of the lithosphere led to intense and voluminous volcanic activity. This affected a large region from Yemen across Ethiopia to northern Kenya during the late Oligocene to early Miocene (Baker et al. 1996; Davidson and Rex 1980; Hofmann 1997; Menzies et al. 1997; WoldeGabriel et al. 1990). At the northern end of the EARS, rifting led to break-up and separation of the Arabian Microplate, thereby creating the oceanic Red Sea and the Gulf of Aden rifts in the early Miocene (ArRajehi et al. 2010 and references therein). Buck (2006) suggested that the genesis of the Afro-Arabian Rift System was driven by magma-assisted extension shortly after the widespread and voluminous basaltic eruptions in the region now occupied by the Red Sea and the Afar Rifts. Following uplift, magmatism, and the inception of rifting at the northernmost part of the EARS in the late Oligocene and early Miocene, plume flow continued southwards beneath the emergent basins and is inferred to have triggered the southward propagation of rifting (Ebinger and Sleep 1998). Today, the Afar Mantle Plume that initiated uplift and magmatism and subsequent rifting along the eastern continental part of the African Plate has dissipated. Passive upwelling is driving the ongoing volcano-tectonic processes along the EARS (Hammond et al. 2013). The Afar Depression represents a proto-oceanic rift that is in transition from continental rifting to seafloor spreading with the ongoing break-up of two additional microplates of the Danakil Block and Somalia from the Nubian Plate (Ebinger 2005; Hammond et al. 2013; Tesfaye and Ghebreab 2013). The volcano-tectonic transition from the Afar Depression to the Main Ethiopian Rift (MER) is manifested by a low-velocity zone associated with shallow passive mantle upwelling and diapiric rise of magma. These processes are responsible for the ongoing rifting and along-axis segmentation (Hammond et al. 2013).

Farther south, rocks of the Mozambique Orogenic Belt underlie the Kenya and Tanzania Rifts of the eastern arm of the EARS. Geochronological data indicate that the belt evolved as a tectono-metamorphic complex between 850 and 500 Ma ago (Shackleton 1993). The meta-sediments were probably deposited between 1200 and 900 Ma ago. The Mozambique Belt trends north to south through Kenya and Tanzania and comprises predominantly paragneisses and orthogneisses at amphibolite to granulite facies (Cahen et al. 1984). Typically known as the Eastern Granulite–Cabo Delgado Nappe Complex of southern Kenya, Tanzania, and Mozambique, the extended crust that formed adjacent to the Mozambique Ocean experienced a 650–620 Ma granulite facies metamorphism that terminated with closure of the Mozambique Ocean (Fritz et al. 2013). Thus, a number of rock assemblages identified as dismembered ophiolites have also been found within the Mozambique Orogenic Belt. Completion of the nappe assembly around 620 Ma is defined as the East African Orogeny, which was followed by phases of post-orogenic extension, with early extension being recorded from 600–550 Ma, approximately contemporaneous with that of the Arabian–Nubian Shield (Fritz et al. 2013). The western front of the belt is characterized by well-defined zones of mylonites and thrusts separating it from the Archean rocks of the Tanzania Craton. The eastern front is less clearly defined. A joint assessment of all available structural, geochronological, and metamorphic evidence led Shackleton (1993) to conclude that two different collisional orogenies occurred, one at about 800 Ma on a suture 60 km SSE of Lake Turkana and the other at about 580 Ma in a zone about 100 km to the west of the first. With respect to the evolution of the rift, it is important to note that exposed, steeply east-dipping gneissic fabrics of the Mozambique Belt along the Nguruman and Elgeyo Escarpments of the southern and northern Kenya Rift facilitated extensional reactivation and fracture propagation associated with Cenozoic Extension (Hetzel and Strecker 1994).

The western branch of the EARS is considered to have been initiated in an area of thinner lithosphere with Paleoproterozoic cover in the Virunga Area and has propagated northwards where it now abuts against thick cratonic lithosphere covered by a thrust belt consisting of gneisses, meta-sediments, and metavolcanics of Neoarchaean to Proterozoic age (e.g. Link et al. 2010). Reactivation of the Tanzanian Ubendian Belt during the Neoproterozoic Pan-African Event (ca. 750 Ma) is suggested to have produced sinistral brittle–ductile shear zones. These later became the preferential locus for the brittle rift faults during development of the western branch (e.g. Lenoir et al. 1994). Ebinger (1989a, b), however, assumes that along-axis segmentation is generally not inherited but rather a result of north and southward along-axis propagation of rift-border-fault segments, thereby linking originally isolated basins. The two dominant extensional stress regimes (WNW–ESE and ENE–WSW, e.g. Aanyu and Koehn 2011) have a strong basement control (e.g. Tiercelin et al. 1988). Nyblade and Brazier (2002) combined the uppermost-mantle P-wave velocities beneath Tanzania with the results of other seismic, gravity, heat-flow, and xenolith studies. They suggest that as the southward-propagating Kenya Rift reached the strong Tanzania cratonic lithosphere margin ca. 12–10 Ma, the Western Rift began to develop northwards in the weaker mobile belt lithosphere on the west side of the Craton. However, a more complex scenario on rift evolution in these areas has emerged over the last years, which clearly shows that extensional processes resulting in thermochronologic cooling signatures and rift basin formation must have begun already during the Paleogene (Bauer et al. 2013; Roberts et al. 2012). In Kenya, for example, the Kerio Basin along the Elgeyo Escarpment has a two-stage history (see below), which began during the Paleogene and resulted in superposed rift basins (Hautot et al. 2000; Mugisha et al. 1997).

2.3 Pre-rift Tectonics

There are several rift-related Quaternary features that are consistent with the influence of deeper mantle processes than the superficial pre-rift structures. For example, deeper magmatic-driven tectonic processes apparently control the variable orientation of the major rift basins along the axis of the EARS, the confinement of dense faulting and volcanism to a narrow width of the axial rift zone (60–80 km) and segmentation, and the dominance of the proximal rift basin at the zone of overlap of two or more rift systems in SW Ethiopia and in the Afar Depression. Most of the major faults that define the MER and the Afar Rifts, however, were described as reactivated Pan-African Mozambique Belt structural expressions related to the Arabian–Nubian Shield (Korme et al. 2004 and references therein). The pre-rift structural features are oriented NW–SE (Red Sea Rift), E–W to ENE–WSW (Gulf of Aden Rift), NE–SW (MER), and N–S to NNE–SSW (Quaternary axial rift zones of the MER). These structural features are closely associated with crystalline basement outcrops that are folded and faulted along NE–SW and NW–SE directions and along N–S- to NNE–SSW-striking suture zones including ophiolites that crop out above the high grade metamorphic rocks (Berhe 1990; Kazmin et al. 1978, 1979 and references therein). No major Paleozoic Era structures are known in Ethiopia. During the Mesozoic, however, thick marine and continental sedimentary rocks formed along NW-trending basins, approximately contemporaneous with the break-up and separation of East Gondwana (Bosellini 1989; Bosellini et al. 1997; Gaina et al. 2013; Merla et al. 1973; Russo et al. 1994 and references therein). In the late Cretaceous to middle Tertiary, low-relief, deep weathering and laterite deposits, and igneous intrusions characterized Africa, including the area transected by the EARS (Burke 1996; Burke and Gunnel 2008). By early to middle Tertiary (45–30 Ma), the Afar Plume encroached the base of the lithosphere in the NE part of the African Continental Plate. This caused a low-density layer, vertical upwelling, and structural and topographic swells, culminating in the earliest basaltic eruptions in southern and SW Ethiopia (Davidson and Rex 1980; Ebinger et al. 1993; Ebinger and Sleep 1998; WoldeGabriel et al. 1991; Yemane et al. 1999).

Pre-rift tectonism in the Kenya and Tanzania sector of the eastern arm of the EARS is closely linked to the tectono-metamorphic history of the Mozambique Belt. Figure 2.1 shows the tectonic setting of the Kenya Rift Valley including a small section of the rift in northern Tanzania. The rift overlaps the boundary between the thick, cold, rigid Tanzania Craton (Archean basement) and the thinner, anisotropic Proterozoic Mozambique Belt (Macdonald 1994). Evidence from seismic and gravity studies shows that the rift developed across a transition zone, thought to represent the sheared Proterozoic boundary between the Archean Tanzania Craton and the mobile Mozambique Belt (Birt et al. 1997). While the model by Nyblade and Pollack (1992) defined the boundary between the Craton and the mobile belt to be a suture 50–100 km west of the Rift Valley, Birt et al. (1997) have suggested that the whole region from the rift westwards to the location of the suspected suture may have been affected by the Proterozoic collision so that the Archean Craton probably extends at depth to the edge of the rift. This suggestion follows the observation that a km-thick layer with velocity of 4.0–4.4 km s^{-2} extends at the surface to the west of the rift for 120 km. Crossley (1979) provides a brief account of the pre-rift history of the region. Accordingly, the Gondwana surface was raised in the Jurassic to form the African continental watershed, and a deeply weathered, well planed, and lateritized surface developed in the late Mesozoic. Further uplift in the early Tertiary caused active erosion of the late Mesozoic surface so that, in the Miocene, the landscape had a relief of 2000 m. The Kenya Rift was initiated as a downwarp, which coincided fairly closely with the axis of the uplift. Pre-rift uplift is implied by paleodrainage reconstruction by King et al. (1972) and King (1978). This is supported by the dispersion patterns and thickness variations within the phonolite lavas of the central Kenya Rift. Striking examples of topographic control are flows travelling eastwards to form the Yatta Phonolites, westwards where they border on the Nyanza Rift, and southwestwards towards Mara Plains (Smith 1994). Smith (1994) pointed out that when faulting and tilting effects are removed, the present-day outcrop of lavas defines a low-relief, elongate shield 400 km in length and 200 km wide, giving a total area of about 80,000 km^2. Smith (1994) concluded that available evidence supports the presence of an elongate area of limited (less than 1 km) crustal uplift that formed prior to initiation of volcanism in the central Kenya Rift. In a subsequent study using the viscosity of the Yatta Flows on the eastern shoulder of the central Kenya Rift, Wichura et al. (2010) demonstrated that the eastward-flowing Yatta Phonolite must have originated at an elevation of 1400 m–13.5 Ma. Importantly, this assessment, combined with the recently established spatiotemporal framework of a 17.5-Ma-old whale bone found west of Lake Turkana, provides two paleoelevation data points that suggest that domal uplift in Kenya must have impacted topography and environmental conditions after 17.5 and prior to 13.5 Ma (Wichura et al. 2015). Veldkamp et al. (2007) reached a similar assessment of high topography based on lava flow directions.

The western branch of the East African Rift developed in Proterozoic mobile belts (e.g. Kokonyangi et al. 2007). The middle Jurassic extensional phase generated basins that appear to be confined to the coastal regions of Kenya, Tanzania, and Mozambique (Morley et al. 1999). It is suggested that from the Jurassic through the Cretaceous an extensive rift system ran across Central Africa from the west to eastern Africa. This was accompanied by episodic active rifting lasting in areas until the early Tertiary, now evident in the northern Kenya Rift region of Lake Turkana. These rifts are considered to form deep basins (6–8 km) that in some places have significant accumulations of hydrocarbons. The topographic low in northern Kenya between the Kenya–Tanzania and Afar Domes coincides with the trend of Cretaceous–Paleogene rifting and may reflect a regional influence of the older rift trend on the younger one. Studies in western Tanzania confirm the extent of the northwest-orientated Paleoproterozoic Ubendian Belt contribution to the Phanerozoic Rift (e.g. Klerkx et al. 1998).

2.4 Cenozoic Volcano-Tectonic Framework of the Ethiopian Rift System

The following sections describe the tectonics, volcanism, sedimentological, and geomorphic features of the three major rift basins in Ethiopia. The Afar Depression, the three sectors of the MER, and the broadly rifted zone of SW Ethiopia form distinct structural and geomorphic features compared with each other and with the rest of the EARS (Fig. 2.1a–c). The characteristic features of the three rift basins are highlighted to show how complex volcano-tectonic processes and interactions led to the genesis of the northern part of the EARS with time.

2.4.1 Afar Rift: Volcano-Tectonic and Sedimentological Records and Hydrological Features

As shown in Fig. 2.1a, the Afar Rift—also known as the Afar, Danakil, or Dallol Depression or the Afar Triangle—covers a large area (~200,000 km^2) in northeastern Ethiopia, eastern Eritrea, and Djibouti. It defines a distinct geomorphic feature created by the interactions of extensional processes in the oceanic Red Sea and the Gulf of Aden Rifts with the MER, resulting in a Triple Junction (CNR-CNRS 1973). The triangular morphology of the Afar Rift is also a product of the interactions of the boundary faults of the three rift systems (WoldeGabriel et al. 2009).

The major chronostratigraphic units of the Afar Rift consist of crystalline basement of the Nubian–Arabian Shield, marine and continental sedimentary units of Mesozoic age, and Tertiary pre- and post-rifting volcanic and sedimentary rocks. The latter are mostly exposed along the western, southern, and eastern rift escarpments, along marginal grabens and the foothills and horsts within the rift floor (Merla et al. 1973; Varet and Gasse 1978; Zanettin and Justin-Visentin 1975). The earliest basaltic eruptions along the rift margins are late Oligocene (24 Ma) in age (Chernet et al. 1998 and references therein). The basin floor is filled with 500–1500-m-thick late-Miocene to Plio–Pleistocene flood basalts, and silicic lavas and tephras, interbedded with fluvial and evaporite deposits that are assigned to the Afar Series (Tefera et al. 1996; Varet and Gasse 1978).

Sedimentological studies and ages of volcanic rocks in the Red Sea and Gulf of Aden Rifts and adjacent regions suggest that rifting commenced between 35 and 27 Ma (Ukstins et al. 2002; Watchorn et al. 1998; Wolfenden et al. 2005). Moreover, late Oligocene (24 Ma) fluvial deposits and interbedded basalts of the Red Series exposed along the foothills of the western rift escarpment of the Afar Depression and the Danakil Block suggest that rifting within the Afar Depression was coeval with the Red Sea and Gulf of Aden Rifts (Tiercelin et al. 1980 and references therein). Hundreds of metres of thick red sandstone, conglomerate, and variegated clays accumulated in a pre-existing basin floored by weathered crystalline basement rocks (Tefera et al. 1996). By the late Oligocene and early Miocene, thermal thinning of the lithosphere due to the expansion of the Afar Plume and flow of plume material beneath initial rift basins is thought to have triggered the southward propagation of rifting (Ebinger and Sleep 1998). Today, the Afar Rift floor is fragmented into several tectonic blocks driven by ongoing passive upwelling, magma-assisted extension, magmatic eruptions, and complex tectonic interactions of the three major rifts and their boundary faults (Barberi and Varet 1977; Beyene and Abdelsalam 2005; WoldeGabriel et al. 2009). Evidence for continuous tectonic and magmatic activity and sedimentation (i.e. fluvial, lacustrine, and evaporites) during the late-Miocene to Plio–Pleistocene is preserved on the basin floor, along the foothills of the escarpments, fault scarps, and canyons within the Afar Rift. Thick evaporite deposits (~1.6 km) of marine reef limestone, red beds, gypsum–anhydrite–halite sequences, and potash-rich strata deformed by tectonic and hydrothermally driven salt-forming processes occur in the eastern and northern part of the basin floor (Holwerda and Hutchinson 1968; Teunis et al. 2008).

Ayenew (2009) described saline lakes of variable sizes and depths in the northwestern and eastern parts of the Afar Rift floor (Fig. 2.1a). The largest lakes are Asale and Afdera, which lie −116 and −102 m below sea level. The Awash River, which flows along the NE-trending axial zone of the MER, empties into Lake Abbe, a hypersaline lake in the southeast-central Afar Rift floor. The lake contains about 50-m-high steam-venting limestone chimneys.

2.4.2 Main Ethiopian Rift

The MER is a 600-km-long, well-defined rift basin that propagates from the Afar Rift in a SW direction. The MER is divided into the northern, central, and southern sectors (WoldeGabriel et al. 1990). Each sector exhibits characteristic geomorphic, tectonic, geological, and hydrological features that are highlighted from the northern to the southern sectors (Fig. 2.1b).

2.4.2.1 Northern Sector: Volcano-Tectonic and Sedimentological Records

The approximately 200-km-long and 80-100-km-wide northern sector covers an area of the MER that extends south from the funnel-shaped SW Afar Rift to the drainage divide of the NE-flowing Awash River and the lakes region in central Ethiopia (Fig. 2.1a and b). The basin, bounded by steeply faulted rift escarpments, widens and gently drops in elevation towards the Afar Depression. The basin floor is generally flat except for the Quaternary axial tectonic and volcanic rift zone, which is topographically higher and rugged compared with the marginal basins because of the many cinder cones and rhyolite domes and calderas. The axial rift zone is asymmetrically located closer to the western rift margin.

Even though pre-rift structural features are mostly concealed beneath Miocene volcanic and sedimentary rocks, evidence for syn-rift tectonic activity within the northern sector of the MER is preserved along the rift escarpments and on the rift floor. The most obvious features are changes in the orientation of the boundary faults along the transition zone to the Afar Rift. For example, the western boundary fault of the northern sector has been evolving from NNE to NNW directions, whereas the opposing eastern margin is shifting from NNE to EW due to boundary fault interaction with the Red Sea and the Gulf of Aden Rifts, respectively (WoldeGabriel et al. 2009). These changes, however, are not reflected in the narrow belt of the NE-trending Quaternary axial tectonic and volcanic rift zone of the rift floor (Fig. 2.1a and b).

Based on the distribution of volcanic and sedimentary rocks along the scarps of the boundary faults, rifting in the northern sector is inferred to have started in the middle Miocene (Juch 1980; Kazmin et al. 1980; Sickenberg and Schönfeld 1975; Wolfenden et al. 2004). During the late Miocene, several silicic volcanoes formed along the eastern marginal graben that was filled with thick ignimbrites and fossiliferous sediments, whereas slightly younger diatomaceous and fluvial sedimentary rocks and intercalated tephra units accumulated along the western marginal graben (Chernet et al. 1998; Kazmin et al. 1980; Kuntz et al. 1975; Renne et al. 1999; Sickenberg and Schönfeld 1975; Tiercelin et al. 1980; WoldeGabriel et al. 1992, 2001, 2009). The Awash River gorge adjacent to the axial rift zone exposes a sequence of basaltic flows and interbedded tephra units, but no sedimentary rocks similar to those found along the marginal graben (WoldeGabriel et al. 1992). Most of the Quaternary volcanic and tectonic activities within the northern sector are confined to the segmented and en-échelon displaced axial rift zone.

2.4.2.2 Central Sector: Volcano-Tectonic and Sedimentological Records and Hydrological Features

The central sector is a closed basin approximately 200 km long and about 75 km wide, which becomes narrower towards the southwest (WoldeGabriel et al. 1990 and references therein). Changes in structural and geomorphic features characterize the transition zone from the northern to the central sectors (Fig. 2.1b). For example, the western margin near Addis Ababa shifted north and forms an embayment that overlaps with the east–west-oriented Ambo Lineament (Chernet et al. 1998; Morton et al. 1979). The western margin is poorly defined south of the embayment until it reaches the Guraghe Mountains (Fig. 2.1b). A similar northward shift of the eastern margin is also noted in the vicinity of Asela. On the rift floor, two narrow (5–10 km wide), densely faulted marginal grabens occur. The Wonji Fault Belt defines the east side and it is marked by several Quaternary cinder cones, basaltic lava fields, and rhyolitic centres and calderas; in contrast, the Debre Zeyt-Silti Zone on the west side consists of basaltic cones, lava fields, and maars (Mohr 1967; WoldeGabriel et al. 1990).

In the central sector, high-angle normal faults and horst and graben structures close to the margins delimit the rift basin (Fig. 2.2a and b). Post-rifting arcuate structure, right-angled spurs, and serrated edges and re-entrants also occur along both rift margins. Based on stratigraphic and structural relations of the various lithologic units exposed along both escarpments, rifting in the central sector started in the middle Miocene (Corti 2009; WoldeGabriel et al. 1990). The thick late-Miocene (10.6–8.3 Ma) basalt sequence and vitrophyre at the base of the Guraghe Mountains along the western margin erupted into an existing basin a few kilometres to the west of the tilted Kella Horst, which consists of crystalline basement, Mesozoic sedimentary units, Oligocene (32 Ma) basaltic flows, and fluvial sedimentary deposits but not the late-Miocene Guraghe Basalts. Additional evidence for middle Miocene or older rifting in the central sector is indicated by the stratigraphic succession of the Shebele Trachyte sequence exposed along the eastern rift shoulder. The NE-trending Wabi Shebele River gorge contains more than 150 m of bedded volcaniclastic sediments that underlie middle Miocene (16.9 Ma) mafic lavas of the Shebele Trachyte (WoldeGabriel et al. 1990). The basin floor in the central sector was apparently more than 100 km wide and became narrower, following the eruptions of the voluminous Plio–Pleistocene Butajira Ignimbrite and Chilalo Trachyte flows (WoldeGabriel et al. 1990). During the Quaternary, tectonic activities and volcanism were confined to the Wonji Fault Belt (e.g. Fig. 2.2c) and to the Debre Zeyt-Silti marginal grabens. The two Quaternary volcanic and tectonic segments merge into a singular axial rift zone in the southern part of the central sector (WoldeGabriel et al. 1990).

The early Miocene (16.9 Ma) sedimentary units in the rift-oriented Wabi Shebele River gorge indicate early sedimentation along an

Fig. 2.2 The photographic images from the central sector of the Main Ethiopian Rift show the eastern (**a**) and western (**b**) rift margins, representing well-defined high-angle normal faults and the rift-bound Lake Abijata (**c**) and a silicic centre along the Quaternary axial rift zone of the Wonji Fault Belt (**d**) north east of Lake Abijata. The photos are from G. WoldeGabriel (**a** and **b**) and from M. Schagerl (**c** and **d**)

extensional basin, which was part of the central sector of the MER (Fig. 2.1b). Localized fluvial deposits (2–10 m thick) were also identified above the Oligocene Kella Basalt along the fault scarp of the Guraghe Mountains, but their age was not constrained. In the rift floor, thick fluvial and lacustrine deposits that are intercalated with Plio–Pleistocene volcanic rocks blanket the rift floor (Gasse and Street 1978). Today, several compartmentalized lakes cover a large area of the basin floor of the central sector (Figs. 2.1b and 2.2d). These lakes are remnants of a larger lake that prevailed during the wet periods of the late Pleistocene (Gillespie et al. 1983; Street 1979; Tudorancea and Taylor 2002 and references therein). Input from perennial streams and groundwater flux keep most of these lakes fresh except for some of the closed caldera and/or Crater Lakes (e.g. Shala, Abijata, Chitu, and Tilo) that are saline (Ayenew 2005, 2009).

2.4.2.3 Southern Sector: Volcano-Tectonic and Sedimentological Records and Hydrological Features

The MER becomes narrower as it propagates southwestwards to its termination. The 150–200-long and 30–65-km-wide southern sector forms unique geomorphic and tectonic features unlike the closed basin of the central and the northeastward-draining basin of the northern sectors (Fig. 2.1b). The uplifted southwestern border of the nested Awasa Caldera Complex forms the boundary between the central and southern sectors. The basin floor of the southern sector is rugged and gradually drops in elevation along the axis of the rift towards the southwest to form Lakes Abaya and Chamo. The basin contains a singular, left-laterally displaced and segmented Quaternary axial rift zone that is asymmetrically placed closer to the western margin. High-angle boundary faults mark the eastern rift escarpment, whereas the west side becomes less apparent largely due to the accumulation of thick tephra deposits from the nested early Pliocene Wagebeta Caldera complex located along the rift margin north–northwest of Soddo (Fig. 2.1b). Moreover, the southern half of the basin floor is bifurcated by the NNE-trending, 90-km-long and 25-km-wide Amaro Horst into the western (Ganjuli or Chamo Basin) and the eastern (Galana) Grabens (Levitte et al. 1974). The Quaternary axial rift zone is confined to the Ganjuli Graben, whereas the Galana Graben is filled with sedimentary deposits and interbedded distal tephra, but it does not host volcanic centres or flows.

Evidence from stratigraphic successions and the distribution of Tertiary volcanic centres suggests that rifting in the southern sector started in the middle Miocene, consistent with the confinement of trachytic (14 Ma) and rhyolite (13 Ma) centres and flows along the foothills of the major boundary faults on both sides of the Ganjuli Graben (Ebinger et al. 1993; Zanettin et al. 1978). Moreover, the stratigraphic sequence at the southern end of the Amaro Horst contains middle Miocene fossiliferous volcaniclastic deposits (17–15 Ma) that are capped by basaltic lavas (12–11 Ma). This suggests that the Amaro Horst was probably uplifted in the late Miocene (Suwa et al. 1991; WoldeGabriel et al. 1991).

The NNE–SSW-trending twin basins of the Galana and Ganjuli Grabens of the southern sector abruptly terminate against uplifted crystalline basement and Miocene basalts of the Konso Highlands (Ebinger et al. 1993, 2000; WoldeGabriel and Aronson 1987; Yemane et al. 1999). Within the termination zone, the rift-oriented faults have steeper dips compared with structures of the crystalline basement of the same trend (Ebinger et al. 1993). The termination of the southern sector may be related to a combination of factors, including truncation by the north-trending tectonic zone of the Kenya Rift, diminishing displacements along major faults that intersect uplifted crystalline basement rocks of the Konso Highlands, and the spreading of the tectonic processes over a broader zone of active and inactive rifts (Bosworth 1992; Moore and Davidson 1978; WoldeGabriel and Aronson 1987).

The basin floor of the southern sector is mostly filled with volcaniclastic deposits and widespread distal tephra beds from rift-bound calderas within the central and southern sectors of the MER (Katoh et al. 2000; WoldeGabriel et al. 2005). Lake Abaya (1160 km^2), the largest freshwater lake in the Ethiopian Rift System, and the mildly saline Lake Chamo (317 km^2) occupy rift-oriented basins closer to the western margin (Ayenew 2009). Both lakes are connected through a narrow channel that carries water from Lake Abaya to Chamo during wet periods.

2.4.3 Broad Rift Zone of SW Ethiopia: Volcano-Tectonic and Sedimentological Records

The north-trending, broad rift zone in SW Ethiopia is the northern extension of the Kenya Rift (Ebinger et al. 2000; Moore and Davidson 1978; WoldeGabriel and Aronson 1987). As shown in Fig. 2.1c, this 300-km-wide tectonic zone consists of three half graben and two symmetrical rift basins. The symmetrical rifts and the half grabens are separated by uplifted crystalline basement ridges that are unconformably capped by ferruginous sandstones (grits) and the oldest Tertiary volcanic rocks associated with the uplift and doming of the Afar Dome (Davidson and Rex 1980; Ebinger et al. 2000; Moore and Davidson 1978). From east to west, the Ririba and Chew Bahir Rifts are symmetrical extensional basins, whereas the Omo, Usno, and Kibish Basins represent half grabens (Fig. 2.1c). Rift faults within the tectonic zone are either parallel or normal to the structural fabric of the crystalline basement, and no evidence for reactivation was noted along these structural elements (Moore and Davidson 1978).

The eastern margin and shoulder of the broad rift region is a zone of tectonic overlap and fault interactions between the NE- and N-striking major structures of the southern MER and the Kenya Rift, respectively. At the zone of overlap, the MER faults are truncated by the dominant

N–NNE-striking Kenya Rift fault systems, which are responsible for the symmetrical Chew Bahir and Ririba Rifts along the eastern part of the broad rift region. In contrast, the N-oriented broad rift zone, and in particular the Chew Bahir Rift closest to the western boundary fault of the southern MER, is truncated by NE-striking fault system of alternating and tilted horst and graben structures (Moore and Davidson 1978; WoldeGabriel and Aronson 1987). Such tectonic influence by the closest boundary fault system of a rift basin is also apparent at the northern end of the EARS, where the NW-trending western boundary and the axial rift zone faults of the Red Sea Rift truncate the major structures of the MER and the Gulf of Aden in the central Afar Rift floor (Fig. 2.1a). In the area of tectonic overlap, the dominant rift faults are inferred to be controlled by deep-seated structures of the closest rift basin instead of the pre-rift shallow reactivated structures.

Even though the earliest volcanism (~45 Ma) started in SW Ethiopia, extensional basins and rifting within the broad rift zone did not start to form until the early to late Miocene (Ebinger et al. 2000; Moore and Davidson 1978). Based on a gravity survey, Mammo (2012) reported more than 4 km of early Pliocene sediments in the N-oriented Omo Basin. In northern Kenya, immediately south of the broad rift zone, late Oligocene to early Miocene deep (~7 km) half graben formed west of Lake Turkana. Later, the tectonic activity migrated eastwards to the Turkana Basin (Morley et al. 1992). The distribution of Plio–Pleistocene volcanic rocks and faulting within the eastern part of the broad rift zone also suggests that rifting and dense faulting migrated eastwards from the Chew Bahir to the Ririba Rifts (Ebinger et al. 2000; WoldeGabriel and Aronson 1987). Geological investigations related to paleoanthropological explorations in the Omo–Turkana Basin of SW Ethiopia and northern Kenya provide information about the Plio–Pleistocene tectono-sedimentary records and localized basaltic eruptions (Brown and McDougall 2011; de Heinzelin 1983).

In summary, the broad rift zone of SW Ethiopia is the least investigated and poorly understood sector compared with other parts of the EARS because of its remoteness. However, the geographic location and the formation of the broad rift zone at the terminus of the NNE-oriented Kenya Rift and the NE-trending MER primarily appear to be the product of the tectonic overlap and subsequent fault interactions of the two rift basins and the eastward migration of faulting. The north-trending crystalline basement ridges within the broad rift zone and at the southern end of the MER probably acted as tectonically rigid blocks, allowing for multiple basins to form between these resistant blocks. In this context more future research will also be necessary to better understand the role of the WNW–ESE-oriented Cretaceous Anza Rift in influencing Cenozoic Extensional processes. The Anza Rift once kinematically connected the Central African Rift Zone and the Lamu Embayment at the passive margin of Kenya and was repeatedly active during the Tertiary (Bosworth and Morley 1994).

2.5 The Kenya Rift

2.5.1 Physiographic Features

The Kenya Rift extends from Lake Turkana in northern Kenya to the northern tip of Lake Natron at the Kenya–Tanzania border (Fig. 2.3). It is a topographically closed area that drained into Lake Turkana during lake-level high stands in the past in the north and Lake Natron to the south (Olago 2013). This rift is associated with high elevations of the East African Dome, rift-shoulder uplift, and high volcanic edifices (e.g. the Aberdare Range) and is located at the equator. It constitutes a unique tectonic and climatic setting with pronounced relief contrasts, highly disparate rainfall amounts, rainfall gradients, and an array of complex half graben along its axis (Bergner et al. 2009). The major fault scarps are the Elgeyo and Nguruman Escarpments that reach 1500 m and the Aberdare Range, a Miocene volcanic edifice situated at the present-day eastern rift shoulder that reach over 2000 m above the rift floor (Schlüter 1997). The

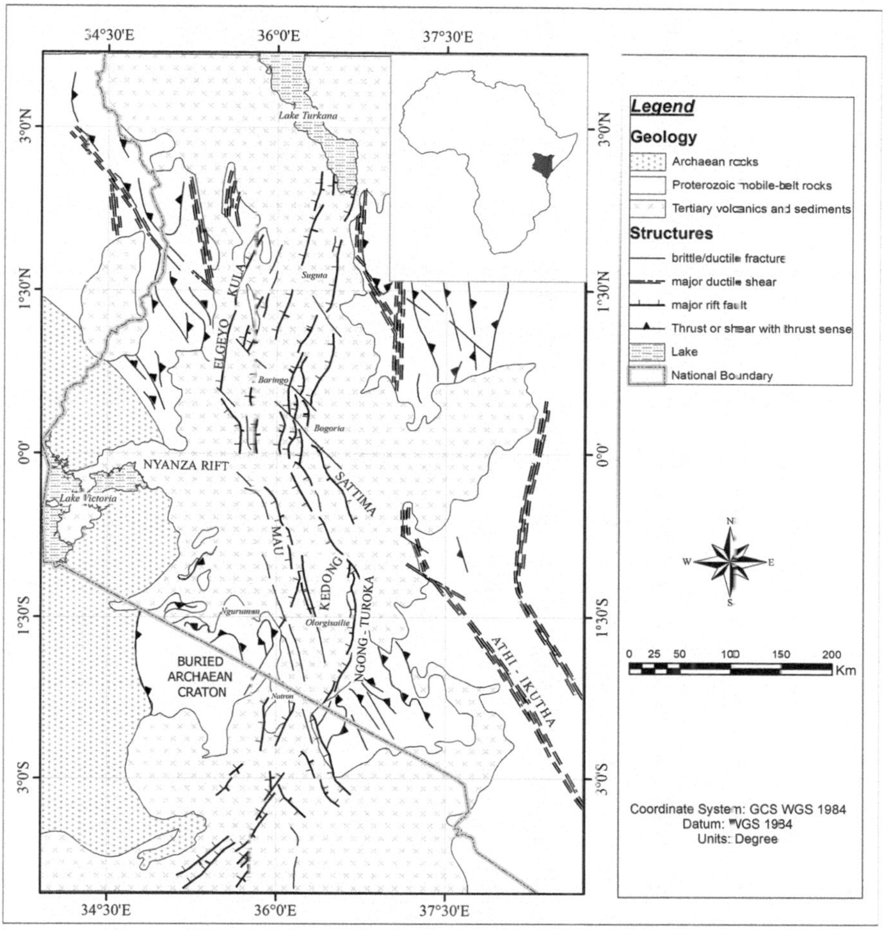

Fig. 2.3 Tectonic setting of the Kenya Rift segments and faults shown in relation to major structures in the Precambrian basement. Modified from Smith (1994)

fault throws are estimated 3000–4000 m (Baker and Wohlenberg 1971). Thus, the present-day general architecture of the EARS is made of a series of relatively narrow (40–70-km-wide) faulted troughs (Tiercelin and Lezzar 2002). Neotectonic faulting and the formation of eruptive centres on the rift floor during the ultimate 400 kyr have given rise to a number of lakes along or close to the axis of the rift. They are virtually isolated today, but may have been connected during more humid episodes in the past. These include Lakes Turkana, Bogoria, Nakuru, Elementaita, Naivasha, and Magadi. The evolution of the rift was evidently characterized by alternating episodes of rifting and volcanism and continuous sedimentation. The complexity of rifting as observed today is to a large extent attributed to the influence of pre-existing structures in the underlying rocks of the Mozambique Mobile Belt (Birt et al. 1997). This influence is reflected in the parallel alignment of faults and gneissic foliations along the Nguruman Escarpment in the south (Crossley 1979) and the Elgeyo Escarpment in the north, where inherited strike-slip zones have been reactivated as extensional transfer faults during Tertiary rifting (Hetzel and Strecker 1994; Mugisha et al. 1997).

The Kavirondo (or Nyanza) Rift is a N80°E-trending rift that branches with the central Kenya

Rift near Lake Bogoria and disappears westwards under Lake Victoria (Chorowicz 2005). Thus, the Lake Bogoria region may represent a Triple Junction between the northern Kenyan, central Kenyan, and Nyanza Rifts, almost at the centre of the Kenya Dome (Chorowicz 2005). Saggerson (1952) asserts that faulting accompanied by tilting began in the Tertiary (but did not specify which epoch) and continued into the Pleistocene. Shackleton (1950) proposed that the Kavirondo Rift was mainly formed during the late Pliocene or earliest Pleistocene; the rift was depressed between faults and flexures, relatively to the regions to the north and south, which were uplifted and tilted. Jones and Lippard (1979), however, based on newer geological evidence, concluded that the Kavirondo Rift was formed during the upper Miocene and that there has been very little volcanic or tectonic activity along it since that time. The floor of the Kavirondo Valley has remained more or less level. This particular rift, so far not studied in as great detail as the Kenya Rift, has only now come under focus due to ongoing exploration work for hydrocarbons. The following sections, therefore, give an overview of the characteristics of the Kenya Rift, including faulting mechanisms and the influence of pre-rift structures on its evolution based on the works of several authors.

2.5.2 Rift Faulting Characteristics

Most of the major faults defining the Kenya Rift Valley are normal or dip-slip faults with dips in excess of 60° (Baker and Wohlenberg 1971). These faults do not extend laterally for many kilometres. Accordingly, the long fault scarps are produced by a number of faults offset from each other in an en-échelon arrangement. Although the escarpment in the central section of the Kenya Rift reaches 2000 m, estimates on the order of 3000–4000 m have been suggested (Baker and Wohlenberg 1971). The grabens in the Kenya Rift are asymmetrical (half grabens). They have a single large fault on the western side (Elgeyo, Mau, Nguruman, etc.) and sets of antithetic faults. The latter originated from the segmentation of a rollover monocline after the major faults along the western margin of the rift had formed. These step faults often produce step-fault platforms or sloping ramps that descend down to the rift floor (Baker and Wohlenberg 1971). The footwall blocks of these structures are usually tilted surfaces sloping away from the Rift Valley, attesting to listric fault geometry. The half-graben characteristics have also been observed in the case of the Anza Graben located in NE Kenya (Dindi 1994). On the graben floor, normal faults are closely spaced and increase in density along the volcano-tectonic axis of the rift (grid faulting, McCall 1967). Here, NNE–SSW- to N–S-striking faults have short lengths, are parallel, and have throws of less than 100 m. Baker (1986) estimated the average spacing of these faults at 1.5 km. Grid faults give rise to minor horsts, graben, and step platforms. Most of these faults offset ~1.4–0.8 Ma lavas, which erupted in the inner trough of the rift (Muirhead et al 2013). Some obliquely striking faults of poorly resolved kinematics and strike-slip faults have been identified in some parts of the Kenya Rift. Kuria et al. (2010), working in the Magadi area, identified four sets of faults (i.e. normal N–S, dextral NW–SE, strike-slip ENE–WSW, and sinistral NE–SW) of different ages. Previous studies had not recognized the existence of sinistral oblique slip NE–SW-trending faults that were created under a regime of E–W extension. The E–W extension has also been confirmed from focal mechanism solutions of earthquake swarms where the four sets of faults intersect. Current E–W extension rates are estimated at ~2–4 mm/year (Katternhorn et al. 2013). Strike-slip faults have also been identified in the area to the south of Lake Turkana (Dunkelman et al 1988). In this context, the 1928 Subukia Earthquake ruptured along a Tertiary dip-slip fault, but with a component of oblique slip (Zielke and Strecker 2009).

The Magadi sector of the rift displays some of the unique characteristics of the half-graben structures in the Kenya Rift. In the Magadi area, the half graben is about 70 km wide, with a high fault escarpment of 1500 m on its western side (Nguruman) and subdued eastern boundary of 200–500 m. The rift floor slopes southwards and has the appearance of a smooth plain of flood lavas, which has been broken by many subparallel minor faults that rarely exceed 200 m in displacement (Baker and Wohlenberg 1971). The major marginal faults form step-fault platforms and south-sloping ramps. The rift floor is cut by numerous normal faults, giving rise to small-scale step faults, horst and graben, and tilted antithetic fault blocks (Baker and Wohlenberg 1971). Faults in the axial zone of the rift floor tend to be more regular in orientation than the main marginal ones. The Magadi sector best exhibits the migration of the zone of fracturing inward, clearly showing a decrease in ages and spacing of faults.

2.5.3 Rift Faulting Mechanisms

The complexity of the Kenya Rift has been attributed to a structural control by the inherited structural trends in the underlying Mozambique Belt. A study of the Precambrian Mozambique Belt in Kenya (Smith and Mosley 1993) demonstrated that a series of late Proterozoic, continental-scale NW–SE- and NS-trending ductile/brittle shear zones exist in the lithosphere beneath the Kenya Rift. According to Pinet and Colletta (1990), inherited features induce discontinuities within the crust, which lead to complex fault patterns of rift systems. Being located within the Mozambique Belt rocks, it is conceivable that the rift architecture was fundamentally influenced by inherited anisotropies and that rheological conditions were conducive to weakening of the lithosphere when mantle upwelling impinged upon the lithosphere–asthenosphere boundary. This ultimately led to rifting in the former collision zone between East and West Gondwana.

2.5.4 Basins and Sedimentation in the Kenya Rift

The basins in the rift have been formed since Mio-Pleistocene times, and these achieved their final morphology between 2 and 0.5 Ma, concomitant with other new lake basins being created as a result of the major tectonic and/or volcanic events that took place during this period (Tiercelin and Lezzar 2002). Present-day lakes such as Lake Baringo or Lakes Nakuru, Elementaita, and Naivasha are the remnants of larger lake domains developed during lower–middle Pleistocene time and are represented by rare sedimentary deposits sometimes uplifted or down-faulted along the inner trough margins (Renaut 1982; Richardson 1966; Richardson and Richardson 1972; Tiercelin and Lezzar 2002; Wilson et al. 2014). There has been a lot of scientific and industry interest in the intercalated volcaniclastic, fluvial, and lacustrine sediments of the Kenya Rift basins. The scientific community is evaluating them to understand tropical paleoclimate and environment, as well as for their rich paleontological, archaeological, anthropological, and evolutionary significance. At the same time, industry is engaged in exploration for oil and gas, as well as for industrial minerals such as trona, phosphate, and diatomite.

2.5.4.1 Turkana, Lokichar, and Suguta Basins

The age of Lake Turkana Basin is given a conservative estimate of 4.3 Ma (by K–Ar and $^{40}Ar/^{39}Ar$ methods), which is recorded from the lowermost tuff bed within the Koobi Fora Basin (McDougall 1985). Nonetheless, volcanic and sediment relationships at the boundary of the Koobi Fora Basin (NE of Lake Turkana) reveal that major Pliocene tholeiitic flood volcanism took place after the initiation of proto-Lake Turkana. Significant lacustrine and fluviatile deposits lie beneath lavas in the upland part of the northern part of the Koobi Fora Basin. This indicates that the age of Lake Turkana may be significantly greater than 4.3 Ma (Watkins 1986). Multifold seismic reflection and geologic mapping indicate that at least 3 km of Neogene sediments occur beneath Lake Turkana. These sediments filled a series of half-graben basins, with centrally located Quaternary volcanic centres, which are linked end-to-end by structural accommodation zones (Dunkelman et al. 1988).

A gravity and seismic survey conducted in 1985 by Amoco Kenya Petroleum Company indicated the presence of two deep-buried, elongate (50 km long; 20 km wide) sedimentary basins, the Lotikipi Basin to the west and the Gatome Basin to the east, separated by the Lokwanamoru Structural High (Thuo 2010). Depth to basement was interpreted to be 4 km and 6 km for the Lotikipi and Gatome Basins, respectively (Desprès 2008; Wescott et al. 1999). The current phase of landscape evolution within the Turkana Basin is dominated by fluvial incision and slope retreat under semiarid conditions. A volumetric analysis dates the present basin landform at 0.11 Ma (Reid and Frostick 1986).

South of the Turkana Basin is the Lokichar Basin with no present-day lake expression. It contains a 7-km-thick sedimentary series of mainly interbedded lacustrine and fluvio-lacustrine sediments of Paleogene to middle Miocene age (Morley et al. 1992). An extensive seismic investigation conducted west of Lake Turkana in the 1980s by the Amoco Kenya Petroleum Company revealed the existence of a string of five large, N–S-trending half graben with up to 6–7 km fill, ranging in age from Palaeocene to late Miocene–Pliocene (i.e. Lokichar, Lothidok, North Lokichar, North Kerio, and Turkana Basins) and linked to the initial phases of rifting (Dunkelman et al. 1988; Morley et al. 1992, 1999; Tiercelin et al. 2004). The pre-Pliocene Turkana Grits clearly reflect intense local deformation, which signals the onset of significant tectonic activity that represents two episodes of local tectonics and deposition (Williamson and Savage 1986). The Plio–Pleistocene sediments of the basin are the product of episodic deposition interspersed with long periods of reworking and, stratigraphically, include a wide range of lithotypes from fine clays to extensive coarse gravel sheets (Reid and Frostick 1986).

The Suguta Valley, directly south of Lake Turkana, hosts Lake Logipi at its extreme northern end, separated from the south end of Lake Turkana by the Volcanic Barrier Complex. Lake Logipi is fed by numerous hot springs located around its northern shores and around the eroded ruff cone of Naperito as well as stream/surface runoff inflows during the wet seasons (Dunkley et al. 1993). Baker (1963) did geological work in the Baragoi area, including a small portion of the Suguta Valley in the northwest sector of his assessment area. He observed a flight of beach terraces on a number of ash cones, coupled with the presence of thick lacustrine and fluviatile deposits. He therefore noted that a paleolake once existed, which was connected to Lake Turkana. Tectonic movements and associated volcanism starting in the Miocene led to the formation of the deep and narrow Suguta Valley (Baker 1963). The sedimentation history in the Suguta Valley comprises recurrent fluvio-lacustrine episodes since the Miocene. These are documented by fluviatile terraces, beach terraces, and lakebed sequences exposed in the river valleys cutting through the rift shoulders, mainly to the east and in the inner graben area (Baker 1963; Casanova et al. 1988; Dunkley et al. 1993; Junginger et al. 2013; Saneyoshi et al. 2006). The Miocene Namurungule Formation, which crops out on the eastern flank, represents basin fill that developed in association with Miocene rifting. It consists mostly of lacustrine delta deposits with minor alluvial fan deposits at its base (Saneyoshi et al. 2006). A combination of steep west-dipping volcanic rocks and pronounced topography on the eastern margin of the northern Suguta Valley has resulted in unstable loading of the valley margin. This possibly led to very significant sediment or rock transport from the rift margins to its axis and may be indicative of rift sedimentation following periods of most rapid structuring (Bosworth and Maurin 1993). The extensive lake phases are well documented in Pleistocene-age sediments: up to 30-m-thick diatomite, silts, and sands that are exposed in the southern and western parts of the Suguta Valley are overlain by flood basalts, documenting a lacustrine phase between 1.0 and 0.7 Ma (Dunkley et al. 1993; Junginger et al. 2013). During the late Quaternary, high lake levels are indicated by terraces marked with deposits such as gastropods (Garcin et al. 2012; Truckle 1976) and stromatolites (Casanova et al. 1988).

2.5.4.2 Kerio, Baringo, and Bogoria Basins

Initiation of the two oldest deep rift basins of the central Kenya Rift, the Kerio and the Baringo Basins, formed in Paleogene time (Hautot et al. 2000; Mugisha et al. 1997; Tiercelin and Lezzar 2002). Exposed lake- and fluvial-type sediments of possible Paleogene age are green laminated shales and sandstones forming the Kimwarer and Kamego Formations that outcrop in the Kerio and Baringo Basins, respectively (Hautot et al. 2000; Renaut et al. 1999). These two formations represent the upper part of a several-km-thick sediment pile of possible Paleogene age that is identified only based on geophysical methods (Hautot et al. 2000; Mugisha et al. 1997). They lie at the base of a nearly entire rift-fill sequence (known as the Tugen Hills sequence) that is considered as one of the best-exposed successions of Neogene sediments in the EARS (Chapman and Brook 1978; Tiercelin and Lezzar 2002). Thick basaltic and phonolites series more or less extensively capped the Kerio and Baringo Basins (Samburu, Sidekh, and Elgeyo Formations) between 23 and 10 Ma (Fig. 2.4) (Chapman and Brook 1978; Hill et al. 1986). During this period, fluvio-lacustrine sedimentation continued to develop in the Kerio Basin of central Kenya, with the deposition of the 400-m-thick Tambach Formation dated between 16 and 14 Ma (Renaut et al. 1999). The Tambach Formation was in turn capped by the Uasin Gishu Phonolites by 14 Ma (Chapman and Brook 1978; Lippard 1973). This was followed by a second phase of rift tectonics that affected the Kerio Basin, initiating the deposition of the Ngorora fluvio-lacustrine formation in an almost 100-km-long and 40-km-wide faulted basin (Tiercelin and Lezzar 2002). The lower half of this thicker Ngorora Formation is only "geophysically" represented, while the upper 400 m of sediments outcrop largely in the Tugen Hills region between the Kerio and Baringo Basins (Chapman and Brook 1978). In the Baringo–Bogoria Basin, a well-dated sequence of diatomites and fluvio-lacustrine sediments document the precessionally forced cycling of an extensive lake system between 2.70 Ma and 2.55 Ma (Wilson et al. 2014). One diatomite unit was studied using the oxygen isotope composition of diatom silica combined with X-ray fluorescence spectrometry and taxonomic assemblage changes. Those approaches reveal a rapid onset and gradual decline of deep-water lake conditions, which exhibit millennial-scale cyclicity of ~1400–1700 years, similar to late Quaternary Dansgaard–Oeschger Events. These conditions are thought to reflect enhanced precipitation coincident with increased monsoonal

Fig. 2.4 Major step faults along the margin of Lake Baringo, Kenya (from Michael Schagerl)

strength and testify to a teleconnection between the high latitudes and East Africa during this period (Wilson et al. 2014).

2.5.4.3 Nakuru, Elementaita, and Naivasha Basins

Late Pleistocene sediments comprising volcanic ash, silts, clays, graded water-laid tuffs, and diatomite beds characterize the sediments of the Lake Nakuru and Elementaita Basins (Thompson and Dodson 1963). The lake sediments occur as lenses mainly consisting of grit, volcanic sand, clays, and silts or as continuous beds with some diatomite beds having thicknesses over 30 m (Odero 1993). The Lake Naivasha Basin was created by normal faulting between 0.9 and 0.7 Ma (Strecker et al. 1990). Its present history began at about 320 kyr BP, when lava flows at Olkaria closed the southern basin outlet between the flanks of the 400-kyr-old Mt. Longonot and the Mau Escarpment (Clark et al. 1990; Scott 1980). The sediments in the basin are generally not exposed, but stratigraphic relations in the Ol Njorowa Gorge in the transition between the Olkaria Volcanic Complex and Mt. Longonot show that no large lake existed before 146 ka BP (Trauth and Strecker 1996). The deposits in the gorge are mainly yellow to buff water-laid tuffs with altered lapilli, diatomite, laminated siltstone, and intercalations of coarse clastic fluvial sediments and pyroclastic deposits indicating intermittent subaerial conditions (Trauth et al. 2003).

2.5.4.4 Olorgesailie, Magadi, and Natron Basins

The Olorgesailie Basin, which has no present-day lake expression, is characterized by two major sedimentary sequences (Behrensmeyer et al. 2002). The Olorgesailie Formation (~1.2–0.5 Ma) consists of laterally extensive diatomites, diatomaceous silts, volcanic sands, gravels, ash, and paleosols that are overlain by a discontinuous series of younger sediments (0.5 Ma to present) that were laid down in and adjacent to a series of major incised river channels (Behrensmeyer et al. 2002). The dominant lithologies include volcanic silts, sands, and gravels that are variably affected by pedogenesis (Owen et al. 2014).

The present Lake Magadi, a saline pan, is the successor to a series of paleolakes that have occupied the basin for probably more than one million years. The oldest sediments, the Oloronga Beds (0.8 to ~0.3 Ma or later), are tuffs, fossiliferous limestone, diatomaceous siltstones, and cherts that were deposited in a large lake (Baker 1958, 1963; Behr 2002). These are overlain by Green Beds (98–40 ka in outcrop although full age range is unknown), lacustrine silts, and zeolitic tuffs that were deposited in an alkaline lake (Behr 2002; Goetz and Hillaire-Marcel 1992). The succeeding High Magadi Beds (HMB) are silts, zeolitic tuffs, rhythmic laminites, and evaporites that formed in a larger alkaline to freshwater paleolake during the late Pleistocene and early Holocene (~23 ka to ~10 ka) (Butzer et al. 1972; Damnati and Taieb 1995; Taieb et al. 1991). At least 40 m of bedded trona has been deposited upon the older sediments since the early to mid-Holocene. The estimate of sediment thickness from the Natron Basin is similar to the 3.5 km depth to basement interpreted in seismic refraction data from the Magadi Basin 30 km north of Natron (Birt et al. 1995). Oloronga Lake developed at >780 ka (Casanova 1987; Casanova and Hillaire-Marcel 1987; Eugster 1980, 1986).

2.6 The Tanzania Rift: Tectonics and Sedimentation in the Tanzania Rift Sector

The eastern arm of the EARS extends through Kenya into northern Tanzania. This is a divergent rift zone, where the eastern rift transitions from a single Rift Valley into a broad zone of tilted fault blocks that radiate a short distance southwards, thus marked by a ragged termination (Veevers 1981). The EARS is poorly defined south of the Kenya Rift (Chorowicz 2005). It diverges from a single, ca. 50-km-wide rift in southern Kenya to a ca. 200-km-wide zone in northern Tanzania, where it comprises three distinct rifts (from west to east: Natron–Manyara–Balangida, Eyasi–Wembere, and the Pangani Rifts) with different orientations (N–S-, NE-, and NW-trending, respectively) (Foster et al. 1997). The individual rift basins are half graben, bounded by a faulted rift escarpment on one side and a flexural warp on the other (Foster et al. 1997). The Natron–Manyara–Balangida and Eyasi–Wembere Basins are shallow (<3 km), and the total extension across the bounding faults is small; K/Ar age determinations on basalts from the western rift basins show that volcanism and sedimentation began in the area at ca. 5 Ma (Foster et al. 1997). Maximum basin depths derived from Euler deconvolution are W. Eyasi, 3.5 km; E. Eyasi, <2 km; Natron, 3.3 km; Manyara, 2.9 km; and Balangida <1 km, decreasing to the south in the Natron–Manyara–Balangida Rift, where faulting occurs across a broader region (Ebinger et al. 1997). Subsequent lacustrine sedimentation at Olduvai Gorge, an incision into a Plio–Pleistocene succession of volcaniclastic, fluvio-lacustrine deposits (Hay 1976), and west of Lake Natron was influenced by minor faulting (Foster et al. 1997). During the same period, Lake Manyara was apparently much larger than today, with sediments composed of interbedded clayey marls and phosphorites, sometimes intercalated with thin clays of volcanic origin (bentonites?) (Schlüter and Kohring 2001). A final major phase of rift faulting occurred at ca. 1.2 Ma and produced the present-day rift escarpments (Foster et al. 1997). Further east, the uplifted flanks of the Pangani Rift are relict features of several tectonic uplift events initiated during the Jurassic, with the final phase of significant tectonic uplift occurring during the early Tertiary (Mbede 2001).

2.7 Western Rift

2.7.1 Physiographic Features

The western arm begins north of Lake Albert and continues southwestwards to Lake Edward, and it then curves south through Lake Kivu and Lake Tanganyika and runs southeastwards through Lake Rukwa, extending south of Lake Malawi trending from NNE–SSW to N–S (Fig. 2.5). It is dominated by large boundary faults and an opposing flexural margin producing mobile asymmetrical full- and half-graben sedimentary (fluvio-deltaic and lacustrine) basins. These basins are typically 80–160 km long, 30–80 km wide, and individually linked along the rift axis (e.g. Lezzar et al. 2002). They are frequently occupied by elongate and narrow, shallow, and deep lakes (up to 1.5 km deep) separated by accommodation or transfer zones. The highest regions rising thousands of metres adjacent to the graben floor are characterized by volcanic ranges (e.g. Virunga Volcanic Field) and steep fault scarps dominated by the Rwenzori Horst block, whose northeastern foothills are still connected to the rift shoulder (Figs. 2.5 and 2.6).

2.7.2 Cenozoic Volcano-Tectonic Framework

From ca. 12 Ma, a shallow downwarp initiated a basin (Lake Obweruka), which became permanent at about 8–7 Ma when the first major rifting episode occurred; the rift shoulders became climatically significant around 4 Ma (Aanyu and Koehn 2011, Pickford et al. 1993). A major tectonic phase for South Kivu is cited at around 10–8 Ma (Ebinger 1989b), with the central and northern basins of Lake Tanganyika subsiding at ca. 12–7 Ma and the southern basin at ca. 4–2 Ma (Cohen et al. 1993). The Malawi Rift started to subside at ca. 9–8 Ma, with major rifting occurring north of Lake Malawi between 6 and 4 Ma; the rift increased in length southwards (Ebinger 1989a; Flannery and Rosendahl 1990). Renewed uplift around 2.6 Ma partitioned the Albertine Rift, with further rifting at 14–12 Ka, diverting the Albert flow northwards into the Nile Basin (Pickford et al. 1993).

2.7.2.1 Volcanism

Volcanic eruptive centres with more potassic magmas appear to be located mainly within accommodation zones, but locally also coincide

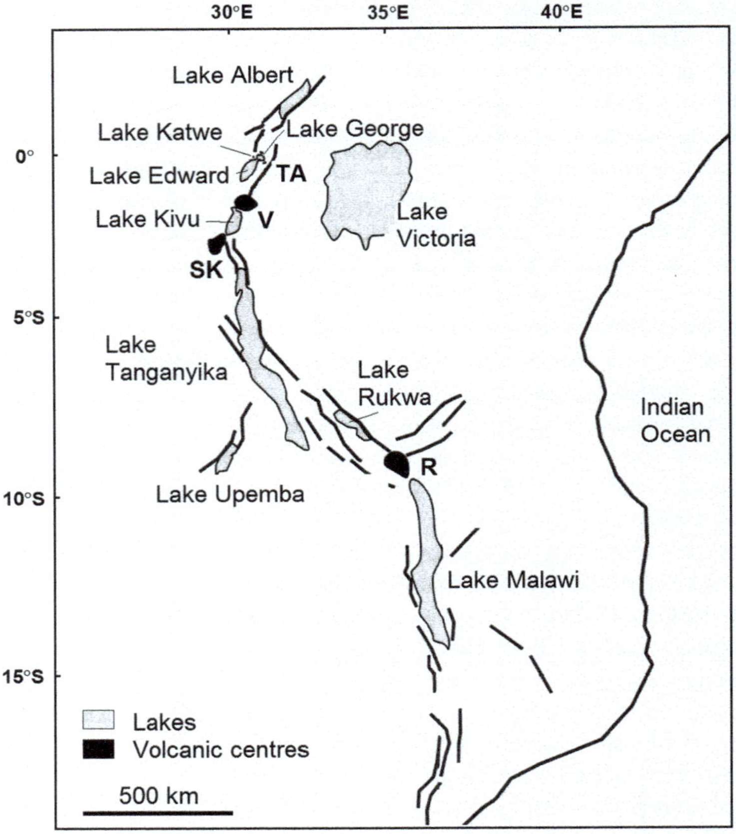

Fig. 2.5 Simplified location map of the Western Rift Valley showing the main lakes and volcanic centres: TA (Toro-Ankole), V (Virunga), SK (South Kivu), and R (Rungwe) [adapted from Morley et al. (1999)]

Fig. 2.6 Panoramic view of the Western Rift in western Uganda showing the rift floor with the Rwenzori Mountains in the background (from Mike Owor)

with border faults and with intra-basinal fissures often parallel to the border faults (Ebinger 1989a). Initial volcanic activity started at ca. 12 Ma in the north and at ca. 7 Ma in the south and is restricted to the areas of Rungwe, Virunga, South Kivu, and Toro-Ankole (Ebinger 1989b). A Miocene age (ca. 11 Ma) has been determined for the onset of the Virunga volcanism (Kampunzu et al. 1998), whereas the Toro-Ankole Volcanoes have been active intermittently during the last 50 ka. This 50 ka period coincides with a distinct phase of Quaternary volcanic activity in the South Kivu (Bukavu) and Virunga Areas (Boven et al. 1998). Late Quaternary volcanism in the Toro-Ankole and the Virunga appears to correlate with periods of tectonic activity in the area (Bishop 1969).

2.7.2.2 Tectonics, Basins, Lakes, and Sedimentation

The Western Rift is marked by a zone of high seismicity with over 11,000 events of local magnitudes from −0.5 to 5.1 and deeper earthquakes down to 30–40 km (e.g. Lindenfeld and Rümpker 2011; Morley et al. 1999). Sedimentary deposits underlie the bulk of the Western Rift basin, i.e. a mixture of recent alluvial material, fluvial deposits, and debris-flow units (Westerhof et al. 2014). Surface uplift associated with the formation of the Edward and Albert Rift basins reversed several major drainage systems, e.g. Katonga, Kagera, and Kafu Rivers, which flow bidirectionally (e.g. Kendall 1969). The Albertine Rift fill is composed of a thick pile (4–7 km) of terrestrial sediments with volcanic intercalations in the southern half (Westerhof et al. 2014). These volcanic flows were divided into an older (Oligocene–)Miocene suite and a younger Plio–Pleistocene–Holocene suite. Sediments in the southern part of the Albertine Rift probably contain a significant proportion of (reworked) volcanic detritus. From surface and subsurface observations, the sedimentary succession is mainly composed of fluvial and lacustrine deposits, which include a range of lithologies, e.g. conglomerates, sandstones, mudstones, cherts, shales, oil shales, and coal-bearing horizons.

At about mid-Miocene (~12 Ma) time, a shallow downwarp caused a basin (Lake Obweruka) that filled up with fluviatile and evaporitic deposits named the Kisegi Formation (Pickford et al. 1993). Lake Obweruka became permanent when the rate of downthrow exceeded sedimentation rates about 8 to 7 Ma; by about 2.6 Ma, it separated into smaller lakes due to renewed uplift and drier conditions. The Lake Edward Basin has as much as 4 km of sediments, which suggests a throw of 7 km from the top of the fault scarp to sub-lake basement (Laerdal and Talbot 2002). The Lake George Basin is largely filled up with lacustrine, alluvial, and fluvial sediments (Bishop 1969). Based on top and bottom unconformities of seismic reflections, Lake Edward, which is connected to Lake George via the Kazinga Channel, a paleo-river valley incised into the flexural margin of the lake, has three depositional sequences (McGlue et al. 2006). Clastic sedimentation on the flexural margin of Lake Edward is strongly modified by inflow from the Kazinga Channel and Lake George during periods of lowered lake level. These authors further suggest that the presence of an incised valley at the margin of the rift controls the stratigraphic development of clinoforms diagnostic of low-stand deltaic sedimentation. The morphology and orientation of these clinoforms point to discharge from Lake George during arid periods via the Kazinga Channel to transport sediment towards the exposed flexural margin of Lake Edward. Dipping reflections in sigmoid–oblique clinoforms indicate that these flows were capable of transporting sediment loads that included coarse grain sizes. Deposition by this process occurred several times during the late Quaternary history of Lake Edward and represents an important modern analogue for linked incised valley–lacustrine–delta sequences in the stratigraphic record.

Ebinger (1989b) cited an age of 5 Ma for fauna in the sediments of Rusizi Basin, which overlie basalts, dated at 6.5–5.0 Ma. The terraces formed by Plio–Pleistocene lacustrine sediments, for instance, in the eastern Kivu and Rusizi Basins, indicate that, during this time, rift basins became narrower due to increased uplift (Ebinger 1989b). The Rukwa Rift is a late Tertiary rift system superimposed on an earlier Karoo Age Rift (Morley et al. 1999). Data from two wells show late Miocene to recent sediments and sandy fluvio-deltaic deposits or shallow-water lacustrine shale. It is further suggested that axial drainage into a relatively short rift segment kept the sediment supply high, hindering the development of extensive deep-water lacustrine shale. Lake Tanganyika has a sedimentary thickness greater than 4 km. Its central and northern basins started to subside at ca. 12–9 Ma and ca. 8–7 Ma, respectively, and the southern basin at ca. 4–2 Ma (Cohen et al. 1993). The basin is subdivided into two subbasins, separated by the transverse Mahali Shoal, which is an active structure located on the Ubende Shear with the deep lake basin mainly developed outside the Ubende Belt (Klerkx et al. 1998). Resulting slope variations due to hypothesized magmatic underplating within the upper drainage basin of north Tanganyika induced strong variations of sedimentary fluxes between the main axial sediment source in the Kivu Rusizi area and the distal depocentres by creating sediment bypass zones. This resulted in the preferential infilling of the most distal depocentres (Lezzar et al. 2002).

High-resolution, airgun-sourced seismic reflection surveys over the offshore regions of five river deltas in Lake Malawi Rift reveal considerable variability in acoustic facies and stratigraphic architecture (Scholz 1995). This variability can largely be attributed to the influences of different structural settings and to a lesser degree to high-amplitude (100–400 m) and high-frequency (1000–100,000 years) fluctuations in lake level. Deltas on flexural and axial margins in the rift lake show well-developed progradational geometries. In contrast, a delta on a steep, accommodation zone margin distributes coarse sediments over a broad depositional apron, rather than concentrating sediment in discrete progradational lobes as on the other deltas. A large border fault margin river delta contains several delta-associated facies, including prograding clinoform packages, fan deltas stacked against a boundary fault, and extensive subaqueous fans. Flexural margin low-stand deltas may be the most prospective for hydrocarbon exploration due to their large, internally well-organized, progradational lobes and their close proximity to deep-water, high-total organic carbon lacustrine source facies.

2.7.2.3 Saline Lakes

Lake Katwe, with an area of about 2.5 km^2, is the largest of eight saline alkaline soda explosion Crater Lakes in the Katwe volcanic field (Morton 1969). Evaporites interbedded with clays extend down at least 40 m below the central part of the lake. Trona and subordinate burkeite are the main evaporite minerals present, with lesser contents of halite and hanksite. Saline springs around the edge of the lake discharge water containing from about 0.5–4 % solutes, which is considered to add about 2000 tons of salts to the lake each year and to be the main source of the salts in the lake. Travertine is associated with many of the springs, and conspicuous mounds of this material are visible at the northeastern end of the lake. These mounds rise 18 m above the present shoreline and were probably formed around springs during an earlier period of higher lake level. Varved chalky beds nearby appear to be contemporaneous with the mounds, and fossil reeds from the base of the varved beds have been dated at 11,000 years B.P. While giving a minimum age for the lake, the dated reeds also suggest that the Katwe Crater was one of the earlier ones in the volcanic field. Tuffs from the mouth of the Semliki River to the west are known, from radiocarbon dating, to be less than 8000 years old.

2.8 Overview of Paleoclimatic Records in the EARS

Burke and Gunnel (2008) discussed the geomorphology, tectonics, and environmental evolution of Africa and suggested that the topographic development and the unique geomorphic features resulted from tectonic processes, which have impacted the African Plate since Jurassic time (180 Ma). One of those geomorphic features was the early Oligocene Afro-Arabian low-elevation and low-relief topography that prevailed prior to the arrival of the Afar Plume in the late Oligocene (31 Ma) and eventually led to the uplift and doming of the Afro-Arabian region (Burke and Gunnel 2008). Climate models and sedimentary records indicate that the Afro-Arabian doming radically influenced the local and regional climate of eastern Africa (Maslin et al. 2014; Prömmel et al. 2013; Sepulchre et al. 2006; WoldeGabriel et al. 2009). According to Sepulchre et al. (2006), atmospheric and biosphere simulations revealed that the doming of eastern Africa altered wind-flow patterns, causing strong aridification and environmental changes during the late Neogene. Prömmel et al. (2013) suggested that tectonic forcing as well as changes in the Earth's orbital processes strongly influenced climate and precipitation in Africa based on regional and global simulations of coupled ocean–atmospheric general circulation models. According to the authors, altered moisture transport due to orbital forcing had a strong impact on precipitation in large parts of Africa during the last interglacial 125,000 years ago.

Within the Afar Rift, sedimentological, paleobiological, and paleosol stable isotope data indicate wet and humid woodland forests prevailing during the late Miocene to early Pliocene (WoldeGabriel et al. 2001, 2009). During the Plio–Pleistocene (<4.0 Ma), paleoenvironmental conditions evolved to open and drier conditions with time. This was due to tectonic forcing related to subsidence of the rift floors and uplift of the rift escarpments and adjacent highlands. Other factors such as sporadic large-volume volcanic eruptions within the Afar Rift and the MER, regional effects from sea-surface temperature fluctuations in the adjacent seas and ocean (i.e. Red Sea, Gulf of Aden, and Indian Ocean), and global effects associated with glacial and interglacial cycles also played a role (Clark et al. 2003; de Heinzelin et al. 1999; de Menocal 2004; de Menocal and Bloemendal 1996; White et al. 1993). Climatic data from late Pleistocene sedimentological records collected in the southwestern (Omo Basin) and northeastern (SW Afar Rift) parts of Ethiopia suggest rapid accumulation of sediments when ancestral Lake Turkana and Yardi Lake, respectively, were at much higher lake levels than at present. This is consistent with cooler and much wetter conditions associated with African monsoons during marine isotope stage (MIS) 6 (190,000–130,000 years ago) (Ayalon et al. 2001).

Sediments from shallow lakes within the Ethiopian Rift System and the adjacent highlands also provide information about Holocene climatic conditions in eastern Africa. For example, cores from a rift-bound shallow lake within the central sector of the MER indicated an early to middle Holocene (12,000–5400 calibrated years before present) wet episode in the region (Chalie and Gasse 2002).

In the Kenya Rift, the hydrology of the basins is mainly influenced by climate, tectonically controlled morphological and volcanic barriers, faults, and local water table variations (Bergner et al. 2009; Clark et al. 1990; Olago et al. 2009). These lakes have been very sensitive to climate changes and variability throughout the Plio–Pleistocene (Maslin et al. 2014; Olago et al. 2007; Street-Perrott et al. 1989; Trauth et al. 2005, 2007). Olaka et al. (2010) explored some of the factors behind the differential sensitivities of the East African rift lakes to past climate changes and found that these can partly be related to morphometric differences arising from their tectono-volcanic histories and not climatic factors alone. Lake sediments in the EARS rift basins suggest that there were three humid periods at 2.7–2.5 Ma, 1.9–1.7 Ma, and 1.1–0.9 Ma. These were superimposed on the longer-term aridification of East Africa and corresponded, respectively, to intensification of the Northern Hemisphere Glaciation, to an important intensification and shift in the east–west zonal atmospheric circulation—the Walker Circulation (Table 2.1); and to the shift from glacial/interglacial cycles every 41,000 years to every ca. 100,000 years (Trauth et al. 2005). A most recent compilation of the relationship between tectonics, climate, and paleolake occurrences is provided in Maslin et al. (2014), building on the compilations by Trauth et al. (2005, 2007). Here, we highlight selected examples, focusing on the Pleistocene–-Holocene periods. The lacustrine phase in paleolake Suguta from 1 to 7 Ma (Dunkley et al. 1993; Junginger et al. 2013) seems to

Table 2.1 Summary of key aspects of the Quaternary evolution of the Kenya Rift [simplified from Olago (2013)]

Age (Ma)	Global climate transitions	East African climate and environment			Kenya Rift system		
					Volcanism	Tectonics	Lake basin formation
Present		General increasing trend of aridity and hydrological variability with precession-linked wet phases and lake level changes	Change in lowland vegetation from woodland to savanna types	Marine transgressions and regressions	Hot springs and fumaroles—Geothermal systems	Minor faulting	Mostly small, shallow rift lakes; desiccation/lowstands during LGM, highstands at ca.10–9 Ka
0.1						Several phases of rift deformation and migration	
0.2							
0.3							
0.4					End of flood volcanism	Formation of axial volcanoes	
0.5							
0.6							Basins formed during Mio-Pliocene times achieve final morphology; Creation of new lake basins
0.7							
0.8	The Mid-Pleistocene Revolution						
0.9							
1.0							
1.1							
1.2							
1.3							
1.4							
1.5							
1.6					Widespread flood volcanism		
1.7							
1.8	Development of Walker Circulation						
1.9							
2.0							
2.1							
2.2							
2.3							
2.4							
2.5							
2.6	Onset of Northern Hemisphere Glaciation						
2.7							
2.8							
2.9							
3.0							

correlate with large lakes in other basins of the EARS (Trauth et al. 2005, 2007). Tiercelin and Lezzar (2002) summarize various studies that indicate a large rift lake existed in Baringo from about 700 ka to less than 200 ka (Cornelissen et al. 1990; Tallon 1976, 1978). Evidence from the vertebrate fauna (Tallon 1978) and the sediment mineralogy from rare outcrops (Renaut 1982; Renaut et al. 1999) indicate that the Kapthurin paleolake was alternatively fresh or saline/alkaline. It started to retreat by ~200 ka, resulting in the development of a wide alluvial plain and the permanence at the northern end of the basin of a small lake prefiguring the present-day Lake Baringo (Fig. 2.4). Other large and quite deep lakes developed in the Kenya Rift during middle to upper Pleistocene including the Nakuru–Naivasha and the Suguta Basins (Tiercelin and Lezzar 2002). In the Olorgesailie Basin farther south, Owen et al. (2008, 2009, 2011) found evidence of five phases of deep-water and eleven periods of shallow-water lacustrine sedimentation from about 1.2 Ma to present.

The lacustrine sedimentary record in the rift lakes and other sites indicates that the changes in precipitation in eastern Africa are driven by the precession cycle of 23,000 years (Olago et al. 2007; Trauth et al. 2001) and its higher precession harmonics (Olago et al. 2000). The best-studied period is the late Pleistocene to Holocene: high lake levels have been recorded for several lakes in eastern Africa at ca. 10,000 years BP (e.g. Street-Perrott et al. 1989). In contrast to the essentially dry present-day conditions in the Suguta Valley, which contains only the shallow, highly alkaline, seasonal Lake Logipi at its northern end, during the AHP the valley contained a large lake (paleo-Lake Suguta) that covered 2200 km^2 and was almost 300 m deep (Garcin et al. 2009; Junginger and Trauth 2013). Lake Suguta overflowed into the Lake Turkana Basin through an overflow sill located at an elevation of 570 m above sea level in the south of the western escarpment, via the Kerio River, at least from 13 to 8.5 cal ka BP, probably interrupted during transient low-stand periods (Garcin et al. 2009, 2012). It was earlier thought that, at this time, Lakes Nakuru, Elementaita, and Naivasha may have united to form a single, deep, freshwater lake (Richardson and Richardson 1972), but more recent evidence suggests that Lake Naivasha was not connected with the Nakuru–Elementaita Basin by surface water but by subsurface linkages (Dühnforth et al. 2006). In comparison with the present-day lake, the paleo-Lake Naivasha at 9000 years BP was three times larger and 150 m deep (versus 9 m for the present-day lake; Bergner et al. 2003). The early Holocene increased precipitation phase is related to a number of factors. These include globally rising temperatures, CO_2, and CH_4 partly related to orbitally controlled changes in solar irradiation (e.g. Chappellaz et al. 1993; Dansgaard et al. 1993, GRIP 1993; Taylor et al. 1993), a significantly higher P:E ratio as compared to today (Bergner et al. 2003, 2009), and a northeastward shift in the Congo Air Boundary (CAB) caused by an enhanced atmospheric pressure gradient between East Africa and India during a Northern Hemisphere insolation maximum (Junginger et al. 2013).

From about the late Miocene, increased uplift and differential warping in the Western Rift led to a relative depression producing a deep graben that formed the paleolake Obweruka in the Western Rift zone (Doornkamp and Temple 1966). Paleolake Obweruka was then cut into half by the upthrusting of the Rwenzori Horst, gradually initiating the formation of Lakes Edward and Albert in around the late Pliocene. Differential warping and uplift parallel to the Western Rift escarpment overcame westward river incision and direction, which after progressive aggradation was reversed. Several authors have demonstrated that during the last glacial maximum (ca. 20 ka), most of the major lakes in the region were hydrologically closed waterbodies (e.g. Kendall 1969; Talbot et al. 2000). Johnson et al. (1996) proposed that Lake Victoria was at least 65 m below its present level at the end of the last glaciation (15–17 ka). This supports the detection of the chemical concentration of calcite in lakebed sediment cores, which characterizes a closed or shallow basin prior to ca. 12.5 ka (Kendall 1969). Paleoenvironmental indicators preserved in lake sediments suggest rapidly rising lake levels during the late Pleistocene, and both Lakes Victoria and Albert were filled up by ca. 10–12 ka (Talbot et al. 2000). By 11.5 ka, it has been inferred from fossil remains that an integrated drainage network similar to the present one was already in existence. Based on $^{87}Sr/^{86}Sr$ isotopic data, Lake Victoria must have overflowed no later than this period. Kendall (1969) used $\delta^{14}C$ and Beuning et al. (2002) used cellulose $\delta^{18}O$ in sediment cores to place the transition from closed to open-basin conditions following the desiccation of the last glacial maximum at ca. 12.5-13 ka. Further evidence for a short period of Lake Victoria fall and low floods in the White Nile directly before ca. 10 ka suggests that the network had been almost certainly episodically disrupted; its level is controlled largely by down cutting of the Victoria–Nile outlet (Kendall 1969; Talbot et al. 2000).

2.9 Discussion

The EARS is a Cenozoic analogue of the major tectonic forces and associated magmatic processes that repeatedly impacted the African Plate along its eastern part during the Mesozoic. The African Plate was strongly influenced by geodynamic processes related to its divergent plate margins, which caused the relatively sluggish motion of the plate since the Jurassic (150 Ma), while the underlying mantle induced magmatic upwelling that triggered major events, including the break-up and separation of Gondwana and related episodic volcanism during the later part of the Mesozoic (Burke 1996; Gaina et al. 2013 and references therein). During the Cenozoic, tectonic forces related to the Afar Plume caused uplift and doming followed by widespread and voluminous volcanism. This culminated in the formation of the EARS and the subsequent break-up and splitting of the Arabian Plate during the early Miocene (ArRajehi et al. 2010; Ebinger and Sleep 1998). The EARS started in the Red Sea–Afar Rift region in the Miocene and is thought to have gradually propagated in a southwesterly direction, creating distinct rift basins along the eastern continental part of the African Plate.

During the Plio–Pleistocene, the EARS appears to have transitioned from plume-powered to passive upwelling-driven volcano-tectonic processes. These were responsible for the narrowing and segmentation of the axial rift zone with time, which may be a prelude to a continental break-up (Buck 2006; Hammond et al. 2013). The EARS was initially controlled by older structural elements that resulted in broad and shallow symmetrical and half graben. With time, however, the diffused volcano-tectonic processes progressively transitioned to a narrow belt marked by a strongly fractured and faulted zone, containing densely packed basaltic and silicic volcanic centres and associated calderas. In Ethiopia, there is a trend in tectonic and volcanic intensity and frequency from the proto-oceanic Afar Rift in the north to the continental rift of broadly rifted half grabens in the south.

In Kenya the influence of axial doming is evident during mid- to late-Miocene time as shown by the radial flow patterns of the initial phonolites (Vedkamp et al. 2012; Wichura et al. 2010) and paleontological data (Wichura et al. 2015). Furthermore, the geometry of the rift in this sector is closely controlled by pre-existing structures in the underlying Mozambique Mobile Belt. Faulting alternated with volcanism, with the centre of activity shifting with time from the boundary faults towards the axial region. As a result of the rifting, a number of lakes of differing sizes and depths have developed in this area through geological time. Today, more than 35 lakes lie in these depressions, having varied morphological, geological, physical, chemical, and biological characteristics (Tiercelin and Lezzar 2002). Rifting, uplift, and volcanism produced geomorphic changes that locally to regionally affected the climate. Coupled with the reorganization of global weather systems and the sensitivity of these lakes to orbitally and intrinsically driven climatic changes, the lake basin sediments offer a wealth of data on the paleohydrology and paleoenvironmental conditions of the region from the Pliocene to present.

The Western Rift was initiated in late Miocene at an area of thinner and weaker mobile belt lithosphere with Paleo–Proterozoic cover west of the strong Tanzania cratonic lithosphere margin and propagated northwards as the southward-propagating Eastern Rift reached the Craton (Nyblade and Brazier 2002). Two dominant extensional stress regimes, i.e. WNW–ESE and ENE–WSW (Aanyu and Koehn 2011), have a strong basement control (Tiercelin et al. 1988) with along-axis segmentation being a result of northward and southward along-axis propagation of rift-border-fault segments, thereby linking originally isolated basins (Ebinger 1989b). Volcanism was initiated southwards from ca. 12 Ma, with more potassic magmas located mainly within accommodation zones, although locally also coinciding with border faults and intra-basinal fissures often parallel to the border faults (Ebinger 1989b). The volume of Western Rift volcanic products differs markedly from the Eastern Rift and ranges from 2000 to 100,000 km^3, which is spread over a number of isolated volcanic provinces of limited extent, representing the only volcanic centres (e.g. Toro-Ankole, Virunga, Kivu, and Rungwe) over a distance of 1200 km (Westerhof et al. 2014 and references therein). The bulk of the Western Rift is underlain by sedimentary (lacustrine, alluvial, and fluvial) deposits of varying thicknesses; the Albertine Rift fill has a thick pile (4–7 km) of terrestrial sediments with volcanic intercalations (Westerhof et al. 2014). The Lake Edward Basin has sediments (up to 4 km), whereas the Lake George Basin is characterized by largely shallow sediments (e.g. Laerdal and Talbot 2002); Kivu and Rusizi lacustrine sediments overlie basalts (Ebinger 1989b). The Rukwa Rift has late Miocene to recent sediments and sandy fluvio-deltaic deposits or shallow-water lacustrine shales (Morley et al. 1999). The Tanganyika Rift has a mixed sedimentary thickness greater than 4 km (e.g. Cohen et al. 1993), while the Malawi Rift has river deltaic sediments of considerable variability in stratigraphic architecture (Scholz 1995).

2.10 Conclusions

Despite major rifting and subsidence within the continental interior in the Mesozoic Era, the EARS formed along the eastern part of the African Plate beginning in the middle Tertiary. Why the EARS formed along the eastern continental margin probably has to do with the major tectonic forces that triggered the separation and drifting of East Gondwana from the African Plate and the subsequent rise of the Réunion Plume in the late Cretaceous, which had a coupled tectonic effect on the African and Indian plates. By the early Tertiary, the Afar Plume encroached the lithosphere beneath what are now the Red Sea and the Afar Rifts, instigating lithospheric thinning, regional doming, and the initial eruptions of Eocene flood basalts in southern Ethiopia. Volcanism in the Afro-Arabian Dome peaked in the late Oligocene to early Miocene and was followed by rifting and the break-up and drifting of the Arabian Microplate and the opening of the oceanic Red Sea and the Gulf of Aden Rifts. Today, the Afar Rift represents a proto-oceanic floor and continues to propagate to the continental sector of the MER, which is characterized by magma-assisted rifting and segmentation.

The present morphology of the eastern arm of the EARS is a result of a sequence of closely related events that include domal uplift, tectonism, faulting, volcanism, erosion, and sedimentation. Faulting and volcanism alternated with each other, with the centre of these two activities shifting with time from the flanks to the axial zone of the rift.

The Western Rift is marked by a zone of high seismicity and is dominated by large boundary faults and an opposing flexural margin. These produced mobile asymmetrical full and half-graben sedimentary basins that are individually linked along the rift axis, separated by accommodation or transfer zones. The Western Rift was initiated in the late Miocene and propagated northwards as the southward-propagating Kenya Rift reached the Tanzanian Craton. Isolated volcanic centres of limited extent include southwards syn-rift late Pliocene to recent volcanic fields (Toro-Ankole) and older, pre-rift, Oligocene to Miocene volcanic fields (Virunga, Kivu, and Rungwe). The Western Rift is largely underlain by sedimentary (fluvial, alluvial, and lacustrine) deposits of varying composition and thickness.

Acknowledgement The authors equally contributed to the geological, tectonic, hydrological, and paleoenvironmental highlights of the various parts of the EARS presented in this overview paper. Giday WoldeGabriel acknowledges the collaborative studies conducted as part of the Middle Awash and the Konso Geological and Paleoanthropological research projects in different parts of the Ethiopian Rift System for his contribution. Michael Owor sends special thanks to all colleagues on this team, the RIFTLINK project, and to the Uganda Petroleum Exploration and Production Department (PEPD) for all information provided. Finally, the authors would like to thank Michael Schagerl for the opportunity to put together a chapter on the Genesis of the East African Rift System. The manuscript greatly benefited from constructive comments and inputs by two anonymous reviewers and the editor, Michael Schagerl.

References

Aanyu K, Koehn D (2011) Influence of pre-existing fabrics on fault kinematics and rift geometry of interacting segments: analogue models based on the Albertine Rift (Uganda), western branch East African Rift System. J Afr Earth Sci 59(2–3):168–184

ArRajehi A, McClusky S, Reikinger R et al (2010) Geodetic constraints on present-day motion of the Arabian plate: implications for Red Sea and Gulf of Aden rifting. Tectonics 29, TC3011

Ayalon A, Bar-Matthews M, Kaufman A (2001) Climatic conditions during marine oxygen isotope stage 6 in the eastern Mediterranean region from the isotopic composition of speleothems of Soreq Cave, Israel. Geology 30:303–306

Ayenew T (2005) Major ions composition of the groundwater and surface water systems and their geological and geochemical controls in the Ethiopian volcanic terrain. SINET Ethiop J Sci 28(2):171–188

Ayenew T (2009) Natural lakes of Ethiopia. Addis Ababa University Press, Addis Ababa

Baker BH (1958) Geology of the Magadi area. Geological Survey of Kenya, Report No. 42, Nairobi, Kenya

Baker BH (1963) Geology of the Baragoi area. Geological Survey of Kenya, Report No. 53, Nairobi, Kenya

Baker BH (1986) Tectonics and volcanism of the southern Kenyan Rift Valley and its influence on sedimentation. In: Frostick LE, Renaut RW, Reid I, Tiercelin J-J (eds) Sedimentation in the African Rifts, Geol Soc Spec Publ 25:267–284

Baker BH, Wohlenberg J (1971) Structure and evolution of the Kenya Rift Valley. Nature 229:538–542

Baker J, Snee L, Menzies M (1996) A brief Oligocene period of flood volcanism in Yemen; implications for the duration and rate of continental flood volcanism at the Afro-Arabian triple junction. Earth Planet Sci Lett 138:39–55

Barberi F, Varet J (1977) Volcanism of Afar: small-scale plate tectonics implication. Geol Soc Am Bull 88:1251–1266

Bauer FU, Glasmacher UA, Ring U, Karl M, Schumann A, Nagudi B (2013) Tracing the exhumation history of the Rewnzori Mountains, Albertine Rift, Uganda, using low-temperature thermochronology. Tectonophysics 599:8–28

Behr HJ (2002) Magadiite and Magadi-chert: a critical analyses of the silica sediments in the Lake Magadi Basin, Kenya. In: Renaut RW, Ashley GM (eds) Sedimentation in continental rifts, SEPM Spec Publ 73:257–274

Behrensmeyer A, Potts R, Deino A, Ditchfield P (2002) Olorgesailie, Kenya: a million years in the life of a rift basin. In: Renaut RW, Ashley GM (eds) Sedimentation in continental rifts, SEPM Spec Publ 73:97–106

Bergner AGN, Trauth MH, Bookhagen B (2003) Paleoprecipitation estimates for the Lake Naivasha Basin (Kenya) during the last 175 k.y. using a lake-balance model. Glob Planet Change 36:117–136

Bergner AGN, Strecker MR, Trauth MH, Deino AL, Gasse F, Blisniuk P, Dühnforth M (2009) Tectonic versus climate influences on the evolution of the lakes in the central Kenya Rift. Quat Sci Rev 28:2804–2816

Berhe SM (1990) Ophiolites in Northeast and East Africa: implications for Proterozoic crustal growth. J Geol Soc London 147:41–57

Beuning KRM, Kelts K, Russell J, Wolfe BB (2002) Reassessment of Lake Victoria-Upper Nile River paleohydrology from oxygen isotope records of lake-sediment cellulose. Geology 30(6):559–562

Beyene A, Abdelsalam MG (2005) Tectonics of the Afar Depression: a review and synthesis. J Afr Earth Sci 41:41–59

Birt C, Maguire PKH, Khan MA, Thybo A (1995) A combined interpretation of the KRISP 94 seismic and gravity data: evidence for a mantle plume beneath the East African Plateau (abstract). Eos Trans AGUnion 76:F607–608

Birt CS, Maguire PKH, Khan MA, Thybo H, Keller GR, Patel J (1997) The influence of pre-existing structures on the evolution of the southern Kenya Rift Valley—evidence from seismic and gravity studies. Tectonophysics 278:211–242

Bishop WW (1969) Pleistocene stratigraphy in Uganda. Geological survey of Uganda, memoir no. X

Bosellini A (1989) The continental margins of Somalia: their structural evolution and sequence stratigraphy. Mem Sci Geol 101:373–384

Bosellini A, Russo A, Fantozzi PL, Assefa G, Tadesse S (1997) The Mesozoic succession of the Mekele Outlier (Tigrai Province, Ethiopia). Mem Sci Geol 49:95–116

Bosworth W (1992) Mesozoic and early tertiary rift tectonics in East Africa. Tectonophysics 209:115–137

Bosworth W, Maurin A (1993) Structure, geochronology and tectonic significance of the northern Suguta Valley (Gregory Rift) Kenya. J Geol Soc London 150:751–762

Bosworth W, Morley CK (1994) Structural and stratigraphic evolution of the Anza Rift, Kenya. Tectonics 236:93–115

Boven A, Pasteels P, Punzalan LE, Yamba TK, Musis JH (1998) Quaternary perpotassic magmatism in Uganda (Toro-Ankole Volcanic Province): age assessment and significance for magmatic evolution along the East African Rift. J Afr Earth Sci 26:463–476

Brown FH, McDougall I (2011) Geochronology of the Turkana depression of northern Kenya and southern Ethiopia. Evol Anthropol 20:217–227

Buck WR (2006) The role of magma in the development of the Afro-Arabian Rift System. In: Yirgu G, Ebinger CJ, Maguire PKH (eds) The Afar Volcanic Province within the East African Rift System, Geol Soc Spec Publ 259:43–54

Burke K (1996) The African plate. South Afr J Geol 99:339–410

Burke K, Cannon JM (2014) Plume-plate interaction. Can J Earth Sci 51:208–221

Burke K, Gunnel Y (2008) The African erosion surface; a continental-scale synthesis of geomorphology, tectonics, and environmental change over the past 180 million years. Geol Soc Am Mem 201:1–66

Butzer KW, Isaac GL, Richardson JL, Washburn-Kamau C (1972) Radiocarbon dating of East African lake levels. Science 175:1069–1076

Cahen L, Snelling NJ, Delhal J, Vail JR (1984) The geochronology and evolution of Africa. Clarendon Press, Oxford, pp xiii, 1–512

Cande SC, Stegman DR (2011) Indian and African Plate motions driven by push force of the Réunion Plume head. Nature 475:47–52

Casanova J (1987) Stromatolites et hauts niveaux lacustres Pléistocènes du bassin Natron-Magadi (Tanzanie-Kenya). Sci Géol Bull 40:135–153

Casanova J, Hillaire-Marcel C (1987) Chronologie et paléohydrologie des hauts niveaux Quaternaires du bassin Natron-Magadi (Tanzanie-Kenya) d'après la composition isotopique (^{18}O, ^{13}C, ^{14}C, U/Th) des stromatolites littoraux. Sci Géol Bull 40:121–134

Casanova J, Hillaire-Marcel C, Page N, Taieb M, Vincens A (1988) Stratigraphie et paléohydrologie des épisodes lacustres du Quarternaire récent du rift Suguta (Kenya). CR Acad Sci Paris 307(Série II):1251–1258

Chalie F, Gasse F (2002) Late Glacial–Holocene diatom record of water chemistry and lake level change from the tropical East African Rift Lake Abiyata (Ethiopia). Palaeogeogr Palaeoclimatol Palaeoecol 187 (3–4):259–283

Chapman GR, Brook M (1978) Chronostratigraphy of the Baringo Basin, Kenya Rift Valley. In: Bishop WW (ed) Geological background to fossil man. Geological Society/Scottish Academic, London, pp 207–223

Chappellaz J, Fung IY, Thompson AM (1993) The atmospheric CH_4 increase the last glacial maximum. 1. Source estimates. Tellus 45B:228–241

Chernet T, Hart WK, Aronson JL, Walter RC (1998) New age constraints on the timing of volcanism and tectonics in the northern Main Ethiopian Rift- southern Afar Rift transition zone, Ethiopia. J Volcanol Geotherm Res 80:267–280

Chorowicz J (2005) The East African Rift System. J Afr Earth Sci 4:379–410

Clark MCG, Woodhall DG, Allen D, Darling G (1990) Geological, volcanological and hydrogeological controls on the occurrence of geothermal activity in the area surrounding Lake Naivasha, Kenya. Ministry of Energy, Nairobi

Clark JD, Beyene Y, WoldeGabriel G, Hart WK, Renne P, Gilbert H, Defleur A, Suwa G, Katoh S, Ludwig K, Boisserie J-R, Asfaw B, White TD (2003) Stratigraphic, chronological, and behavioral contexts of Pleistocene *Homo sapiens* from Herto, Ethiopia. Nature 423:747–752

CNR-CNRS-Afar Team (1973) Geology of northern Afar (Ethiopia). Rev Geogr Phys Geol 2:343–390

Cohen A, Soreghan M, Schotz C (1993) Estimating the age of formation of lakes: an example from Lake Tanganyika, East African Rift. Geology 21:511–514

Cornelissen E, Boven A, Dabi A, Hus J, Ju Yong K, Keppens E, Langohr R, Moeyersons J, Pasteels P, Pieters M, Uytterschaut H, Van Noten F, Workineh H (1990) The Kapthurin formation revisited. Afr Archaeol Rev 8:23–75

Corti G (2009) Continental rift evolution: from rift initiation to incipient breakup in the Main Ethiopian Rift, East Africa. Earth Sci Rev 96:1–53

Crossley R (1979) The Cenozoic stratigraphy and structure of the western part of the rift valley in southern Kenya. J Geol Soc London 136:393–405

Damnati B, Taieb M (1995) Solar and ENSO signatures in laminated deposits from Lake Magadi (Kenya) during the Pleistocene/Holocene transition. J Afr Earth Sci 21:373–382

Dansgaard W, Johnsen SJ, Clausen HB et al (1993) Evidence for general instability of past climate from a 250 kyr ice core record. Nature 364:218–220

Davidson A, Rex DC (1980) Age of volcanism and rifting in southwestern Ethiopia. Nature 283:657–658

de Heinzelin J (1983) The Omo Group. In: de Heinzelin (ed) Archives of the international Omo research expedition, Belgique, Annales, Serie in 8, Sciences Geologiques No. 85, p 365

de Heinzelin J, Clark JD, White T, Hart W, Renne P, WoldeGabriel G, Beyene Y, Vrba E (1999)

Environment and behavior of 2.5-million-year-old Bouri Hominids. Science 284:625–629

de Menocal PB (2004) African climate change and faunal evolution during the Pliocene-Pleistocene. Earth Planet Sci Lett 220:3–24

de Menocal PB, Bloemendal J (1996) Plio-Pleistocene climatic variability in subtropical Africa and the paleoenvironment of hominid evolution: a combined data-model approach. In: Vrba ES, Denton GH, Partridge TC, Burckle LH (eds) Plaeoclimate and evolution, with emphasis on human origins. Yale University Press, New Haven, CT, pp 262–288

Desprès A (2008) Evolution Tectono-Sédimentaire des Bassins de Rift Crétacé-Paléogène du Nord du Kenya. Master 2, Université de Rennes 1, France

Dindi EW (1994) Crustal structure of Anza Graben from gravity and magnetic investigations. Tectonophysics 236:359–371

Doornkamp JC, Temple PH (1966) Surface, drainage and tectonic instability in part of southern Uganda. Geogr J 132:238–252

Dühnforth M, Bergner AGN, Trauth MH (2006) Early Holocene water budget of the Nakuru-Elmenteita basin, central Kenya Rift. J Paleolimnol 36:281–294

Dunkelman TJ, Karson JA, Rosendahl BR (1988) Structural style in the Turkana Rift, Kenya. Geology 16:258–261

Dunkley PN, Smith M, Allen DJ, Darling WG (1993) The geothermal activity and geology of the northern sector of the Kenya Rift Valley. British Geological Survey Research Report SC/93/1

Eagles G, König M (2008) A model of plate kinematics in Gondwana breakup. Geophys J Int 173:703–717

Ebinger CJ (1989a) Geometric and kinematic development of border faults and accommodation zomes, Kivu-Rusizi Rift, Africa. Tectonics 8:117–133

Ebinger CJ (1989b) Tectonic development of the western branch of the East African Rift System. Geol Soc Am Bull 101:885–903

Ebinger C (2005) Continental breakup: the East African perspective. Astron Geophys 46:2.16–2.21

Ebinger CJ, Sleep NH (1998) Cenozoic magmatism throughout east Africa resulting from impact of a single plume. Nature 395:788–791

Ebinger CJ, Yemane T, WoldeGabriel G, Aronson JL, Walter RC (1993) Late eocene-recent volcanism and faulting in the southern Main Ethiopian Rift. J Geol Soc London 150:99–108

Ebinger CJ, Djomani YP, Mbede E et al (1997) Rifting Archaean lithosphere: the Eyasi–Manyara–Natron Rifts, East Africa. J Geol Soc London 154:947–960

Ebinger CJ, Yemane T, Harding DJ et al (2000) Rift deflection, migration, and propagation: linkage of the Ethiopian and Eastern Rifts, Africa. Geol Soc Am Bull 112:163–176

Eugster HP (1980) Lake Magadi, Kenya and its precursors. In: Nissenbaum A (ed) Hypersaline brines and evaporites, Dev Sedimentol 28:195–232

Eugster HP (1986) Lake Magadi, Kenya: a model for rift valley hydrochemistry and sedimentation. In: Frostick LE, Renaut RW, Reid I, Tiercelin J-J (eds) Sedimentation in the African Rifts, Geol Soc Spec Publ 25:177–189

Flannery J, Rosendahl B (1990) The seismic stratigraphy of Lake Malawi, Africa: implications for interpreting geological processes in lacustrine rifts. J Afr Earth Sci 10(3):519–548

Foster A, Ebinger C, Mbede E, Rex D (1997) Tectonic development of the northern Tanzanian sector of the East African Rift System. J Geol Soc London 154:689–700

Fritz H, Abdelsalam M, Ali KA, Bingen B, Collins AS, Fowler AR, Ghebreab W, Hauzenberger CA, Johnson PR, Kusky TM, Macey P, Muhongo S, Stern RJ, Viola G (2013) Orogen styles in the East African Orogen: a review of the neoproterozoic to Cambrian tectonic evolution. J Afr Earth Sci 86:65–106

Gaina C, Müller RD, Brown B, Ishihara T, Ivanov S (2007) Breakup and early seafloor spreading between India and Antarctica. Geophys J Int 170:151–169

Gaina C, Torsvik TH, van Hinsbergen DJJ, Medvedev S, Werner SC, Labails C (2013) The African plate: a history of oceanic crust accretion and subduction since the Jurassic. Tectonophysics 604:4–25

Garcin Y, Junginger A, Melnick D, Olago DO, Strecker MR, Trauth MH (2009) Late Pleistocene-Holocene rise and collapse of Lake Suguta, northern Kenya Rift. Quat Sci Rev 28(9–10):911–925

Garcin Y, Melnick D, Olago D, Strecker R, Tiercelin J-J (2012) East African mid-Holocene wet-dry transition recorded in palaeo-shorelines of Lake Turkana, northern Kenya Rift. Earth Planet Sci Lett 331:322–334

Gasse F, Street FA (1978) Late Quaternary lake-level fluctuations and environments of the northern rift valley and Afar region (Ethiopia and Djibouti). Palaeogeogr Palaeoclimatol Palaeoecol 24:279–325

Gibbons AD, Whittaker JM, Müller RD (2013) The breakup of East Gondwana: assimilating constraints from Cretaceous ocean basins around India into a best-fit tectonic model. J Geophys Res Solid Earth 118:808–822

Gillespie R, Street-Perrot AF, Switsur R (1983) Postglacial arid episodes in Ethiopia have implications for climate prediction. Nature 306:680–683

Goetz C, Hillaire-Marcel C (1992) U-series disequilibria in early diagenetic minerals from Lake Magadi sediments, Kenya: dating potential. Geochim Cosmochim Acta 56:1331–1341

GRIP (Greenland Ice Core Project Members) (1993) Climate instability during the last interglacial period recorded in the GRIP ice core. Nature 364:203–207

Hammond JOS, Kendal J-K, Stuart GW, Ebinger CJ, Bastow ID, Keir D, Ayele A, Belachew M, Goitom B, Ogubazghi G, Wright TJ (2013) Mantle upwelling and initiation of rift segmentation beneath the Afar Depression. Geology 41:635–638

Hautot S, Tarits P, Whaler K, Le Gall B, Tiercelin J-J, Le Turdu C (2000) Deep structure of the Baringo Rift

Basin (central Kenya) from three-dimensional magnetotelluric imaging: implications for rift evolution. J Geophys Res 105:493–518
Hay RL (1976) Geology of the Olduvai Gorge. University of California Press, Berkeley, CA
Hetzel R, Strecker M (1994) Late Mozambique Belt structures in western Kenya and their influence on the evolution of the Cenozoic Kenya Rift. J Struct Geol 16:389–401
Hill A, Curtis G, Drake R (1986) Sedimentary stratigraphy of the Tugen Hills, Baringo, Kenya. In: Frostick LE, Renaut RW, Reid I, Tiercelin J-J (eds) Sedimentation in the African Rifts, Geol Soc Spec Publ 25:285–295
Hofmann C (1997) Timing of the Ethiopian flood basalt event and implications for plume birth and global change. Nature 389:834–841
Holwerda JG, Hutchinson RW (1968) Potash-bearing evaporites in the Danakil area, Ethiopia. Econ Geol 63:124–150
Johnson TC, Scholz CA, Talbot MR, Kelts K, Ricketts RD, Ngobi G, Beuning K, Ssemmanda I, McGill JW (1996) Late Pleistocene desiccation of Lake Victoria and rapid evolution of cichlid fishes. Science 273:1091–1093
Jones WB, Lippard SJ (1979) New age determinations and the geology of the Kenya Rift-Kavirondo Rift junction, W Kenya. J Geol Soc London 136:693–704
Juch D (1980) Tectonics of the southeastern escarpment of Ethiopia. Accad Naz Lincei Rome 47:407–418
Junginger A, Trauth MH (2013) Hydrological constraints of paleo-Lake Suguta in the Northern Kenya Rift during the African humid period (15–5 ka BP). Glob Planet Change 111:174–188
Junginger A, Roller S, Olaka LA, Trauth MH (2013) The effects of solar irradiation changes on the migration of the Congo air boundary and water levels of paleo-Lake Suguta, northern Kenya Rift, during the African humid period (15–5 ka BP). Palaeogeogr Palaeoclimatol Palaeoecol 396:1–16
Kampunzu AB, Bonhomme MG, Kanika M (1998) Geochronology of volcanic rocks and evolution of the Cenozoic western branch of the East African Rift System. J Afr Earth Sci 26:441–461
Katoh S, Nagaoka S, WoldeGabriel G, Renne PR, Snow MG, Beyen Y, Suwa G (2000) Chronostratigraphy and correlation of the Plio–Pleistocene tephra layers of the Konso formation, southern Main Ethiopian Rift. Quat Sci Rev 19:1305–1317
Katternhorn SA, Muirhead JD, Dindi E, Fischer TP, Lee H, Ebinger CJ (2013) Contribution of transverse structures, magma and crustal fluids to continental rift evolution: the East African Rift in southern Kenya. Eos Trans AGU 94, Fall Meet. Suppl, Abstract T12B-02
Kazmin V, Shifferaw A, Balicha T (1978) The Ethiopian basement: stratigraphy and possible manner of evolution. Geol Rundsch 60:531–546
Kazmin V, Shifferaw A, Teferra M, Berhe SM, Chowaka S (1979) Precambrian structures of western Ethiopia. In: Proceedings 5th conference of African Geology. Annals of the geological survey of Egypt, Cairo 9, 1–8
Kazmin V, Berhe SM, Nicoletti M, Petrucciani C (1980) Evolution of the northern part of the Ethiopian Rift. Accad Naz Lincei Rome 47:275–292
Kendall RL (1969) An ecological history of the Lake Victoria basin. Ecol Monogr 39(2):121–176
King BC (1978) Structural evolution of the Gregory rift valley. In: Bishop WW (ed) Geological background to fossil man. Geological Society/Scottish Academic, London, pp 29–54
King BC, Le Bas MJ, Sutherland DS (1972) The history of the alkaline volcanoes and intrusive complexes of eastern Uganda and western Kenya. J Geol Soc London 128:173–205
Klerkx J, Theunissen K, Delvaux D (1998) Persistent fault controlled basin formation since the Proterozoic along the Western Branch of the East African Rift. J Afr Earth Sci 26(3):347–361
Kokonyangi JW, Kampunzu AB, Armstrong RA, Arima M, Yoshida M, Yoshida M, Okudaira T (2007) U-Pb SHRIMP dating of detrital zircons from the Nzilo Group (Kibaran Belt): implications for the source of sediments and Mesoproterozoic evolution of central Africa. J Geol 115:99–113
Korme T, Acocella V, Abebe B (2004) The role of pre-existing structures in the origin, propagation and architecture of faults in the Main Ethiopian Rift. Gondwana Res 7:467–479
Kuntz K, Kreuzer H, Muller P (1975) Potassium-Argon age determination of the trap basalt of southeastern part of the Afar Rift. In: Pilger A, Rösler A (eds) Afar depression of Ethiopia, vol 2. E. Schweizerbart, Verlag, Stuttgart, pp 370–374
Kuria ZN, Woldai T, van der Meer FD, Barongo JO (2010) Active fault segments as potential earthquake sources: inferences from integrated geophysical mapping of the Magadi fault system, southern Kenya Rift. J Afr Earth Sci 57:345–359
Laerdal T, Talbot MR (2002) Basin neotectonics of Lakes Edward and George, East African Rift. Palaeogeogr Palaeoclimatol Palaeoecol 187:213–232
Lenoir J-L, Kuster D, Liegeois JP, Utke A, Haider A, Matheis G (1994) Origin and regional significance of late Precambrian and early Palaeozoic granitoids in the Pan-African belt of Somalia. Geol Rundsch 83:624–641
Levitte D, Columba J, Mohr PA (1974) Reconnaissance geology of the Amaro Horst, southern Ethiopian rift. Geol Soc Am Bull 85:417–422
Lezzar KE, Tiercelin JJ, Turdu CL, Cohen AS, Reynolds DJ, Gall BL, Scholz CA (2002) Control of normal fault interaction on the distribution of major Neogene sedimentary depocenters Lake Tanganyika, East African rift. AAPG Bull 86(6):1027–1059
Lindenfeld M, Rümpker G (2011) Detection of mantle earthquakes beneath the East African Rift. Geophys J Int 186:1–5
Link K, Koehn D, Barth MG, Tiberindwa JV, Barifaijo E, Aanyu K, Foley SF (2010) Continuous cratonic crust

between the Congo and Tanzania blocks in western Uganda. Int J Earth Sci 99(7):1559–1573

Lippard SJ (1973) The petrology of phonolites from the Kenya Rift. Lithos 6:217–234

Macdonald R (1994) Petrological evidence regarding the evolution of the Kenya Rift Valley. Tectonophysics 236:373–390

Mammo T (2012) Analysis of gravity field to reconstruct the structure of Omo Basin in SW Ethiopia and implications for hydrocarbon potential. Mar Petrol Geol 29:104–114

Maslin MA, Brierley CM, Milner AM, Shultz S, Trauth MH, Wilson KE (2014) East African climate pulses and early human evolution. Quat Sci Rev 101:1–17

Mbede EI (2001) Tectonic setting and uplift analysis of the Pangani Rift Basin in northern Tanzania using apatite fission track thermochronology. Tanz J Sci 27A:23–38

McCall GJH (1967) Geology of the Nakuru-Thompson's Falls-Lake Hannington area. Geological Survey of Kenya Report 78

McDougall I (1985) K-Ar and $^{40}Ar/^{39}Ar$ dating of the hominid-bearing Plio-Pleistocene sequence at Koobi Fora, Lake Turkana, northern Kenya. Geol Soc Am Bull 96:159–175

McGlue MM, Scholz CA, Karp T, Ongodia B, Lezzar KE (2006) Facies architecture of flexural margin lowstand delta deposits in Lake Edward, East African Rift: constraints from seismic reflection imaging. J Sediment Res 76:942–958

Menzies M, Gallagher K, Yelland A, Hurford AJ (1997) Volcanic and nonvolcanic rifted margins of the Red Sea and Gulf of Aden: crustal cooling and margin evolution in Yemen. Geochim Cosmochim Acta 61:2511–2527

Merla G, Abbate E, Azarolli A, Canuti P, Sagri M, Tacconni P (1973) Geological map of Ethiopia and Somalia, 1:2,000,000 and comments with map of major land forms. CNR, Florence

Mohr PA (1967) The Ethiopian Rift System. Geophys Observatory Bull 12:27–56

Moore JM, Davidson A (1978) Rift structure in southern Ethiopia. Tectonophysics 46:159–173

Morley CK, Wescott WA, Stone DM, Harper RM, Wigger ST, Karanja FM (1992) Tectonic evolution of the northern Kenyan Rift. J Geol Soc London 149:333–348

Morley CK, Ngenoh DK, Ego JK (1999) Introduction to the East African Rift System. In: Morley CK (ed) Geoscience of rift systems—evolution of East Africa. AAPG Stud Geol 44:1–18

Morton MH (1969) The Lake Katwe Crater. Uganda Geological Survey and Mines Department report

Morton WH, Rex DC, Mitchell JG, Mohr PA (1979) Riftward younging of surface volcanic units in the Addis Ababa region, Ethiopian Rift Valley. Nature 280:284–288

Mugisha F, Ebinger CJ, Strecker M, Pope D (1997) Two-stage rifting in the Kenya rift: implications for half-graben models. Tectonophysics 278:63–81

Muirhead JD, Kattenhorn SA, Dindi EW, Gama R (2013) The evolving contribution of border faults and intra-rift faults in early-stage east African Rifts: insights from the Natron (Tanzania) and Magadi (Kenya) Basins. Eos Trans AGU 94, Fall Meet. Suppl., Abstract T21B-2543

Nyblade AA, Brazier RA (2002) Rift system Precambrian lithospheric controls on the development of the East African Rift System. Geology 30:755–758

Nyblade AA, Pollack HN (1992) A gravity model for the lithosphere in western Kenya and northeastern Tanzania. Tectonics 212:257–267

Odero DO (1993) A hydrogeological study of Njoro Area. Unpublished MSc. Thesis, University of Nairobi

Olago DO (2013) Quaternary Evolution. In: Paron P, Olago D, Omuto CT (eds) Kenya: a natural outlook—geoenvironmental resources and hazards. Elsevier B.V, Amsterdam, pp 31–38

Olago DO, Odada EO, Street-Perrott FA, Perrott RA, Ivanovich M, Harkness DD (2000) Long-term temporal characteristics of palaeomonsoon dynamics in equatorial Africa. Glob Planet Change 26 (1–3):159–171

Olago DO, Umer M, Ringrose S, Huntsman-Mapila P, Sow EH, Damnati B (2007) Palaeoclimate of Africa: an overview since the last glacial maximum. In: Otter L, Olago DO, Niang I (eds) Global change processes and impacts in Africa: a synthesis. East African Educational Publishers Ltd, Nairobi, pp 1–32

Olago DO, Opere A, Barongo J (2009) Holocene paleohydrology, groundwater and climate change in the lake basins of the central Kenya Rift. Hydrolog Sci J 54(4):765–780

Olaka LA, Odada EO, Trauth MH, Olago DO (2010) The sensitivity of East African rift lakes to climate fluctuations. J Paleolimnol 44:629–644

Owen RB, Potts R, Behrensmeyer AK, Ditchfield P (2008) Diatomaceous sediments and environmental change in the Pleistocene Olorgesailie formation, southern Kenya Rift Valley. Palaeogeogr Palaeoclimatol Palaeoecol 269:17–37

Owen RB, Renaut RW, Scott JJ, Potts R, Behrensmeyer AK (2009) Wetland sedimentation and associated diatoms in the Pleistocene Olorgesailie Basin, southern Kenya Rift Valley. Sediment Geol 222:124–137

Owen RB, Renaut RW, Potts R, Behrensmeyer AK (2011) Geochemical trends through time and lateral variability of diatom floras in the Pleistocene Olorgesailie formation, southern Kenya Rift Valley. Quat Res 76:167–179

Owen RB, Renaut RW, Behrensmeyer AK, Potts R (2014) Quaternary geochemical stratigraphy of the Kedong–Olorgesailie section of the southern Kenya Rift valley. Palaeogeogr Palaeoclimatol Palaeoecol. doi:10.1016/j.palaeo.2014.01.011

Pickford M, Senut B, Hadoto D (1993) Geology and palaeobiology of the Albertine rift valley Uganda-Zaire, volume I: geology, occasional publication 24. CIFEG, International Center for Training and Exchanges in Geosciences

Pinet B, Colletta B (1990) Probing into extensional sedimentary basins: comparison of recent data and derivation of tentative models. Tectonophysics 173:185–197

Prömmel K, Cubasch U, Kaspar F (2013) A regional climate model study of the impact of tectonic and orbital forcing on African precipitation and vegetation. Palaeogeogr Palaeoclimatol Palaeoecol 369:154–1

Reid I, Frostick LE (1986) Slope processes, sediment derivation and landform evolution in a rift valley basin, northern Kenya In: Frostick LE, Renaut RW, Reid I, Tiercelin J-J (eds) Sedimentation in the African Rifts, Geol Soc London Spec Publ 25:99–111

Renaut RW (1982) Late quaternary geology of the Lake Bogoria Fault-Trough, Kenya Rift Valley. Unpublished PhD thesis, University of London

Renaut RW, Ego J, Tiercelin J-J, Le Turdu C, Owen RB (1999) Saline, alkaline palaeolakes of the Tugen Hills-Kerio Valley region, Kenya Rift Valley. In: Andrews P, Banham P (eds) Late cenozoic environments and hominid evolution: a tribute to Bill Bishop. Geological Society, London, pp 41–58

Renne PR, WoldeGabriel G, Hart WK, Heiken G, White TD (1999) Chronostratigraphy of the Miocene–Pliocene Sagantole Formation, Middle Awash Valley, Afar Rift, Ethiopia. Geol Soc Am Bull 111:869–885

Richardson JL (1966) Changes in level of Lake Naivasha, Kenya, during postglacial times. Nature 209:290–291

Richardson JL, Richardson AE (1972) History of an African Rift lake and its climatic implications. Ecol Monogr 72:499–534

Roberts EM, Stevens NJ, O'Connor PM, Dirks PMO, Gottfried MD, Clyde WC, Armstrong RA, Kemp AIS, Hemming S (2012) Initiation of the western branch of the East African Rift coeval with the eastern branch. Nat Geosci. doi:10.1038/NGE01432

Russo A, Assefa G, Atnafu B (1994) Sedimentary evolution of the Abay River (Blue Nile) Basin. N Jb Geol Palaonl Mh 5:291–308

Saggerson EP (1952) Geology of the Kisumu District. Rep. No. 21, Geol. Surv. Kenya, Nairobi, Kenya

Saneyoshi M, Nakayama K, Sakai T, Sawada Y, Ishida H (2006) Half graben filling processes in the early phase of continental rifting: the Miocene Namurungule formation of the Kenya Rift. Sediment Geol 186:111–131

Schlüter T (1997) Geology of East Africa. In: Bender F, Jacobshagen V, Lüttig G (eds) Beiträge zur Geologie der Erde, Band 27, Gebrüder Borntraeger. Berlin, Stuttgart, p 484

Schlüter T, Kohring R (2001) Palaeopathological fish bones from phosphorites of the Lake Manyara area, northern Tanzania—fossil evidence of a physiological response to survival in an extreme biocenosis. Environ Geochem Health 24:131–140

Scholz CA (1995) Deltas of the Lake Malawi Rift, East Africa: seismic expression and exploration implications. AAPG Bull 79(11):1679–1697

Scott SC (1980) The geology of Mt. Longonot Volcano, central Kenya; a question of volumes. Philos Trans R Soc Lond 296:437–465

Sepulchre P, Ramstein G, Fluteau F, Schuster M, Tiercelin J-J, Brunet M (2006) Tectonic uplift and eastern Africa aridification. Science 313:1419–1423

Shackleton RM (1950) A contribution to the geology of the Kavirondo Rift Valley. Q J Geol Soc 106:345–392

Shackleton RM (1993) Tectonics of Mozambique Belt in East Africa. In: Prichard HM, Alabaster T, Harris NBW, Neary CRE (eds) Magmatic processes and plate tectonics, Geol Soc Spec Publ 76:345–362

Sickenberg O, Schönfeld M (1975) The Chorora formation—Lower Pliocene limnical sediments in the southern Afar. In: Pilger A, Rosler A (eds) Afar depression of Ethiopia, vol 1. Schweizerbart Verlag, Stuttgart, pp 277–284

Smith M (1994) Stratigraphic and structural constraints on mechanisms of active rifting in the Gregory Rift, Kenya. Tectonophysics 236:3–22

Smith M, Mosley P (1993) Crustal heterogeneity and basement influence on the development of the Kenya Rift, East Africa. Tectonics 12:591–606

Strecker MR, Blisniuk PM, Eisbacher GH (1990) Rotation of extension direction in the central Kenya Rift. Geology 18:299–302

Street FA (1979) Late Quaternary lakes in the Ziway-Shala Basin, Ethiopia. Palaeoecol Afr 15:99–110

Street-Perrott FA, Marchand DS, Roberts N, Harrison SP (1989) Global lake-level variations from 18,000 to 0 years ago: a palaeoclimatic analysis. US Department of Energy, Washington, DC

Suwa G, White T, Asfaw B, WoldeGabriel G, Yemane T (1991) Miocene faunal remains from the Burji Soyama area, Amaro Horst, southern sector of the Main Ethiopian Rift. Palaeontol Afr 28:23–38

Taieb M, Barker P, Bonnefille R, Damnati B, Gasse F, Goetz C, Hillaire C, Icole M, Masault M, Roberts N, Vincens A, Willamson D (1991) Histoire paléohydrologique du lac Magadi (Kenya) au Pleistocène superieur. C R Acad Sci Sér 2(313):339–346

Talbot MR, Williams MAJ, Adamson DA (2000) Strontium isotope evidence for late Pleistocene reestablishment of an integrated Nile drainage network. Geology 28(4):343–346

Tallon PWJ (1976) The stratigraphy, palaeoenvironments and geomorphology of the Pleistocene Kapthurin Formation, Kenya. Unpublished PhD thesis, University of London

Tallon PWJ (1978) Geological setting of hominid fossils and Acheulian artefacts from the Kapthurin Formation, Baringo District, Kenya. In: Bishop WW (ed) Geological background to fossil man. Geological Society/Scottish Academic, London, pp 361–378

Taylor KC, Hammer CU, Alley RB, Clausen HB, Dahl-Jensen D, Gow AJ, Gundestrup NS, Kipfsstuhl J, Moore JC, Waddington ED (1993) Electrical conductivity measurements from the GISP2 and GRIP Greenland ice cores. Nature 366:549–552

Tefera M, Chernet T, Haro W (1996) Explanation of the Geological Map of Ethiopia, vol 3. Ethiopian Institute of Geological Surveys, Addis Ababa

Tesfaye S, Ghebreab W (2013) Simple shear detachment fault system and marginal grabens in the southernmost of the Red Sea Rift. Tectonophysics 608:1268–1279

Teunis H, Baker E, Mark T, Boyd KP (2008) Modern analog for interlayered evaporites and seaward dipping volcanic reflectors; Danakil Depression, northern Afar, East Africa. AAPG 2008 Annual Convention and Exhibition

Thompson AO, Dodson RG (1963) Geology of the Naivasha Area. Report No. 55, Geological Survey of Kenya, Nairobi, Kenya

Thuo P (2010) Stratigraphic, petrographic and diagenetic evaluation of cretaceous/paleogene potential reservoir sandstones of Western Turkana, Kenya. Implications on the petroleum potential of Northwestern Kenya. PhD Thesis, Universite de Bretagne Occidental, France

Tiercelin JJ, Lezzar K-E (2002) A 300 million years history of rift lakes in Central and East Africa: an updated broad review. In: Odada EO, Olago DO (eds) The East African Great Lakes: limnology, palaeolimnology and biodiversity, Adv Glob Change Res 12:3–60

Tiercelin JJ, Taieb M, Faure H (1980) Continental sedimentary basins and volcano-tectonic evolution of the Afar. Accad Naz Lincei Rome 47:491–504

Tiercelin J, Chorowicz J, Bellon H, Richert J, Mwanbene J, Walgenwitz F (1988) East African Rift System: offset, age and tectonic significance of the Tanganyika Rukwa-Malawi intracontinental transcurrent fault zone. Tectonophysics 148:241–252

Tiercelin J-J, Potdevin J-L, Morley CK, Talbot MR, Bellon H, Rio A, Le Gall B, Vetel W (2004) Hydrocarbon potential of the Meso-Cenozoic Turkana depression, northern Kenya. I. Reservoirs: depositional environments, diagenetic characteristics, and source rock–reservoir relationships. Mar Petrol Geol 21:41–62

Trauth MH, Strecker MR (1996) Late pleistocene lake-level fluctuations in the Naivasha Basin, Kenya. In: Johnson TC, Odada EO (eds) The limnology, climatology and paleoclimatology of the East African Lakes, the international decade for the East African Lakes (IDEAL), Gordon and Breach, Newark, NJ, p 549–557

Trauth MH, Deino A, Strecker MR (2001) Response of the East African climate to orbital forcing during the last interglacial (130–117 ka) and the early last glacial (117–60 ka). Geology 29(6):499–502

Trauth MH, Deino AL, Bergner AGN, Strecker MR (2003) East African climate change and orbital forcing during the last 175 kyr BP. Earth Planet Sci Lett 206:297–313

Trauth MH, Maslin MA, Deino A, Strecker MR (2005) Late Cenozoic moisture history of East Africa. Science 309:2051–2053

Trauth MH, Maslin MA, Deino AL, Bergner ML, Strecker MR, Bergner AGN, Dühnforth M (2007) High- and low-latitude controls and East African forcing of Plio-Pleistocene East African climate and early human evolution. J Hum Evol 53:475–486

Truckle P (1976) Geology and late Cainozoic lake sediments of the Suguta Trough, Kenya. Nature 263:380–383

Tudorancea C, Taylor WD (2002) Ethiopian Rift Valley lakes. Backhuys Publishers, Leiden, p 289

Ukstins IA, Renne PR, Wolfenden E, Baker J, Ayalew D, Menzies M (2002) Matching conjugate volcanic rifted margins: $^{40}Ar/^{39}Ar$ chronostratigraphy of pre- and syn-rift bimodal flood volcanism in Ethiopia and Yemen. Earth Planet Sci Lett 198:289–306

Varet J, Gasse F (1978) Geology of Central and Southern Afar (Ethiopia and Djibouti Republic). In: Gasse F (ed) Chapter IV on sedimentary formation. CNRS, France

Vedkamp A, Schoorl JM, Wijbrans JR, Claessens L (2012) Mount Kenya volcanic activity and the late Cenozoic landscape reorganization in the upper Tana fluvial system. Geomorphology 145–146:19–31

Veevers JJ (1981) Morphotectonics of rifted continental margins in embryo (East Africa), youth (Africa-Arabia), and maturity (Australia). J Geol 89(1):57–82

Veldkamp A, Buis E, Wijbrans JR, Olago DO, Boshoven EH, Maee M, van den van Saparoea (2007) Late Cenozoic fluvial dynamics of the Tana, Kenya, an uplift dominated record. Quat Sci Rev 26 (22–24):2897–2912

Watchorn F, Nichols GJ, Bosence DWJ (1998) Rift related sedimentation and stratigraphy, southern Yemen (Gulf of Aden). In: Purser BH, Bosence DWJ (eds) Sedimentation and tectonics of rift basins: Red Sea–Gulf of Aden. Chapman and Hall, London, p 663

Watkins RT (1986) Volcano-tectonic control on sedimentation in the Koobi Fora sedimentary basin, Lake Turkana. In: Frostick LE, Renaut RW, Reid I, Tiercelin J-J (eds) Sedimentation in the African Rifts, Geol Soc Spec Publ 25:85–95

Wescott WA, Wigger ST, Stone DM, Morley CK (1999) Geology and geophysics of the Lotikipi Plain. In: Morley CK (ed) Geoscience of rift systems-evolution of East Africa, American Association of Petroleum Geologists, Stud Geol 44:55–65

Westerhof AB, Härmä P, Isabirye E, Katto E, Koistinen T, Kuosmanen E, Lehto T, Lehtonen MI, Mäkitie H, Manninen T, Mänttäri I, Pekkala Y, Pokki J, Saalmann K, Virransalo P (2014) Geology and geodynamic development of Uganda with explanation of the 1:1,000,000-scale geological map. Geol Surv Final 55:2–405

White TD, Suwa G, Hart WK, Walter RC, WoldeGabriel G, de Heinzelin J, Clark JD, Asfaw B, Vrba E (1993) New pliocene hominids from Maka, Ethiopia. Nature 366:261–265

Wichura H, Bousquet R, Oberhänsli R, Strecker M, Trauth MH (2010) The mid-Miocene East African Plateau: a pre-rift topographic model inferred from the emplacement of the phonolitic Yatta lava flow, Kenya. Geol Soc Spec Publ 357:285–300

Wichura H, Jacobs LL, Lin A, Polcyn MJ, Manthi FK, Winkler DA, Strecker MR, Clemens M (2015) A 17-my-old whale constrains onset of uplift and climate change in East Africa. Proc Natl Acad Sci USA. doi:10.1073/pnas.1421502112

Williamson PG, Savage RJG (1986) Early rift sedimentation in the Turkana Basin, northern Kenya. In: Frostick LE, Renaut RW, Reid I, Tiercelin J-J (eds) Sedimentation in the African Rifts, Geol Soc Spec Publ 25:267–283

Wilson KE, Maslin MA, Leng MJ, Kingston JD, Deino AL, Edgar RK, Mackay AW (2014) East African lake evidence for Pliocene millennial-scale climate variability. Geology. doi:10.1130/G35915.1

WoldeGabriel G, Aronson JL (1987) Chow Bahir Rift: a "failed" rift in southern Ethiopia. Geology 15:430–433

WoldeGabriel G, Aronson JL, Walter RC (1990) Geology, geochronology, and rift basin development in the central sector of the Main Ethiopian Rift. Geol Soc Am Bull 102:439–458

WoldeGabriel G, Yemane T, Suwa G, White T, Asfaw B (1991) Age of volcanism and rifting in the Burji-Soyoma area, Amaro Horst, southern sector of the Main Ethiopian Rift; geo- and bio-chronologic data. J Afr Earth Sci 13:437–447

WoldeGabriel G, White TD, Suwa G, Semaw S, Beyene Y, Asfaw B, Walter R (1992) Kesem-Kebena: a newly discovered paleontological research area in Ethiopia. J Field Archaeol 19:471–493

WoldeGabriel G, Haile-Selassie Y, Renne PR, Hart WK, Ambrose SH, Asfaw B, Heiken G, White TD (2001) Geology and paleontology of the late Miocene Middle Awash valley, Afar Rift, Ethiopia. Nature 412:175–178

WoldeGabriel G, Hart WK, Katoh S, Beyene Y, Suwa G (2005) Correlation of Plio–Pleistocene tephra in the Ethiopian and Kenyan Rift Basins: temporal calibration of geological features and hominid fossil records. J Volcanol Geotherm Res 147:81–108

WoldeGabriel G, Hart WK, Renne PR, Haile-Selassie Y, White TD (2009) Stratigraphy of the Adu-Asa formation. In: Haile-Selassie Y, WoldeGabriel G (eds) *Ardipithecus kadabba*: late Miocene evidence from the Middle Awash, Ethiopia. University of California Press, Berkeley, CA, pp 27–61

Wolfenden E, Yirgu G, Ebinger C, Deino A, Ayalew D (2004) Evolution of the northern Main Ethiopian Rift: birth of a triple-junction. Earth Planet Sci Lett 24:213–228

Wolfenden E, Ebinger C, Yirgu G, Renne PR, Kelley SP (2005) Evolution of a volcanic rifted margin: Southern Red Sea, Ethiopia. Geol Soc Am Bull 117:846–864

Yatheesh V, Kurian PJ, Bhattacharya GC, Rajan S (2013) Morphotectonic architecture of an India-Madagascar breakup related anomalous terrace complex on the southwest continental margin of India. Mar Petrol Geol 46:304–318

Yemane T, WoldeGabriel G, Tesfaye S, Berhe SM, Durary S, Ebinger CJ, Kelley S (1999) Temporal and geochemical features of Tertiary volcanic rocks in the southern Main Ethiopian Rift and the adjacent volcanic fields. Acta Vulcanol 11:99–119

Zanettin B, Justin-Visentin E (1975) Tectonical and volcanological evolution of the western Afar margin (Ethiopia). In: Pilger A, Rosler A (eds) Afar depression of Ethiopia, vol. 1. Schweizerbart Verlag, Stuttgart, p 300–309

Zanettin B, Nicoletti M, Petrucciani C (1978) The evolution of the Chencha Escarpment and the Ganjuli Graben (Lake Abaya) in southern Ethiopian rift. NJB Geol Palaont Mon 8:473–490

Zielke O, Strecker MR (2009) Recurrence of large earthquakes in magmatic continental rifts: insights from a palaeoseismic study along the Laikipia-Marmanet Fault, Subukia Valley, Kenya Rift. BSSA 99(1):61–70

Morphometry and Physical Processes of East African Soda Lakes

3

John M. Melack and Sally MacIntyre

Abstract

The altitudes of the saline lakes in the East African Rift System (EARS) vary from below sea level in the Danakil Depression (−155 m) to 2000 m in the Ethiopian highlands; most lakes are located above 1000 m. These lakes are among the largest (Turkana) and deepest (Shala) saline lakes in the world, though many formed by volcanism or lahars are small. High concentrations of phytoplankton or suspended sediments lead to high light attenuation, with Secchi visibilities usually less than 1 m. In the shallow, saline lake strong stratification usually develops during the day with mixing to the bottom at night. Seasonal variations in stratification and mixing have been observed in the deep lakes. Topography alters winds and mixing in lakes within volcanic craters. Chemically stratified, saline lakes occur throughout the EARS, and meromixis was documented in Lake Sonachi. Seasonal variations in thermal stratification and horizontal and vertical gradients in salinity occur in Lake Turkana. Only brief visits, occasional year-round studies and long lapses in the study of soda lakes of the EARS leave large gaps in our understanding of temporal variations in their limnology and their responses to climatic and human-caused changes in their hydrology.

3.1 Introduction

Physical characteristics of lakes include their location, altitude, area and depth, optical conditions and stratification and mixing. These characteristics reflect geologic and climatic history and influence ecological and chemical features of the lakes. We examine the physical characteristics of East African soda lakes

J.M. Melack (✉)
Department of Ecology, Evolution and Marine Biology, University of California at Santa Barbara, Santa Barbara, CA 93106, USA

Bren School of Environmental Science and Management, University of California at Santa Barbara, Santa Barbara, CA 93106, USA
e-mail: john.melack@lifesci.ucsb.edu

S. MacIntyre
Department of Ecology, Evolution and Marine Biology, University of California at Santa Barbara, Santa Barbara, CA 93106, USA
e-mail: sally@eri.ucsb.edu

M. Schagerl (ed.), *Soda Lakes of East Africa*, DOI 10.1007/978-3-319-28622-8_3

(EASL) by summarizing the available literature and offering perspectives based on current understanding of physical processes. Data on hydrodynamics of these lakes are scarce and largely based on studies in the 1970s and 1980s.

The semiarid EARS extends from northern Tanzania into the Danakil Depression in Ethiopia and Djibouti. Tectonic and volcanic activities have formed the lake basins (Table 3.1). Small lakes within volcanic craters are scattered throughout and include the Basotu lakes (Tanzania; Downie and Wilkinson 1962) and Bishoftu lakes (Ethiopia; Prosser et al. 1968). The Momela Lakes were formed by a lahar that originated on Mt. Meru (Tanzania; Hecky 1971). Both broad, shallow lakes and large, deep lakes occupy tectonic basins.

No rivers flow to the ocean from the EARS; hence, the lakes are within endorheic basins. As a consequence, most of the lakes are saline because solutes are concentrated by evaporation, and the lakes vary considerably in depth, size and salinity as a function of variations in rainfall. These variations occur seasonally, interannually and over longer time periods. Early geological and archaeological expeditions noted relict shorelines above the extant lakes, and subsequent analyses based on radiocarbon dating of sediment cores and use of microfossils to infer salinities and water levels indicated a wetter

Table 3.1 Morphometric characteristics of lakes and mode of origin of basin

Name	Altitude (m asl)	Area (km^2)	Mean depth (m)	Maximum depth (m)	Basin origin
Assal[a,b]	−155	54	7.4	40	Tectonic
Abhe[b]	240	350	8.6	37	Tectonic
Arenguade[c]	1900	0.54	18.5	32	Volcanic crater
Kilotes[c]	2000	0.77	2.6	6.4	Volcanic crater
Biete Mengest[c]	1850	1.03	17.5	38	Volcanic crater
Bishoftu[c]	1870	0.9	55	87	Volcanic crater
Pawlo[c]	1870	0.6	38	65	Volcanic crater
Chitu[b]	1600	0.8		21	Volcanic crater
Shala[b,d]	1558–1567	329–409	88	266	Tectonic
Abiyata[b]	1578	176	7.6	14	Tectonic
Langano[b]	1582	241	17	48	Tectonic
Turkana[e]	427	8860	28	106	Tectonic
Bogoria[f,g]	963	33	5.4	9	Tectonic fault scarp
Nakuru[h]	1759	36–49	0.4–3.6	0.6–4.6	Tectonic
Elmenteita[i]	1776	20	0.7–1.1	0.85–1.35	Tectonic
Sonachi[i]	1891	0.18–0.16	3.8–4.0	6.1–7.0	Volcanic crater
Oloidien[j]	1890	5.5	5.6	8.4	Tectonic
Magadi[a]	660, 683	108			Tectonic
Magad[k]	1722	17		2	Volcanic caldera
Manyara[k]	960	413		3.7	Tectonic fault scarp
Natron[a]	675	1000			Tectonic
Eyasi[a]	1030	1050			Tectonic
Reshitani[k]	1448	0.2		29	Lahar
Big Momela[k]	1448	0.9		31	Lahar
Embagai[l]				75	Volcanic crater
Tulusia[l]				13	Lahar
Lekandiro[l]				11	Lahar
Gidamuniud[l,m]	1600	0.02		9	Volcanic crater

Sources: (a) Serruya and Pollingher (1983), (b) Wood and Talling (1988), (c) Prosser et al. (1968), (d) Melack (1983), (e) Spigel and Coulter (1996), (f) Melack et al. (1981), (g) World Lake Database (1999), (h) Vareschi (1982), (i) Melack (1976), (j) Gaudet and Melack (1981), (k) Melack and Kilham (1974), (l) MacIntyre and Melack (1982), (m) Kilham and Cloke (1990) climate in the early to mid-Holocene (Livingstone 1975; Halfman and Johnson 1988; Street and Grove 1976; Verschuren 1996).

3.2 Morphometry

Morphometric data are available for only a subset of EASL (Table 3.1; see Schagerl and Renaut, Chap. 1). Given the changes in water level caused by variations in rainfall and evaporation, the morphometric data summarized in Table 3.1 represent only depths and sizes when the lakes were studied. The altitudes of the lake surfaces vary from below sea level in the Danakil Depression (−155 m) to 2000 m in the Ethiopian highlands; most lakes are located above 1000 m. Melack (1981) provides a compilation of tropical soda lakes throughout Africa with morphometric divisions as shallow (<2 m), intermediate depths (between 2 and 15 m) and deep (>15 m) and as small (<5 km^2) and large (>5 km^2).

Several EASL are among the largest and deepest saline lakes in the world, although they are not listed as such in Hammer (1986). With the Aral and Caspian Seas excluded, Lake Turkana would be within the top five by area and top ten by depth; Shala (Ethiopia) would be within the top five by maximum and mean depth. In contrast, the volcanic crater lakes and those formed by a lahar are small (0.015–1.03 km^2), though can be fairly deep (up to 87 m, Bishoftu, Ethiopia).

3.3 Optical Conditions

The EASL often have high concentrations of phytoplankton or suspended sediments, leading to high attenuation of incoming light with several limnological and ecological implications. Thermocline depth, vertical nutrient fluxes, rates of photosynthesis and predation rates are all related to penetration of light through the water column. We summarize estimates of underwater transparency made with Secchi discs and attenuation of light measured with underwater sensors. Photosynthetically available radiation (PAR, 400–700 nm) was measured with a filtered, cosine-corrected silicon photodiode (Biggs et al. 1971). Light attenuation was determined in three spectral regions (495, 540 and 650 nm; 50 % bandwidth was 75, 50 and 75 nm, respectively) with filtered photoresistors and ohmmeter (Melack and Kilham 1974) or in four spectral regions (460, 540, 630 and 685 nm) with filtered selenium barrier-layer sensors and microammeter (Talling et al. 1973). The visual colour apparent to a human observer was also noted, and turbidity, expressed as Jackson Turbidity Units (JTUs), was measured with a Hach DR-EL portable instrument. JTUs are an inverse measure of the length of a column of water needed to completely obscure a light source viewed through it, i.e., the lower the value, the clearer the water.

Among the EASL, Secchi disc visibilities varied from 0.1 to 1.0 m with the exception of Lake Turkana (Table 3.2). The abundant phytoplankton common in these lakes is a primary and persistent cause of the shallow Secchi disc visibilities as indicated by Jenkin's (1936) observation at Lake Nakuru in 1929 matching those in the 1970s. P. Kilham (personal communication), as part of his chemical survey of Africa lakes (Kilham 1971), recorded turbidites, as JTUs, in several eastern Rift Valley saline lakes: Big Momela (58), Lekandiro (115), Tulusia (34), Reshitani (40), Small Momela (44) and Embagai (40). All these values are high and indicative of the suspended particles, most likely phytoplankton based on visual observations.

As noted by Talling et al. (1973), in very dense suspensions of phytoplankton, such as those found in Ethiopian lakes Arenguade and Kilotes, it is difficult to obtain high accuracy in measurements of underwater light. Even so, the attenuation coefficients reported for these two lakes are extraordinary (Table 3.2). Other lakes with abundant phytoplankton had attenuation of PAR or visible wavelengths usually 1 and 10 m^{-1}. The very high attenuation in Lake Manyara was caused by suspended sediments. The elevated attenuation values enhance the heating of the water by

Table 3.2 Secchi disc visibility (m), underwater light attenuation (m^{-1}) and visual colour

Lake	Secchi	PAR	460	495	540	630	650	685	Visual colour
Arenguade[a]	–	–	110	–	60	30	–	25	Green
Kilotes[a]	–	–	30–50	–	10–20	8–20	–	7–18	Green
Shala[b]	–	–	3	–	1.3	1	–	1	Clear
Abiyata[b]	–	–	9.5	–	4.6	4.8	–	4	Green
Langano[b]	–	–	8.5	–	6.5	3.1	–	3	Reddish brown
Turkana[c]	1–4.8	–	–	–	–	–	–	–	–
Bogoria[d,e]	0.28–0.60	–	–	–	–	–	–	–	–
Nakuru[f]	0.15	–	–	10	12	–	15	–	Green
Nakuru[g]	–	3.6–17.0	–	–	–	–	–	–	Green
Elmenteita[f]	0.17	–	–	7	9	–	30	–	Green
Elmenteita[d,i]	0.12–1	1.9–3.6	–	–	–	–	–	–	Green to yellow
Elmenteita[e]	0.15	9.2	–	–	–	–	–	–	Green
Sonachi[e,h]	0.3–0.55	1.9–3.5	–	–	2.3–3.8	–	–	–	Yellowish green
Oloidien[j]	0.5–1.0	1.3–2.4	–	–	1.7–2.9	–	–	–	–
Magad[f]	0.1	–	–	4	4	–	4	–	–
Manyara[f]	–	–	–	16	100	–	80	–	–
Reshanti[f]	–	–	–	4	5	–	7	–	Green
Big Momela[f]	0.25	–	–	4	6	–	5	–	Green
Embagai[k]	–	–	–	1.8	2.6	–	2.5	–	–
Gidamuniud[k]	–	–	–	1.2	1.1	–	0.9	–	–

Sources (a) Talling et al. (1973), (b) Wood et al. (1978) (estimated from Fig. 3.2), (c) Källqvist et al. (1988), (d) Melack (1981), (e) Melack et al. (1981), (f) Melack and Kilham (1974), (g) Vareschi (1982), (h) Melack (1982), (i) Melack (1976), (j) Melack (1979), (k) P. Kilham, personal communication

insolation, with concomitant strong stratification during the day.

3.4 Stratification and Mixing

The EASL are generally warm, including those in highlands of Ethiopia. At temperatures near and above 20 °C, warming of the upper waters in the day and cooling at night cause appreciable changes in density and lead to pronounced cycles of diel stratification and mixing. The high evaporation at warm temperatures contributes to surficial cooling and nocturnal convective mixing. Furthermore, since seasonal variations in temperature are slight, temperature gradients between surface and bottom waters are not large. As a result, shallow lakes are usually polymictic with daily or frequent stratification and mixing to the bottom. Seasonal stratification occurs in deep lakes. General reviews of stratification and mixing in tropical lakes are provided in Beadle (1981), Livingstone and Melack (1984), MacIntyre (2012, 2013), Talling and Lemoalle (1998) and Spigel and Coulter (1996), and a more comprehensive discussion is in MacIntyre et al. (2002, 2014).

3.4.1 Shallow Lakes

In shallow, saline lakes, strong stratification usually develops during the day with mixing to the bottom at night (Melack 2009; Melack and Kilham 1974; Vareschi 1982). Time-series observations in Lake Elmenteita, Kenya (Fig. 3.1), obtained over 13 months, indicate a high degree of predictability, based on information theoretic indices, in stratification dynamics (Melack 2009).

As the wind passes across the surface of a stratified lake, the thermocline usually tilts, and velocity shear develops within it. Whether shear-induced instabilities will form in the thermocline and become turbulent can be predicted from the Richardson Number, $Ri = -g\ \rho^{-1}\ (\delta\rho/\delta z)$

Fig. 3.1 Lake Elmenteita with evaporative soda deposits in foreground. Photograph by J. Melack, June 1973

$(\delta u/\delta z)^{-2}$ where g is gravity, ρ is density, u is velocity and z is depth. If Ri drops below 0.25, the fluid will become turbulent and mix. In one of the few analyses of Ri for a tropical lake, MacIntyre (1981) and MacIntyre and Melack (1995) report vertical profiles of current speeds, measured with a warm-bead thermistor sensor (MacIntyre 1986), and temperature in Lake Nakuru. Temperatures taken through the day document morning stratification followed by afternoon and evening mixing, as expected from previous studies noted above. The stability of the stratification, determined from the Brunt-Väisälä frequency, $N^2 = -g\,\rho^{-1}\,\delta\rho/\delta z$, indicated strong stratification during portions of the day. Temperature fluctuations in the stratified regions suggested high-frequency internal wave motions. Currents, measured on November 29, 1976, at 0.07, 0.11, 0.37 and 0.95 m, were less than 2.5 cm s^{-1} except at 0.11 m. At 0.11 m, they were highest at noon, reaching 5 cm s^{-1}, and decreased throughout the afternoon. Richardson numbers were either negative, implying heat-loss-augmented wind mixing due to winds of 3–5 m s^{-1} during the afternoon, or close to the critical value for turbulent mixing above the thermocline and much larger than the critical value across the thermocline. The high values imply reduced flux between the upper and lower water column during the day. The persistent stratification enabled anoxia to develop in the lower water column during the day, which would then enable phosphorus to be in ionic form and support growth of autotrophs. The reduced mixing also enabled buoyant Cyanobacteria to control their vertical position in the water column.

Intermittent winds can induce surface seiches. Based on the morphometry of Lake Nakuru, Vareschi (1982) calculated a theoretical period of ~1.3 h. Following a windy period, an automatic stage recorder documented a period of 1.1 h and an amplitude of 4.5 cm. While conducting diel measurements at Lake Elmenteita, occasional solitary, surface waves of ~ 10 cm amplitude were observed. We inferred that these waves were surface seiches.

3.4.2 Crater Lakes

Prosser et al. (1968) noted that the variation in the height of the crater rim surrounding a crater lake is likely to influence the extent to which wind reaches the lakes and moderates stratification and mixing of the water. Subsequently, Melack (1978) proposed an index of the exposure to wind-induced mixing as the ratio $D\,H^{-1}$, where D is the maximum diameter of the lake and H is the minimum height of the crater rim above the lake. Wind is an important factor, as the energy imparted to a lake by wind causes currents and shear and internal and surface waves, and contributes to evaporative cooling and convective mixing. The ratio explained 31 % of the variation in the depth of anoxic or hypoxic water measured in a number of African crater lakes. Scatter in the relation is expected given the multiple processes that determine stratification and because mixing depth can vary on timescales from hours to months to years.

Stratification and mixing in the upper waters of tropical crater lakes occur on a diel basis (MacIntyre 1981; Melack 1976, 1982; Talling et al. 1973). Within sheltered Lake Sonachi (Fig. 3.2), diurnal heating occurred in the upper metre with cooling restricted to the upper 3 m in May 1973 (Fig. 3.3). The extent of heating and cooling offshore varied with solar radiation, wind speed and its timing during day and night, light attenuation and lateral transport due inshore and offshore differences in temperature. For example, on November 1 and 2, 1976, daily insolation was 356 and 321 W m^{-2}, respectively. Underwater attenuation of PAR was 3 m^{-1}; hence, the 1 % light level was at 1.5 m. Wind, measured on a mid-lake platform (Fig. 3.4), ranged from 0.5 to 7.5 m s^{-1} with significant diel variations (Fig. 3.5). Crater walls rise 33 to 117 m above the lake (Fig. 3.6), and the lakeshore is fringed by *Acacia xanthophloea* trees. As the lake heated in the morning, thermal stratification developed and peaked in mid-afternoon with temperatures differing by as much as 4 °C in the first metre (Fig. 3.7). After nocturnal cooling, stratification was minimal around dawn. Temperature oscillations in thermally stratified regions ranged in amplitude from 0.05 to 1.60 °C and probably indicate short-period internal waves. By late afternoon, cooler temperatures occurred near the surface than in the water below; these decreases are indicative of surficial cooling leading to convective mixing. Thermal stability, calculated according to Idso (1973), was least in the early morning and increased with increases in the lake's heat content (MacIntyre and Melack

Fig. 3.2 Lake Sonachi in foreground; only the southern portion of the volcanic crater around Lake Sonachi is shown. Aerial photograph by J. Melack, January 1971

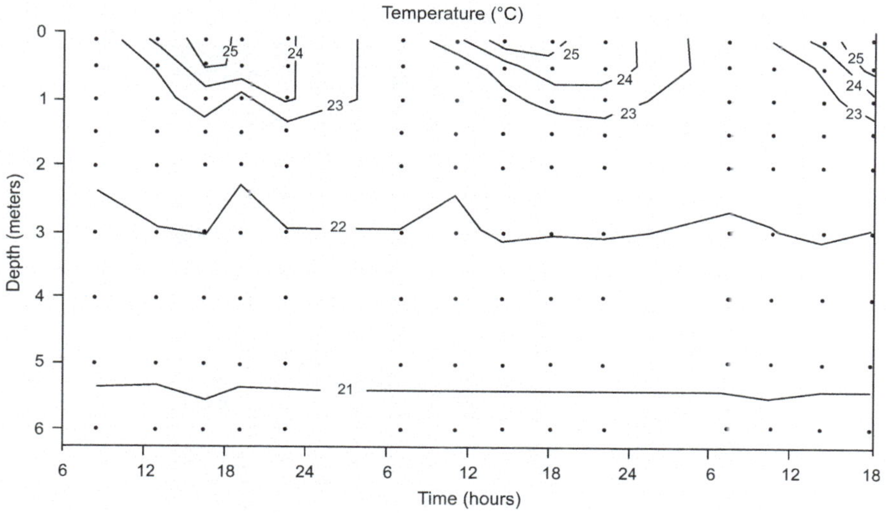

Fig. 3.3 Time-depth diagram of temperature in Lake Sonachi, 5–7 March 1973. From Melack (1976)

Fig. 3.4 Research platform with mast for current sensors and pole for wind sensor and housing for recorder, Lake Sonachi. Photograph by J. Melack, November 1976

1982). On November 1, mid-afternoon (1420 h), thermal stability was 8.7 J m^{-2} and on November 2 at 1500 h was 10.5 J m^{-2}.

Conditions in Lake Sonachi in the morning and early afternoon indicate that most of the turbulent kinetic energy imparted by the wind was consumed by work against the buoyancy produced by solar heating. In other words, stable stratification and Richardson's numbers between 0.05 and 100 developed during morning and early afternoon despite morning winds reaching 6 m s^{-1}. In fact, the strongly stratified region reached the surface (MacIntyre 1981). Brunt-Väisälä frequencies were high (0.03–0.12 s^{-1})

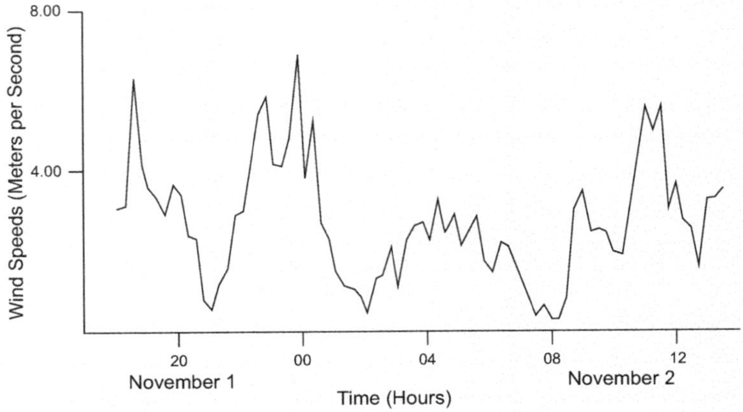

Fig. 3.5 Wind speeds (m s^{-1}), Lake Sonachi, November 1–2, 1976. Wind speeds were measured mid-lake, 1.9 m above the water with a 3-cup anemometer and chart recorder and averaged for 15-min periods. From MacIntyre (1981)

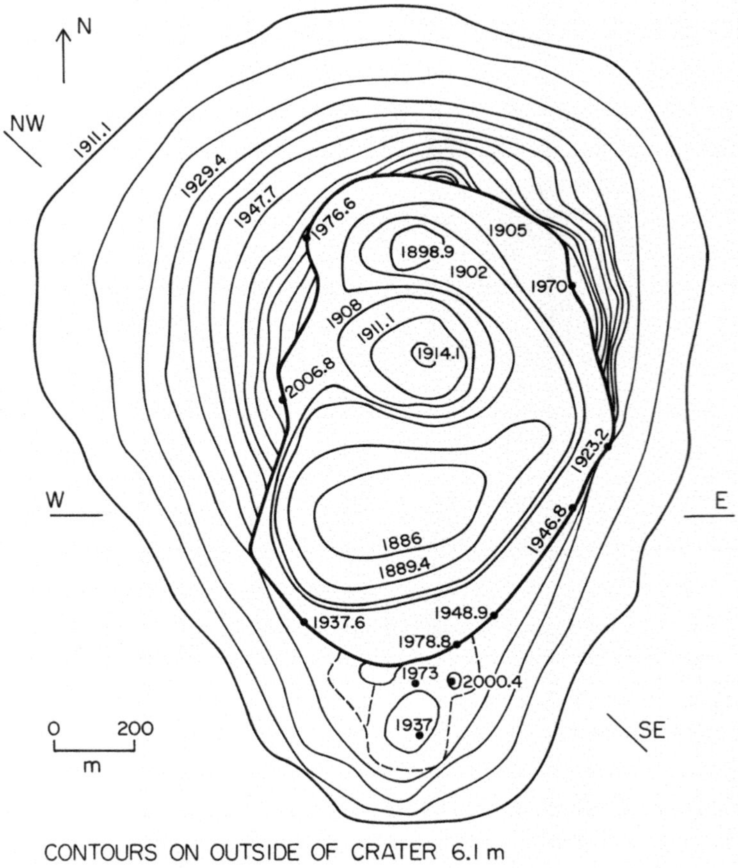

Fig. 3.6 Volcanic crater and surrounding hillside around Lake Sonachi derived from Kenya Ministry of Works drawing #44,330 at 1:1000 scale. Lake lies between 1886 and 1889.4 m contours in lower half of crater. From MacIntyre (1981)

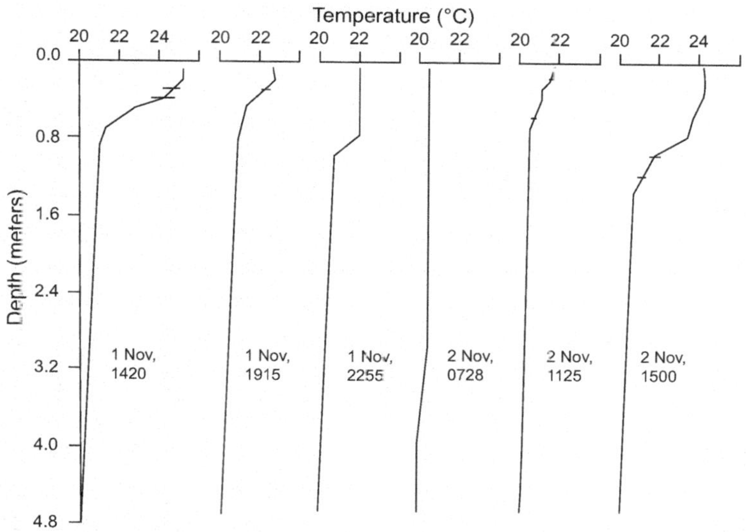

Fig. 3.7 Temperature profiles, Lake Sonachi, 1–2 November 1976. Horizontal bars indicate range of temperature oscillations. Measurements made at 0.1 m intervals in first metre, at 0.2 m intervals in the second metre and at 0.5 m intervals until 4.5 m. From MacIntyre (1981)

near the surface, and temperature oscillations pointing to high-frequency internal waves were present. *Ri* computed for 0.3 m intervals occasionally decreased below the critical value of 0.25. Values below 0.25 indicate that conditions were sometimes conducive to overturning internal waves and intermittent turbulence in the stratified upper region. The temperature inversion between 0.3 and 0.4 m at 1125 h on November 2, 1976, provides an example of a shear instability (Fig. 3.3). Overall, however, the stabilizing effect of heating exceeded that from wind-induced shear, and near-surface mixing did not begin until later in the afternoon when heat loss began to dominate the surface energy budget.

Seasonal variations in stratification and mixing have been observed in some crater lakes. For example, Wood et al. (1976) documented seasonal holomixis in Lake Babogaya (formerly Lake Pawlo), one of the Bishoftu crater lakes. With clear skies in January and February, heat loss by long-wave radiation increased to maximal values. Latent heat fluxes are expected to have increased with the concomitant decrease in relative humidity, and sensible heat fluxes are expected to have increased as air temperatures dropped. Thus, nocturnal cooling would have been increased relative to that during other seasons and led to the destratification and mixing of this lake.

MacIntyre (1981) analysed how crater topography moderates the wind field impacting lakes within volcanic craters and used Lake Sonachi as an example; we summarize this study and relevant theory here. Airflow is affected by atmospheric stratification within and above the crater and by the trees that fringe the lake. Possible patterns of airflow can surmised based on studies of airflows in valleys or behind hills. Bell and Thompson (1980) used experimental data and theory to show that when the Froude number (F) was below 1.3, air in a valley was stagnant. $F = U/Nh$, where U is the wind speed upstream of a hill, N is the Brunt-Väisälä frequency and h is the height of a hill or depth of a valley. F is the ratio of forces that generate mixing to buoyancy forces that resist mixing. Stratification of air within a crater is also influenced by the diel temperature variations of the land within the crater, by the lake and by the relative proportion of the crater occupied by the lake or land. Lake Sonachi occupies only about 20 % of the area within the crater. During the day, heat from the warming land leads to unstable stratification of the atmosphere over the land, while at night the cool ground leads to stable stratification. Lakes tend to be cooler than the air during mid-day, and thus the air is stably stratified and warmer than overlying air at night and the atmosphere unstably stratified. Thus, the proportion of land to water within the crater will determine whether the atmosphere is unstable or stable; this is a first-order control as to when winds penetrate into the crater.

Airflows over a crater lake for stratified or neutral atmospheric conditions inferred from laboratory studies (Hunt et al. 1978; Hunt and Snyder 1980) are illustrated in Fig. 3.7. These conditions would apply during the day when the surface area of the lake is large enough relative to the overall surface area of the crater such that its temperature dominates atmospheric stability. Small boundary layer separations directly behind the crater rim or the trees fringing the lake are not depicted. At $F = 0.2$ (Fig. 3.8a), air within the crater is stagnant, a small hydraulic jump forms near the crater rim, and the boundary layer separation in the lee of the crater extends approximately 100 m over the lake. At $F = 0.4$ (Fig. 3.8b), air within the crater remains stagnant, and the vertical eddies in the wake of the hydraulic jump probably do not reach the lake. At $F = 0.6$, eddies are likely to reach the lake's surface. For F greater than 0.4, lee waves of about the same height as the hill over which the air is flowing may form; unstable regions called rotors can occur in association with lee waves. Bell and Thompson (1980) observed rotors at Froude numbers between 0.8 and 1.2. Above the critical Froude number of 1.3, flow passes though the crater and lee waves form with a rotor (Fig. 3.8c). As wind speed increases and F exceeds 1.7, air sweeps through the crater except for a small region of backflow (Fig. 3.8d). Measured wind speeds often fluctuate rapidly, probably due to interactions with trees and irregular terrain. For example, horizontal fluctuations

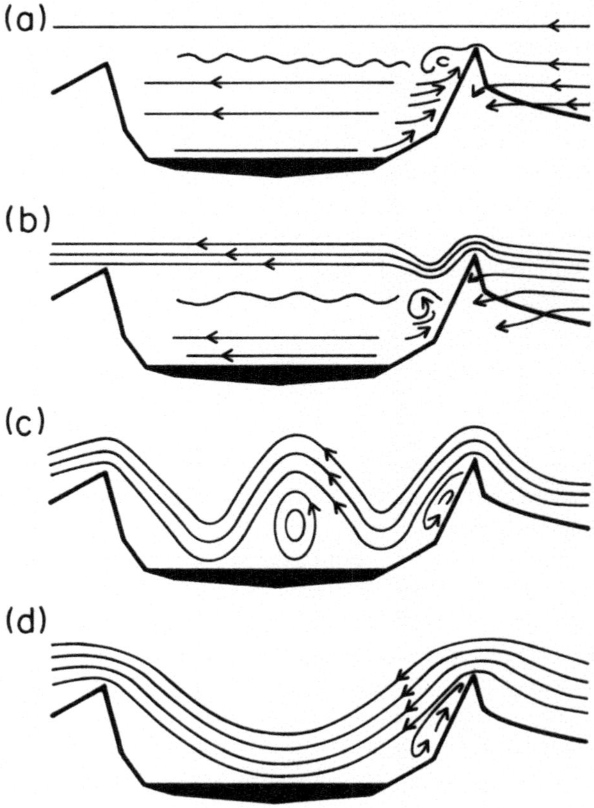

Fig. 3.8 Sketches of air flow over Lake Sonachi at different Froude numbers. Explanation in text. From MacIntyre (1981)

in wind speed were found to be five times greater over a wooded ridge than over smooth terrain (Panofsky et al. 1978).

During the periods when wind speeds outside the crater relative to stratification are too low for them to penetrate, the temperature differences over the lake and over the land within the crater will likely generate lake-land breezes. For Lake Sonachi, the larger land mass suggests that the atmosphere as a whole was unstable during the day and outside winds could penetrate. At night, however, the atmosphere within and outside the crater was likely stable. Hence, when outside winds were light, the greater cooling of the air over the land inside the crater likely generated airflows over the lake. This mechanism would enable heat loss by evaporation at night and contribute to nocturnal mixing. Even in craters in which the lake's surface area is proportionately larger, the greater cooling of the crater walls at night could induce downslope flows and promote heat loss and mixing within the lake.

Given the wide range of crater and lake dimensions among the crater lakes in eastern Africa, winds will impart a wide range of energies to mixing processes. Among the Basotu lakes, the longest dimension of the lakes (L_m) varies from 68 to 700 m; the maximum height of the crater above the lake (H_m) ranges from 12 to 96 m (Kilham and Cloke 1990). The ratio of these two terms (H_m/L_m) ranges from 0.017 to 0.50. Among the Bishoftu lakes, L_m varies from 900 to 1600, H_m from about 40 to about 200 m and the ratio from 0.04 to 0.22 (Prosser et al. 1968). For Lake Sonachi, $L_m = 650$ m, $H_m = 117$ m, and the ratio = 0.18. Among the 16 volcanic crater lakes associated with the western Rift Valley in Uganda and examined by Melack (1978), L_m ranged from 75 to 3750 m, H_m from 5 to 215 m and the ratio from 0.0013 to 0.6. The analysis of controls on airflows over crater lakes indicates why the predictive ability is low with the ratio D:H. Further improvement would likely be achieved by including the ratio of lake surface area to crater area and the ratio of daytime to night-time air temperatures. This is because a larger range implies a greater probability of formation of gravity currents, which would cause a land breeze over the lake at night.

3.4.3 Chemically Stratified Lakes

Chemically stratified, saline lakes occur throughout the EARS (MacIntyre and Melack 1982). One consequence of chemical stratification is meromixis, i.e., incomplete vertical mixing over at least several years. The water in the upper portion is called the mixolimnion, while that below the chemocline is called the monimolimnion. To demonstrate meromixis requires repeated measurements over many years, and one of the few lakes in eastern Africa with such data is Lake Sonachi (Kenya). MacIntyre and Melack (1982) described an 8-year period that included a weakening and strengthening of chemical stratification, without complete vertical mixing, as the amount of rainfall varied. Based on an analysis of sediment cores from Lake Sonachi, Verschuren (1999) and Verschuren et al. (1999) inferred meromictic and holomictic periods spanning a few years to several decades over the prior 175 years.

Several aspects of MacIntyre and Melack (1982) are generally relevant to studies of saline lakes. Water density and its changes with temperature are critical when examining stratification and meromixis. Because the solutes dissolved in the waters of soda lakes differ from seawater and are in much higher concentration than in freshwaters, density must be calculated based on physical chemical principles or measured (Boehrer et al. 2010; Jellison and Melack 1993; MacIntyre and Melack 1982; Millero et al. 1976). Because it is usual to measure profiles of electrical conductivity, relationships between the densities and conductivities are needed. Calculating the stability of the stratification, as determined following Idso (1973), and its partition into chemical and thermal components are useful metrics to evaluate the likely persistence of meromixis. In the case of Lake Sonachi, the chemical stability varied from a high of 200 J m^{-2} to a low of

0.7 J m^{-2}, when the lake almost mixed. During the period with lowest chemical stability in 1976, maximum daily thermal stability ranged from 1.95 to 10.5 J m^{-2}.

A rare type of thermal stratification with the warmest water at mid-depth (heliothermy) can occur in small, chemically stratified lakes, and this was observed in Lake Mahega, a crater lake in the Rift Valley of Uganda (Kilham and Melack 1972; Melack and Kilham 1972). The lake had strong chemical stratification, a dense accumulation of pigmented phytoplankton and Bacteria at 1 m and a Secchi disc transparency of 0.33 m, which allowed solar energy to reach the turbid layer. As a result, the water temperature increased from 31 °C at 0.1 m to 40 °C at 1 m and decreased to 37 °C at 3.5 m. Though this type of stratification has not been observed in EASL, the Basotu crater lakes are likely candidates.

3.4.4 Large, Deep Lakes

Lake Turkana, the largest lake in the EARS, is 256 km long with a mean width of 35 km (Spigel and Coulter 1996). It receives almost all its inflow from the Omo River draining from Ethiopia. Its local climate is arid—with precipitation on the lake of 360 mm y^{-1} and evaporation from the lake of 2340 mm y^{-1}—and hot with mean annual air temperatures of 30 °C (Spigel and Coulter 1996). The lake has a salinity of about 2.5‰ (Yuretich and Cerling 1983), with horizontal and vertical gradients in salinity attributed to inputs of less saline Omo River water (Ferguson and Harbott 1982). The larger central basin has a maximum depth of 73 m, while the smaller southern basin has a maximum depth of 106 m. A closed-basin lake with no surface outlet, its water level varies, e.g., levels declined approximately 5 m from 1972 to 1988 (Källqvist et al. 1988). During the southwest monsoon, which typically occurs from May or June through August or September, strong southerly winds can exceed 15 m s^{-1} with persistent winds from 5 to 11 m s^{-1}; these are superimposed on a diel pattern of strong winds in the afternoon and night (Ferguson and Harbott 1982).

Seasonal changes in near-surface waters are about 2 °C, and near-bottom temperatures range from 25.5 to 26.4 °C (Ferguson and Harbott 1982). Time series of vertical profiles of temperature and dissolved oxygen reported by Källqvist et al. (1988) for 1987–1988 indicate prolonged periods of stratification. Temperature and conductivity profiles shown by Spigel and Coulter (1996) for the monsoon-influenced period in 1987 and 1988 vary with depth but lack well-defined mixed layers and thermoclines. A transect of temperature profiles during August and October indicated holomixis in the southern lake and stratification in the north (Ferguson and Harbott 1982). While this pattern perhaps reflects upwelling induced by the strong southerly winds, freshwater inputs from the Omo River are largest from June through September and enter the northern lake as a sediment-laden overflow; this likely enhances stratification. An additional factor increasing the likelihood of a stratified northern versus southern lake is the large difference in the depth of the 1 % light level: 0.5–3.7 m 25 km from the Omo River mouth versus 10.2–13 m 230 km to the south (Ferguson and Harbott 1982; Spigel and Coulter 1996). Based a survey of 11 stations in the northern basin in January 1990, Halfman (1996) reported water temperatures slightly below 27 °C at 40 m and near-surface heating under calm conditions to just over 28 °C.

Lake Shala is a large, deep soda lake within a graben in the Ethiopian Rift Valley. Lake levels were low during the last glacial maximum and high 8000–9000 years B.P. (Beadle 1981; Street and Grove 1976). As summarized by Melack (1983), no chemical stratification was noted by Loffredo and Maldura (1941) or Baxter et al. (1965), but chemical stratification was observed when sampled by Baumann et al. (1975). Both Vatova (1940) and Baxter et al. (1965) reported thermal stratification with near-surface temperatures from 23 to 26 °C and near-bottom temperatures of about 21 °C.

3.5 Temperature Trends and Climate Changes

Though long-term records of water temperatures in EASL are not available, observations of skin temperatures have been reported for three lakes in the region (Schneider and Hook 2010). Following the terminology of Donlon et al. (2002), the skin temperature is that measured by a radiometer within a thin layer (~500 μm) on the water side of the air-water interface. Schneider and Hook (2010) used thermal infrared imagery obtained by the satellite-borne Along Track Scanning Radiometer and Advanced Very High Resolution Radiator series of sensors to examine trends in skin temperature from 1985 to 2009. Night-time values were averaged for the months of January, February and March to capture the warming periods in equatorial Africa. Temperature trends expressed as °C y^{-1} of 0.02 (L. Turkana), 0.05 (L. Eyasi) and 0.02 (L. Abaya) were found. Except for L. Eyasi, which often has shallow water distributed over a salt pan, these rates are less than the global average rate of 0.045 ± 0.011 °C y^{-1}. The lower values may result from the increased evaporative heat losses that accompany heating in warm tropical lakes.

3.6 Future Directions

Only brief visits, occasional year-round studies and long lapses in the study of EASL leave large gaps in our understanding of temporal variations in their limnology and their responses to climatic and human-caused changes in their hydrology. Though recent studies or analyses of ecological aspects are available (e.g., Burian et al. 2014; Krienitz and Kotut 2010; Schagerl and Oduor 2008, see Schagerl and Burian, Chap. 12), few investigations have focused on physical limnology since the early efforts by MacIntyre described above. With the availability of automatic recording systems, robust sensors, remote sensing techniques and models, it should be possible to initiate or continue studies in these lakes. As an example, Tebbs et al. (2013) developed an algorithm to monitor chlorophyll levels in Lake Bogoria and lakes with similar phytoplankton using an orbiting sensor (Landsat). Examples of approaches appropriate to eastern African saline lakes include the applications in a large, saline lake of microstructure profilers to calculate turbulence (MacIntyre et al. 1999, 2009; MacIntyre and Melack 2009), multi-station, time series of temperature and conductivity to document changing density gradients and vertical mixing (Jellison and Melack 1993) and three-dimensional modelling supported by meteorological data and temperature loggers (Vidal et al. 2013). These activities are of increasing urgency as pollutants from urban and agricultural activities and water diversion schemes, compounded by climate changes, are altering inflows to eastern African saline lakes (Dagnachew Legesse et al. 2004; Hadgembes 2006; Melack 1996; Odada et al. 2004; Velpuri and Senay 2012). International partnerships with researchers at universities in Ethiopia, Kenya and Tanzania should be nurtured and collaborative activities and local infrastructure and capacity developed further.

References

Baumann A, Förstner U, Rodhe R (1975) Lake Shala: water chemistry, mineralogy and geochemistry of sediments in an Ethiopian rift lake. Geol Rundsch 64:593–609

Baxter RM, Prosser MV, Talling JF, Wood RB (1965) Stratification in tropical African lakes at moderate altitudes (1500 to 2000 m). Limnol Oceanogr 10:510–520

Beadle LC (1981) The inland waters of tropical Africa. Longman Group, New York

Bell RC, Thompson RO (1980) Valley ventilation by cross winds. J Fluid Mech 96:757–767

Biggs WW, Edison AR, Eastin JD, Brown KW, Maranville JW, Clegg MD (1971) Photosynthesis light sensor and meter. Ecology 52:125–131

Boehrer B, Herzsprung P, Schultze M, Millero FJ (2010) Calculating density of water in geochemical lake stratification models. Limnol Oceanogr Methods 8:567–574

Burian A, Kainz MJ, Schagerl M, Yasindi A (2014) Species-specific separation of lake plankton reveals

divergent food assimilation patterns in rotifers. Freshw Biol. doi:10.1111/fwb.12345
Dagnachew Legesse C, Vallet-Coulomb CV, Gasse FC (2004) Analysis of the hydrological response of a tropical terminal lake, Lake Abiyata (main Ethiopian Rift Valley) to changes in climate and human activities. Hydrol Process 18:487–504
Donlon CJ, Minnett PJ, Gentemann C, Nightingale TJ, Barton IJ, Ward B, Murray MJ (2002) Toward improved validation of satellite sea surface temperature measurements for climate research. J Climate 15:353–369
Downie C, Wilkinson C (1962) The explosion craters of Basotu, Tanganyika Territory. Bull Volcan 14:389–420
Ferguson ADJ, Harbott BJ (1982) Geographical, physical and chemical aspects of Lake Turkana. In: Hopson AJ (ed) Lake Turkana: a report on the findings of the Lake Turkana project 1972–1975. Overseas Development Administration, London, pp 1–107
Gaudet JJ, Melack JM (1981) Major ion chemistry in a tropical African lake basin. Freshw Biol 11:309–333
Hadgembes T (2006) Review of environmental threats to East African saline lakes and their likely conservation strategies. Available via http://www.academia.edu/3528211/. Accessed Dez 2014
Halfman JD (1996) CTD-transmissometer profiles from Lakes Malawi and Turkana. In: Johnson TC, Odada EO (eds) The limnology, climatology and paleoclimatology of the East African lake. Gordon and Breach Publishing, Amsterdam, pp 169–182
Halfman JD, Johnson TC (1988) High resolution record of cyclic climate change during the past 4 ka from Lake Turkana, Kenya. Geology 16:496–500
Hammer UT (1986) Saline lake ecosystems of the world. Dr. W Junk Publishers, Dordrecht
Hecky RE (1971) The paleolimnology of the alkaline, saline lakes on the Mt. Meru lahar. Dissertation, Duke University, Durham, NC
Hunt JCR, Snyder WH (1980) Experiments on stably and neutrally stratified flow over a model three-dimensional hill. J Fluid Mech 96:671–704
Hunt JCR, Snyder WH, Lawson RE (1978) Flow structure and turbulent diffusion around a three-dimensional hill. EPA-600/4-78-041. National Technical Information Service, Springfield, VA
Idso SB (1973) On the concept of lake stability. Limnol Oceanogr 18:681–683
Jellison R, Melack JM (1993) Meromixis in hypersaline Mono Lake, California I. Vertical mixing and density stratification during the onset, persistence, and breakdown of meromixis. Limnol Oceanogr 38:1008–1019
Jenkin PM (1936) Reports of the Percy Sladen expedition to some Rift Valley lakes in Kenya in 1929. VII. Summary of ecological results, with special reference to the alkaline lakes. Annu Mag Nat Hist 18:133–181
Källqvist T, Lien L, Liti D (1988) Lake Turkana limnological study 1985–1988. Norwegian Institute of Water Research, Report 0–85313, Blindern, Norway
Kilham P (1971) Biogeochemistry of African lakes and rivers. Dissertation, Duke University, Durham, NC
Kilham P, Cloke PL (1990) The evolution of saline lake waters: gradual and rapid biogeochemical pathways in the Basotu Lake District, Tanzania. Hydrobiologia 197:35–50
Kilham P, Melack JM (1972) Primary northupite deposition in Lake Mahega, Uganda. Nature 238:123
Krienitz L, Kotut K (2010) Fluctuating algal food populations and the occurrence of lesser flamingos (*Phoeniconaias minor*) in three Kenyan Rift Valley Lakes. J Phycol 46:1088–1096
Livingstone DA (1975) Later quaternary climatic change in Africa. Annu Rev Ecol Syst 6:249–280
Livingstone DA, Melack JM (1984) Some lakes of subsaharan Africa. In: Taub FB (ed) Lake and reservoir ecosystems. Ecosystems of the world, vol 23. Elsevier, Amsterdam, pp 467–497
Loffredo S, Maldura CM (1941) Risultati generali delle ricerche di chimica limnologica sulle acque dei laghi dell'Africa orientale italiana esplorati dalla Missione ittiologica. In: Brunelli G et al (eds) Esplorazione dei laghi della Fossa Galla, vol I, Collezione scientifica e documentaria dell'Africa Italiana III. Airoldi, Verbania, pp 181–200
MacIntyre S (1981) Stratification and mixing in shallow tropical African lakes. Dissertation, Duke University, Durham, NC
MacIntyre S (1986) Flow measuring system for small lakes. Limnol Oceanogr 31:900–906
MacIntyre S (2012) Stratification and mixing in tropical African lakes. In: Bengtsson L, Herschy RW, Fairbridge RW (eds) Encyclopedia of lakes and reservoirs. Springer Science + Business Media B.V, Dordrecht, pp 737–743. doi:10.1007/978-1-4020-4410-6
MacIntyre S (2013) Climate variability, mixing dynamics, and ecological consequences in the African Great Lakes. In: Goldman CR, Kumagai M, Robarts RD (eds) Effects of global warming on freshwater ecosystems of the world: impacts and mitigation for ecosystems and societies. Wiley, Chichester, pp 311–336
MacIntyre S, Melack JM (1982) Meromixis in an equatorial African soda lake. Limnol Oceanogr 27:595–609
MacIntyre S, Melack JM (1995) Vertical and horizontal mixing in lakes: linking littoral, benthic and pelagic habitats. J N Am Benthol Soc 14:599–615
MacIntyre S, Melack JM (2009) Mixing dynamics in lakes across climatic zones. In: Likens GE (ed) Encyclopedia of ecology of inland water, vol 2. Elsevier, Oxford, pp 603–612
MacIntyre S, Romero JR, Kling GW (2002) Spatial-temporal variability in mixed layer deepening and lateral advection in an embayment of Lake Victoria, East Africa. Limnol Oceanogr 47:656–671
MacIntyre S, Clark JF, Jellison RJ, Fram JP (2009) Turbulent mixing induced by non-linear internal waves in Mono Lake, CA. Limnol Oceanogr 54:2255–2272

MacIntyre S, Romero JR, Silsbe GM, Emery BM (2014) Stratification and horizontal exchanges in Lake Victoria, East Africa. Limnol Oceanogr 59: 1805–1838

MacIntyre S, Flynn KM, Jellison R, Romero JR (1999) Boundary mixing and nutrient flux in Mono Lake, CA. Limnol Oceanogr 44:512–529

Melack JM (1976) Limnology and dynamics of phytoplankton in equatorial African lakes. Dissertation, Duke University, Durham, NC

Melack JM (1978) Morphometric, physical and chemical features of the volcanic crater lakes of western Uganda. Arch Hydrolbiol 34:430–453

Melack JM (1979) Photosynthetic rates in four tropical African fresh waters. Freshw Biol 9:555–571

Melack JM (1981) Photosynthetic activity of phytoplankton in tropical African soda lakes. Hydrobiologia 81:71–85

Melack JM (1982) Photosynthetic activity and respiration in an equatorial African soda lake. Freshw Biol 12:381–400

Melack JM (1983) Large, deep salt lakes: a comparative limnological analysis. Hydrobiologia 105:223–230

Melack JM (1996) Saline and freshwater lakes of the Kenyan rift valley. In: McClanahan TR, Young TP (eds) East African ecosystems and their conservation. Oxford University Press, New York, pp 171–190

Melack JM (2009) Diel variability and community metabolism in African soda lakes. In: Oren A, Palacios P, Wurtsbaugh W (eds) Saline lakes around the world: unique systems with unique values, vol XV, Natural resources and environmental Issues. S.J. and Jessie E. Quinney Natural Resources Research Library, Logan, pp 153–160

Melack JM, Kilham P (1972) Lake Mahega: a mesothermic, sulphato-chloride lake in western Uganda. Afr J Trop Hydrobiol Fish 2:141–150

Melack JM, Kilham P (1974) Photosynthetic rates of phytoplankton in East African alkaline, saline lakes. Limnol Oceanogr 19:743–755

Melack JM, Kilham P, Fisher TR (1981) Responses of phytoplankton to experimental fertilization with ammonium and phosphate in an African soda lake. Oecologia 52:321–326

Millero FJ, Lawson D, Gonzalez A (1976) The density of artificial river and estuarine waters. J Geophys Res 81:1177–1179

Odada E, Raini J, Ndetei R (2004) Experiences and lessons learned brief, Lake Nakuru. Available via http://www.worldlakes.org/uploads/Nakuru_6.22.04.pdf. Accessed Dez 2014

Panofsky HA, Egolf CA, Lipschutz R (1978) On characteristics of wind direction fluctuations in the surface layer. Bound Lay Meterol 15:439–446

Prosser MV, Wood RB, Baxter RM (1968) The Bishoftu crater lakes: a bathymetric and chemical study. Arch Hydrobiol 65:309–324

Schagerl M, Oduor SO (2008) Phytoplankton community relationship to environmental variables in three Kenyan Rift Valley saline-alkaline lakes. Mar Freshw Res 59:125–136

Schneider P, Hook SJ (2010) Space observations of inland water bodies show rapid surface warming since 1985. Geophys Res Lett 37, L22405. doi:10.1029/2010GL045059

Serruya C, Pollingher U (1983) Lakes of the warm belt. Cambridge University Press, Cambridge

Spigel RH, Coulter GW (1996) Comparison of hydrology and physical limnology of the East African Great Lakes: Tanganyika, Malawi, Victoria, Kivu and Turkana (with reference to some North American Great Lakes). In: Johnson TC, Odada EO (eds) The limnology, climatology and paleoclimatology of the East African lake. Gordon and Breach Publishing, Amsterdam, pp 103–140

Street FA, Grove AT (1976) Environmental and climatic implications of late Quaternary lake-level fluctuations in Africa. Nature 261:385–390

Talling JF, Lemoalle J (1998) Ecological dynamics of tropical inland waters. Cambridge University Press, Cambridge

Talling JF, Wood RB, Prosser MV, Baxter RM (1973) The upper limit of photosynthetic productivity by phytoplankton: evidence from Ethiopian soda lakes. Freshw Biol 3:53–76

Tebbs EJ, Remedios JJ, Harper DM (2013) Remote sensing of chlorophyll-a as a measure of cyanobacterial biomass in Lake Bogoria, a hypertrophic, saline-alkaline, flamingo lake, using Landsat ETM+. Remote Sens Environ 135:92–106

Vareschi E (1982) The ecology of Lake Nakuru (Kenya) III. Abiotic factors and primary production. Oecologia 55:81–101

Vatova A (1940) Notizie idrografiche e biologiche sui Laghi dell'Africa Orientale Italiana. Thalassia 4:25

Velpuri NM, Senay GB (2012) Assessing the potential hydrological impact of the Gibe III Dam on Lake Turkana water level using multi-source satellite data. Hydrol Earth Syst Sci 16 3561–3578

Verschuren D (1996) Comparative paleolimnology in a system of four shallow tropical lake basins. In: Johnson TC, Odada EO (eds) The limnology, climatology and paleoclimatology of the East African lake. Gordon and Breach Publishing, Amsterdam, pp 559–572

Verschuren D (1999) Influence of depth and mixing regime on sedimentation in a small, fluctuating tropical soda lake. Limnol Oceanogr 44:1103–1113

Verschuren D, Cocquyt C, Tibby J, Roberts CN, Leavitt PR (1999) Long-term dynamics of algal and invertebrate communities in a small, fluctuating tropical soda lake. Limnol Oceanogr 44:1216–1231

Vidal J, MacIntyre S, McPhee-Shaw EE, Shaw WJ, Monismith SG (2013) Temporal and spatial variability of the internal wave field in a lake with complex morphometry. Limnol Oceanogr 58:1557–1580

Wood RB, Talling JF (1988) Chemical and algal relationships in a salinity series of Ethiopian inland waters. Hydrobiologia 158:29–67

Wood RB, Prosser MV, Baxter RM (1976) The seasonal pattern of thermal characteristics of four of the Bishoftu crater lakes, Ethiopia. Freshw Biol 6:519–530

Wood RB, Prosser MV, Baxter RM (1978) Optical characteristics of the Rift Valley lakes, Ethiopia. Ethiop J Sci 1:73–85

World Lakes Database (1999) International Lake Environment Committee Foundation. http://wldb.ilec.or.jp/. Accessed Jan 2015

Yuretich RF, Cerling TE (1983) Hydrogeochemistry of Lake Turkana, Kenya: mass balance and mineral reactions in an alkaline lake. Geochim Cosmochim Acta 47:1099–1109

Geochemistry of African Soda Lakes

4

Daniel M. Deocampo and Robin W. Renaut

Abstract

Soda lakes are those in which sodium and carbonate species dominate the dissolved ions. They form in hydrologically closed lake basins, where inflow is balanced primarily by evaporation. They are important habitats, closely tied to water resource issues in arid lands, and they support important economic activities. Soda lakes differ from other saline lakes in that their drainage basins typically are dominated by volcanic terrains that produce very alkaline dilute inflow. In East Africa, soda lakes are commonly associated with volcanism and tectonic depressions, mostly along the East African Rift. Upon evaporative concentration, salinity increases, but many solutes are also lost to solid phases through the actions of abiotic mineral precipitation and biological utilization. Thus, soda lake waters tend to be dominated by sodium, bicarbonate and/or carbonate, with potentially significant sulfate and chloride concentrations as well. Alkaline earths (calcium and magnesium) are almost always negligible in concentration in evolved soda brines. Hydrothermal input is an important contributor of silica in soda lake waters and may also play an important role in deep subsurface environments (far below the sediment-water interface) by enriching CO_2, enhancing silicate hydrolysis and generating high alkalinity.

4.1 Introduction

Soda lakes are those saline lakes in which the dissolved ions are dominated by sodium and carbonate species. They are an important type of lake in many regions around the globe, providing unique biological habitats, affecting surface and groundwater resources as well as impacting human communities and numerous aspects of economic activity. Through geological

D.M. Deocampo (✉)
Department of Geosciences, Georgia State University, 33 Gilmer Street SE, Atlanta, GA, USA
e-mail: deocampo@gsu.edu

R.W. Renaut
Department of Geological Sciences, University of Saskatchewan, 114 Science Place, Saskatoon, SK S7N 5E2, Canada
e-mail: robin.renaut@usask.ca

M. Schagerl (ed.), *Soda Lakes of East Africa*, DOI 10.1007/978-3-319-28622-8_4

time, such lakes have had enormous impacts on the global carbon budget and hydrocarbon accumulation (Horsfield et al. 1994). The world's largest sources of industrial sodium carbonate minerals are fossil soda lakes in North America and China (Dyni et al. 1995). Understanding the nature and evolution of modern soda lakes is therefore important not only for understanding their environmental roles, but also for assessing their social and economic values. In modern times, soda lakes are common to East Africa, a region in which various geological and climatological parameters favour their development, as discussed below.

Soda lakes are characterized by elevated pH and dominance of sodium and carbonate species in the cation and anion dissolved solutes, respectively. The geochemical key to the occurrence of such lakes is a great excess of conservative cations (in dilute waters: Na^+, K^+, Ca^{2+}, Mg^{2+}) over conservative anions (Cl^-, SO_4^{2-})—in other words, very high alkalinity. Because charge balance is maintained through the carbonic acid system and gas exchange of CO_2 with the atmosphere, high alkalinity results in high bicarbonate (HCO_3^-) and carbonate (CO_3^{2-}) content. It is not uncommon for carbonate alkalinity concentrations to exceed five or more times the sum of chloride and sulfate concentrations. Most such waters have pH values over 9, but in some cases when aqueous CO_2 builds up due to relatively low degassing rates, pH can be as low as 6 despite the high alkalinity (Deocampo and Ashley 1999). Most dilute waters do not have very high alkalinity and thus require some combination of water-rock interaction (weathering), evaporative concentration and solute loss to produce highly alkaline water. Thus, the development of high alkalinity in surface waters is closely linked to hydrology, climate and regional geology.

4.2 Hydrology and Climate

As with all saline lakes, hydrological closure is generally required for solute concentrations to rise sufficiently to generate the great alkalinities observed in soda lakes (see Schagerl and Renaut, Chap. 1). Such conditions are commonly seen in East Africa, in crater lakes as well as within the East African Rift itself. One possible exception to hydrological closure is when saline groundwater is directly discharged into lacustrine environments—it may either be brines discharging from a shallow groundwater circulation system or may involve deeper hydrothermal systems. In such cases, soda brines may be present at the surface even before significant residence time in a lake leads to further evaporative concentration.

In hydrologically closed lakes (i.e. internally drained), evaporation is the dominant pathway for water to exit the basin (Rosen 1994). Such settings generally occur only in tectonically active regions, such as adjacent to uplifted regions, within continental rifts or within intermontane or volcanic (i.e. caldera) basins (Deocampo and Jones 2014). The spatial distribution of soda lakes around the world shows that, without exception, they are associated with modern or ancient volcanic activity (Pecoraino et al. 2015). Whereas hydrologically closed regions may produce saline lakes, those that occur in volcanically active tectonic settings may furthermore produce carbonate-rich saline lakes, i.e. soda lakes (Fig. 4.1). Aside from the Pacific Northwest and the East African Rift, the only other major volcanic region on Earth that has widespread hydrological closure is the central and southern Andes in South America. The saline waters developed there (e.g. on the Altiplano) tend to be lower in carbonate and sulfate rich, however, due to a greater presence of volcanic sulfur in the region (Pecoraino et al. 2015; Risacher et al. 2003).

As in other saline lakes, three hydrological conditions must be met for soda lakes to form (Fig. 4.2): (1) outflow from the lake to streams or groundwater must be minimal compared to evaporation; (2) at steady-state equilibrium (which may never effectively be reached because of the dynamics of climate), evaporation equals inflow; and (3) inflow must be great enough to maintain water in the lake and avoid desiccation. The fundamental control on the geochemistry of

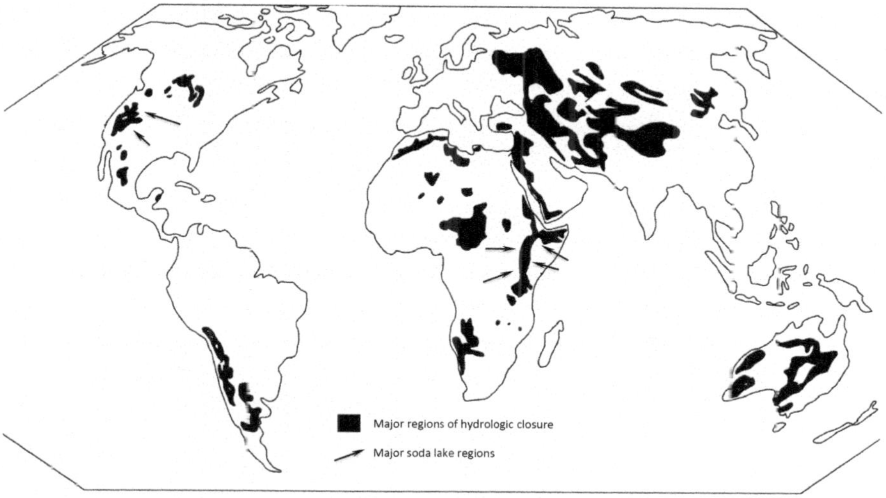

Fig. 4.1 Global distribution of major areas of internal drainage and hydrological closure (from Deocampo and Jones 2014). *Arrows* indicate the two major regions of the world where soda lakes are the dominant form—on the eastern flanks of the Pacific Northwest in the rain shadow of the Cascade Range and throughout the East African Rift. Although local features may favour soda lake development in other parts of the world, the East African Rift is an entire region in which tectonic, volcanic and climate conditions allow soda lakes to dominate

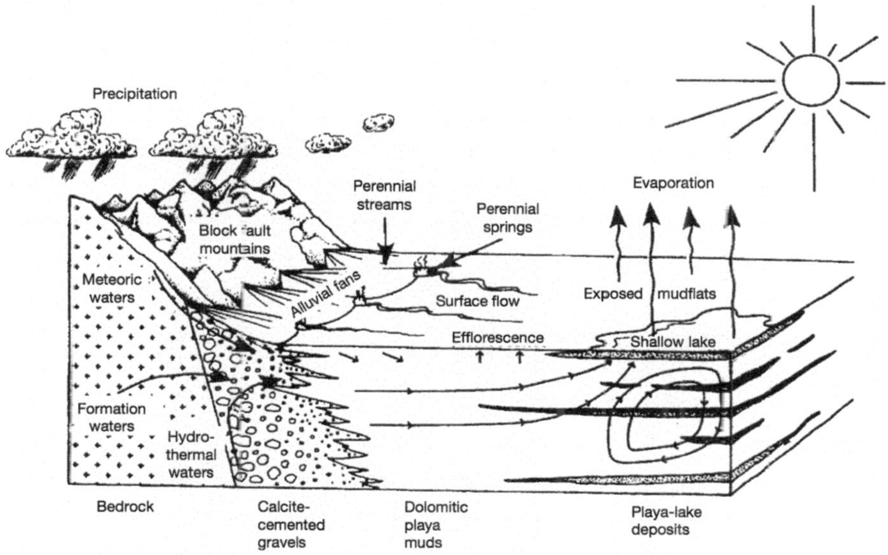

Fig. 4.2 Schematic block diagram showing major hydrological features of closed basins, after Eugster and Hardie (1975) and Deocampo and Jones (2014). *Flow lines* indicate areas of potentially salinity driven density circulation and interaction between fresh meteoric input and circulating brines (Duffy and Al-Hassan 1988; Rosen 1994)

these lakes is the geology-weathering reactions in the drainage basins surrounding these lakes that determine whether, upon evaporation, lake waters evolve towards a carbonate-rich (i.e. "soda") endmember or towards more neutral or even acidic compositions.

For hydrologically closed lakes, change in surface area is the principal lake response to climatically driven changes in inflow and evaporation rates (Almendinger 1990). The resulting change in the volume of water storage in the lake and the associated change in the lake level (see Schagerl and Burian, Chap. 12) are therefore functions of the bathymetry and gradients in the basin. Lake volume and water level changes in high-gradient basins will therefore be much greater than those in gently sloping basins, even for an identical surface area response (Fig. 4.2). When such a lake approaches steady state, a constant lake volume is approached, but solute concentrations rise as water is lost to evaporation. In the ideal case, solute concentrations may also eventually reach a steady state because thermodynamically driven mineral precipitation and biologically mediated mineralization act as solute sinks, removing solids from solution. These processes can be complex and heavily affected by stratification, seasonality and multiple inflows.

4.3 Solutes and Alkalinity

The dominant ions and dissolved species (e.g. aqueous silica as H_4SiO_4 or as ions) in soda lakes originate from three main sources (Fig. 4.3): (1) rain falling directly on the surface of the lake and drainage basin, (2) weathering reactions between runoff and groundwater, bedrock, surface soils and sediments and (3) hydrothermal fluids and their interactions with subsurface rocks.

Much of the rainfall in East Africa is linked to monsoonal winds blowing from the Indian Ocean and the north-south seasonal migration of the Intertropical Convergence Zone (ITCZ) (Johnson

Fig. 4.3 Main sources of solutes in East African soda lakes. (**a**) Convection rainstorm falling on Lake Baringo, Kenya. (**b**) Spheroidally weathered basalt (Kaparaina Basalt), Tugen Hills, Kenya. (**c**) Hydrothermally altered trachyte (clay minerals, sulfur, opaline silica), Koibobei, southern Lake Bogoria

and Odada 1996). The chemical composition of the rains therefore reflects marine aerosols, dissolution of airborne dust (derived locally or regionally) and dissolved atmospheric gases (mainly CO_2). Atmospheric carbon dioxide dissolves to form carbonic acid, much of which dissociates into bicarbonate and hydrogen ions:

$$H_2O + CO_2 = H_2CO_3$$

$$H_2CO_3 = HCO_3^- + H^+$$

Most of the precipitation in East Africa falls in the tropical and subtropical highlands, but some falls in the rain shadow on the rift floors. Although the isotopic composition of the rain is well known (e.g. Levin et al. 2009), few chemical analyses of East African rain have been presented (e.g. Kanuga 1982; Renaut 1982; Visser 1961). Table 4.1 shows analyses of rain that fell at Lake Bogoria, Kenya; Table 4.2 compares ionic ratios in one analysis with those of seawater. Ionic ratios in the rainwater are similar to those in seawater, implying that marine aerosols are the source of many ions. Calcium and bicarbonate are in excess, possibly reflecting dissolution of carbonate dust. Much of the chloride in the soda lakes originates from rainfall, showing the effects of atmospheric processes on lake chemistry. Although the marine influence on rainwater composition is likely perennial, seasonal changes potentially affect the abundance of carbonate dust available for dissolution by atmospheric water.

The effects of marine sulfo-chloride aerosols have been well documented in saline lakes near the ocean, for example, in the Bahamas (Dix et al. 1999). This is also an important factor in dryland lake basins where efflorescent crusts of minerals such as trona and halite (NaCl) commonly form on lake flats (Nielson 1999). If they are eroded and mobilized by wind action (e.g. "dust devils"), they can become redeposited in local water bodies and contribute strongly to the dissolved fraction (Fig. 4.4). Although this may not be a significant factor with surface brines that are already highly evolved, it can strongly influence the chemistry of more dilute waters such as those in freshwater streams and marginal wetlands (Deocampo 2004, 2005).

Table 4.1 Rainfall (RF1, 5 August 1977 and RF2, 11 August 1977) analyses, Loburu Delta, Lake Bogoria

mg/L	RF 1	RF2
Na	2.34	2.19
K	0.33	0.31
Ca	2.68	2.75
Mg	0.21	0.24
HCO_3	5.99	5.85
Cl	4.12	4.81
SO_4	0.68	1.24
pH	6.23	6.12

4.3.1 Chemical Weathering as a Source of Solutes

Most soda lakes in East Africa lie in drainage basins in which Neogene to recent volcanic rocks, their weathering products and the associated volcaniclastic sediments are the dominant lithology in outcrops (Abbate et al. 2015; Dawson 2008; MacDonald 2003). Older basement rocks, composed of ancient cratonic (typically metamorphic) and younger mobile-belt rocks, are also common, but igneous rocks dominate the catchments of most soda lakes. These volcanic rocks vary in composition from basaltic to trachytic to phonolitic and locally include carbonatites composed of eruptive Ca-Na carbonates. Carbonatite ashes are particularly potent alkali sources—within 24 h of an eruption at Ol Doinyo Lengai in northern Tanzania. Zaitsev and Keller (2006) found samples composed of nahcolite ($NaHCO_3$), trona ($Na_3H(CO_3)_2{\cdot}2H_2O$), sylvite (KCl), halite, kalicinite ($KHCO_3$) and villiaumite (NaF). These deposits weather rapidly to produce secondary nahcolite, trona, thermonatrite ($Na_2CO_3{\cdot}H_2O$), pirssonite ($CaCO_3{\cdot}Na_2CO_3{\cdot}2H_2O$), gaylussite ($CaCO_3{\cdot}Na_2CO_3{\cdot}5H_2O$) and calcite ($CaCO_3$); much of this material is completely soluble, thus releasing alkalis and carbonate rapidly into surface- and groundwaters. The impact on the alkalinity and pH of such waters is so rapid and strong that it has been recognized as imposing a "mock aridity" on landscapes and ecosystems, because highly reactive alkaline volcaniclastics mimic the

Table 4.2 Ratios of ions in Lake Bogoria rainwater samples (RF1 and RF2) compared with those in seawater. Alk = HCO_3- + CO_3^{2-}

Ionic ratio	RF1	RF2	Sea water (SW)	RF (mean)/SW
Ca:Na	1.20	1.30	0.038	32.9
K:Na	0.14	0.14	0.036	3.89
Mg:Na	0.09	0.11	0.120	0.83
Cl:Na	1.80	2.20	1.800	1.11
SO_4:Cl	0.17	0.26	0.140	1.54
Alk:Cl	1.50	1.20	0.008	169.00

The data imply that the rainwater originates from marine aerosols, but the excess Ca and carbonate in RF1 implies dissolution of incorporated carbonate dust

Fig. 4.4 The Ngoitokitok freshwater marsh on the margin of Lake Makat, Ngorongoro Crater, Tanzania. Sodium-carbonate dust is being reworked from the lake flats by wind action, providing a potential source of cations (and hence alkalinity) to the wetlands (Deocampo 2004)

effect of arid-land evaporative processes (Harris and Van Couvering 1995). The basaltic rocks are relatively rich in calcium, whereas the more alkaline rocks are variably dominated by sodium-rich silicate minerals and abundant K-feldspar. Quartz is absent in many of the volcanic rocks.

The intensity and rates of chemical weathering vary with location, but both are highest in elevated regions that receive the highest rainfall, including rift margins and central volcanoes. The lowest rates are in areas of rain shadow such as rift floors and in semiarid regions of modest topography.

4.3.2 Acquisition of Alkalinity

Other than atmospheric input, weathering reactions taking place in upslope surficial soils and groundwater environments dominate the geochemistry of dilute inflow to saline lakes. The nature of these reactions depends on the mineral content of the local geology and hydrological factors that determine the residence time required for water-rock interactions (Yechieli and Sivan 2011). These reactions produce key initial solute ratios that are important in determining the subsequent geochemistry of evolved brines; for example, soda lakes may develop only in basins where the dilute inflow has $Ca^{2+} < HCO_3^-$. This condition is favoured by silicate hydrolysis, such as the idealized hydrolysis reaction describing the weathering of sodium-rich plagioclase feldspar:

$$\underset{\text{(Na-feldspar)}}{2NaAlSi_3O_{8(s)}} + 2H^+_{(aq)} + 9H_2O_{(l)}$$

$$\rightarrow \underset{\text{(kaolinite)}}{Al_2Si_2O_5(OH)_{4(s)}} + 2Na^+_{(aq)} + 4H_4SiO_{4(aq)}$$

The above example reaction shows that the hydrolysis of silicate minerals consumes hydrogen ions from solution and releases alkali cations. This has the effect of both raising the pH and increasing the alkalinity. Moreover, the rate of hydrolysis reactions is enhanced by aqueous interactions with the glassy components of volcanic ashes that are more reactive—because they lack a crystalline structure and have a higher surface area—than the crystalline other fractions of ash.

Other weathering reactions release ions into solution, but do not necessarily contribute to alkalinity, and some reactions can even reduce alkalinity or add acidity. The strongest such action is the oxidation of sulfide minerals, such as pyrite (FeS_2), which involves the oxidation of both ferrous iron and sulfide. This can be described conceptually in a two-step reaction, with the first step releasing ferrous iron and sulfide:

$$\underset{\text{(pyrite)}}{4FeS_{2(s)}} + 14O_{2(g)} + 4H_2O_{(l)}$$

$$\rightarrow 4Fe^{2-}_{(aq)} + 8SO_4{}^{2-}_{(aq)} + 8H^+_{(aq)}$$

Ferrous iron is not stable in an oxygen-rich environment, and so it would rapidly oxidize as well:

$$4Fe^{2+}_{(aq)} + O_{2(g)} + 4H^+_{(aq)} \rightarrow 4Fe^{3+}_{(aq)} + 2H_2O_{(l)}$$

Because Fe^{3+} is only minimally soluble in most waters, the vast majority of Fe^{3+} produced in this way will precipitate into solid phases, generally Fe-oxyhydroxides (Jones 1986). Thus, the oxidation of sulfides contributes sulfate and hydrogen ions into solution, reducing alkalinity and lowering the pH.

When multiple weathering reactions occur during soil development, some reactions may combine to enhance or limit the development of alkalinity. For example, limestone dissolution combined with sulfide oxidation can produce dilute waters where $Ca^{2+} >> HCO_3^-$. Thus, the mineralogy of the host rocks in soils of the basin is crucial for determining water chemistries.

Other major weathering reactions also affect the geochemical composition of inflowing waters, although they do not have a large impact on the alkalinity that is so important for the development of soda lakes. For example, dissolution of simple chloride or sulfate salts adds to the dissolved solids, but because these add conservative cations and anions in equal proportion, they do not affect alkalinity. Similarly, the dissolution of carbonate minerals such as calcite does add to alkalinity, but it does not affect the Ca^{2+}/HCO_3^- ratio because the two components are added in equal proportions.

The general chemistry of the weathering reactions can be illustrated by comparing bedrock composition with the adjacent in situ residue (saprolite) following weathering. Table 4.3 compares an example of trachyphonolite bedrock west of Lake Bogoria (Hannington Trachyphonolite Formation: Griffiths and Gibson 1980) with its adjacent saprolite. Assuming that Al (or Ti) remains immobile (Retallack 2001) and that little material has been added to the weathering profile (e.g. from dust or airfall tephra), it is possible to compare losses and gains during chemical weathering. In this example, all major oxides and some silica show losses relative to Al and Ti, largely reflecting silicate hydrolysis. Ferrous iron oxides decrease, while ferric iron increases because of oxidation. The losses broadly reflect the composition of local river waters, in which Na > K and Ca > Mg. Water has been gained (added to the mineral component) during neoformation of clay minerals and Fe-oxyhydroxides.

Table 4.3 Chemical analyses of a fresh Hannington trachyphonolite (Hannington Trachyphonolite Formation of Griffiths and Gibson 1980) and its overlying in situ saprolite

	(1)	(2)	(3)	(4)
SiO_2	56.40	53.80	46.00	−18.5
TiO_2	0.52	0.61	0.52	–
Al_2O_3	16.00	18.70	16.00	–
Fe_2O_3	6.70	10.60	0.07	+35.2
FeO	3.20	0.78	0.67	−79.3
MnO	0.64	0.77	0.66	+3.1
MgO	0.91	0.71	0.61	−33.0
CaO	1.60	0.74	0.63	−60.6
Na_2O	6.52	2.89	2.47	−62.2
K_2O	4.51	3.02	2.58	−42.8
H_2O	2.76	6.63	5.67	+105.0
P_2O_5	0.07	0.05	0.04	−42.9
CO_2	–	0.42	–	–
Total	99.9	99.8	84.9	–

Oxides are reported in weight %. Samples taken on midslope site, ~ 150 m west of Chemurkeu, Lake Bogoria (1) Fresh trachyphonolite, (2) trachyphonolite saprolite. (3) = (2) recalculated assuming the amount of Al_2O_3 in (1). (TiO_2 gives an almost identical result.); (4) Gains (+) and losses (−) in chemical weathering assuming constant Al_2O_3 (i.e. immobility in the profile because of its low solubility), which are differences in composition between (1) and (3), expressed as a percentage of the original composition (1)

4.3.3 Hydrothermal Processes as Sources of Solutes

Hydrothermal processes have not usually been considered as major factors in explaining the origin of solutes in saline lakes (Eugster and Hardie 1978; Eugster and Jones 1979). Nonetheless, thermal fluids are increasingly recognized as possible contributors in lakes in tectonic basins, especially in volcanic rifts, and in saline crater lakes (e.g. Benvenuti et al. 2013; Earman et al. 2005; Pecoraino et al. 2015). Hot springs flow into most Kenyan and Ethiopian rift lakes and into the rift lakes of northern Tanzania; volcanic gases, mainly CO_2, also discharge directly into some rift lakes. Even where springs are not evident in the lake itself, thermal spring waters may feed streams in the drainage basin (Renaut and Jones 2011). In some lakes (e.g. Lake Turkana), the proportion of thermal inflow is minor; in others, however, hot springs contribute much of the annual recharge (e.g. Lakes Bogoria, Logipi, Magadi, Nasikie Engida). In Lake Magadi and Nasikie Engida, Kenya, for example, perennial saline, alkaline hot springs (up to 86 °C) contribute most of the annual inflow (Jones et al. 1977).

Recirculated groundwaters discharge at the surface as low-enthalpy (warm) springs with temperatures <45 °C or as higher-enthalpy springs and geysers with temperatures up to local boiling point. The warm springs are relatively dilute, Na-HCO_3 or Na-Ca-HCO_3 waters. The higher-temperature fluids are more saline and silica enriched, however, and may contain high levels of volcanogenic gases such as CO_2 and species of sulfur. These hot fluids commonly interact with bedrock during subsurface circulation. Some are acidic and produce extensive hydrothermal alteration, thereby adding solutes to fluids that discharge at the surface (Darling et al. 1996; Gizaw 1996; Pecoraino et al. 2015). Bicarbonate-rich hot waters are formed mainly when steam and gases (mainly CO_2) condense in shallow groundwaters. Such fluids normally have near-neutral pH when discharged or may be alkaline. The initial acidity of rising CO_2-rich fluids is neutralized by hydrolytic reactions with bedrock that consume protons (H^+), leaving waters dominated by Na^+ and HCO_3^- (Nicholson 1993; Renaut and Jones 2011). If shallow boiling occurs as pressure is relieved on approach to the surface, CO_2 and other gases are exsolved and the pH may rise rapidly. Magmatic CO_2 can also be discharged directly in some lakes (Benvenuti et al. 2013).

4.4 Geochemical Evolution

Once acquired, solutes in surface and subsurface waters that flow towards the lakes undergo fractionation processes that progressively change the overall composition and ionic ratios in the inflow waters. Their salinity usually increases unless diluted by rains. The water becomes partly or fully depleted of some ions, while others increase in relative abundance or remain in solution until the fluids reach the open lakes. Alkalinity generally increases with progressive evaporative concentration in hydrologically closed basins. If alkalinity is considered as an excess of conservative cation charge over conservative anion charge, then it follows simply that if evaporation is the only process at work, then excess of cation charge will increase upon evaporative concentration (Deocampo and Jones 2014). The proportion of carbonate species tends to drop significantly even for the most alkaline soda lakes, however. This is apparent in examination of basin-wide surface- and groundwaters within a watershed such as the Ngorongoro Crater, Tanzania (Fig. 4.5). As in most other soda lake basins, dilute waters in Ngorongoro acquire alkalinity through the hydrolysis of volcaniclastic sediment—in this case the feldspathic glasses from trachytic to trachyandesitic ashes of the Ngorongoro Volcanic Highlands (Deocampo 2004; Mollel and Swisher 2008). As waters become more evaporatively concentrated towards the centre of the basin, the absolute concentration of carbonate species rises from ~150 mg L^{-1} (~3 meq L^{-1}) to over 4000 mg L^{-1} (~80 meq L^{-1}). Despite the enormous rise in alkalinity due to evaporative concentration, the proportion of anion charge attributable to carbonate species falls from 95 % in dilute inflow to ~73 % in Lake Makat (as sampled in 1999; Deocampo 2004). The drop in the proportion of carbonate corresponds to a matching drop in Ca^{2+} and Mg^{2+} concentrations (Fig. 4.5), to the point that alkaline earth carbonates are at extremely low levels in evolved brines.

The same pattern is evident in regional analyses spanning multiple basins, such as major rivers and lakes of the Great Basin in the western USA. (Fig. 4.6). Thus, even the most "soda" lakes with very high alkalinity, such as Pyramid Lake and Soda Lake, have anion compositions with high proportions of chloride.

Upon evaporative concentration, the precipitation of alkaline earth carbonates is usually the

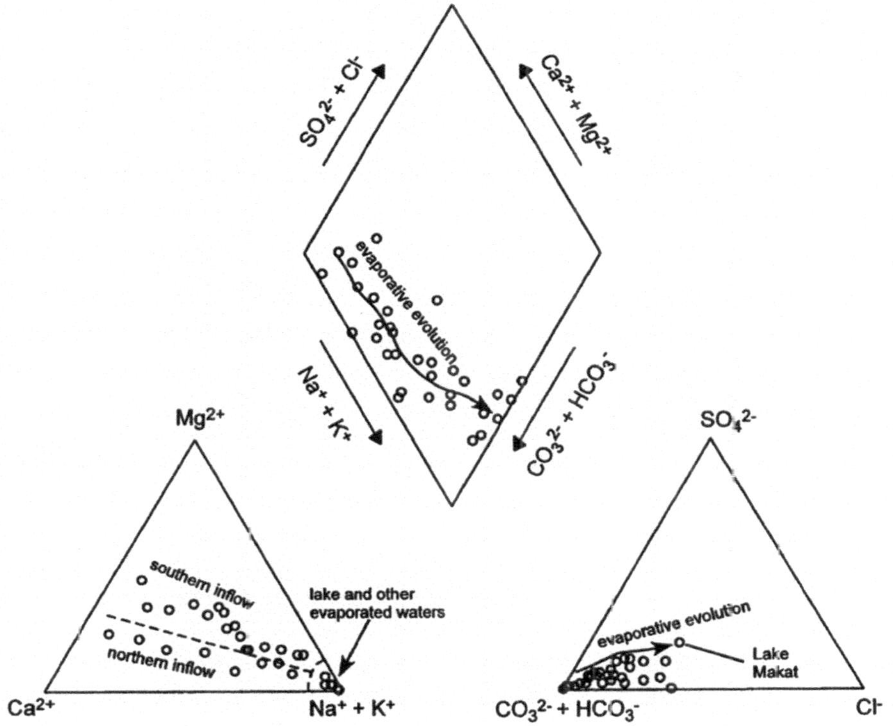

Fig. 4.5 Piper diagram of the geochemical evolution of surface waters in Ngorongoro Crater, Tanzania (Deocampo 2004)

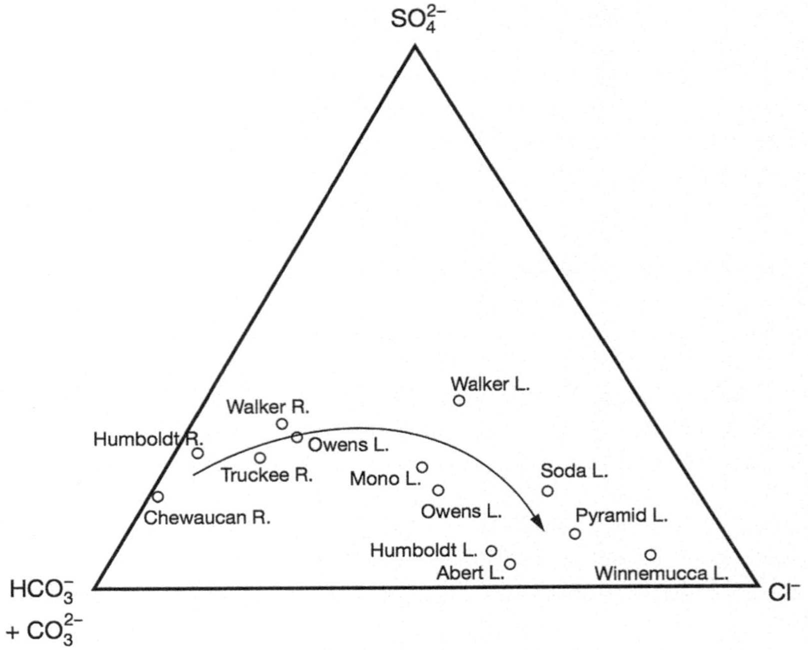

Fig. 4.6 Anion compositional trends in the Great Basin, USA, from Hutchinson (1957). *Arrow* indicates generally increasing total dissolved solids, from dilute inflow sources to saline and soda lakes such as Pyramid Lake, Nevada

first thermodynamically driven (i.e. abiotic) mineralization process to affect the geochemistry of lake waters. This can occur at relatively dilute concentrations and is often associated with either the physical margins of (calcium-free) alkaline brines or, in the central basin, with an influx of Ca-bearing freshwater into the basin, such as during monsoonal rains. Calcite precipitation is the first of the "chemical divides" of Eugster and Hardie (1978)—critical mineral precipitation events that determine the subsequent fate of brine geochemistry. The chemical divide stipulates that once a mineral is supersaturated in an evaporating solution, it will precipitate until one of the solutes (whichever one is less concentrated in the dilute solution) is effectively consumed. This is why the Ca/HCO_3 ratio is so important in dilute waters: waters in which $Ca > HCO_3$ will effectively lose all carbonate once calcite precipitates, whereas waters in which $Ca < HCO_3$ will effectively lose all calcium. The latter is associated with soda lake development. Later in the evaporative evolution of waters, as salinity increases, other chemical divides come into play, such as gypsum (a divide between Ca and SO_4), magnesite ($MgCO_3$—a divide between Mg and CO_3) and others (Fig. 4.7) (Deocampo and Jones 2014).

The importance of initial Ca/HCO_3 ratios in dilute inflow is highlighted in Fig. 4.8. In the case of the "high Ca" path, waters evolve into sulfo-chloride brines nearly devoid of carbonate—examples of these include the salt lakes of the Northern Great Plains of North America (Last 1992; Last and Ginn 2005). In the case of the "low Ca" path, precipitation of alkaline earth carbonates eliminates Ca (and possibly Mg) from solution. Any Mg remaining in solution after carbonate precipitation is subsequently eliminated by sulfate precipitation. Thus, the East African soda lakes have very little Ca or Mg in ambient waters.

Waters flowing into the East African lakes have variable compositions, but those in the volcanic regions of Kenya, Ethiopia and Tanzania are usually Na- or Ca-dominated among cations in dilute waters, reflecting silicate hydrolysis of the abundant feldspars and glass in parent bedrock (Deocampo and Jones 2014; Renaut et al. 2013). Potassium is commonly the

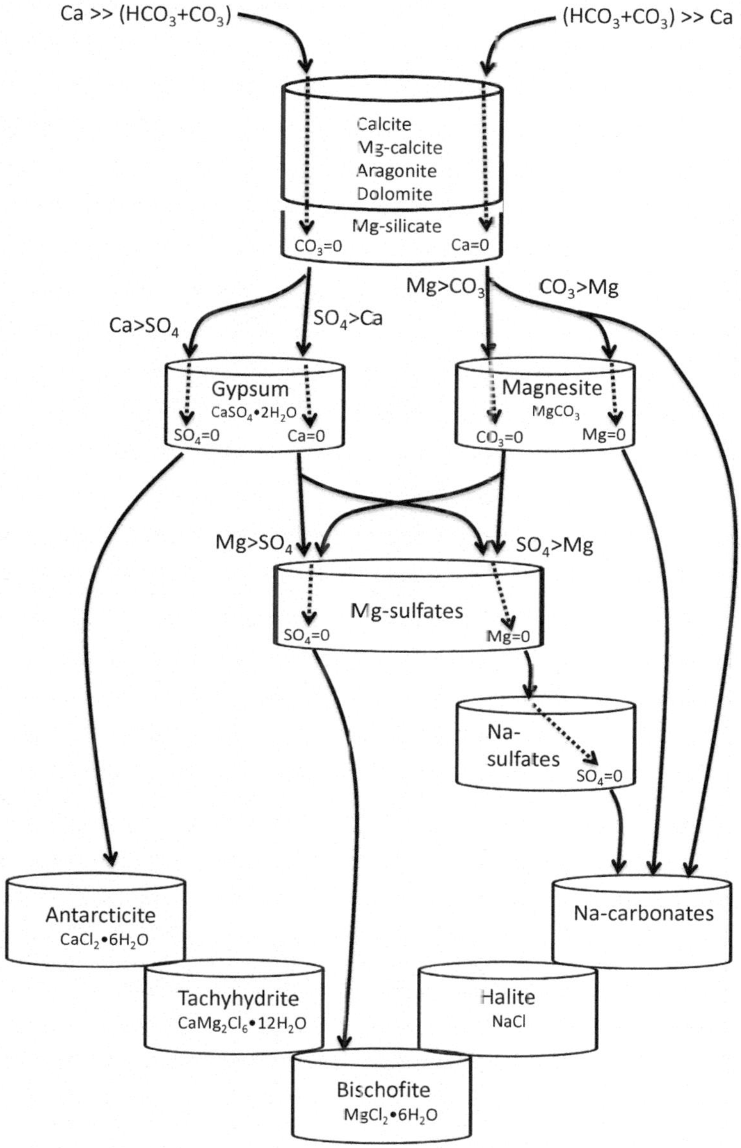

Fig. 4.7 Simplified flow chart of evaporative evolution of a closed-basin brine, with greater evaporative concentration towards the bottom of the diagram, from Deccampo and Jones (2014), following Eugster and Hardie (1978). Solute labels refer to total charged aqueous species (in equivalents), and "CO_3" refers to all aqueous carbonate species. Flow paths do not account for the many environmental complexities that impact geochemical evolution, such as hydrologic fluctuations, solute recycling and leakage and others as discussed in the text

subsidiary cation. Mg is rarely important because ultramafic rocks and Mg-rich basement (e.g. serpentinite) are uncommon, but Mg can be high in basaltic terrains. The anions in inflow are dominated by carbonate species, with variable but commonly high chloride concentrations and variable sulfate. Sulfate is highest where the drainage basin is partly floored by sulfide-bearing metamorphic basement rocks or contains sedimentary rocks that formed in anoxic settings where pyrite formed and became available for oxidation. Gypsum ($CaSO_4{\cdot}2H_2O$) and anhydrite ($CaSO_4$) are rare in bedrock except where seawater mixed with continental waters, for example, in the Afar region. Near volcanic centres, some sulfur can originate from magmatic sources (as H_2S, SO_2 or other species). Sulfate-rich waters are rare but contribute to runoff or lake

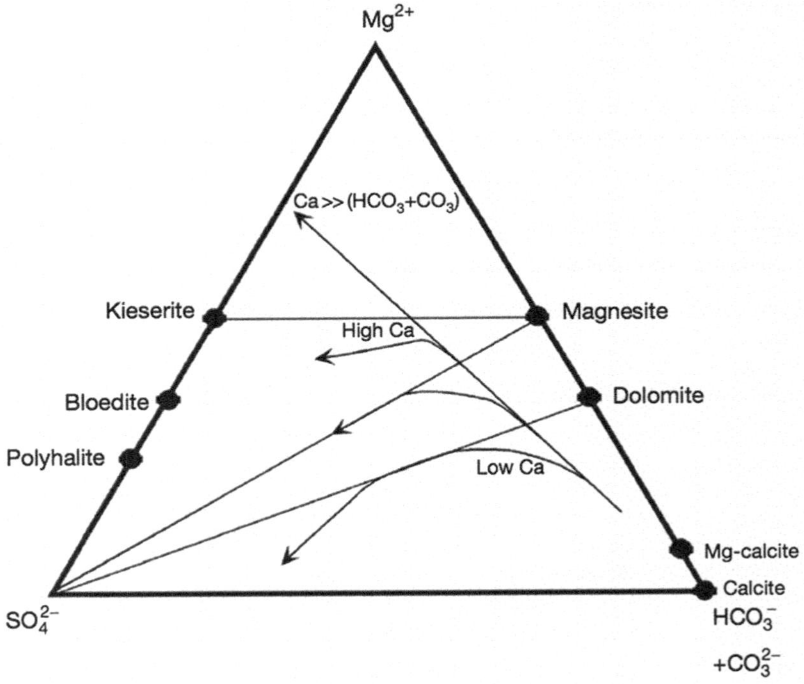

Fig. 4.8 Ternary diagram of solute compositions in idealized evaporatively concentrated waters, from Deocampo and Jones (2014). The path taken and the ultimate fate of the brine depend on how rapidly Ca^{2+} is depleted, relative to Mg^{2+}. Kieserite ($MgSO_4{\cdot}H_2O$) represents all the hydration states of $MgSO_4$

waters where they derive from condensates. Silica and fluoride are also high in dilute waters, reflecting the dominance of the volcanic rocks.

Early precipitation of the alkaline earths, Ca and Mg, as solid phases is a critical stage in development of sodium carbonate brines (Deocampo and Jones 2014; Eugster and Jones 1979; Kilham and Cloke 1980; Renaut and Gierlowski-Kordesch 2010). Such precipitates can take many forms in lacustrine settings or in the surrounding watershed (Fig. 4.4a–h). Calcium carbonate commonly forms at the land surface and in near-surface sediments and soils; at spring vents and in their proximal outflow channels where fed by ambient to warm, dilute meteoric waters; in hydrothermal systems where they precipitate from deeply circulating hot groundwaters, both at the surface and in the subsurface; and, at times, in lakes by various abiotic and biological processes. Additionally, organisms—vertebrates, invertebrates and microbes—remove calcium directly to form skeletal matter (invertebrate shells and skeletal components such as fish bones and otoliths) or indirectly by metabolic and passive mediation (microbialites and biologically induced precipitation in lakes) and early diagenetic processes (e.g. bacterial formation of carbonate concretions and nodules in sediments). Precipitation of calcium (with or without magnesium) carbonate removes both alkaline earths and carbonate species from the fluid. If, however, the carbonate concentration (HCO_3^- and CO_3^{2-}) exceeds that of alkaline earths, then the action of the chemical divide will result in depletion of Ca (and Mg). Although many processes are involved, the result is sodium-rich alkaline fluids that have been depleted of Ca and Mg, flowing towards closed-basin lakes from both rivers and groundwaters.

Throughout the drier parts of East Africa but especially on the semiarid rift valley floors, calcite often precipitates in near-surface sediments and soils. There, it forms cement and veins in siliciclastic sediments and permeable volcanic rocks and thinly encrusts sand and gravel clasts and/or bedrock with variable lateral continuity (Deocampo 2010). In places, enough $CaCO_3$ precipitates to form beds of calcrete, locally up to ~50 cm thick, composed of variably cemented sediments, isolated nodules, pisoids millimetres to centimetres in diameter, cemented nodules and laminated limestone (Hay and Reeder 1978) (Fig. 4.9a and b). Most calcretes reflect periods when climate, hydrology and tectonics remained fairly stable. These near-surface carbonates and calcretes form by several methods including precipitation from evaporation of downward-percolating runoff and precipitation from seasonally fluctuating shallow groundwaters near the water table, by replacement of existing rocks and through pedogenic processes. In the latter, plants and associated biota, either directly or indirectly, induce calcium carbonate precipitation, for example, photosynthesis or evapotranspiration. In places, evidence for plants is preserved in the form of rhizoliths (root casts and related structures) (Fig. 4.9a; Cohen 1982; Owen et al. 2008). These soil and groundwater processes are regionally important in removing Ca and Mg from dilute surface- and groundwaters moving towards lakes.

Groundwater derived from runoff circulates deeply along faults and fractures on the rift floors (Fig. 4.9c). Some groundwater is derived from the elevated rift margins that receive high rainfall; other waters originate locally. These fluids can be discharged into rivers or lakes after shallow circulation, moving laterally or axially along the rift floors, with fractured volcanic rocks, porous volcanic ash layers, paleosols and permeable siliciclastic sediments serving as aquifers. Other fluids move into deeper aquifers, and some may discharge hundreds of kilometres from where the rains fell on the land surface (Kaufman et al. 1990). Those that are locally derived, or which have not circulated to great depths, may discharge at the surface as ambient-temperature or warm springs or may seep into lakes. The deeper fluids, which have sometimes reached depths of several kilometres, flow at the surface as hot springs or form steam vents and fumaroles. These deeper fluids are usually more saline than surface waters, and some are enriched by volcanic gases, including CO_2. These deeper hot fluids commonly interact with the volcanic bedrock, which changes their chemical composition through hydrothermal alteration processes. In the East African Rift, such deep, hot fluids contribute to waters in many soda lakes.

As in dilute lakes, the concentrations of nutrients (i.e. N, P) in soda lakes are controlled by primary productivity and biogeochemical cycling (Cohen 2003). Therefore, they tend not to rise in concentration due to evaporative concentration (as opposed to the major solutes). Indeed, potentially lower abundances of heterotrophs in soda lakes can lead to very high primary productivity and preservation of organic matter in lacustrine bottom waters and sediment (Deocampo and Jones 2014; Renaut and Tiercelin 1994). As a result, stratification enhanced by salinity gradients commonly produces anoxic bottom waters, leading to metal sulfide precipitation (Domagalski et al. 1990).

Calcium carbonate is precipitated locally at springs in the form of tufa (from dilute ambient and warm (<45 °C) Na-Ca-HCO_3 waters) and as travertine (from higher-enthalpy hot springs). In porous tufa, plants, mosses and microbial remains are commonly encrusted by calcite, usually linked to photosynthetic removal of CO_2. In travertine, calcite and aragonite precipitates from very hot waters undergoing exsolution of carbon dioxide at the land surface (Fig. 4.9e) (Lee et al. 2013; Renaut et al. 2013). Some precipitation occurs subaqueously on lake floors, forming chimney-like mounds of carbonate (Dekov et al. 2014). Ca-depleted spring outflow then flows into the drainage system or directly into lakes.

Ca and Mg carbonates also form in the lakes themselves by several methods, depending on the lake status at that time (Deocampo 2010; Hargrave et al. 2014). Alkaline earth carbonates precipitate when a lake is relatively dilute or when dilute river- or groundwaters mix with existing saline, alkaline waters. In open lakes with through-flowing waters, the lake waters may remain undersaturated with respect to $CaCO_3$, although Ca may be removed by organisms as skeletal components (Fig. 4.9f). Once saturated, however, $CaCO_3$ may precipitate in the water column by several processes. The

Fig. 4.9 Sites where alkaline earths (mainly calcite) are removed from waters flowing into soda lakes. (**a**, **b**) Soils and near-surface sediments. (**a**) shows rootcasts, (Loboi Plain, Kenya). (**b**) Cemented calcite nodules forming calcrete (Dry Lagoon, Magadi, Kenya). (**c**) Calcite precipitated from groundwater along fault plane cutting basaltic scoria cone (Karosi, Baringo, Kenya). (**d**) Pleistocene tufa precipitated around reeds in outflow from calcareous spring (Pleistocene, Olorgesailie, Kenya). (**e**) Travertine (calcite) that has precipitated from boiling waters at the vent of a perpetually spouting hot spring, Loburu, Lake Bogoria, Kenya. (**f**) Fossil oyster shells (*Etheria* sp.) (early Holocene lake marginal sediments, Lake Turkana, Kenya). (**g**) Large oncoids (microbialites) along paleoshoreline (early Holocene, Galana Boi Formation, near Ileret, Lake Turkana, Kenya). (**h**) Calcite stromatolites encrusting shoreline gravel and boulders, southern Lake Bogoria, Kenya

effects of biotic activity are commonly evident less in the uptake of major solutes than in strong control of water column pH. For example, evaporative concentration and solar heating may be enough for carbonate crystals to nucleate, but photosynthetic removal of CO_2 by plankton can strongly promote nucleation in surface waters. These crystals then settle through the water column which, if supersaturated, enables accumulation of micritic mud on the lake floor. Ca and Mg carbonates also precipitate around lake margins in efflorescent crusts and in beach rock (e.g. NE Lake Turkana: Johnson et al. 1991; Lake Eyasi: Deocampo 2002), where carbonate precipitates as cements in littoral sands.

Microbialites (stromatolites, thrombolites) form only when the lake is sufficiently saturated for $CaCO_3$ to precipitate from ambient waters (Fig. 4.9g and h) (Casanova 1994; Deocampo 2010). With evaporative concentration and without replenishment in a closed basin, Ca^{2+} becomes depleted as the lake volume shrinks and the formation of stromatolitic limestones ceases. The carbonates are therefore preserved within a narrow vertical range or "window" around the lake margins (Renaut et al. 2013), and they do not form when the salinity is high in soda lakes. When dilute inflows from surface- or groundwaters reach the margins of such a lake, the incoming alkaline earths may cause intense carbonate precipitation in the zone of mixing; if the inflow is focused on one side of the lake, this may result in an abundance of sedimentary carbonate on only that side of the lake.

Mixing of dilute, Ca-bearing inflow with alkaline sodic waters can sometimes induce precipitation of mixed salts such as pirssonite and gaylussite (Deocampo 2010; Frost and Dickfos 2007). The effects of the chemical divide on dilute waters are so strong that simply evaporated dilute inflow cannot produce salts such as these—they require an admixture of dilute Ca-bearing water into soda brine. These can form plumes of minute crystals that are blown onshore and accumulate in lake marginal muds, such as those recorded at Lake Bogoria, Nasikie Engida and other soda lakes.

At high salinities, Natron or trona commonly precipitates from sodium carbonate brines, with Natron precipitation generally restricted to lower temperatures (Jones et al. 1977; Renaut and Long 1989). If waters have high pCO_2, far above the levels that occur through simple equilibration with ambient atmospheric CO_2, then nahcolite may precipitate (Garrels and Christ 1965). Under modern atmospheric CO_2 levels, nahcolite is therefore generally restricted to hypercarbonic brines, for example, due to biogenic or volcanogenic CO_2 enrichment (Earman et al. 2005; Mees et al. 1998).

4.5 Conclusions

Soda lakes form in hydrologically closed basins where dilute inflow has very high alkalinity, typically associated with both weathering of volcanic rocks and discharge of hydrothermal fluids. Many of the geochemical processes observed in East Africa are also known to occur in soda lakes throughout the world; however, soda lakes tend to form in East Africa because the unique tectonic, volcanic, hydrothermal and climatological settings all favour soda lake formation. The high alkalinity of soda lakes is generally attributable to silicate hydrolysis of easily weathered volcanic rocks in the drainage basin, a process that may be enhanced by the presence of high-CO_2 hydrothermal waters. Because dilute inflow in these settings has a $Ca^{2+} < HCO_3^-$ ratio, upon calcium carbonate precipitation, Ca and Mg are effectively removed from the waters. The carbonate sediments associated with this process commonly form extensive cements, calcrete, nodules, pisoids, laminites and other deposits. Because Ca and Mg are removed early in the process of brine evolution, soda lake cations are dominated by Na. Thus, freshwater input—such as groundwater or surface runoff—is usually required for precipitation of carbonates within a soda lake. At more extreme salinities, evaporites such as sodium carbonates may form, as well as double salts of sodium-calcium carbonate if a freshwater source of calcium is introduced.

References

Abbate E, Bruni P, Sagri M (2015) Geology of Ethiopia: a review and geomorphological perspectives. In: Billi P (ed) Landscapes and landforms of Ethiopia. Springer, Dordrecht, pp 33–64

Almendinger JE (1990) Groundwater control of closed-basin lake levels under steady-state conditions. J Hydrol 112:293–318

Benvenuti M, Bonini M, Tassia F, Corti G, Sania F, Agostinia A, Manettia P, Vasellia O (2013) Holocene lacustrine fluctuations and deep CO_2 degassing in the northeastern Lake Langano Basin (Main Ethiopian Rift). J Afr Earth Sci 77:1–10

Casanova J (1994) Stromatolites from the East African Rift: a synopsis. In: Bertrand-Sarfati J, Monty C (eds) Phanerozoic stromatolites II. Springer, New York, pp 193–226

Cohen AS (1982) Paleoenvironments of root casts from the Koobi Fora Formation, Kenya. J Sediment Petrol 52:401–414

Cohen AS (2003) Paleolimnology. Oxford University Press, Oxford

Darling WG, Gizaw B, Arusei MK (1996) Lake-groundwater relationships and fluid-rock interaction in the East African Rift Valley: isotopic evidence. J Afr Earth Sci 22:423–431

Dawson JB (2008) The Gregory Rift Valley and Neogene–Recent volcanoes of northern Tanzania. Mem Geol Soc Lond 33:1–98

Dekov VM, Egueh NM, Kamenov GD, Bayon G, Lalonde SV, Schmidt M, Liebetrau V, Munnik F, Fouquet Y, Tanimizu M, Awaleh MO, Guirreh I, Le Gall B (2014) Hydrothermal carbonate chimneys from a continental rift (Afar Rift): mineralogy, geochemistry, and mode of formation. Chem Geol 387:87–100

Deocampo DM (2002) Sedimentary processes and lithofacies in lake-margin groundwater-fed wetlands in East Africa. In: Renaut RW, Ashley GM (eds) Sedimentation in continental rifts, vol 73. SEPM (Society for Sedimentary Geology) Special Publication, Tulsa, pp 295–308

Deocampo DM (2004) Hydrogeochemistry in the Ngorongoro Crater, Tanzania, and implications for land use in a World Heritage Site. Appl Geochem 19:755–767

Deocampo DM (2005) Evaporative evolution of surface waters and the role of aqueous CO_2 in magnesium silicate precipitation: Lake Eyasi and Ngorongoro Crater, northern Tanzania. S Afr J Geol 108:493–504

Deocampo DM (2010) The geochemistry of continental carbonates. In: Alonso-Zarza AM, Tanner LH (eds) Carbonates in continental settings, geochemistry, diagenesis, and applications, Developments in Sedimentology, vol 62. Elsevier, Amsterdam, pp 1–59

Deocampo D, Ashley GM (1999) Siliceous islands in a carbonate sea: modern and Pleistocene records of spring-fed wetlands in Ngorongoro Crater and Oldupai Gorge, Tanzania. J Sediment Res 69(5):974–979

Deocampo DM, Jones BF (2014) Geochemistry of saline lakes. In: Holland HD, Turekian KK (eds) Treatise on geochemistry, vol 7, 2nd edn. Elsevier, Oxford, pp 437–469

Dix GR, Patterson RT, Park LE (1999) Marine saline ponds as sedimentary archives of Late Holocene sea level and climate variation: an example along a carbonate platform-margin, Bahamas. Palaeogeogr Palaeoclimatol Palaeoecol 150:223–246

Domagalski JL, Eugster HP, Jones BF (1990) Trace metal geochemistry of Walker, Mono, and Great salt lakes. In: Spencer RJ, Chou IM (eds) Fluid-mineral interactions: a tribute to H.P. Eugster. Geochemical Society Special Publication 2, San Antonio, pp 315–354

Duffy CJ, Al-Hassan S (1988) Groundwater circulation in a closed desert basin: topographic scaling and climatic forcing. Water Res Res 24:1675–1688

Dyni JR, Wiig SV, Grundy WD (1995) Trona resources in southwest Wyoming. Nonrenew Res 4:340–352

Earman S, Phillips FM, McPherson BJOL (2005) The role of "excess" CO_2 in the formation of trona deposits. Appl Geochem 20:2217–2232

Eugster HP, Hardie LA (1975) Sedimentation in an ancient playa-lake complex: the Wilkins Peak member of the Green River formation of Wyoming. Geol Soc Am Bull 86:319–334

Eugster HP, Hardie LA (1978) Saline lakes. In: Lerman A (ed) Lakes: chemistry, geology, physics. Springer, New York, pp 237–293

Eugster HP, Jones BF (1979) Behavior of major solutes during closed-basin brine evolution. Am J Sci 279:609–631

Frost RL, Dickfos M (2007) Hydrated double carbonates—a Raman and infrared spectroscopic study. Polyhedron 26:4503–4508

Garrels RM, Christ CL (1965) Solutions, minerals, and equilibria. Freeman, Cooper & Company, San Francisco

Gizaw B (1996) The origin of high bicarbonate and fluoride concentrations in waters of the Main Ethiopian Rift Valley, East African Rift system. J Afr Earth Sci 22:391–402

Griffiths PS, Gibson IL (1980) The geology and petrology of the Hannington Trachyphonolite formation, Kenya Rift Valley. Lithos 13:43–53

Hargrave JE, Hicks M, Scholz CA (2014) Lacustrine carbonates from Lake Turkana, Kenya: a depositional model of carbonates in an extensional basin. J Sediment Res 84:224–237

Harris J, Van Couvering J (1995) Mock aridity and the paleoecology of volcanically influenced ecosystems. Geology 23:593–596

Hay RL, Reeder RJ (1978) Calcretes of Olduvai Gorge and the Ndolanya Beds of northern Tanzania. Sedimentology 25:649–673

Horsfield B, Curry DJ, Bohacs K, Littke R, Rullkötter J, Schenk HJ, Radke M, Schaefer RG, Carroll AR, Isaksen G, Witte EG (1994) Organic geochemistry of freshwater and alkaline lacustrine sediments in the

Green River formation of the Washakie Basin, Wyoming, USA. Org Geochem 22:415–440
Hutchinson GE (1957) A treatise on limnology, vol 1. Wiley, New York
Johnson TC, Halfman JD, Showers WJ (1991) Paleoclimate of the past 4000 years at Lake Turkana, Kenya, based on the isotopic composition of authigenic calcite. Palaeogeogr Palaeoclimatol Palaeoecol 85:189–198
Johnson TC, Odada E (eds) (1996) Limnology, climatology and paleoclimatology of the East African Lakes. Gordon and Breach, Toronto
Jones BF (1986) Clay mineral diagenesis in lacustrine sediments. In: Mumpton FA (ed) Studies in diagenesis, vol 1578. U.S. Geological Survey Bulletin, Denver, pp 291–300
Jones BF, Eugster HP, Rettig SL (1977) Hydrochemistry of the Lake Magadi basin, Kenya. Geochim Cosmochim Acta 41:53–72
Kanuga KK (1982) Chemical composition of rains over Kenya. J Weather Modif 14:38–42
Kaufman A, Margaritz M, Paul M, Hillaire-Marcel C, Hollos G, Boaretto E, Taieb M (1990) The ^{36}Cl ages of the brines in the Magadi-Natron basin, East Africa. Geochim Cosmochim Acta 54:2827–2833
Kilham P, Cloke PL (1980) The evolution of saline lake waters: gradual and rapid biogeochemical pathways in the Basotu Lake District, Tanzania. Hydrobiologia 197:35–50
Last WM (1992) Chemical composition of saline and subsaline lakes of the northern Great Plains, western Canada. Int J Salt Lake Res 1:47–76
Last WM, Ginn FM (2005) Saline systems of the Great Plains of western Canada: an overview of the limnogeology and paleolimnology. Saline Systems 1:10. doi:10.1186/1746-1448-1-10
Lee RKL, Owen RB, Renaut RW, Behrensmeyer AK, Potts R, Sharp WD (2013) Facies, geochemistry and diatoms of late Pleistocene Olorgesailie tufas, southern Kenya Rift. Palaeogeogr Palaeoclimatol Palaeoecol 374:197–217
Levin NE, Zipser EJ, Cerling TE (2009) Isotopic composition of waters from Ethiopia and Kenya: insights into moisture sources for eastern Africa. J Geophys Res 114, D23306. doi:10.1029/2009JD012166
MacDonald R (2003) Magmatism of the Kenya Rift Valley: a review. Trans R Soc Edinb Earth Sci 93:239–253
Mees F, Reyes E, Keppens E (1998) Stable isotope chemistry of gaylussite and nahcolite from the deposits of the crater lake at Malha, northern Sudan. Chem Geol 146:87–98
Mollel GF, Swisher CC III (2008) The Ngorongoro Volcanic Highland and its relationships to deposits at Olduvai Gorge and East African Rift volcanism. J Hum Evol 63:274–283
Nicholson K (1993) Geothermal fluids: chemistry and exploration techniques. Springer, Berlin
Nielson JM (1999) East African Magadi (trona): fluoride concentration and mineralogical composition. J Afr Earth Sci 29:423–428
Owen RA, Owen RB, Renaut RW, Scott JJ, Jones B, Ashley GM (2008) Mineralogy and origin of rhizoliths on the margins of saline, alkaline Lake Bogoria, Kenya Rift Valley. Sediment Geol 203:143–163
Pecoraino G, D'Alessandro W, Inguaggiato S (2015) The other side of the coin: geochemistry of alkaline lakes in volcanic areas. In: Rouwet D et al (eds) Volcanic lakes, Advances in volcanology. Springer-Verlag, Berlin, pp 219–237
Renaut RW (1982) Late Quaternary geology of the Lake Bogoria Fault–Trough, Kenya Rift Valley. Dissertation, University of London
Renaut RW, Gierlowski-Kordesch EG (2010) Lakes. In: James NP, Dalrymple R (eds) Facies models 4. Geological Association of Canada, St. Johns, pp 541–575
Renaut RW, Jones B (2011) Hydrothermal environments, terrestrial. In: Reitner J, Thiel V (eds) Encyclopedia of geobiology, Encyclopedia of earth sciences series. Springer, Berlin, pp 467–479
Renaut RW, Long PR (1989) Sedimentology of the saline lakes of the Cariboo Plateau, interior British Columbia. Sediment Geol 64:239–264
Renaut RW, Tiercelin JJ (1994) Lake Bogoria, Kenya Rift Valley—a sedimentological overview. In: Renaut RW, Last WM (eds) Sedimentology and geochemistry of modern and ancient saline lakes. SEPM Special Publication 50, pp 101–124
Renaut RW, Owen RB, Jones B, Tiercelin JJ, Tarits C, Ego JK, Konhauser KK (2013) Impact of lake level changes on the formation of thermogene travertine in continental rifts: evidence from Lake Bogoria, Kenya Rift Valley. Sedimentology 50:428–468
Retallack GJ (2001) Soils of the past. An introduction to paleopedology. Unwin Hyman, Boston
Risacher F, Alonso H, Salazar C (2003) The origin of brines and salts in Chilean salars: a hydrochemical review. Earth Sci Rev 63:249–293. doi:10.1016/S0012-8252(03)00037-0
Rosen MR (1994) The importance of groundwater in playas: a review of playa classifications and the sedimentology and hydrology of playas. In: Rosen MR (ed) Paleoclimate and basin evolution of playa systems. Geological Society of America Special Paper 289, pp 1–18
Visser SA (1961) Chemical composition of rainwater in Kampala, Uganda and its relation to meteorology and topographical conditions. J Geophys Res 66:3759–3965
Yechieli Y, Sivan O (2011) The distribution of saline groundwater and its relation to the hydraulic conditions of aquifers and aquitards: examples from Israel. Hydrogeol J 19:71–81
Zaitsev AN, Keller J (2006) Mineralogical and chemical transformation of Oldoinyo Lengai natrocarbonatites, Tanzania. Lithos 91:191–207

Part II

Organisms and Their Interactions

Organisms inhabiting soda lakes have developed special adaptations to cope with high pH and salinity. The biodiversity of Bacteria and Archaea is extremely high in the hypo- and mesosaline range; for other groups, a low diversity is assumed because the extreme conditions exclude many organisms. Those adapted to the saline alkaline environment thrive and reach densities that are ranked amongst the world's highest. The productivity of soda lakes is comparable to tropical rain forests, and, often, the cyanoprokaryote *Arthrospira fusiformis* dominates the water column. This species is known to the public under the trademark "Spirulina" and sold as a food supplement. The East African soda lakes are renowned for their huge flocks of flamingos, and hundreds of thousands of tourist visit these lakes to enjoy this spectacle, but the systems also host many other fascinating organisms. The following chapters review the different groups, from viruses, Archaea and Bacteria and micro- and macrophytes to animals such as protozoans, insects, fish and birds. Interactions of the food web compartments, the erratic community shifts, biodiversity and productivity are also discussed.

Bacteria, Archaea and Viruses of Soda Lakes

5

William D. Grant and Brian E. Jones

Abstract

The soda lake environment presents microorganisms with many combinations of physical and chemical extremes, amongst them extremes of alkaline pH, salinity (NaCl), temperature and incident radiation and water and oxygen availability. Despite this harsh regime, soda lakes are fertile habitats for an enormous diversity of alkaliphilic prokaryotic microbiota, which in turn is a host to a range of alkaliphilic viruses. The main driver for these ecosystems is the prodigious rate of photosynthetic primary production provided by alkaliphilic Cyanobacteria and anoxygenic phototrophic Bacteria, which deliver fixed carbon for a vast array of aerobic and anaerobic chemoorganotrophic Bacteria and Archaea. Many of these microbes are unique to soda lakes although the soda lake microorganisms are found within all the main prokaryotic evolutionary divisions and amongst all the major trophic groups and often have relatives or counterparts in freshwater and salt lake ecosystems.

Due mainly to their often difficult accessibility and a general lack of familiarity with soda lakes, they were little explored by microbiologists until the 1980s, when systematic studies began on the alkaline lakes of the East African Rift System (EARS) in Kenya. The investigations in East Africa stimulated and inspired examination of other soda lakes around the globe. Accordingly, as of today (June 2015), around 200 Bacteria and Archaea, many of them novel or unique to soda lakes, have been reported. Based on analyses in the EARS soda lakes, the microorganisms involved in the cycling of carbon, together with nitrogen and sulphur, in soda lakes

W.D. Grant (✉)
Department of Infection, Immunity and Inflammation,
University of Leicester, Leicester LE1 9HN, UK
e-mail: wdg1@leicester.ac.uk

B.E. Jones
DuPont Industrial Biosciences/Genencor International
BV, Leiden 2333CN, The Netherlands
e-mail: bejones1948@gmail.com

M. Schagerl (ed.), *Soda Lakes of East Africa*, DOI 10.1007/978-3-319-28622-8_5

worldwide can now be construed with some confidence, revealing a complex trophic web of microbial interactions. This chapter provides a reference for the microbiota of the EARS soda lakes, as determined by culture-dependent and culture-independent methods, in relation to that in soda lakes around the world.

5.1 Introduction

Soda lakes are highly alkaline aquatic environments. Often impermanent in nature, they are widely distributed around the world (Table 5.1). Most soda lakes are confined to arid or semiarid subtropical latitudes in continental interiors or rain shadow zones in areas that include the East African Rift System (EARS), regions of California and Nevada, the Inner Mongolia Autonomous Region stretching to Tibet and the Kulunda Steppe in South Siberia (Russia).

As the name implies, these lakes contain large amounts of soda (sodium carbonate, Na_2CO_3, or complexes of this salt), formed by evaporative concentration. All of these soda lakes share certain features in common, and there seems to be universal agreement that the most important characteristic is the lack of alkaline earth cations, Ca^{2+} and Mg^{2+}, in the rocks of the water-catchment area. The amount of Ca^{2+} in the local geology (and to a lesser extent Mg^{2+}) is critical in determining the final pH of brines. Waters high in Na^+, Cl^- and HCO_3^-/CO_3^{2-} evaporate down, and when the concentration of HCO_3^-/CO_3^{2-} greatly exceeds that of any Ca^{2+} and Mg^{2+}, these cations precipitate as insoluble carbonates, leaving behind an alkaline brine that develops as a consequence of a shift in the $CO_2/HCO_3^-/CO_3^{2-}$ equilibrium towards CO_3^{2-}. This causes the development of a soda (Na_2CO_3) lake with pH values usually between 10 and 11, occasionally > pH 12. Frequently, NaCl co-concentrates with Na_2CO_3, producing an alkaline and saline brine. The genesis of such lakes is described in detail in WoldeGabriel et al. (Chap. 2).

Some of the African lakes have, however, some claim to being unique, in that, unlike most of the other sites, they are standing bodies of brine with constant replenishment from surface or subsurface sources. Moreover, many of these soda lakes are in areas of active volcanism and as such are frequently associated with alkaline hot springs, making the sites complex biological systems. Total salinities in the East African soda lakes (EASL) vary with the rainy season and range from around 5 % (w/v) total salts in the more dilute lakes such as lakes Elmenteita, Bogoria and Nakuru, made up mainly from roughly equal amounts of NaCl and Na_2CO_3 at a pH of about 11, to saturated brines like Lake Magadi at > 30 % (w/v) total salts, again dominated by NaCl and Na_2CO_3 at pH values approaching 12. Measurement of such high pH in the presence of large amounts of Na^+ presents technical problems, so there is some doubt about exactly how alkaline the most concentrated lakes actually are. Calculations of pH based on ionic composition suggest pH 12 is probably the upper pH limit, making these lakes amongst the most alkaline naturally occurring environments on the surface of the Earth.

Microorganisms that inhabit soda lakes have to be of necessity alkaliphilic or at least alkalitolerant. The term 'alkaliphile' is usually reserved for microorganisms growing optimally, or very well, at pH values above 9, often with pH optima for growth around 10, showing little or no growth at near neutral pH values. Such organisms inhabit the more dilute lakes, whereas 'haloalkaliphiles', which are generally held to be alkaliphiles with an additional obligate requirement for at least 1.5–2 M NaCl for optimum growth, are characteristically found in the hypersaline soda lakes like Lake Magadi and Lake Natron. Occasionally, there is a preference for sodium carbonates over sodium chloride, and these organisms are sometimes additionally

Table 5.1 Worldwide distribution of soda lakes and soda deserts (modified from Grant and Sorokin 2011)

Africa	
Libya	Lake Fezzan
Egypt	Wadi Natrun
Sudan	Malha Crater Lake, Dariba Lakes
Ethiopia	Lake Aranguadi, Lake Kilotes, Lake Abijatta, Lake Shala, Lake Chitu, Lake Hertale, Lake Metahara
Kenya	Lake Bogoria, Lake Nakuru, Lake Elmenteita, Lake Magadi, Lake Simbi, Crater Lake (Lake Sonachi), Lake Turkana
Tanzania	Lake Natron, Lake Embagi, Lake Magad, Lake Manyara, Lake Kusare, Lake Tulusia, El Kekhotoito, Momela Lakes, Lake Lekandiro, Lake Reshitani, Lake Lgarya, Lake Ndutu, Lake Balangida, Basotu Crater lake
Uganda	Lake Rukwa North, Lake Katwe, Lake Mahega, Lake Kikorongo, Lake Nyamunuka, Lake Murumuli, Lake Munyanyange, Lake Nunyampaka.
Chad	Lake Bodu, Lake Rombou, Lake Dijikare, Lake Momboio, Lake Yoan
Botswana	Lake Ngami, Makgadikgadi Pan
The rest of the world	
Canada	Lake Manito, Goodenough Lake
United States	Alkali Valley, Albert Lake, Lake Lenore, Soap Lake, Big Soda Lake, Owens Lake, Mono Lake, Searles Lake, Deep Springs, Rhodes Marsh, Harney Lake, Summer Lake, Surprise Valley, Pyramid Lake, Walker Lake, Alkali Lake, Borax Lake, Union Pacific Lakes (Green River), Ragtown Soda Lakes
Mexico	Lake Texcoco, Lake Alchichica
Venezuela	Lagunilla Valley
Chile	Antofagasta
Hungary	Lake Feher, Kelemen-szék
Russia	Kulunda Steppe, Kunkuskaya Steppe, Altai Steppe, Tanatar Lakes, Transbaikal, Karakul, Araxes Plain, Chita, Barnaul, Slavgorod, Buryatia, Tuva, Lake Shuluutai, Lake Verkhnee Beloe, Lake Nizhnee Beloe, Lake Ostozhe, Lake Doroninskoe, Lake Khatyn, Lake Malye Kasytui
Turkey	Lake Van, Lake Salda
Iran	Lake Urmia
India	Lonar Lake, Sambhar Lake
China	Various 'nors'; Qinhgai Hu, Sui-Yuan, Heilungkiang, Kirin, Jehol, Chahar, Shansi, Shensi, Kansu; Lake Chahannor, Lake Chagannor, Lake Zabuye, Bange Lake, Lake Baer, Lake Wudunao, Lake Hamatai, Xiarinaoer Lake, Dali Lake, Lake Xiaokule, Lake Bagaejinor, Haoji Lake
Mongolia	Lake Horsemeno, Lake Dzun Ulziin
Australia	Lake Corangamite, Red Rock Lake, Lake Werowrap, Lake Chidnup

designated as 'natronophilic' (Banciu and Sorokin 2013).

5.2 Soda Lake Bacteria

Soda lakes and soda deserts have been the object of microbiological analyses for well over a century. A pioneering morphological and culture study was carried out by Isachenko in 1931–1933 on lakes of the Kulunda Steppe region of Siberia, although the original publications in Russian did not receive much attention for many years until English translations became available (Isachenko 1951). Although there appear to be some inconsistencies in the English translation, it is clear from this account and references to the Russian literature that microbiological studies on Kulunda Steppe lakes date back to the end of the nineteenth century. It was about the same time that geochemical and microbiological studies were also reported on the Wadi Natrun (Egypt) lakes (Schweinfurth and Lewin 1898 cited by Oren 2013; Sickenberger 1892). It can be argued that the African lakes were the first soda lakes to receive modern, detailed microbiological analyses, although in recent years there has been significant interest in other sites, particularly the Kulunda Steppe region of Russia, but also the Inner Mongolian lakes and the unusual impact crater lakes, Lonar Lake in India and Mono Lake in California.

Soda lake microbiology has been reviewed several times over the years, from the first general review by Grant and Tindall (1986). Thus, there are general reviews by Grant (1992, 2003, 2006), Grant and Mwatha (1989), Grant and Jones (2000), Jones and Grant (2000) and Jones et al. (1994, 1998a, b). Antony et al. (2013) have reviewed the microbiology of Lonar Lake in India and have included general aspects of soda lake microbiology. Very recently, Sorokin et al. (2014a) have updated our current understanding of microbial diversity and biogeochemical cycling in soda lakes. There are reviews concerned with specific aspects such as applications (Grant et al. 1990; Grant and Horikoshi 1992), sulfur oxidisers (Sorokin and Kuenen 2005a, b; Sorokin et al. 2006), aerobic chemoorganotrophs and chemolithotrophs (Grant and Sorokin 2011) and the anaerobe community (Zavarzin et al. 1999; Zhilina and Zavarzin 1994), and Grant et al. (2011) specifically reviewed the microbiology of several Inner Mongolian soda lakes.

One of the striking facts about soda lakes is that, despite inhospitable caustic conditions, these environments are extremely productive because of high ambient temperatures, high light intensities and effectively unlimited supplies of CO_2 via the $HCO_3^-/CO_3^{2-}/CO_2$ equilibrium. Primary production rates in excess of 10 g C m^{-2} day^{-1} have been recorded (the mean primary productivity of streams and lakes of the world is about 0.6 g C m^{-2} day^{-1}), making these the most productive aquatic environments anywhere in the world (Melack and Kilham 1974; Oduor and Schagerl 2007; Talling et al. 1973; Whittaker and Likens 1973. It is generally considered that photosynthetic primary productivity, mainly the result of the dense blooms of Cyanobacteria (up to 13,000 cyanobacterial filaments mL^{-1}) (Grant 1992), supports the rest of the microbial community. These Cyanobacteria blooms are usually dominated by what used to be known as *Spirulina* (now *Arthrospira*) spp., but in different lakes and also depending on seasonal factors, *Cyanospira* spp. and unicellular forms, which might be *Synechococcus* or *Chroococcus*, may also be common (Jones and Grant 2000). The cyanobacterial population of the African lakes is considered in detail in Krienitz and Schagerl (Chap. 6).

Anoxygenic phototrophic Bacteria are also capable of forming visible blooms in soda lakes. In early studies based on morphological data, Isachenko (1951) and Jannasch (1957) observed visible accumulations of anoxygenic phototrophs in Kulunda Steppe lakes and the Wadi Natrun lakes in Egypt. Anoxygenic phototrophs at the Wadi Natrun sites were later confirmed and studied by Imhoff et al. (1978, 1979, 1981) and Imhoff and Trüper (1977). The first recorded culture-based study on an East African lake was by Grant et al. (1979), who isolated an *Ectothiorhodospira* sp. from the relatively dilute Lake Bogoria (Kenya). The ubiquitous and probably universal presence of these particular anaerobic phototrophs in soda lakes was confirmed by the detection of *Ectothiorhodospira vacuolata* in both East African and Wadi Natrun lakes (Imhoff et al. 1981). There are now a considerable number of other anoxygenic phototroph isolates recorded from soda lakes, including African soda lakes. Grant and Tindall (1986) recorded phototrophs in saltern brines at hypersaline Lake Magadi as revealed by the presence of typical peaks in absorption spectra of the brines (Fig. 5.1), where there is a pronounced absorption maximum at around 800 nm characteristic of bacteriochlorophyll, microscopic examination again suggesting *Ectothiorhodospira* or more likely *Halorhodospira* spp. The absorption maximum at the shorter wavelengths is attributable to a mix of bacteriochlorophyll and a larger component of archaeal carotenoids.

Since those early studies, there have been further investigations on other soda lakes emphasising the importance of the anoxygenic phototrophs. Sorokin et al. (2004b) did a wide-ranging study of several geochemically significant physiological groups of Bacteria in more than 20 alkaline salt lakes of the northeast Mongolian steppe with water salinity from 3 to

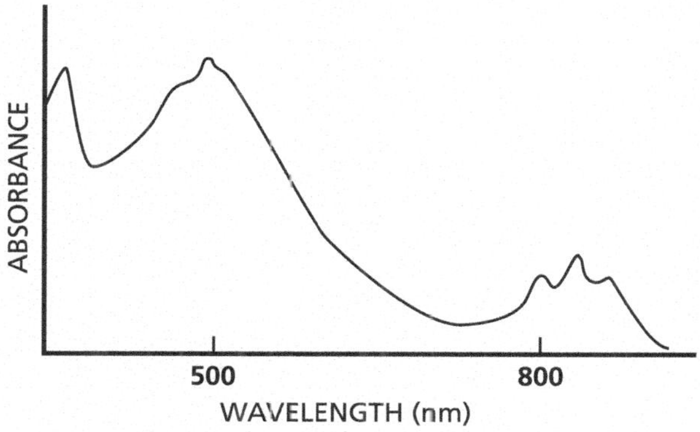

Fig. 5.1 Absorption spectrum of Lake Magadi saltern brine (redrawn from Grant and Tindall 1986)

390 g L^{-1} and pH values ranging from 9.0 to 10.6. Active and diverse microbial communities have been found in most of the lakes. Purple sulfur Bacteria of the *Ectothiorhodospiraceae* family represented the largest population of anaerobic phototrophic Bacteria. Salinity was the key factor in determining the activity and the composition of the microbial communities. The most diverse and active prokaryotic populations, including aerobic and anaerobic phototrophs, were found in lakes with salinity less than 60 g L^{-1}. Kompantseva et al. (2007) looked at the structure of the anaerobic phototrophic communities of soda lakes of the southeastern Transbaikal region. Here, communities were characterised by a wide diversity and evenness of species composition. Purple sulfur Bacteria of the families *Ectothiorhodospiraceae* and *Chromatiaceae* were predominant. Purple non-sulfur Bacteria of the family *Rhodobacteraceae*, green filamentous Bacteria *Oscillochloris* sp. and heliobacteria were also detected. According to this study, no less than 15 species of anaerobic phototrophic Bacteria occur in the studied lakes. Russian scientists (Gorlenko et al. 2010; Bryantseva et al. 2000c) looked at the cultivable microbial communities of the stratified soda Lake Doroninskoe (Transbaikal region) particularly in respect of phototrophic and sulfur cycle organisms. Bacteriochlorophyll *a*-containing aerobic anoxygenic phototrophic *Roseinatronobacter* sp. capable of thiosulfate oxidation was isolated from samples collected from the aerobic zone. Purple sulfur Bacteria belonging to the species *Thioalkalicoccus limnaeus*, *Ectothiorhodospira variabilis* and *Ectothiorhodospira shaposhnikovii* were isolated from samples of deep silt sediments. These Bacteria were the major anaerobic phototrophs in the shoreline algal–bacterial mat. *Ectothiorhodospira shaposhnikovii*, purple non-sulfur Bacteria of the genera *Rhodobacter* and *Roseococcus* sp. were isolated from the samples collected from mud. These observations around the world indicate that anoxygenic phototrophic Bacteria must make a significant, if as yet unquantified, contribution to the primary productivity in soda lakes.

5.2.1 Aerobic Chemoorganotrophs (African Soda Lakes)

There are now more than 150 descriptions of bacterial species that have been isolated from soda lakes or other soda-rich environments around the world (Table 5.2). Table 5.2 does not include Cyanobacteria since these oxygenic phototrophs are considered in Krienitz and Schagerl (Chap. 6). Isolates are numerically dominated by members of the *Proteobacteria* and *Firmicutes* with roughly equal numbers of aerobes and anaerobes. The vast majority have relatives living in less extreme environments, although there are two isolates that, to date, represent distinct lineages with no other representatives in culture.

As noted above, the first culture-based studies on African soda lakes largely relate to anoxygenic phototrophic Bacteria, but it is also clear from reading an early review of soda lake

Table 5.2 Isolates of soda lake Bacteria

Soda lake Bacteria		
Species	Original location	Reference
Aerobic organotrophs		
Alcalilimnicola halodurans[a]	Lake Natron, Tanzania	Yakimov et al. (2001)
Alkalilimnicola ehrlichii[a*]	Mono Lake, CA, USA	Hoeft et al. (2007)
Alkalibacillus haloalkaliphilus[b*]	Sambhar Lake, India	Fritze (1996), Jeon et al. (2005)
Alkalimonas amylolytica[a]	Lake Chahannor, Inner Mongolia	Ma et al. (2004b)
Alkalimonas delamerensis[a]	Lake Elmenteita, Kenya	Ma et al. (2004b)
Amphibacillus haojiensis[b]	Haoji soda lake, Inner Mongolia	Zhao et al. (2004)
Amphibacillus jilinensis[b*]	Soda lake, Jilin, China	Wu et al. (2010)
Bacillus alkalicola[b]	Lake Zabuye, Tibet	Zhai et al. (2014)
Bacillus alkalidiazotrophicus[b]	Soda soils, Mongolia	Sorokin et al. (2008f)
Bacillus alkalisediminis[b]	Soda ponds, Kiskunság, Hungary	Borsodi et al. (2011)
Bacillus arsenicoselenatis[b]	Mono Lake, CA, USA	Blum et al. (1998)
Bacillus aurantiacus[b]	Soda lake, Hungary	Borsodi et al. (2008b)
Bacillus beveridgei[b]	Mono Lake, CA, USA	Baesman et al. (2009)
Bacillus bogoriensis[b]	Lake Bogoria, Kenya	Vargas et al. (2005)
Bacillus caseinilyticus[b]	Lonar Lake, India	Reddy et al. (2015b)
Bacillus chagannorensis[b]	Lake Chagannor, Inner Mongolia	Carrasco et al. (2007b)
Bacillus daliensis[b]	Dali Lake, Inner Mongolia	Zhai et al. (2012)
Bacillus daqingensis[b]	Soda soil, Daqing, China	Wang et al. (2014)
Bacillus locisalis[b]	Soda lakes, China, Kenya, Tanzania	Marquez et al. (2011), Reddy et al. (2015d)
Bacillus lonarensis[b]	Lonar Lake, India	Reddy et al. (2015a)
Bacillus selenitireducens[b]	Mono Lake, CA, USA	Blum et al. (1998)
Belliella kenyensis[b]	Lake Elmenteita, Kenya	Akhwale et al. (2015a)
Bogoriella caseilytica[c]	Lake Bogoria, Kenya	Groth et al. (1997)
Cecembia lonarensis[h*]	Lonar Lake, India	Kumar et al. (2012)
Cellulomonas bogoriensis[c]	Lake Bogoria, Kenya	Jones et al. (2005)
Cohaesibacter haloalkalitolerans[d]	Lonar Lake, India	Sultanpuram et al. (2013)
Dietzia natronolimnaea[c]	Lake Nakuru, Kenya	Duckworth et al. (1998)
Georgenia satyanarayanai[c]	Lonar Lake, India	Srinivas et al. (2012)
Halolactibacillus alkaliphilus[b]	Xiarinaoer Lake, Inner Mongolia	Cao et al. (2008)
Halomonas alkalitolerans[a]	Soda meadow, Heilongjiang, China	Wang et al. (2011)
Halomonas campaniensis[a]	Alkaline mineral pool, Italy	Romano et al. (2005b)
Halomonas campisalis[a]	Alkali Lake, WA, USA Lake Magadi, Kenya	Mormile et al. (1999), Boltyanskaya et al. (2004)
Halomonas chromatireducens[a]	Soda soil, Kulunda, Russia	Shapovalova et al. (2009)
Halomonas kenyensis[a]	Lake Magadi, Kenya	Boltyanskaya et al. (2007)
Halomonas magadiensis[a]	Lake Magadi, Kenya	Duckworth et al. (2000)
Halomonas mongoliensis[a]	Soda lake, Mongolia	Boltyanskaya et al. (2007)
Indibacter alkaliphilus[h*]	Lonar Lake, India	Kumar et al. (2010a)
Jonesia quinghaiensis[c]	Soda lake, Qinghai, China	Schumann et al. (2004)
Litoribacter ruber[h]	Soda lake, Yongsheng, China	Tian et al. (2010)
Lunatimonas lonarensis[h]	Lonar Lake, India	Srinivas et al. (2014)
Marinospirillum alkaliphilum[a]	Haoji Lake, Inner Mongolia	Zhang et al. (2002a)
Marinospirillum celere[a]	Mono Lake, CA, USA	Namsaraev et al. (2009)
Mongolicoccus roseus[h]	Lake Horsemeno, Mongolia	Liu et al. (2012)
Mongoliitalea lutea[h]	Lake Dugerno, Mongolia	Yang et al. (2012)
Nesterenkonia aethiopica[a]	Lake Abijatta, Ethiopia	Delgado et al. (2006)

(continued)

Table 5.2 continued: Soda lake Bacteria

Species	Original location	Reference
Nesterenkonia suensis[a]	Alkaline salt pan, Botswana	Govender et al. (2013)
Nitrincola lacisaponensis[a]	Soap Lake, WA, USA	Dimitriu et al. (2006)
Paracoccus bogoriensis[d]	Lake Bogoria, Kenya	Osanjo et al. (2009)
Roseinatronobacter thiooxidans[d]	Soda lake, Kunkuskaya Steppe, Russia	Sorokin et al. (2000b)
Roseinatronobacter monicus[d]	Mono Lake, CA, USA	Boldareva et al. (2007)
Roseococcus suduntuyensis[d]	Lake Shuluutai, Russia	Boldareva et al. (2009a)
Salinicoccus alkaliphilus[b]	Baer Lake, Inner Mongolia	Zhang et al. (2002b)
Salinicoccus halitifaciens[b]	Lonar Lake, India	Ramana et al. (2013)
Salinicoccus kekensis[b]	Lake Xiaokule, China	Gao et al. (2010)
Salisediminibacterium haloalkalitolerans[b]	Lonar Lake, India	Reddy et al. (2015d)
Salisediminibacterium halotolerans[b]	Xiarinaoer Lake, Inner Mongolia	Jiang et al. (2012)
Salsuginibacillus halophilus[b]	Xiarinaoer Lake, Inner Mongolia	Cao et al. (2010)
Salsuginibacillus kocurii[b]	Lake Chagannor, Inner Mongolia	Carrasco et al. (2007a)
Streptomyces alkaliphilus[c]	Lake Elmenteita, Kenya	Akhwale et al. (2015b)
Streptomyces alkalithermotolerans[c]	Lonar Lake, India	Reddy et al. (2015c)
Texcoconibacillus texcoconensis[b]	Lake Texcoco, Mexico	Ruiz-Romero et al. (2013a)
Anaerobic organotrophs		
Alkalibacter saccharofermentans[b]	Lake Nizhnee Beloe, Russia	Garnova et al (2004)
Alkaliflexus imshenetskii[h*]	Lake Verkhnee Beloe, Russia	Zhilina et al. (2004)
Alkaliphilus peptidofermentans[b]	Lake Verkhnee Beloe, Russia	Zhilina et al. (2009b)
Alkalitalea saponilacus[b]	Soap Lake, WA, USA	Zhao and Chen (2012)
Amphibacillus fermentum[b]	Lake Magadi, Kenya	Zhilina et al. (2001a)
Amphibacillus tropicus[b]	Lake Magadi, Kenya	Zhilina et al. (2001a)
Anaerobacillus alkalilacustre[b]	Lake Khatyn, Russia	Zavarzina et al. (2009)
Anaerobranca gottschalkii[b]	Lake Magadi, Kenya	Prowe and Antranikian (2001)
Anaerovirgula multivorans[b]	Owens Lake, CA, USA	Pikuta et al. (2006)
Anoxynatronum sibiricum[b]	Lake Nizhnee Beloe, Russia	Garnova et al. (2003)
'Candidatus *Contubernalis alkalaceticum*'[b]	Lake Khatyn, Kulunda Steppe, Russia	Zhilina et al. (2005c)
'Candidatus *Syntrophonatronum acetioxidans*'[b]	Bitter Lake, Kulunda Steppe, Russia	Sorokin et al. (2014c)
Chitinivibrio alkaliphilus[g*]	Soda lake, Kulunda Steppe, Russia	Sorokin et al. (2014b)
Geoalkalibacter ferrihydriticus[f]	Lake Khatyn, Kulunda Steppe, Russia	Zavarzina et al. (2006)
Halanaerobium hydrogenoformans[b*]	Soap Lake, WA, USA	Brown et al. (2011)
Halarsenatibacter silvermanii[b]	Searles Lake, CA, USA	Blum et al. (2009)
Halonatronum saccharophilum[b]	Lake Magadi, Kenya; Mongolia	Zhilina et al. (2001b)
Natranaerobaculum magadiense[b]	Lake Magadi, Kenya	Zavarzina et al. (2013)
Natranaerobius thermophilus[b]	Wadi Natrun, Egypt	Mesbah et al. (2007b)
Natranaerobius trueperi[b]	Wadi Natrun, Egypt	Mesbah and Wiegel (2009)
Natranaerovirga hydrolytica[b]	Soda lake, Kulunda, Russia	Sorokin et al. (2012e)
Natranaerovirga pectinivora[b]	Soda lake, Kulunda, Russia	Sorokin et al. (2012e)
Natroniella acetigena[b]	Lake Magadi, Kenya	Zhilina et al. (1996b)
Natroniella sulfidigena[b]	Wadi Natrun, Kulunda Steppe	Sorokin et al. (2011c)
Natronincola ferrireducens[b]	Lake Verkhnee Beloe, Russia	Zhilina et al. (2009a)
Natronincola histidinovorans[b]	Lake Magadi, Kenya	Zhilina et al. (1998)

(continued)

Table 5.2 continued: Soda lake Bacteria

Species	Original location	Reference
Natronincola peptidovorans[b]	Lake Verkhnee Beloe, Russia	Zhilina et al. (2009a)
Natronobacillus azotifigens[b]	Soda soils, Siberia, Libya, Egypt, Mongolia	Sorokin et al. (2008d)
Natronocella acetinitrilica[a]	Soda lake, Kulunda, Russia	Sorokin et al. (2007b)
Natronoflexus pectinivorans[b]	Soda lake, Kulunda, Russia	Sorokin et al. (2011b)
Natronovirga wadinatrunensis[b]	Wadi Natrun, Egypt	Mesbah and Wiegel (2009)
Nitriliruptor alkaliphilus[c]	Soda lakes, Kulunda, Russia	Sorokin et al. (2009)
Nitritalea halalkaliphila[h*]	Lonar Lake, India	Kumar et al. (2010b)
Proteinivorax tanatarense[b]	Lake Tanatar III, Altai, Russia	Kevbrin et al. (2013)
Romboutsia sedimentorum[b]	Alkaline–saline lake, Daqing, China	Song et al. (2015)
Ruminiclostridium alkalicellulosi[b*]	Lake Verkhnee Beloe, Russia	Zhilina et al. (2005a), Yutin and Galperin (2013)
Spirochaeta africana[e]	Lake Magadi, Kenya	Zhilina et al. (1996a)
Spirochaeta alkalica[e*]	Lake Magadi, Kenya	Zhilina et al. (1996a)
Spirochaeta americana[e]	Mono Lake, CA, USA	Hoover et al. (2003)
Spirochaeta asiatica[e]	Lake Khatyn, Russia	Zhilina et al. (1996a)
Spirochaeta dissipatitropha[e]	Searles Lake, CA, USA	Pikuta et al. (2009)
Spirochaeta sphaeroplastigenens[e]	Lonar Lake, India	Reddy et al. (2013)
Tindallia californiensis[b]	Mono Lake, CA, USA	Pikuta et al. (2003a)
Tindallia magadiensis[b]	Lake Magadi, Kenya	Kevbrin et al. (1998)
Tindallia texcoconensis[b]	Lake Texcoco, Mexico	Alazard et al. (2007)
Sulfur-oxidisers		
Thioalkalibacter halophilus[a]	Soda lakes, Kulunda Steppe, Siberia	Banciu et al. (2008)
Thioalkalivibrio denitrificans[a]	Lake Bogoria, Kenya	Sorokin et al. (2001b)
Thioalkalivibrio halophilus[a]	Soda lake, Altai Steppe, Russia	Banciu et al. (2004)
Thioalkalivibrio jannaschii[a]	Mono Lake, CA, USA	Sorokin et al. (2002b)
Thioalkalivibrio nitratireducens[a*]	Wadi Natrun, Egypt	Sorokin et al. (2003)
Thioalkalivibrio nitratis[a]	Lake Nakuru, Kenya	Sorokin et al. (2001b)
Thioalkalivibrio thiocyanodenitrificans[a]	Mixed samples, Russia, Egypt	Sorokin et al. (2004a)
Thioalkalivibrio paradoxus[a]	Mixed samples, Kenya and Egypt	Sorokin et al. (2002a)
Thioalkalivibrio sulfidiphilus[a]	Lake Elmenteita, Kenya	Sorokin et al. (2012b)
Thioalkalivibrio thiocyanoxidans[a*]	Mixed samples, Siberia, Kenya, Egypt	Sorokin et al. (2002a)
Thioalkalivibrio versutus[a*]	Lake Bogoria, Kenya; soda lakes, Siberia	Sorokin et al. (2001b)
Thioalkalimicrobium aerophilum[a*]	Soda lakes, Siberia and Kenya	Sorokin et al. (2001b)
Thioalkalimicrobium cyclicum[a*]	Mono Lake, CA, USA	Sorokin et al. (2002b)
Thioalkalimicrobium sibericum[a]	Soda lakes, Siberia	Sorokin et al. (2001b)
Thioalkalimicrobium microaerophilum[a]	Soap Lake, WA, USA	Sorokin et al. (2007c)
Thioalkalispira microaerophila[a]	Wadi Natrun, Egypt	Sorokin et al. (2002c)
Thiohalophilus thiocyanoxidans[a*]	Soda lake, Kulunda, Russia	Sorokin et al. (2007e)
Sulfate reducers		
Desulfitispora alkaliphila[b]	Soda lakes, Kulunda, Russia	Sorokin and Muyzer (2010a)
Desulfobotulus alkaliphilus[f]	Soda lakes, Kulunda, Russia	Sorokin et al. (2010a)

(continued)

Table 5.2 continued: Soda lake Bacteria

Species	Original location	Reference
Desulfobulbus alkaliphilus[f]	Soda lake, Kulunda, Russia	Sorokin et al. (2012d)
Desulfohalophilus alkaliarsenatis[f]	Searles Lake, CA, USA	Blum et al. (2012)
Desulfonatronobacter acidivorans[f]	Soda lake, Kulunda, Russia	Sorokin et al. (2012d)
Desulfonatronospira delicata[f]	Wadi Natrun, Egypt	Sorokin et al. (2008a)
Desulfonatronospira thiodismutans[f*]	Soda lakes, Kulunda, Russia	Sorokin et al. (2008a)
Desulfonatronovibrio magnus[f]	Soda lakes, Kulunda, Russia	Sorokin et al. (2011a)
Desulfonatronovibrio halophilus[f]	Soda lake, Kulunda, Russia	Sorokin et al. (2012g)
Desulfonatronovibrio hydrogenovorans[f]	Lake Magadi, Kenya	Zhilina et al. (1997)
Desulfonatronovibrio thiodismutans[f]	Soda lakes, Kulunda, Russia	Sorokin et al. (2011a)
Desulfonatronum thioautotrophicum[f]	Soda lakes, Kulunda, Russia	Sorokin et al. (2011a)
Desulfonatronum buryatense[f]	Alkaline lake, Buryatia, Russia	Ryzhmanova et al. (2013)
Desulfonatronum cooperativum[f]	Lake Khatyn, Russia	Zhilina et al. (2005b)
Desulfonatronum lacustre[f*]	Lake Khatyn, Russia	Pikuta et al. (1998)
Desulfonatronum thiodismutans[f]	Mono Lake, CA, USA	Pikuta et al. (2003b)
Desulfonatronum thiosulfatophilum[f]	Soda lake, Kulunda, Russia	Sorokin et al. (2011a)
Desulfonatronum zhilinea[f]	Lake Alginskoe, Transbaikal, Russia	Zakharyuk et al. (2015)
Desulfuribacillus alkaliarsenatis[f]	Soda lake, Kulunda, Russia	Sorokin et al. (2012f)
Sulphur reducers		
Desulfurispira natronophila[g]	Soda lakes, Kulunda, Russia	Sorokin and Muyzer (2010b)
Desulfurivibrio alkaliphilus[f*]	Wadi Natrun, Egypt	Sorokin et al. (2008b)
Dethiobacter alkaliphilus[b]	Soda lake, Mongolia	Sorokin et al. (2008b)
Desulfurispirillum alkaliphilum[j]	Mixed sample, bioreactor	Sorokin et al. (2007a)
Sulfurospirillum alkalitolerans[k]	Mixed sample, bioreactor	Sorokin et al. (2013)
Nitrifiers		
Nitrobacter alkalicus[d]	Soda lakes, Kenya, Russia	Sorokin et al. (1998)
Nitrosomonas halophila[n]	Soda lakes, Mongolia	Sorokin et al. (2001c)
Anoxygenic phototrophic Bacteria		
'Candidatus *Chloroploca asiatica*'[m]	Soda lakes, Transbaikal, Russia, and Mongolia	Gorlenko et al. (2014)
Ectothiorhodosinus mongolicus[a]	Lake Dzun Ulziin, Mongolia	Gorlenko et al. (2004)
Ectothiorhodospira vacuolata[a]	Kenyan and Egyptian soda lakes	Imhoff et al. (1981)
Ectothiorhodospira haloalkaliphila[a]	Wadi Natrun, Egypt	Imhoff and Süling (1996)
Ectothiorhodospira magna[a]	Soda lake, Transbaikal, Russia	Bryantseva et al. (2010)
Ectothiorhodospira shaposhnikovii[a]	Soda lakes, Transbaikal, Russia	Cherni et al. (1969), Bryantseva et al. (2010)
Ectothiorhodospira variabilis[a]	Wadi Natrun, Egypt; Siberia, Mongolia	Gorlenko et al. (2009)
Halorhodospira abdelmalekii[a]	Wadi Natrun, Egypt	Imhoff and Trüper (1977), Imhoff and Süling (1996)
Halorhodospira halochloris[a*]	Wadi Natrun, Egypt	Imhoff and Trüper (1977), Imhoff and Süling (1996)
Halorhodospira halophila[a*]	Wadi Natrun, Egypt	Imhoff and Süling (1996)
Heliorestis acidaminivorans[b]	Wadi Natrun, Egypt	Asao et al. (2012)
Heliorestis baculata[b]	Lake Ostozhe, Russia	Bryantseva et al. (2000b)
Heliorestis convoluta[b]	Wadi Natrun, Egypt	Asao et al. (2006)
Heliorestis daurensis[b]	Soda lake, Siberia	Bryantseva et al. (2000a)

(continued)

Table 5.2 continued: Soda lake Bacteria

Species	Original location	Reference
Rhodobaca barguzinensis[d]	Soda lake, Barguzin, Russia	Boldareva et al. (2008)
Rhodobaca bogoriensis[d*]	Lake Bogoria, Kenya	Milford et al. (2000)
Rhodobaculum claviforme[d]	Transbaikal steppe lakes, Russia	Bryantseva et al. (2015)
Rhodovulum steppense[d]	Steppe soda lake, Russia	Kompantseva et al. (2010)
Rhodovulum tesquicola[d]	Steppe soda lake, Russia	Kompantseva et al. (2012)
Roseibacula alcaliphilum[d]	Lake Doroninskoe, Russia	Nuyanzina-Boldareva and Gorlenko (2014)
Roseinatronobacter monicus[d]	Mono Lake, CA, USA	Boldareva et al. (2007)
Rubribacterium polymorphum[d]	Soda lake, Barguzin, Russia	Boldareva et al. (2009b)
Thioalkalicoccus limnaeus[a]	Soda lakes, Russia	Bryantseva et al. (2000c)
Thiocapsa imhoffii[a]	Soap Lake, WA, USA	Asao et al. (2007)
Thiorhodospira sibirica[a]	Lake Malye Kasytui, Russia	Bryantseva et al. (1999)
Methylotrophs		
Methylohalomonas lacus[a]	Soda lake, Kulunda, Russia	Sorokin et al. (2007d)
Methylomicrobium alcaliphilum[a*]	Soda lake, Tuva, Russia	Kalyuzhnaya et al. (2008)
Methylomicrobium buryatense[a*]	Soda lakes, Transbaikal, Russia	Kalyuzhnaya et al. (2001)
Methylomicrobium kenyense[a]	Soda lakes, Kenya	Kalyuzhnaya et al. (2008)
Methylonatrum kenyense[a]	Lake Magadi, Kenya	Sorokin et al. (2007d)
Methylophaga alcalica[a]	Soda lake, Mongolia	Doronina et al. (2003)
Methylophaga lonarensis[a*]	Lonar Lake, India	Antony et al. (2012b)
Methylophaga natronica[a*]	Soda lake, Transbaikal, Russia	Doronina et al. (2003a)
Chemolithotrophic hydrogenotrophs		
Fuchsiella alkaliacetigena[b]	Lake Tanatar III, Altai, Russia	Zhilina et al. (2012)
Fuchsiella ferrireducens[b]	Lake Tanatar III, Altai, Russia	Zhilina et al. (2015)

[a]*Gammaproteobacteria*, [b]*Firmicutes*, [c]*Actinobacteria*, [d]*Alphaproteobacteria*, [e]*Spirochaeta*, [f]*Deltaproteobacteria*, [g]novel lineage, [h]*Bacteroidetes*, [j]*Chrysiogenetes*, [k]*Epsilonproteobacteria*, [m]*Chloroflexi*, [n]*Betaproteobacteria*. [*]Partial or complete genome sequence available
Isolates from African lakes in bold-type face (Updated June 2015)

microbiology (Grant and Tindall 1986) following Isachenko's specific monograph, that alkaliphilic aerobic organotrophic Bacteria are abundant in the more dilute EASL. Viable counts on media described by Grant and Tindall (1980) reveal a remarkably constant population over a 12-month period of 10^5–10^6 mL^{-1}, with total counts indicating 10^7–10^8 mL^{-1} in lakes Elmenteita, Nakuru and Bogoria (Mwatha 1991). Systematic studies of EASL chemoorganotrophs did not begin until the late 1980s and were hampered not only by accessibility of material for study but also by the limitations of the technology of the time. Numerous aerobic chemoorganotrophic Bacteria were isolated at pH 9–10, but there was a lack of a means of comparison to existing Prokaryotes. Early studies relied upon phenotypic characteristics and the application of numerical taxonomy (Jones et al. 1994, 1995, 1998a; Jones and Grant 2002). Other than clustering similar strains, however, it did not throw a lot of light on the identity of the Bacteria. It was clear that all the isolates represented novel taxa, many of which appeared unique to soda lakes. Subsequently, many of the isolates from a number of studies where organisms had both been isolated by enrichment procedures and by direct plating on agar-based media were subject to phylogenetic analysis by 16S rRNA gene sequencing (Duckworth et al. 1996), which confirmed that the alkaliphilic phenotype is not monophyletic and that the EASL contain a wide range of physiological groups. In particular, many representatives of *Gammaproteobacteria* and both *Firmicutes* and *Actinobacteria* from amongst the Gram-positive Bacteria were isolated. Some isolates were classified by polyphasic characterisation, which revealed novel taxa (e.g. *Alkalimonas*; Ma et al. 2004b) or new species within existing taxa (e.g. *Cellulomonas bogoriensis*; Jones et al. 2005), but these were limited to a small number by a lack of resources or the need for a more thorough taxonomic revision (e.g. *Halomonas*; Duckworth et al. 2000). It is still surprising perhaps that even after 20 years, some of these early chemoorganotrophic bacterial isolates remain unidentified despite the numerous reports of novel Prokaryotes from EASL and elsewhere. However, a reanalysis of the Duckworth et al. (1996) data and other isolates in relation to the current, now much expanded database of 16S rRNA gene sequences does indicate some likely affiliations (Tables 5.3 and 5.4).

Despite the growing interest in soda lakes during the past three decades, there have been few attempts at a systematic study of the dynamics of bacterial populations. The difficulties of regular access to many of these locations are the probable cause of this hiatus. Many of the East African lakes are now readily accessible, but the only available systematic studies date from 1988 (Mwatha 1991). Regular observations, which included bacterial type and numbers in relation to physical and chemical parameters, were performed on a monthly basis from October 1988 to September 1989 at lakes Elmenteita, Nakuru (Table 5.5) and Bogoria (data not shown).

The different Bacteria in the monthly samples were selected based on colony characteristics and differentiated by SDS-PAGE. A vigorous attempt was made to characterise the isolates by numerical taxonomy and chemotaxonomy methods, but the limited framework of reference as noted previously hampered adequate interpretation of the data. Since then, many, but not all, of the Mwatha isolates have been identified (BE Jones, unpublished) (Tables 5.3 and 5.4), and we are able to reconstruct some of the ecological observations performed by Mwatha (1991). Bacterial culture was generally carried out under aerobic conditions on a relatively rich medium, which inevitably biases recovery of chemoorganotrophs.

Each of the lakes exhibited unique characteristics, which is probably a reflection of local conditions. Bacterial numbers varied by several orders of magnitude during the study year but showed no statistically significant correlation with rainfall, which had little influence on water conductivity or alkalinity in these highly buffered systems. There was also no apparent correlation between algal blooms (as measured by chlorophyll *a*) and a subsequent increase in bacterial numbers, although the Cyanobacteria were clearly influenced by changes in salinity and alkalinity. There was a small correlation between water conductivity and alkalinity and bacterial numbers, but its significance may be masked by the dynamics of dominant species rather than by total numbers. This is more clearly seen using abstract mapping procedures (UNIRAS/UNIMAP) as shown in Fig. 5.2, where there are evidently at least two separate chemoorganotrophic populations in Lake Nakuru dominating at different water conductivities and alkalinity, while at Lake Bogoria there appeared to be at least three separate bacterial populations at different combinations of salinity and alkalinity (data not shown).

Amongst the main factors affecting microbial population dynamics was the input of nutrients. Phosphate concentration was a major influence on bacterial numbers and also total nitrogen, as can be seen at Lake Elmenteita (Fig. 5.3). Conventional statistical analysis failed to highlight the effects of phosphate and nitrogen limitations on total bacterial numbers, probably because in each of the lakes different bacterial [species] populations responded to different threshold values of total phosphate and total nitrogen. Nitrogen limitation is less of a problem at Lake Nakuru and Lake Bogoria, which are usually home to large resident populations of grazing

Table 5.3 Identities of aerobic chemoorganotrophic isolates (from Duckworth et al. 1996)

Isolate (GenBank Acc. No.)	Identification	Reference	Comments
Lake Nakuru isolate 19 N1 (X92149) Lake Magadi isolate 21M1^T (X92150) Lake Bogoria isolate 24B1 (X92151)	*Halomonas magadiensis* (99–100 %)	Duckworth et al. (2000)	Valid name
Lake Nakuru isolate 28 N1 (X92144)	*Halomonas nitritophilus* (99 %)	Gilvanova et al. unpublished	Possible identity to GenBank AJ309564
Crater Lake isolate 29C1 (X92148)	*Halomonas campisalis* (96 %)	Joshi et al. (2007)	Closest identity to Lonar Lake strain MCM B-365
Lake Bogoria isolate WB5 (X92136)	*Halomonas campisalis* (99 %)	Mormile et al. (1999)	Denitrifying halomonad
Lake Elmenteita isolate 1E1^T (X92130) Lake Elmenteita isolate 2E1 (X92131)	*Alkalimonas delamerensis* (100 %)	Ma et al. (2004b)	Valid name
Lake Nakuru isolate 20 N1 (X92128)	*Aliidiomarina sanyensis* (97 %)	Wang et al. (2013)	GenBank JX981923
Lake Bogoria isolate 10B1 (X92127) Crater lake isolate 11C1 (X92129)	*Idiomarina* sp. (98 %)		GenBank KM013952
Lake Natron isolate 97NT4 (X92132)	Unknown		
Lake Elmenteita isolate 45E3 (X92133) Lake Nakuru isolate 52 N3 (X92134)	*Pseudomonas salegens* (98 %)	Amoozegar et al. (2014)	GenBank KF830707
Lake Bogoria isolate 66B4 (X92158) Lake Nakuru isolate 18 N1 (X92159) Lake Bogoria isolate 64B4 (X92160)	*Bacillus pseudofirmus* (97–99 %)	BE Jones (unpublished)	GenBank NR_026139
Lake Elmenteita isolate WE4 (X92161) Lake Elmenteita isolate 3E1 (X92162)	*Bacillus saliphilus* (98 %)	Romano et al. (2005a)	Confirmed identity (BE Jones, unpublished)
Lake Natron isolate 103NT4 (X92163) Lake Elmenteita isolate WE1 (X92164)	*Bacillus locisalis* (98 %)	Marquez et al. (2011)	Valid name
Little Lake Magadi isolate 95LM4 (X92165) Lake Natron isolate 105NT4 (X92166)	*Salsuginibacillus halophilus* (97 %)	Cao et al. (2010)	Probable identity (GenBank NR_114275)
Lake Bogoria isolate 9B1 (X92167)	*Bacillus aurantiacus* (98 %)	Borsodi et al. (2008b)	BE Jones (unpublished) GenBank AJ605772
Lake Nakuru isolate 17 N1 (X92168)	*Bacillus* sp. nov.	BE Jones (unpublished)	
Lake Nakuru isolate WN16 (X92169)	*Alkalibacterium* sp. (97 %)	Ntougias and Russell (2001)	BE Jones (unpublished) GenBank AF143512
Lake Bogoria isolate 69B4^T (X92152)	*Cellulomonas bogoriensis* (100 %)	Jones et al. (2005)	Valid name

(continued)

Table 5.3 (continued)

Isolate (GenBank Acc. No.)	Identification	Reference	Comments
Lake Bogoria isolate WB3 (X92153)	*Nesterenkonia pannonica* (99 %)	BE Jones (unpublished)	Probable identity to GenBank FM179648, Borsodi et al. (2008a), unpublished
Lake Elmenteita isolate WE2 (X92154) Crater Lake isolate 13C1 (X92155)	*Nesterenkonia* sp. (99 %)	BE Jones (unpublished)	Probable close identity to GenBank EU432579
Little Lake Naivasha isolate 14LN1 (X92156) Little Lake Naivasha isolate 15LN1^T (X92157)	*Dietzia natronolimnaea* (100 %)	Duckworth et al. (1998)	Valid name
Little Lake Magadi isolate 93dLM4 (X92170)	New genus	D. Sorokin (pers. comm.)	Sulphur reducer
Lake Magadi isolate 82 M4 (X92171)	*Natronolimnobius innermongolicus* (98 %)	BE Jones (unpublished)	Probable sp. identity, GenBank AB125108, Itoh et al. (2005)
Lake Magadi isolate 89 M4 (X92172) Little Lake Magadi isolate 93lLM4 (X92173) Lake Magadi isolate 86 M4 (X92175)	*Natronococcus occultus* (98 %)	Tindall et al. (1984)	Probable identity to type strain SP4, GenBank NR_104253
Lake Natron isolate 98NT4 (X92174)	*Natrialba* sp. (97 %)	Kamekura et al. (1997)	Probable genus identity, GenBank NR_113522

Table 5.4 Unpublished soda lake isolates and their identity (BE Jones, unpublished results)

Isolate	Origin	Identity	Comments and GenBank Acc. Nos.
23 M1	Lake Magadi, Kenya (W.D. Grant, 1987 as strain M1)	*Nesterenkonia* sp. (99 %)	EU432579
60E4	Lake Elmenteita, Kenya (BE Jones, 1988 as strain K6.8b)	*Bacillus* sp. (99 %)	Closely related to *B. saliphilus* (NR_025554) and various soda lake isolates (e.g. AY347312, JX945788)
72C4	Crater lake, Kenya (BE Jones, 1988 as strain K27.7)	*Nesterenkonia* sp. (99 %)	Closely related to isolates from Lonar Lake, India (KJ417902); Xinjiang, China (EU432579); Mono Lake, California (AF140001)
80LN4	Little Lake Naivasha (BE Jones, 1988 as strain K33.2)		
WN10	Lake Nakuru, Kenya (W.E. Mwatha, 1988 as strain N1)	*Alkalibacterium pelagium* (99 %)	Ishikawa et al. (2009) (NR_11424)
WN12	Lake Nakuru, Kenya (W.E. Mwatha, 1988 as strain N12)	*Amphibacillus jilinensis* (99 %)	Wu et al. (2010) (NR_116553)

flamingos whose contribution to carbon and nitrogen recycling is likely to be hugely significant. However, the abstract mapping approach clearly indicated a competition for resources, especially phosphate, between Cyanobacteria and chemoorganotrophs, since maximum bacterial numbers were seen at high phosphate levels when chlorophyll was low.

Table 5.5 Monthly data at lakes Nakuru and Elmenteita: October 1988 to September 1989 from Mwatha (1991)

Month	Rainfall (mm)	pH	Conductivity (mS cm^{-1})	Alkalinity (meq L^{-1})	CFU mL^{-1} ($\times 10^5$)	Chlorophyll (μg mL^{-1})	TOC (μg mL^{-1})	Total nitrogen (μg mL^{-1})	Total phosphate (μg mL^{-1})	Ammonium (μg mL^{-1})	Sulfate (μg mL^{-1})
Lake Nakuru											
Oct. 1988	71.7	10.5	19	110.5	2.9	53.7	4.2	3.8	3.00	0.46	280
Nov.1988	58.2	10.7	19	106.0	1.4	65.8	3.9	4.0	2.90	0.46	280
Dec. 1988	23.3	10.4	17	102.0	23.0	83.6	3.3	7.8	3.25	0.50	300
Jan. 1989	28.6	10.5	16	100.4	17.0	4.0	5.5	5.8	5.50	0.41	173
Feb. 1989	67.5	10.6	30	126.8	2.6	2.0	1.3	4.0	1.25	0.44	300
Mar. 1989	54.9	10.8	24	112.5	6.4	121.0	5.1	3.0	1.94	0.42	255
Apr. 1989	97.4	11.0	24	105.8	1.5	160.0	6.8	2.8	1.94	0.42	345
May 1989	82.4	11.0	24	102.0	1.5	228.5	6.0	2.8	1.82	0.38	255
Jun. 1989	59.4	10.9	25	102.0	0.85	307.8	7.0	3.0	1.80	0.29	195
Jul. 1989	111.8	10.9	26	91.0	1.1	160.0	7.0	3.8	2.00	0.42	260
Aug. 1989	80.0	10.9	30	71.2	4.5	62.5	6.8	3.5	2.33	0.45	230
Sep. 1989	203.1	11.2	27	91.2	12.0	45.0	7.1	5.5	2.32	0.44	240
Lake Elmenteita											
Oct. 1988	88.0	10.4	13	135.0	3.0	40.0	4.5	3.8	2.70	0.27	360
Nov.1988	37.1	11.0	13	135.0	5.1	64.0	4.8	4.7	2.65	0.35	360
Dec. 1988	10.7	10.8	17	133.0	1.2	181.0	5.4	5.6	3.20	0.43	380
Jan. 1989	34.8	10.5	16	130.0	1.5	46.0	6.0	5.8	2.50	0.40	373
Feb. 1989	43.6	10.5	19	128.0	0.024	8.0	2.5	2.4	1.20	0.44	360
Mar. 1989	72.8	10.5	16	132.0	0.3	86.0	4.5	4.4	1.80	0.29	360
Apr. 1989	89.3	10.6	18	138.0	1.2	46.0	4.5	4.8	1.60	0.30	343
May 1989	117.7	10.6	17	134.0	0.12	66.0	5.3	4.2	1.80	0.34	340
Jun. 1989	60.0	10.6	17	102.0	0.12	56.0	6.5	4.0	1.20	0.49	340
Jul. 1989	72.7	10.5	17	126.0	0.16	76.0	8.0	2.8	2.50	0.35	350
Aug. 1989	95.9	10.5	18	109.2	0.13	58.1	9.5	3.8	2.75	0.35	310
Sep. 1989	48.8	10.9	19	93.6	1.4	46.1	8.3	6.6	2.40	0.37	225

CFU = [bacterial counts] colony forming units. TOC = total organic carbon

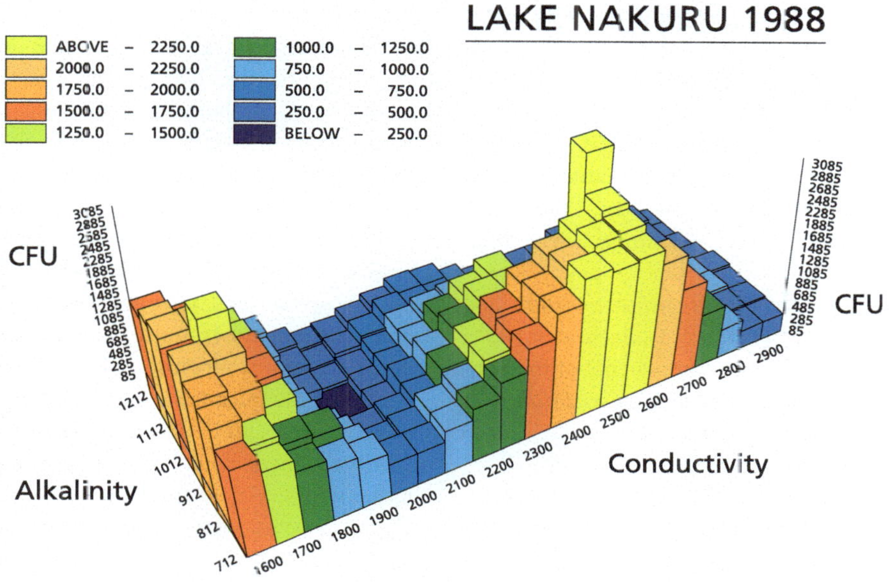

Fig. 5.2 Lake Nakuru: bacterial numbers in relation to alkalinity and conductivity. Conductivity [(mS cm^{-1}) × 10^{-2}], alkalinity [(meq L^{-1}) × 10^{-1}) and bacterial numbers [(CFU × 10^{2}) mL^{-1}]

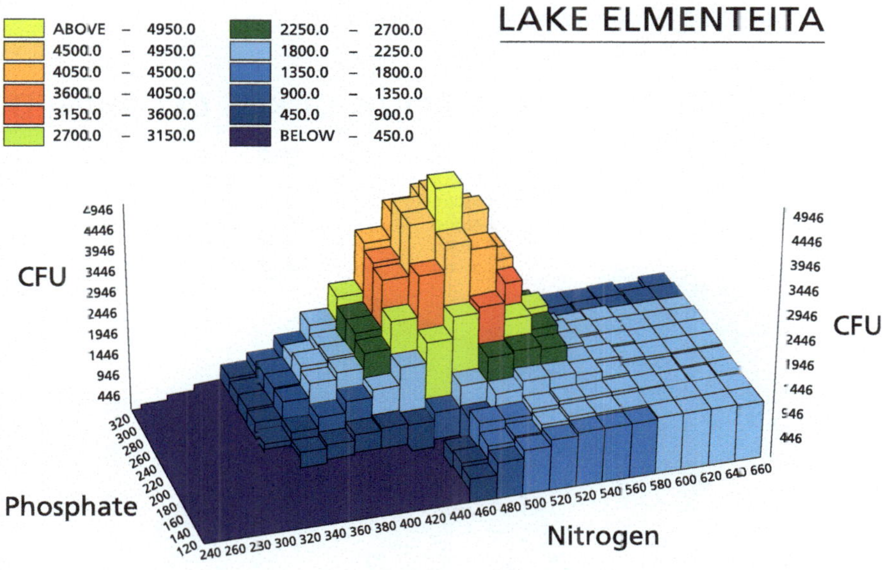

Fig. 5.3 Lake Elmenteita: bacterial numbers in relation to phosphate and nitrogen concentrations. Nitrogen \$6# [(μg mL^{-1}) × 10^{-2}], phosphate [(μg mL^{-1}) × 10^{-2}] and bacterial numbers [(CFU × 10^{2}) mL^{-1}]

Halomonas spp. (Tables 5.3 and 5.4) are ubiquitous and account for at least 30 % of the aerobic chemoorganotrophs in the cultivated population. Under favourable circumstances *Halomonas* spp. provide up to 70 % of the strains isolated but not necessarily the same species at any one time. In the summer months at Lake Nakuru, *Halomonas* spp. were at their lowest level and other Bacteria such as *Amphibacillus jilinensis* (WN12, Table 5.4) at 20 % achieve significant numbers. *Alkalibacterium pelagium* (WN10, Table 5.3) are present at Nakuru in low numbers throughout the year, but it is another *Alkalibacterium* species (WN16, Table 5.3) that came to prominence in February 1989 when water conductivity and alkalinity were high (Table 5.5).

In Lake Elmenteita, *Bacillus locisalis* and *Bacillus saliphilus* (Table 5.3), each at <10 %, were amongst the minor components of the population, which was dominated by a *Halomonas* sp (isolate WE5, Duckworth et al. 2000) (15–60 %) and an unknown isolate (30–70 %) in a reciprocal relationship.

The most abundant strains at Lake Bogoria are *Halomonas* spp.; *H. magadiensis* and *H. campisalis* were present in low numbers throughout the year, occasionally providing 30 % of the strains isolated. The most abundant (10–80 %) strain of *Halomonas* (WB2, Duckworth et al. 2000) became particularly prominent when phosphate levels were high in combination with high conductivity and alkalinity. *Nesterenkonia* sp. was a minor, incidental component of the population, also at Lake Elmenteita. Combined bacterial data for lakes Nakuru, Elmenteita and Bogoria (north and south) indicate several distinct bacterial populations whose dynamics were influenced by an intricate combination of chemical and physical factors, the most important of which were phosphate and nitrogen levels (Fig. 5.4), water conductivity (salinity) and alkalinity, and limited through competition by phototrophs.

While these [reconstructed] data offer a useful historical perspective on the population dynamics of the Kenyan soda lakes, they suffer several shortcomings. For example, they provide data only on aerobic chemoorganotrophs that were readily isolated on a particular rich medium. The data do not include the contributions by anaerobic Bacteria or the considerable numbers of Bacteria peculiar to the sulfur cycle. This study would benefit from being repeated using modern tools of pyrosequencing and metagenomics.

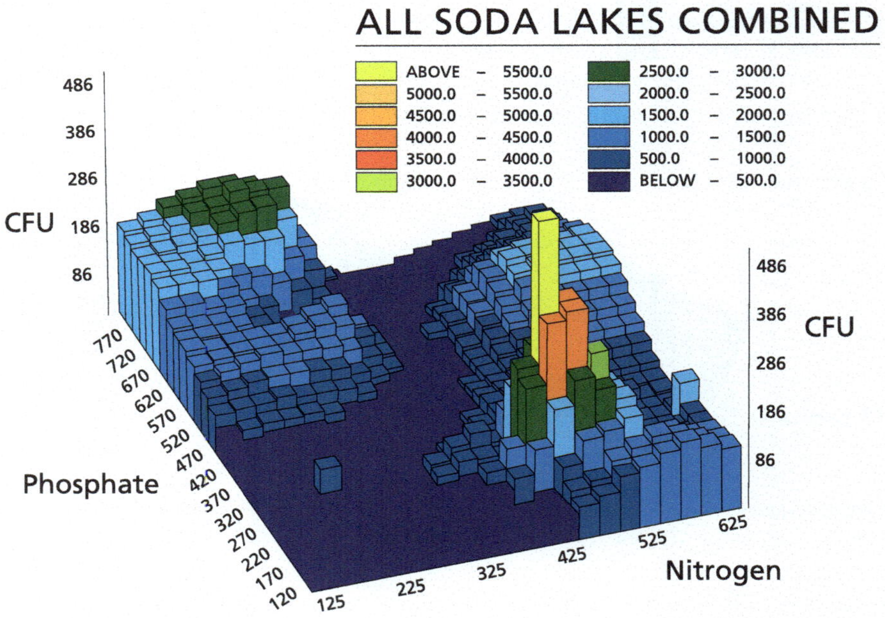

Fig. 5.4 Bacterial numbers (combined data all lakes) showing different populations in relation to phosphate and nitrogen concentrations. Nitrogen [(μg mL^{-1}) × 10^{-2}], phosphate [(μg mL^{-1}) × 10^{-2}] and bacterial numbers [(CFU × 10^{3}) mL^{-1}]

It is clear from the range of media used by Duckworth et al. (1996) that isolates from soda lakes were likely to have commercial potential in that they secrete many extracellular hydrolytic enzymes, including proteinases, cellulases and lipases capable of functioning at high pH and possibly high temperature and salt concentration. As such they are of interest to the detergent industry in particular (Horikoshi 2006). In the past, two different cellulases derived from Gram-positive African soda lake isolates were marketed for use in laundry and textile processes, and there have been attempts to directly access enzyme genes from DNA extracted from African soda lake samples (reviewed by Grant and Heaphy 2010). Cellulases stable at high pH and salt can also be used to release sugars from recalcitrant lignocellulose in agricultural waste for the production of bioethanol (Zhang et al. 2011; Zhu 2008). In addition, cyclodextrin glucanotransferase (CGTase) is another industrially important enzyme, used in the commercial production of cyclodextrins. Al-Sharawi et al. (2013) and Ibrahim et al. (2012) report the isolation of a new and potent CGTase from alkaliphilic Bacteria including *Amphibacillus* spp. collected from Egyptian soda lakes. The possibility of using Wadi Natrun soda lake isolates for the bioremediation of aromatics and chloroaromatics was explored by Hassan et al. (2012), and chromium bioremediation has also been recorded by an *Amphibacillus* sp. from the same site (Ibrahim et al. 2011). Besides the discovery of novel hydrolases, nitrile hydratases are important industrial enzymes that catalyse the hydration of a broad scope of nitrile compounds into commercially more valuable amides (e.g. acrylamide), and a nitrile hydratase activity was isolated in the soda lake actinobacterium *Nitriliruptor alkaliphilus* (Sorokin et al. 2009; 2007b). General reviews on the industrial use of enzymes from alkaliphiles, including soda lake Bacteria, are presented by Ito (2011) and Horikoshi (2011).

The only other comparable culture-based study of EASL is that of Mwirichia et al. (2010a), where a culture analysis of Lake Elmenteita (Kenya) was carried out. Here, the authors used lake brine as the basis of various enrichment media. This study had the benefit of a relatively recent 16S rRNA gene database for comparisons to be made and showed that the majority of the isolates were affiliated to the class *Gammaproteobacteria* and to the *Firmicutes* division genus *Bacillus*. Isolates from the genus *Halomonas* (class *Gammaproteobacteria*) and *Bacillus* constituted 37 and 31 % of the total sequenced isolates, respectively. Other groups recovered were related to *Marinospirillum*, *Idiomarina*, *Vibrio*, *Enterococcus*, *Alkalimonas*, *Alkalibacterium*, *Amphibacillus*, *Marinilactibacillus* and the *Actinobacteria* Gram-positive division genera *Nocardiopsis* and *Streptomyces*. These isolates did not have close affiliation to the organisms studied by Duckworth et al. (1996), presumably a reflection of the different media used in the study.

A much more limited study was carried out by Vargas et al. (2004), who looked at lipase-producing isolates from Lake Bogoria (Kenya). Ten of the isolates were Gram-negative, nine of which were closely related to the *Pseudomonas* group and one to the *Halomonas* group sharing high similarity profile with *Halomonas desiderata*. The remaining Gram-positive isolates were closely related to the *Bacillus* cluster and were grouped with *B. halodurans*, *B. alcalophilus* and *B. licheniformis*.

In a similar restricted study, Martins et al. (2001) isolated alkaliphilic starch utilisers from Ethiopian soda lakes. All isolates except one were motile endospore-forming rods and were found to be closely related to the *Bacillus* cluster, being grouped with *B. pseudofirmus*, *B. cohnii*, *B. vedderi* and *B. agaradhaerens*. The one exception had nonmotile coccoid cells and was closely related to *Nesterenkonia halobia*.

Tourova et al. (1999) carried out a phylogenetic analysis on 13 strains of alkaliphilic obligately or facultatively anaerobic Bacteria isolated from two soda lakes with pH 10 – the highly mineralised Lake Magadi (Kenya) and the slightly mineralised Lake Nizhnee Beloe (south-eastern Transbaikal region). 16S rRNA gene sequence analysis showed that all the strains belonged to the phylogenetic branch of *Firmicutes*. Other anaerobes affiliated with

Clostridium spp. and the *Firmicutes* order *Halanaerobiales*. The strains investigated were sufficiently phylogenetically distinct to represent new taxa.

5.2.2 Aerobic Chemoorganotrophs (The Rest of the World's Soda Lakes)

There have been a number of culture-based studies of other soda lakes around the world, usually with a view to isolating examples of particular genera using specific enrichment procedures for particular physiological attributes. The organisms are, in general, biochemically active, hydrolysing a wide range of polymers, including cellulose, chitin, starch, saccharides, sugars, polypeptides and amino acids under aerobic conditions, sometimes linked to the reduction of nitrate (Boltyanskaya et al. 2007) and even chromate (Shapovalova et al. 2009). There are, however, a few instances of a more general approach similar to that taken by Duckworth et al. (1996) on the EARS lakes. Most studies have concentrated on chemoorganotrophs using relatively rich media under aerobic conditions.

The cultivable bacterial diversity of Lonar Lake in India was studied by Joshi et al. (2008) and shown to be comparable to the EASL. Phylogenetic analysis of 196 strains indicated that most of the Lonar Lake isolates were related to the phylum *Firmicutes*, with different genera: *Bacillus*, *Paenibacillus*, *Alkalibacillus*, *Exiguobacterium*, *Planococcus*, *Enterococcus* and *Vagococcus*. Seven strains were Gram-negative, with different genera: *Halomonas*, *Stenotrophomonas* and *Providencia* affiliated to *Gammaproteobacteria*, *Alcaligenes* to *Betaproteobacteria* and *Paracoccus* to *Alphaproteobacteria*. Only five isolates were *Actinobacteria* with various genera: *Cellulosimicrobium*, *Dietzia*, *Arthrobacter* and *Micrococcus*. Methylotrophic Bacteria growing on methanol and methylamine were also recorded in Lonar Lake as well as in the lakes of the Kulunda Steppe (Antony et al. 2012b; Sorokin et al. 2000a).

Sahay et al. (2012) were interested in the agricultural and industrial potential of Bacteria from Sambhar Lake in India. A majority of 93 isolates (53 %) showed similarity with the phylum *Firmicutes*, which was followed by *Proteobacteria* (40 %) and *Actinobacteria* (6 %). The isolates were further evaluated for the production of extracellular enzymes such as amylase, cellulase, protease and xylanase and plant growth-promoting attributes such as phosphorus solubilisation. Rusznyak et al. (2007) studied biofilm communities on reed stems in Hungarian soda lakes, where they isolated 260 strains and 16S rRNA gene sequence analyses identified members of the genera *Agrobacterium*, *Paracoccus*, *Halomonas*, *Pseudomonas*, *Bacillus*, *Planococcus* and *Nesterenkonia*. Sorokin et al. (2008c) looked at nitrogen fixation rates in hypersaline soda solonchak soils collected in several locations of Central Asia and in Egypt. Cultures were dominated by Gram-positive spore-forming Bacteria. Most of the isolates (eight) were affiliated with the aerotolerant fermentative haloalkaliphilic bacterium *Amphibacillus tropicus* and a single strain clustered with the obligately anaerobic haloalkaliphile *Bacillus arseniciselenatis*. Diazotrophy had never been recognised previously in these groups of Gram-positive Bacteria. Sorokin et al. (2012a) recorded chitin utilisation dominated by several obligately haloalkaliphilic members of the genus *Marinimicrobium* in Kulunda Steppe lakes. Additionally, a new lineage of fermentative haloalkaliphilic Bacteria specialised to exclusively utilise chitin as growth substrate was isolated and has recently been described as *Chitinivibrio alkaliphilus* (Sorokin et al. 2014b).

Aerobic methanotrophic Bacteria that resemble *Methylomicrobium* spp. were first recorded in East African lakes (Sorokin et al. 2000a) but have also been isolated from saline and alkaline lakes of southeast Siberia, Mongolia and North America (Khmelenina et al. 2000; Kalyuzhnaya et al. 2008), and methane oxidation has also been recorded in Mono Lake (Carini et al. 2005).

The capacity for aerobic carboxydotrophy in soda lakes was also studied by Sorokin et al. (2010b), mostly in the Kulunda Steppe lakes, but one culture was obtained from the Wadi Natrun. Using CO as the only substrate enabled the isolation of strains related to *Alkalispirillum*/*Alkalilimnicola*.

5.2.3 Anaerobic Chemoorganotrophs

Many chemoorganotrophic anaerobes have been cultured from African soda lakes and soda lakes elsewhere around the world. Viable counts (at 37 °C) on the black, anoxic soda lake sediment slurries (Elmenteita, Bogoria) indicate they contain more than 10^6 cfu mL^{-1} chemoorganotrophic alkaliphilic anaerobes (Jones et al. 1998a, b). In the soda lake environment, obligate anaerobes are probably 'secondary organotrophs' utilising the products of primary hydrolysis provided by aerobes and facultative anaerobes. Fermentative anaerobes are composed of two subgroups: those that degrade polymers and the secondary heterotrophs that utilise the resulting monomers. Soda lake fermentative anaerobes are dominated by haloalkaliphilic *Firmicutes*, such as members of the genera *Anoxynatronum*, *Anaerovirgula*, *Alkaliphilus*, *Natranaerobius* and *Natranaerobaculum* and certain species of *Anaerobranca*, *Spirochaeta* and *Anaerobacillus* (Bowers et al. 2009; Grant and Sorokin 2011; Mesbah and Wiegel 2012; Zavarzin et al. 1999; Zavarzin and Zhilina 2000). The black lake sediments clearly imply the presence of sulfate-reducing Bacteria (SRB), and there are now a significant number of these in culture (Table 5.2), considered elsewhere in this chapter. Like the aerobic chemoorganotrophs, the anaerobic chemoorganotrophs have mostly been isolated by enrichment procedures on specific substrates. Again, like the aerobic types, the anaerobic isolates are in general biochemically reactive, fermenting a range of polymers such as chitin (Sorokin et al. 2012a), cellulose (Zhilina et al. 2005a) and pectin (Sorokin et al. 2012d) with many examples fermenting sugars and amino acids to end products such as acetate, formate, lactate and H_2. Nitrate reduction, nitrogen fixation, iron reduction and arsenic reduction are also recorded (Baesman et al. 2009; Oremland et al. 2004; Zavarzina et al. 2006; Zhilina et al. 2009a, b). There are also isolates recorded as enriched on odd substrates like isobutyronitrile and acetonitrile (Sorokin et al. 2007b, 2009).

Dimitriu et al. (2008) looked at spatial and temporal patterns in the anaerobic microbial diversity of Soap Lake in Washington State (USA). Late-summer samples harboured the highest prokaryotic diversity. Most-probable-number assays targeting anaerobic microbial guilds were performed to compare summer and fall samples. In both seasons, the anoxic samples appeared to be dominated by lactate-oxidising sulfate-reducing Prokaryotes. High numbers of lactate- and acetate-oxidising iron-reducing Bacteria, as well as fermentative microorganisms, were also found. Apart from this study, there seems to be no large general culture-based studies of anaerobic chemotroph composition and distribution in soda lakes, although Jones et al. (1998b) record a number of different anaerobes from East African soda lakes, including Lake Magadi that affiliated with different *Clostridium* subgroups.

5.2.4 Chemolithotrophs

In addition to the vigorous cycling of carbon, there are also active sulphur and nitrogen cycles in soda lakes, where chemolithotrophs play key roles in the interconversions of different valence states of these elements. The study of these cycles has been mainly carried out not in African soda lakes, but in Russian soda lakes, mainly of the Kulunda Steppe region (reviewed by Grant and Sorokin 2011), although the first isolations of aerobic chemolithotrophic sulphur-oxidising Bacteria (Sorokin et al. 2001b) and sulfate-reducing Bacteria (Zhilina et al. 1997) included isolations from African soda lakes. The first chemolithotrophic nitrifying Bacteria (Sorokin et al. 1998) were also simultaneously co-isolated from African and Russian soda lakes. The role of *Ectothiorhodospira* and *Halorhodospira* spp. in the carbon cycle and their early isolations from African soda lakes

(Imhoff et al. 1978, 1979, 1981) has already been noted. These Bacteria also use reduced sulphur species as electron donors for photosynthesis, generating elemental sulphur and finally sulfate.

There are now a considerable number of soda lake chemolithotrophic sulfur oxidisers that have been described, mostly derived from Russian lakes. These oxidise a variety of reduced sulphur compounds including thiocyanate (Sorokin et al. 2002a; Sorokin and Kuenen 2005a, b). Some chemolithotrophic strains are denitrifying (Sorokin et al. 2003).

There has been some industrial interest in using sulfur oxidisers in the sustainable removal of sulphide from bio- and industrial waste gases, and the genome of *Thioalkalivibrio sulfidophilus* has been sequenced because of its relevance in this regard (Muyzer et al. 2011). The whole cells of these haloalkaliphiles have been used for the sustainable removal of toxic sulphur compounds from wastewater (de Graaff et al. 2011), gas streams (Sorokin et al. 2008e; van den Bosch et al. 2007) and paper mill effluent (Janssen et al. 2009).

The sulfur cycle is completed by sulfate-reducing Bacteria and around 18 different sulfate-reducers are recorded (Table 5.2), again mostly from Russian lakes although very similar organisms are found in the African lakes. An early study on sulfur reduction in the Wadi Natrun lakes was that of Abd-el-Malek and Ritz (1963). Some of the soda lakes in the USA and the Kulunda Steppe (Sorokin et al. 2012e) harbour sulfate reducers that will also reduce arsenate (Blum et al. 2012). Mono Lake in the USA has a complete arsenic cycle in that there are arsenite-oxidising facultative chemolithotrophs that grow aerobically or at the expense of nitrate reduction (Hoeft et al. 2007; Oremland et al. 2009, 2004, 2005). There are even magnetotactic Bacteria recorded in lakes in the USA, including Mono Lake, that are related to sulfate-reducing *Desulfonatronum* spp. (Lefèvre et al. 2011). In common with other aquatic environments, soda lakes harbour sulphur-reducing Bacteria, some of which are capable of the disproportionation of sulphur (Poser et al. 2013; Sorokin et al. 2008a, 2012e). The majority of Sulfidogenic Bacteria are probably not hydrogenotrophic (Grant and Sorokin 2011), although hydrogenotrophic sulfate reducers are described in Kenyan soda lakes by Zavarzin et al. (1999). Non-sulphur hydrogenotrophs are present in some lakes, including Kenyan lakes, such as *Alkalilimnicola* spp. (Hoeft et al. 2007; Sorokin et al. 2014a) and the more recently described *Fuchsiella alkaliacetigena* (Zhilina et al. 2012).

Both groups of nitrifying Bacteria have been described in soda lakes. Ammonification of nitrogenous substrates is widely carried out by the chemotrophic anaerobes in the lakes (Kevbrin et al. 1998; Zhilina et al. 2009a). Ammonia-oxidising organisms with a phylogenetic relationship to *Nitrosomonas* spp. are described by Sorokin (1998; Sorokin et al. 2001a) in African, Russian and Mongolian lakes, and the nitrite-oxidising *Nitrobacter alkalicus* is also present at several sites (Grant and Sorokin 2011). Sorokin et al. (2014a) point out that the high pH of the lakes is likely to cause volatilisation of NH_3 from the system so the N-cycle in soda lakes may be somewhat different from other lake systems. The capacity for nitrate reduction is widely distributed amongst the chemoorganotrophs with organisms such as *Halomonas* spp. (Duckworth et al. 2000; Jones and Grant 2000; Shapovalova et al. 2009), and there are additionally chemolithotrophs such as *Thioalkalivibrio nitratireducens* (Sorokin et al. 2003) that carry out this process. The nitrogen cycle is completed by nitrogen fixation capability, again widely distributed amongst the chemoorganotrophs, particularly the anaerobes (Sorokin et al. 2008d; Zavarzina et al. 2006, 2009). However, it is commonly assumed that Cyanobacteria are responsible for much of the nitrogen fixation in African soda lakes (Grant and Jones 2000; Jones and Grant 2000; Grant 2003) although the best-known bloom-forming species *Arthrospira platensis* is not recorded as having this attribute. It is also assumed that many of the anoxygenic phototrophic Bacteria participate in this part of the nitrogen cycle, backed up by the demonstration of the *nifH* gene in several soda lake anoxygenic phototroph representatives (Tourova et al. 2007, 2014).

5.3 Soda Lake Archaea

A quite different population of Prokaryotes is present in the concentrated brines of the very saline soda lakes and soda deserts throughout the world. Brines and soda flats are often characteristically red due to extensive blooms of microorganisms (10^6–10^7 mL^{-1}). The organisms most commonly cultured are both profoundly halophilic and alkaliphilic and are members of the haloarchaeal lineage. Although mass developments presumed to be Haloarchaea were documented in Wadi Natrun lakes (Imhoff et al. 1979), the first recorded isolation of such organisms was from Lake Magadi in Kenya (Tindall et al. 1980), where the authors devised an appropriate alkaline high salt medium. This was essentially devoid of divalent cations in order to mimic the chemistry of the lake brines and alkaline salterns where alkaline brines are essentially devoid of Mg^{2+} since the anion is removed by precipitation of insoluble $MgSO_4$ as the alkaline brine develops. Later, Soliman and Trüper (1982) isolated another strain from one of the Wadi Natrun lakes in Egypt and described *Halobacterium* (now *Natronomonas*) *pharaonis*. Tindall et al. (1984) extended the Lake Magadi study, proposing two new genera, *Natronobacterium* and *Natronococcus*, describing several new species, assigning the Soliman and Trüper isolate to *Natronobacterium* at that time. Alkaliphilic Haloarchaea continued to be classified in these two genera until the reclassification of these and other isolates by Kamekura et al. (1997) on the basis of 16S rRNA gene sequences. There are now more than 20 validly named alkaliphilic Haloarchaea in nine different genera (Table 5.6), rare examples of which, like the genus *Halorubrum*, harbour non-alkaliphilic members. Table 5.6 reveals that in recent years most of the isolates derive from other parts of the world, notably Inner Mongolia, Tibet and China. Some of the unnamed isolates described in Duckworth et al. (1996) can now be assigned to particular genera (Table 5.3).

Hypersaline environments are relatively low in oxygen due to reduced oxygen solubility (2 ppm in saturated NaCl, compared with 7 ppm in seawater) and have already been noted to have anaerobic anoxygenic Bacteria such as *Halorhodospira* spp. coexisting with these alkaliphilic Haloarchaea. There seems to be no alkaliphilic equivalent of the ubiquitous *Salinibacter* that coexists with Haloarchaea in neutral salt environments (Oren 2011). Haloarchaeal blooms in neutral solar salterns are known to promote crystallisation of halite and almost certainly also do so in alkaline salterns (Table 5.6).

Alkaliphilic Haloarchaea are aerobic heterotrophs (Grant and Jones 2000; Grant 2003), occasionally using other electron acceptors such as nitrate (Itoh et al. 2005; Xin et al. 2001). They presumably develop to high densities in hypersaline sites by deriving carbon and cell protein from phototrophic primary productivity by other haloalkaliphilic organisms like the anoxygenic phototrophic Bacteria and probably less salt-tolerant Cyanobacteria that form blooms during seasonal dilution in rainy seasons. The extremely alkaline conditions would appear to favour the use of sodium ions rather than protons as the coupling ion in energy generation, but analysis of the genome sequence of the alkaliphilic haloarchaeon *Natronomonas pharaonis* clearly identifies protons as the coupling ion between respiratory chain and ATP synthesis (Falb et al. 2005). Alkaliphilic Haloarchaea also seem not to have the capacity for photoheterotrophy since they, unusually amongst Haloarchaea, lack the light-mediated bacteriorhodopsin proton pump which is coupled with energy generation in other Haloarchaea, although they do possess other photoactive retinal pigments (Bivin and Stoeckenius 1986). It is worth noting that Grant and Ross (1986) suggested the possibility of anaerobic growth linked to Sulfidogenesis, and a recent (Sorokin et al. 2012c) review of Sulfidogenesis in the lakes of the Kulunda Steppe also points towards this possibility.

Table 5.6 Isolates of soda lake Archaea

Species	Original location	Reference
Halophiles		
Halalkalicoccus tibetensis	Lake Zabuye, Tibet	Xue et al. (2005)
Halobiforma nitratireducens	Unnamed lake, China	Xin et al. (2001), Hezayen et al. (2002)
Halobiforma sp. BNMIPTR	Sambhar Lake, India	Gupta et al. (2015)
Halorubrum alkaliphilum	Xinjiang Province, China	Feng et al. (2005)
Halorubrum luteum	Lake Chagannor, Inner Mongolia	Hu et al. (2008)
Halorubrum tibetense	Lake Zabuye, Tibet	Fan et al. (2004)
Halorubrum vacuolatum	Lake Magadi, Kenya	Mwatha and Grant (1993), Kamekura et al. (1997)
Haloglobiforma haloterrestri	Lake Chahannor, Inner Mongolia	Hezayen et al. (2002)
Halostagnicola bangensis	Lake Bange, Tibet	Corral et al. (2015)
Natronobacterium gregoryi*	Lake Magadi, Kenya	Tindall et al. (1984)
Natronobacterium texcoconense	Lake Texcoco, Mexico	Ruiz-Romero et al. (2013c)
Natronococcus occultus*	Lake Magadi, Kenya	Tindall et al. (1984)
Natronococcus amylolyticus	Lake Magadi, Kenya	Kanai et al. (1995)
Natronococcus roseus	Lake Chagannor, Inner Mongolia	Corral et al. (2013)
Natronomonas pharaonis*	Lake Magadi, Kenya	Tindall et al. (1984), Kamekura et al. (1997)
Natrialba magadii*	Lake Magadi, Kenya	Tindall et al. (1984), Kamekura et al. (1997)
Natrialba hulunbeirensis	Soda lake, Hulunbeir, Inner Mongolia	Xu et al. (2001)
Natrialba chahannaoensis	Lake Chahannor, Inner Mongolia	Xu et al. (2001)
Natronolimnobius baerhuensis	Lake Baer, Inner Mongolia	Itoh et al. (2005)
Natronolimnobius innermongolicus	Lake Baer, Inner Mongolia	Itoh et al. (2005)
Natronorubrum bangense	Bange Lake, Tibet	Xu et al. (1999)
Natronorubrum tibetense	Bange Lake, Tibet	Xu et al. (1999)
Natronorubrum sulfidifaciens	Aiding Lake, China	Cui et al. (2007)
Natronorubrum texcoconense	Lake Texcoco, Mexico	Ruiz-Romero et al. (2013b)
Methanogens		
Methanobacterium alcaliphilum	Wadi Natrun, Egypt	Worakit et al. (1986)
Methanocalculus sp. LA2	Lonar Lake, India	Surakasi et al. (2007)
Methanocalculus natronophilus	Lake Tanatar II, Russia.	Zhilina et al. (2013)
Methanoculleus sp. LH2	Lonar Lake, India	Surakasi et al. (2007)
Methanohalophilus oregonensis	Alkali Lake, USA	Boone (2001), Liu et al. (1990)
Methanolobus sp.	Soda lakes, Kulunda Steppe, Russia	Sorokin et al. (2015a, b)
Methanosaeta sp.	Lake Tanatar VI, Kulunda Steppe, Russia	Sorokin et al. (2015c)
Methanosalsum zhilinaeae*	Lake Magadi, Kenya; Wadi Natrun, Egypt	Mathrani et al. (1988), Boone and Baker (2001)
Methanosalsum natrophilum	Soda lakes, Kulunda Steppe, Russia	Sorokin et al. (2015a, b)

Isolates from African lakes in bold type (updated June 2015)
*Partial or complete genome sequence available

There is evidence of methanogenesis in soda lakes worldwide, including African lakes, although relatively few methanogens have been isolated and validly published (Table 5.6). Hydrolysis products of complex polymers to acetic acid, hydrogen and CO_2 by anaerobic decomposition (Grant et al. 1998; Grant and Sorokin 2011; Zavarzin et al. 1999) provide potential substrates for methanogens, although many of the methanogens isolated to date from hypersaline environments are methylotrophic, utilising compounds such as methanol and methylamine (Oren 2011). Obligately methylotrophic methanogens seem to have a distinct ecological advantage over other groups of methanogens as their absolute requirement for one-carbon (C1) compounds such as methanol and methylated amines avoids competition with sulfate-reducing Bacteria (Oremland et al. 1982). There are, however, examples of hydrogenotrophic methanogens such as *Methanocalculus natronophilus* (Zhilina et al. 2013) and *Methanobacterium alcaliphilum* (Worakit et al. 1986). Culture-based studies have isolated organisms affiliating with *Methanosarcina*, *Methanocalculus* and *Methanoculleus* spp. from Lonar Lake in India (Surakasi et al. 2007; Thakker and Ranade 2002). Other methylotrophic methanogens belonging to the genus *Methanohalophilus* (Kevbrin et al. 1997; Liu et al. 1990; Mathrani et al. 1988) isolated from several soda lake environments including Lake Magadi have now been reclassified as *Methanolobus oregonensis* and *Methanosalsum zhilinae* (Boone 2001; Boone and Baker 2001).

5.4 Direct Molecular Analysis of Soda Lake Microbiota

5.4.1 African Soda Lakes

There are relatively few reports of culture-independent analyses of African soda lakes, although arguably, the first account of such an analysis applied to a soda lake anywhere relates to Lake Magadi in Kenya (Grant et al. 1999). In this analysis, the lake brine was indirectly studied by examination of an adjacent alkaline saltern pan used to harvest salt from the source lake brine. This study examined the archaeal population by retrieval of 16S rRNA genes from DNA preparations that were obtained from cells filtered from saltern brine on site, stabilised in ice-cold buffer and extracted within 4 h. Sequences around 95 % related to known isolates from the site such as the haloalkaliphiles *Natronobacterium gregoryi* and *Halorubrum vacuolatum* were detected, but the study is noticeable for the retrieval of two sequences that were only 88–90 % related to any then known Haloarchaea. It was more than a decade later, following the development of direct metagenomic sequencing applied to hypersaline environments (Andrei et al. 2012; Ghai et al. 2011; Narasingarao et al. 2011), that a new class of as yet uncultivated Archaea, the *Nanohaloarchaea*, was described and to which these Lake Magadi sequences were shown to affiliate as the first recorded examples of this group.

Rees et al. (2004) extracted DNA from water and sediment samples taken from five soda lakes of the Kenyan–Tanzanian Rift Valley, the hypersaline Lake Magadi, and four more dilute lakes, namely, Crater Lake (Lake Sonachi), Lake Elmenteita, Lake Nakuru and Lake Bogoria. 16S rRNA genes were amplified by the polymerase chain reaction (PCR) and microbial diversity also studied using denaturing gradient gel electrophoresis (DGGE) of 16S rRNA gene amplicons. Several of the amplicon sequences had high identities, up to 99 %, with 16S rRNA genes of organisms previously isolated from soda lakes. Others were only distantly related, with identities as low as 82 %. Phylogenetic analysis of the more closely related amplicons indicated that sequences were related to haloarchaeal, *Bacillus/Clostridium*, *Rhodobacter*, *Thioalkalivibrio/Methylobacter*, *Bacteroidetes* groups and the enterobacteria/*Aeromonas/Vibrio* part of the *Gamma*-3 subdivision of the *Proteobacteria*.

Mwirichia et al. (2010b) studied haloarchaeal diversity in Lake Elmenteita and despite the relatively low salinity of this site recovered a considerable number of haloarchaeal sequences. Comparative sequence analysis of the archaeal clones affiliated them to a wide range of genera including *Natronococcus*, *Halovivax*, *Halobiforma*, *Halorubrum* and *Halalkalicoccus*. The highest percentage (46 %) of the clones, however, belonged to uncultured members of the domain Archaea in the order *Halobacteriales*.

Mwrichia et al. (2011) also followed up their previous culture-based study of Lake Elmenteita Bacteria using culture-independent methods, detecting up to 37 orders in the domain Bacteria. Most diversity was detected in *Firmicutes* with many *Bacillus*, *Lactobacillus* and *Clostridium* types recorded in sediments, whereas *Actinobacteria* were much less well represented. *Proteobacteria* dominated water samples with methylotrophs and non-sulphur phototrophs (*Alphaproteobacteria*), hydrogenotrophic Bacteria and methylotrophs (*Betaproteobacteria*), nitrifiers, sulfur oxidisers and anoxygenic phototrophs (*Gammaproteobacteria*) being represented. Sediments also had sulfate reducers (*Deltaproteobacteria*). Other sequences, mostly from sediments, were related to the *Bacteroidetes* group of Bacteria and *Spirochaetes*, indicating in common with other studies on the East African lakes great phylogenetic complexity.

Three Wadi Natrun lakes in Egypt have been studied by Wiegel and colleagues (Mesbah et al. 2007a). These analyses showed that the bacterial community was diverse. Nearly 42 % of the sequences that were retrieved were less than 90 % similar to previously described sequences or organisms in culture. The bacterial community in both the water and sediments of the lakes was dominated by clones affiliated with the low G + C *Firmicutes*, *Gammaproteobacteria* and *Bacteroidetes* groups, patterns that have been observed in previously described East African ecosystems (Rees et al. 2004). The bacterial community composition of the water and sediment of the same lake and of different lakes was significantly different. Anoxygenic phototrophic Bacteria in *Alpha*- and *Gammaproteobacteria* such as *Ectothiorhodospira*, *Halorhodospira* spp. and members of the *Rhodobacterales* were frequently detected in all lakes, in agreement with the historical observations of blooms of these Bacteria. Chemolithotrophic sulfur oxidisers and sulfate-reducing Bacteria were also widely detected. Examples of *Firmicutes* include members of the *Clostridiales* including *Alkalibacillus* and *Alkaliphilus* and members of the *Halanaerobiales*, mostly in sediments. *Spirochaetes*, *Verrucomicrobia* and *Chloroflexi* were detected only in Lake Hamra (Egypt), together with an absence of archaeal sequences, possibly reflecting the more oxygenated and dilute nature of this lake. The archaeal community composition of the other Wadi Natrun lakes appeared less diverse: all sequences fell into the phylum *Euryarchaeota*. A few sequences were related to the methanogen *Methanohalophilus oregonensis* and the haloarchaeon *Natronomonas pharaonis*, but the vast majority were largely unrelated to previously documented sequences, with the exception of some similarity to several of the unusual Lake Magadi salt pan sequences recorded by Grant et al. (1999), although not the nanohaloarchaeal sequences. An earlier study by Grant et al. (2004) looked at anaerobic enrichment cultures set up from sediment samples and soda soil samples obtained from lakes at the same site. Clones derived from lake sediments most closely matched *Clostridium* spp., *Natronincola histidinovorans*, *Halocella cellulolytica*, *Bacillus* spp. and the *Bacteroidetes* phylum. Similar clones were identified in the soda soils, but *Bacillus* spp. predominated. Many of the clones were most closely related to organisms already identified in alkaline or saline environments.

5.4.2 The Rest of the World's Soda Lakes

Following on from these pioneering studies on African soda lakes, other authors have recorded similar analyses on Inner Mongolian soda lakes; Qinghai Lake, China; soda lakes in the USA, Hungary, the Transbaikal region of Russia and the Kulunda Steppe region; Lonar Lake in India; Lake Texcoco in Mexico; and Lake Van in Turkey. Some of these studies have used conventional 16S rRNA gene clone library analyses to measure the overall diversity of Bacteria and Archaea, whereas others have looked at genes defining particular physiological groups such as autotrophs, sulfur oxidisers, aerobic methylotrophs, methanogens and phototrophic Bacteria.

Borsodi et al. (2013) looked at one of the shallow soda ponds located in the Kiskunság National Park, Hungary, and found that most of the total of 157 16S rRNA gene sequences affiliated with the anoxic phototrophic Bacteria in the genera *Rhodobaca* and *Rhodobacter* (*Alphaproteobacteria*), *Ectothiorhodospira* (*Gammaproteobacteria*) and *Heliorestis* (*Firmicutes*). Phylotypes related to the phylum *Bacteroidetes* formed the second most abundant group. Clones related to the mainly anaerobic and alkaliphilic Bacteria of the genera *Anoxynatronum* (*Firmicutes*), *Spirochaeta* (*Spirochaetes*) and *Desulfonatronum* (*Deltaproteobacteria*) were also found, revealing a very similar pattern to strains isolated from the African lakes.

Lopez-Garcia et al. (2005) carried out an analysis of the carbonate-rich microbialite columns produced by ingress of calcium-enriched waters into the soda Lake Van in Turkey. Although not a direct examination of the brine, molecular diversity studies based on 16S rRNA gene amplification revealed the presence of bacterial lineages affiliated to the *Alpha*-, *Beta*- and *Gammaproteobacteria*, the Cyanobacteria, the *Bacteroidetes* group, *Actinobacteria* and *Firmicutes*. Cyanobacteria and *Bacteroidetes* members were detected only in surface layers. The most abundant and diverse lineages were the *Firmicutes*. Again, there was considerable similarity to soda lake sites elsewhere in the world. The authors speculate that these heterotrophic Bacteria may play a crucial role in the formation of Lake Van Microbialites by locally promoting carbonate precipitation.

Wani et al. (2006) looked at the prokaryotic diversity of Lonar Lake (India). Most of the bacterial 16S rRNA gene clones were classified as *Firmicutes* (34 %). The other clones were grouped into *Proteobacteria* (29.5 %), *Actinobacteria* (6.8 %), *Deinococcus-Thermus* (4.5 %), *Bacteroidetes* (13.3 %), *Planctomycetes* (6.8 %), Cyanobacteria (4.5 %) and *Spirochaetes* (2.3 %). In the case of the archaeal 16S rRNA gene library, analysis of 25 randomly selected clones revealed the presence of 13 distinct phylotypes; five phylotypes were associated with *Crenarchaeota* and eight with *Euryarchaeota*. Most of the *Euryarchaeota* sequences were related to methanogens.

Ma et al. (2004a) looked at bacterial diversity in the Inner Mongolian Baer soda lake. Here the picture was slightly different in that the 16S rRNA gene library was dominated by Gram-negative Bacteria (24 % *Alphaproteobacteria*, 31 % *Betaproteobacteria*, 33 % *Gammaproteobacteria* and 2 % *Deltaproteobacteria*), with a low percentage of clones corresponding to Gram-positive Bacteria. Forty cloned sequences were similar to that of known bacterial isolates (>97 % sequence similarity), represented by the species of the genera *Brevundimonas*, *Comamonas*, *Alcaligenes*, *Stenotrophomonas* and *Klebsiella*. Eighteen cloned sequences showed less affiliation with known taxa (<97 % sequence similarity) and may represent novel taxa.

Jan-Roblero et al. (2004), in the first molecular study at this site, studied lagoons at Lake Texcoco in Mexico to determine any differences between an original lagoon and a restored lagoon. The restored Facultativa lagoon harboured mainly *Gamma*- and *Betaproteobacteria*, *Firmicutes* and several members of the *Halobacteriaceae*. The original Nabor Carrillo lagoon mainly included typical halophilic and alkaliphilic *Firmicutes*, *Gammaproteobacteria* and *Betaproteobacteria* similar to those found in other soda lakes. The lake site was revisited in 2009 (Valenzuela-Encinas et al. 2009) after a period of flooding of adjacent alkaline–saline soil. Three different soil sites were chosen on the basis of conductivity. 16S rRNA gene clone identification of the phyla *Proteobacteria*, *Actinobacteria*, *Acidobacteria*, Cyanobacteria, *Bacteroidetes*, *Firmicutes* and *Chloroflexi* showed that the bacterial communities were different. Species and genera of the *Rhizobiales*, *Rhodobacterales* and *Xanthomonadales* orders of the *Alpha*- and *Gamma*-subdivisions of *Proteobacteria* were found at all three sites. Species and genera of the *Rhodospirillales*, *Sphingobacteriales*, *Clostridiales*, *Oscillatoriales* and *Caldilineales* were found only in the most saline soil; *Sphingomonadales*, *Burkholderiales*, *Pseudomonadales* and *Actinomycetales* in the intermediate salinity soil; and *Myxococcales* and

Actinomycetales in the least saline soil. This study clearly showed the influence of the chemical composition on microbial populations, with only the most saline site bearing much resemblance to typical soda lake microbiota.

Ghozlan et al. (2006) analysed the microbial composition of Qinghai Lake in China in relation to depth by characterisation of 16S rRNA gene libraries. Whereas the *Alpha-*, *Beta-* and *Gammaproteobacteria* and *Bacteroidetes* were dominant at the water–sediment interface, *Firmicutes* became the predominant group in the anoxic sediments. They concluded that the presence of *Actinobacteria* and *Acidobacteria/Holophaga* in significant proportions in the Qinghai Lake sediments appeared to be unique. The archaeal diversity was much lower, and clone sequences could be grouped in the *Euryarchaeota* and *Crenarchaeota* domains. The archaeal clones were not related to any known cultures but to sequences previously found in methane-rich sediments.

Humayoun et al. (2003) also did a depth-distribution study, but in this case on Mono Lake (USA). 16S rRNA gene sequences fell into *Alpha-* and *Gammaproteobacteria* (6 % and 10 %, respectively), *Bacteroidetes* (19 %), *Actinobacteria* (25 %) and *Firmicutes* (*Bacillus* and *Clostridium*, 19 %). Twelve percent were identified as chloroplasts. The remaining 9 % represented *Beta-* and *Deltaproteobacteria*, *Verrucomicrobiales* and candidate divisions. They noted elevated diversity in anoxic bottom water samples relative to oxic surface water samples.

In addition to the general molecular analyses described above, there have been a number of analyses designed to look at particular physiological groups using group-specific 16S rRNA gene amplification or the amplification of genes characteristic of the physiology of the targeted group. Thus the methylotroph population of Lonar Lake was studied by group-specific 16S rRNA gene amplification (Antony et al. 2010; Surakasi et al. 2010) and methylotrophs in Mono Lake by the amplification of methane monooxygenase genes, whereby Lin et al. (2004, 2005) used a combination of ^{13}C-labelled CH_4 uptake, 16S rRNA gene amplification and methane monooxygenase gene amplification to study methylotrophs in Transbaikal soda lake sediments. All came to the conclusion that strains of *Methylomicrobium*, *Methylobacter*, *Methylomonas* and *Methylothermus* had assimilated the labelled methane. Sulfate-dependent methane oxidation in the same lake was studied by amplifying genes involved in sulphide reduction together with archaeal 16S rRNA genes, leading to the conclusion that in this particular lake, Archaea were not involved in the process (Scholten et al. 2005).

Sorokin et al. (2007b) looked at the sulphur-oxidising Bacteria in Soap Lake using specific amplification of 16S rRNA gene sequences and cultivation studies, showing that both approaches revealed the dominance of Bacteria belonging to the genus *Thioalkalimicrobium.*

Tourova et al. (2013) used amplification of a key enzyme of the Sox Pathway, sulfate thiohydrolase, to study sulphur oxidation in salt and soda lakes in southwestern Siberia and Egypt, showing that all clone libraries obtained from soda lakes containing the gene belonged to the genus *Thioalkalivibrio* in agreement with cultivation results. Earlier, Tourova et al. (2011) used RuBisCO genes as molecular markers to assess autotrophic microbial communities in soda lakes of the Kulunda Steppe, showing that diversity of autotrophic Bacteria in the studied sediment horizons was low with haloalkaliphilic Cyanobacteria and sulphur-oxidising Bacteria of the genus *Halorhodospira* predominant. In contrast, in lake waters, halophilic chemoautotrophic *Halothiobacillus* and *Thioalkalivibrio* were found along with photoautotrophic Bacteria of the genus *Ectothiorhodosinus* and Cyanobacteria. Kovaleva et al. (2011) carried out a similar analysis in lake sediments from soda lakes of the Kulunda Steppe (Altai, Russia) and from hypersaline alkaline lakes of Wadi Natrun (Egypt), showing that hypersaline lakes were dominated by members of the extremely haloalkaliphilic sulphur-oxidising *Ectothiorhodospiraceae*, the chemolithotrophic *Thioalkalivibrio* and the phototrophic *Halorhodospira*. The less saline soda lakes from the Kulunda Steppe showed domination of members of a novel phototrophic

Chromatiales lineage and another only distantly related to the *Thiomicrospira* cluster.

Asao et al. (2011) used a photosynthesis reaction centre protein gene to carry out an analysis of phototrophs in Soap Lake and showed that a significant diversity of purple Bacteria inhabit this soda lake with close relatives of several of the phylotypes obtained from cultured species such as alkaliphilic purple sulfur Bacteria of the families *Chromatiaceae* and *Ectothiorhodospiraceae* detected. Several other more divergent phylotypes were also identified.

Recently, Antony et al. (2012a), using a combination of 16S rRNA gene library analyses and functional gene analyses, looked at the methanogen population in Lonar Lake (India). 16S rRNA gene sequences related to potentially novel Archaea within the orders *Methanosarcinales* and *Methanomicrobiales* were obtained together with a significant fraction of sequences representing uncultivated euryarchaeotes. To identify the active methylotrophic Archaea involved in methanogenesis, mRNA transcripts of methyl-coenzyme M reductase enzyme genes were retrieved from methanol-consuming and methane-emitting sediment microcosms at two different time points. Reverse-transcription PCR, qPCR, DGGE fingerprint and clone library analysis showed that the active Archaea were closely related to *Methanolobus* spp.

The question of whether there are significant differences in microbial populations between different hypersaline lakes and if so what determines this was approached in the study of Pagaling et al. (2009). They used cultivation-independent methods to investigate the prokaryotic biogeography of the water column in six salt lakes in Inner Mongolia, China, and a salt lake in Argentina. These lakes had different salt compositions and pH values and were at variable geographic distances, on both local and intercontinental scales, which allowed the exploration of the microbial community composition within the context of both contemporary environmental conditions and geographic distance. Biotic similarity matrices were used in combination with environmental similarity matrices and a distance matrix in the Mantel test to discover which factors significantly influenced biotic similarity. It was shown that archaeal biogeography was influenced by contemporary environmental factors alone (Na^+, CO_3^{2-} and HCO_3^- ion concentrations; pH and temperature). Bacterial biogeography was influenced both by contemporary environmental factors (Na^+, Mg^{2+}, HCO_3^- ion concentrations and pH) and also by geographic distance. In essence, lakes with similar chemistries relatively close to each other have similar populations, but great geographical distances between lakes have an effect on at least some of the microbial population structure when comparisons are made.

5.5 The Thermal Sites

5.5.1 Thermophilic Alkaliphiles

The EARS is an area of active volcanism and some of the soda lakes are fed by heated groundwater rising from deep aquifers along established fault lines (McCall 2010). The waters become rapidly enriched in carbonate on reaching the surface, although the composition of the springs is more dilute than that of the lakes that they feed. The warm and hot springs vary with the nature of the location and lake water levels. Some springs are submerged (e.g. at Lake Elmenteita and Lake Bogoria), but on the shores of Lake Bogoria, these are often strong flows from boiling (96 °C at this altitude) pools and 'geysers' especially at the Loburu Delta (Renaut et al. 2013) (Fig. 5.5), while the Magadi–Natron basin is characterised by weakly flowing 'soda seeps' with a higher conductivity (total dissolved salts) and a south-to-north temperature gradient along the Nguruman Escarpment ranging from 30 °C in the south to 84 °C at the northern end of Little Lake Magadi (Nasikie Engida). There are also hot springs in Ethiopia (Lake Shala and Lake Chitu) and thermal sites on the northern shores of Lake Natron (Tanzania). Collectively these springs and their drainage channels provide remarkable ecosystems with fascinating gradients of pH, salinity and temperature that have attracted almost no microbiological studies.

Fig. 5.5 Hot springs at the Loburu Delta, Lake Bogoria, Kenya (December 1996)

Fig. 5.6 'Stromatolite-like' algal mats in a warm spring drainage channel at the southern end of Lake Magadi. The red mat at the centre of the picture had been inverted to reveal the underside

The weakly flowing warm springs in the Magadi basin at the foot of the Nguruman Escarpment are complex environments characterised by extensive 'algal' mats whose true nature has never been explored and are part of an intricate food web involving at least a brine fly and a species of spider (BE Jones and WD Grant, personal observations). The 'algal' mats in the run-off channels from warm springs at the southern end of Lake Magadi are altogether quite different, resembling 'proto-stromatolites' (Fig. 5.6). All these mats must make a considerable contribution to the carbon input into these hypersaline soda lakes, but this has never been quantified and requires further research

Evidence of true [relict] stromatolites has been demonstrated at Lake Bogoria (McCall 2010), and cyanobacterial mats undergoing calcification have been recorded (Renaut et al. 2013) associated with the shoreline hot springs. Many of the hot springs on the shores of Lake Bogoria vigorously emerge at temperatures close to 96 °C through unconsolidated sediments into incised pools 2 m or so deep or spouting from vents surrounded by travertine terraces (Renaut et al. 2013) (Fig. 5.5) and discharge into narrow, sinuous drainage channels into the lake.

A noticeable feature of the travertine terraces and drainage channels is the extensive, thick microbial mats that range in colour from bright orange close to the hot spring vent (Fig. 5.5) to green and dark brown or black gelatinous mats or streaming filaments in the cooler waters. The upper temperature limit for these mats appears to be 68–70 °C. They are composed of *Chloroflexus* spp., a member of the anoxygenic phototrophic green non-sulphur Bacteria and filamentous or coccoid Cyanobacteria (Grant and Tindall 1986 and unpublished observations) that have been assigned to *Oscillatoria*, *Phormidium*, *Pseudanabaena*, *Spirulina*, *Arthrospira*, *Leptolyngbya*, *Planktotrichoides*, *Synechococcus*, *Xenococcus* and *Nostocales* spp. (Dadheech et al. 2013; Krienitz et al. 2003). A recent metagenomic study based on cyanobacterial 16S rRNA gene sequencing suggested that almost 15 % of the Lake Bogoria hot spring Cyanobacteria could not be assigned to any known taxon (Dadheech et al. 2013).

In the past we have speculated that the microbial mediation of geochemical processes at Lake Bogoria hot springs and EASL is likely to be considerable (Jones et al. 1994). However, Renaut et al. (2013) discounted the involvement of the *Chloroflexus*–Cyanobacteria associations in the travertine formation at the Lake Bogoria hot springs since this occurs at >80 °C, which is above the maximum temperature for photoautotrophic [cyano]bacteria, although microbial involvement in calcite formation at lower temperatures could not be ruled out. Mineralisation of Cyanobacteria mats such as *Phormidium* during wetter periods when lake levels are high has been implicated in the formation of stromatolites (Casanova 1986; Renaut et al. 2013), and biomediated silica precipitation is responsible for silicified microbial crusts and microstromatolites, also observed in the Lake Bogoria hot springs (Renaut et al. 1998).

Renaut et al. (1998, 2013) have speculated on the involvement of hyperthermophiles in biomineralisation processes in the Lake Bogoria hot springs; however the lack of evidence may be due to the shortage of thermophilic, alkaliphilic isolates either from EASL or indeed elsewhere (Table 5.7). Thermophilic and alkaliphilic microbes, which might also be halophilic and anaerobic, are often referred to as polyextremophiles, and these have been extensively reviewed in recent years (Bowers and Wiegel 2011; Bowers et al. 2009; Canganella and Wiegel 2014; Kumar and Satyanarayana 2013; Mesbah and Wiegel 2008, 2011, 2012; Wiegel 2000, 1998).

Many of these organisms described as polyextremophiles are not very remarkable; many of them neither represent the most extreme thermophilic nor the most alkaliphilic of microorganisms, and it appears that pH and temperature optimum are inversely related (Mesbah and Wiegel 2008). Thermoalkaliphiles (or alkalithermophiles) having temperature optima >65 °C with pH optima >9.5 are not common, and this is thought to reflect physiological constraints (van de Vossenberg et al. 1995), but it could equally be due to a lack of exploration. Here we only consider optimum temperatures for growth >50 °C and optimum growth pH >8.5 (Table 5.7). The number of well-characterised thermophilic and alkaliphilic organisms originating from soda lake environments appears quite restricted (Table 5.7). Although other highly alkaline natural environments exist, for example, due to the geochemical process of serpentinisation (Suzuki et al. 2014), they have an altogether different chemistry and do not harbour the same organisms as soda lakes. In addition, serpentinisation sites are not hot. An exceptional geothermally heated alkaline environment is the Rehai geothermal site at Tengchong, China (Pagaling et al. 2012), but again the geochemistry is different from EARS hot soda springs, and

Table 5.7 Thermophilic alkaliphiles

Genus and species	Origin	pH_{opt}	T_{opt} (°C)	Comments	Reference
Archaea					
Natrialba hulunbeirensis	Soda lake, Inner Mongolia, China	9.0	50	Aerobe, halophilic	Xu et al. (2001)
'Natronolimnobius aegyptiacus'	Wadi Natrun, Egypt	9.5	55	Aerobe, halophilic	Mesbah and Wiegel (2012)
Thermococcus acidaminovorans	Marine hydrothermal vent, Vulcano, Italy	9	85	Anaerobes. Optimum pH may be considerably lower at the optimum growth temperature (Wiegel 1998)	Dirmeier et al. (1998)
Thermococcus alkaliphilus	Marine hydrothermal vent, Vulcano, Italy	9	85		Keller et al. (1995)
Thermococcus fumicolans	Deep sea hydrothermal vent, North Fiji basin	8.5	85		Godfroy et al. (1996)
Bacteria					
Peptoclostridium paradoxum [basonym *Clostridium*]	Sewage plant, USA	9.3	56	Anaerobe	Li et al. (1993), Yutin and Galperin (2013)
Peptoclostridium thermoalcaliphilum [basonym *Clostridium*]	Sewage plant, USA	9.2	51	Anaerobe	Li et al. (1994), Yutin and Galperin (2013)
Anaerobranca horikoshii	Hot spring, Yellowstone NP, USA	8.5	57	Anaerobe, alkalitolerant	Engle et al. (1995)
Anaerobranca gottschalkii	Lake Bogoria, Kenya	9.5	55	Anaerobe, alkalitolerant	Prowe and Antranikian (2001)
Thermosyntropha lipolytica	Lake Bogoria, Kenya	8.5	63	Anaerobe, alkalitolerant	Svetlitshniyi et al. (1996)
Thermobrachium celere	Ubiquitous	8.2	66	Anaerobe, alkalitolerant	Engle et al. (1996)
'Thermopallium natronophilum'	Lake Bogoria, Kenya	8.8–9.5	70	Anaerobe	This chapter
Natranaerobius thermophilus	Wadi Natrun, Egypt	9.5	53	Anaerobe, halophilic	Mesbah et al. (2007)
Natranaerobius trueperi	Wadi Natrun, Egypt	9.5	52	Anaerobe, halophilic	Mesbah and Wiegel (2009)
'Natranaerobius grantii'	Lake Magadi, Kenya	9.5	46	Anaerobe, halophilic, thermotolerant	Bowers et al. (2008)
'Natranaerobius jonesii'	Lake Magadi, Kenya	10.5	66	Anaerobe, halophilic	Bowers et al. (2008)
'Caloramator halophilus'	Salt flats, Nevada, USA	9.2	64	Facultative aerobe	Mesbah and Wiegel (2011)
Natronovirga wadinatrunensis	Wadi Natrun, Egypt	9.9	51	Anaerobe, halophilic	Mesbah and Wiegel (2009)
Halonatronum saccharophilum	Lake Magadi, Kenya	8.5	55	Anaerobe, halophilic	Zhilina et al. (2001)
Dichotomicrobium thermohalophilum	Solar Lake, Israel	8.5	50	Aerobe, halophile	Hirsh and Hoffman (1989)

(continued)

Table 5.7 (continued)

Genus and species	Origin	pH_{opt}	T_{opt} (°C)	Comments	Reference
Caldalkalibacillus thermarum	Alkaline hot spring, Tengchong, China	8.5	60	Aerobe, obligate alkaliphile	Xue et al. (2006)
Caldalkalibacillus (Thermalkalibacillus) uzonensis	Hot spring, Kamchatka, Russia	8.3	52	Aerobe, alkalitolerant	Zhao et al. (2006, 2008)

Microorganisms originating from African lakes are noted in bold type

a quite different spectrum of microorganisms has been recorded at the Tengchong site (Hedlund et al. 2012; Pagaling et al. 2012; Xue et al. 2006). All these observations suggest that the Rift Valley thermoalkaliphiles are endemic or at least restricted in their biogeography, but there are too few similar habitats known and too few studies performed to be able to draw firm conclusions.

5.5.2 A Novel Alkaliphilic Member of the *Thermotogales*

Duckworth et al. (1996) reported the isolation and 16S rRNA gene sequence of two thermophilic and alkaliphilic strains from the hydrothermal springs at Lake Bogoria that were clearly related to the genera *Fervidobacterium* and *Thermotoga* in the bacterial order *Thermotogales* for which the name '*Thermopallium natronophilum*' was proposed. However, other than the occasional mention in conference abstracts (Meijer et al. 1996; Wiegel and Kevbrin 2004), no formal description of the new genus was ever published. This may in part be due to the commercial interest since much of the information was published in the patent literature (Jones et al. 1997). The strain was deposited as DSM 9460 for the purpose of patent procedures and investigated for hydrolytic enzymes (Thompson 1998).

Samples were collected anaerobically in December 1992 (Duckworth et al. 1996) (Table 5.8) and enriched on a low conductivity (5–11 mS cm^{-1})-modified *Thermus* medium (TM), pH 9.5 (Table 5.9).

Positive enrichments were obtained only on the low-salt samples 7A and 9A, and pure strains were isolated from single colonies by plating out on TM medium plus 2.5 g L^{-1} starch–agar at 70 °C under N_2. The morphology of the cells is shown in Fig. 5.7, and a brief description of the new natronophilic thermophile is provided in Table 5.9.

Phylogenetic analysis based on the sequencing of the gene coding for the small subunit rRNA indicated that these strains are more closely related to *Fervidobacterium* than to *Thermotoga* (Table 5.10) and are probably representatives of a new genus within the *Thermotogales*. The full sequence of the 16S rRNA gene of '*Thermopallium natronophilum*' strain Tg9A is deposited at EMBL under accession number X91822 and GenBank under U37021. However, the taxonomy of the *Thermotogales* has recently been revised (Bhandari and Gupta 2014), and it is unclear how strain Tg9A fits in as the bacterium is not readily available.

Table 5.8 Anaerobic soda lake samples for enrichment

Sample	Source	Temp °C	pH	Conductivity (mS cm^{-1})	Equiv. salt (%)	Equiv. Na_2CO_3 (gL^{-1})
7A	Bogoria (hot water channel where it enters the lake)	66	9.5	8	0.5	8.5
8A	Bogoria (hot water channel)	85	10	18	1.1	19
9A	Bogoria (geyser pool)	96	8.5	4.8	0.3	5.1
17A	Little Magadi (hot pool)	60–80	9.8	35	2.1	37
18A	Little Magadi (hot water stream)	60–80	9.8	37	2.2	39

Table 5.9 Description of *Thermopallium natronophilum*

Enrichment medium (g L^{-1}): tryptone, 2.0; yeast extract, 1.0; NaCl, 2.0; $Na_2S.9H_20$, 0.5; Na_2SiO_3, 0.004; Castenholz salts (Williams and Da Costa 1992), 1.27 g L^{-1}, at 70 °C under a head space gas phase of $N_2/CO_2/H_2$ (94:5:1 by volume).

Morphology. Straight rod-shaped bacterium with rounded ends surrounded by a sheath-like outer structure ballooning over both ends of the cell similar to *Thermotoga* (Huber and Stetter 1992) Fig. 5.7. During growth the outer sheath-like structure often becomes spherical at one end of the cell. The cells usually occur singly or in short chains up to 3 cells long. During growth cells may become curved or irregular in shape, sometimes forming aggregates. In stationary phase the cells become coccoid with one or more cells within a large spherical body. Forms round, shiny, whitish, translucent colonies on alkaline nutrient agar containing carbonate. Gram reaction, negative; KOH reaction, negative; aminopeptidase reaction, negative. Strict anaerobe.

In the presence of sodium dodecylsulphate 1 % (w/v) both the cells and sheath-like structure disintegrate within a few seconds. There is no effect observed with lysozyme (10 mg mL^{-1}).

Thermophile. Temperature growth range: 52 °C–78 °C.

No growth at 50 °C or 79 °C.

Maximum cell yield at 63 °C–64 °C.

Maximum growth rate at 70 °C.

Alkaliphile. pH range for growth: 7.2 to >10.5.

pH optimum for growth: 8.8–9.5.

'Natronophile', requires HCO_3^- or CO_3^{2-} for growth. No growth is obtained on media adjusted to alkaline pH values with NaOH.

Requires low conductivity/low salt growth medium.

Optimum NaCl concentration for growth = 1 % w/v.

At NaCl concentrations above 2 % w/v cell yield and growth rate is severely reduced.

No growth above 5 % w/v NaCl.

Nutrition. Grows (70 °C) on glucose, galactose, maltose, xylose (weak), acetate, pyruvate (weak), glycine, glycerol, cellulose (weak), xylan, starch, casein, gelatine, tryptone, yeast extract and olive oil. No growth on ribose, formate, lactate, propionate, glutamate or ethanol.

Antibiotics (100 μg mL^{-1}). Sensitive (70 °C) to ampicillin, chloramphenicol, polymyxin B, streptomycin, tetracycline. Not sensitive to cycloheximide or gentamycin.

Growth is inhibited by H_2, which may be relieved by adding sulphur (S^0) but not by Na_2S.

No growth under 100 % CO_2 either with or without S^0. Grows best under N_2.

G + C content: 36.3±0.9 mol% (n=2) (HPLC).

Habitat. Found in low conductivity (<15 mS cm^{-1}) alkaline (carbonate) hot springs.

5.5.3 Metagenome Study of Alkaline Hot Springs

The numerous hot springs at the foot of the Nguruman Escarpment at the northern end of Little Lake Magadi (Nasikie Engida) coalesce to form a substantial perennial stream of water feeding this endorheic lake. The springs appear to be permanent, but their position is ephemeral as the braided water channels vary from year to year probably in response to changing ground-water levels. The springs emerge from the gravel

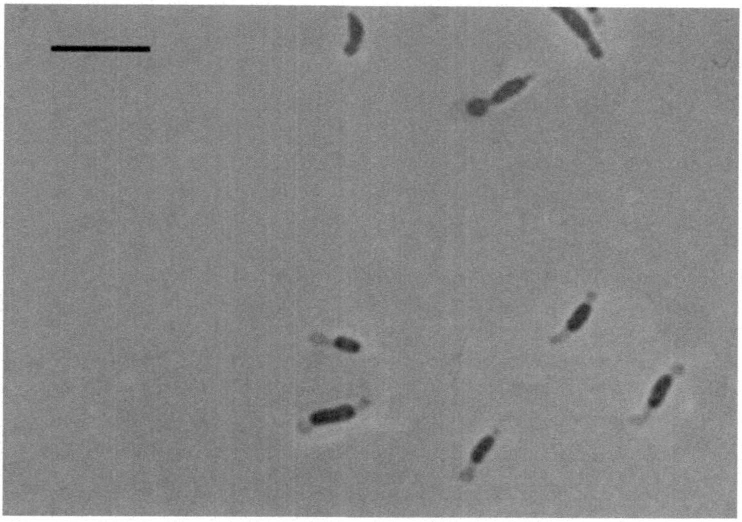

Fig. 5.7 Phase-contrast photomicrograph showing the morphology of '*Thermopallium natronophilum*'. *Bar* represents 10 μm

Table 5.10 16S rRNA sequence homology (%) comparison between *Thermopallium natronophilum* strain Tg9A and some other members of the order *Thermotogales*

	Thermopallium natronophilum	*Fervidobacterium islandicum*	*Fervidobacterium nodosum*	*Thermosipho africanus*	*Thermotoga maritima*
Thermopallium natronophilum	100				
Fervidobacterium islandicum	90.0	100			
Fervidobacterium nodosum	89.9	95.2	100		
Thermosipho africanus	85.9	87.4	86.1	100	
Thermotoga maritima	83.6	85.1	84.1	88.6	100

substratum to form hot pools with temperatures 80°–84 °C which overflow into water channels exhibiting interesting thermal and salinity gradients. We were unable to isolate the '*Thermopallium*' strains from these springs, probably because the salinity is too high (Table 5.8), and we find no molecular signature for them in the metagenome (Table 5.11).

Even so, to the experienced observer (WD Grant and BE Jones, personal observations), there is ample macroscopic evidence of microbial life in and around these hot soda springs. There are green, yellow and red algal mats of various types covering the sediments or as green filaments attached to the coarse gravel. The upper temperature limit for these photosynthetic microbes appears to be 68 °C. At higher temperatures up to 76 °C, there are thick bacterial masses of long 'grey filaments' anchored to gravel and small rocks to prevent wash away in the swiftly flowing water. In the past we have attempted to cultivate these 'grey filaments', but without success. The analysis of extracted DNA was similarly unsuccessful. However, a recent, but incomplete, metagenomic analysis of a 80 °C hot pool at Little Lake Magadi (DA Cowan and BE Jones, unpublished data) gives a clue to the identity of these 'grey filaments', since molecular signatures related to *Thermocrinis* and *Hydrogenobacter* 16S rRNA

Table 5.11 Some 16S rRNA gene sequences detected in the metagenome of a Little Lake Magadi hot spring (D. A. Cowan and B. E. Jones, unpublished results)

Identity [Genbank Acc. No.]
Alkalimonas delamerensis 1E1 (94 %) [NR_044879]
Anaerobacillus sp., *A. alkalilacustris* Z-0521 (97 %) [DQ675454], *A. alkalidiazotrophicus* (95 %) [DQ675454]. *Anerobacillus arseniciselenatis* (92 %) [AJ865469]
Aquificales sp. (94 %), *Thermocrinis* sp. (90 %), *Hydrogenobacter thermophiles* (90 %) [DQ791382]
Halomonas sp. (92 %) [JX240570], *H. nitritophilus* (92 %) [AJ309564], *H. campisalis* (91 %) [AY935689]
Lake Nakuru isolate 20 N1 (99 %) [X92128] *A. haloalkalitolerans?, Aliidiomarina sanyensis* (100 %) [JX981923]
Marinobacter sp. (92 %) [KF908999], *M. alkaliphilus* (95 %) [EU440994]
Natroniella sp. (94 %), [*N. sulfidgena* (94 %), *Halonatrum saccharophilum* (94 %)] [JX240706]
Neptunomonas sp. (80 %) [KC160633]
Nitrincola sp. (98 %) [FN395250], *N. lacisaponensis* (96 %) [AY567473]
Thermus sp. (90 %) [AY699382]
Uncultured bacterial clone (hot spring Taiwan) (95 %) [AB696025]
Uncultured bacterial clone from alkaline hot spring Papua New Guinea related to *Acetothermia* candidate div. OP1 (91 %) [JF935228]
Uncultured bacterial clone (*Bacilliaceae*) (93 %) [EF422410]
Uncultured bacterial clone from Inner Mongolia Xiarinur soda lake (94 %) [GU083678]
Uncultured candidate div. OP1 (*Acetothermia*) (91 %) [JN882362]
Candidatus *Contubernalis alkalaceticum* clone Z-7904 (94 %) [DQ124682]
Uncultured bacterial clone Soap Lake, USA (90 %) [EU645189]

genes were recovered (Table 5.11). Most of the currently described members of *Thermocrinis* and *Hydrogenobacter* were isolated from low ionic strength hydrothermal systems with a fairly neutral pH in continental solfatara fields (Huber and Eder 2006), conditions quite different from those encountered in the Little Lake Magadi hot springs (Table 5.7) with pH 9.5–10 and conductivity >35 mS cm^{-1}. Given these circumstances, it is highly likely that the 'grey filaments' represent a novel alkaliphilic taxon within the *Aquificales*, and many of the physiological features of the organism can be predicted. It is probable that the bacterium is a microaerophilic, chemolithotrophic primary producer fixing CO_2 using the reductive TCA cycle, using H_2 as sole electron donor and maybe additional sulphur and thiosulfate (Eder and Huber 2002; Stöhr et al. 2001). Since the origins of primary productivity in the trona and concentrated brines of the hypersaline lakes of the Magadi–Natron basin are generally unclear (Jones et al. 1998), these 'grey filaments' could be important for our understanding of trophic relationships in these concentrated brine lakes. However, the organism needs to be isolated for study in pure culture. The same can be said of all the other signature sequences recovered in the metagenome (Table 5.11).

5.6 Soda Lake Viruses

Suttle (2007) noted that viruses are the most pervasive biological entities in aquatic systems, the majority being bacteriophages. Accordingly, it is not unreasonable to suppose that the viral population in soda lakes must make a significant contribution to microbial interactions and development, given the often recorded dense blooms of Prokaryotes, particularly Cyanobacteria and Haloarchaea. Very little attention, however, has been paid to the soda lake environment, although there is a considerable body of work on hypersaline virioplankton, particularly haloarchaeal bacteriophages, in neutral hypersaline lakes (reviewed by Baxter et al. 2011). Techniques applied to neutral hypersaline sites include virus isolation by cultivation, phase partition and direct genome isolation by pulsed-field electrophoresis. Population studies have employed filtration concentration techniques together with transmission electron microscopy, and, recently, a metagenomic approach has been applied to these environments. Counts of virioplankton at 3–6 × 10^9 mL^{-1} have been recorded as compared to cell counts of around 6 × 10^7 mL^{-1} of the brine (Baxter et al. 2011), and a considerable number of different morphotypes have been observed, some fusiform types characteristic of haloarchaeal host types and others typical head and tail phages.

Studies on African soda lakes are restricted to investigations of factors involved in population fluctuations of Cyanobacteria and Lesser Flamingos (*Phoeniconaias minor*). Recently, a cyanophage was described that infects *Arthrospira platensis* in mass cultures (Jacquet et al. 2013). Based on morphology and molecular investigation, it was proposed that the virus belongs to the cyanopodovirus group with a capsid and short tail. Fluctuations in cyanobacterial (*Arthrospira* spp.) mass seem to be the overwhelming factor influencing flamingo distribution (Kaggwa et al. 2013a; Krienitz and Kotut 2010) and the cause of episodes of disappearance. Peduzzi et al. (2014) screened for virioplankton and, specifically, cyanophage infection in Lake Nakuru (Kenya) over a period of several months in 2009. Virioplankton numbers ranged from 1.2 × 10^9 to 7.0 × 10^9 mL^{-1}. Flamingo numbers corresponded with the peak of cyanobacterial biomass in the summer months. During the crash in cyanobacterial biomass in late summer (and the departure of flamingos), visible signs of cyanophage infection were detected by transmission electron microscopy in almost 25 % of cells, leading the authors to conclude that a virus-mediated loss of the most important food source triggered the collapse of the flamingo population. In another study, Kaggwa et al. (2013b) correlated morphological change in the *Arthrospira* populations in the Kenyan lakes Nakuru and Bogoria in part with cyanophage numbers and speculate that morphotype patterns may be useful in predicting the best outcomes for commercial '*Spirulina*' production.

Rather more work has been carried out on virus populations in soda lakes elsewhere in the world, notably Mono Lake in California. Brum et al. (2005) and Brum and Steward (2010) looked at spatial and temporal variability of Prokaryotes, viruses and viral infections in the lake and morphologically characterised some of the viruses using electron microscopy. Different viral assemblages were detected in different layers of the lake and fusiform viruses were not found, in contrast to neutral hypersaline sites. The authors conclude that the virioplankton was different from other aquatic environments. Viral numbers were comparable to those observed in neutral hypersaline lakes and Prokaryote mortality ranged from 3.7 % to 16 % over the year, with viral numbers correlated with Prokaryote numbers, although not with algal/cyanobacterial numbers as judged by chlorophyll *a* measurements. Jiang et al. (2004) used pulsed-field gel electrophoresis to reveal a diverse viral community in the same lake with genome sizes ranging from 14 to 400 kb with most in the 30–60 kb size range, typical of viruses infecting Prokaryotes. Pinkart and Storrie-Lombardi (2007) presented preliminary data demonstrating the presence of a diverse group of phage integrases in Soap Lake (Washington State) and speculate on the role of bacteriophages in horizontal gene transfer in soda lakes.

Krammer et al. (2008) deduced the potential influence of viruses in shallow Austrian soda ponds, measuring isotope incorporation by the prokaryotic population following simulated rainwater dilution that compared the addition of distilled water and artificial lake water with filtered lake water, concluding that viral lysis was reduced by dilution with rainwater.

Grant et al. (2011), in their review of several Inner Mongolian soda lakes, describe two lytic viruses isolated from the weakly mineralised soda lake, Lake Bagaejinor. The host is a *Halorubrum* sp. One of these viruses (BJ1) has had its sequence determined and it has very low sequence identity to any previously described virus (Pagaling et al. 2007). BJ1 has an icosahedral head and tail morphology and most likely a linear double-stranded DNA genome exhibiting terminal redundancy. Its genome sequence has

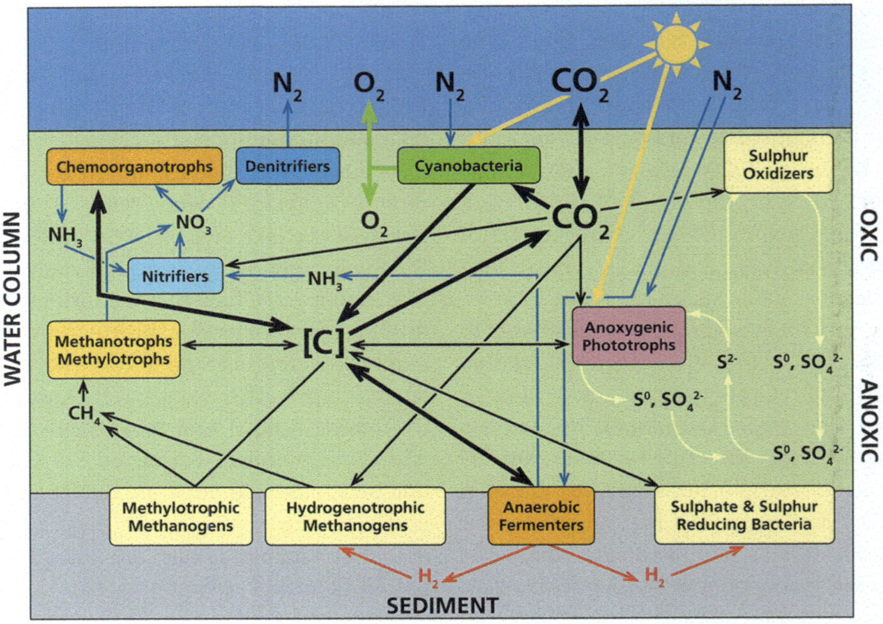

Fig. 5.8 Trophic relationships and element cycling in the soda lake ecosystem (after Grant 2006)

42,271 base pairs with a G + C content of ~65 mol%. The genome of BJ1 is predicted to encode 70 ORFs, including one for a tRNA. Fifty of the 70 ORFs had no identity to database entries; 20 showed sequence identity matches to archaeal and to haloarchaeal viruses. ORFs possibly coding for an origin of replication complex, integrase, helicase and structural capsid proteins were identified. Evidence for viral integration was obtained. The second virus BJ2 has been partially sequenced: 44 contigs containing a total of 97,602 base pairs with a G + C content of 51 mol%. It has no discernible sequence identity to the BJ1 virus. This virus is more fully described by Pagaling (2007).

5.7 Conclusions

It is clear that the soda lake community has attained sufficient diversity to represent all the main branches of the complete phylogenetic tree of the Prokaryotes and additionally all the major functional trophic groups required to close up major biogeochemical cycles. Although we do not know in detail all the aspects of the microbial ecosystem in soda lakes, it is possible to make predictions regarding the roles played by some of the groups in the cycling of nutrients since there are parallels with other aquatic systems. Despite the conditions imposed by high alkalinity (and sometimes high salinity), soda lakes contain large numbers of Cyanobacteria, fermentative and autotrophic haloanaerobes including spirochaetes and sulfate-reducing Bacteria, many aerobic chemoorganotrophs including *Firmicutes* and *Actinobacteria* Gram-positives, *Proteobacteria*, aerobic methylotrophs, sulphur cycle and nitrogen cycle chemolithotrophs, thermophiles, aerobic Haloarchaea and anaerobic methanogenic Archaea, with concomitant active carbon, nitrogen and sulphur cycling under aerobic and anaerobic conditions. Accordingly, the current understanding of trophic relations amongst the microbial community in African soda lakes (and other soda lakes around the world) is predicted and depicted with some confidence in Fig. 5.8.

Acknowledgements Both authors contributed equally to the preparation and writing of this chapter. We thank all the past students and colleagues for their collaboration with our studies in soda lakes. In particular, we express our sincere appreciation of the collaboration with the late Prof. Wanjiru Mwatha, some of whose unpublished PhD studies we present here posthumously. Our soda lake research was supported financially by numerous grants from the European Commission, and we are grateful to Gist-Brocades BV; Genencor International, Inc.; and Danisco A/S for their interest and support.

References

Abd-el-Malek Y, Ritz SG (1963) Bacterial sulphate reduction and the development of alkalinity. III experiments under natural conditions in the Wady Natrun. J Appl Bacteriol 26:20–26

Akhwale JK, Göker M, Rohde M, Schumann P, Klenk H-P, Boga HI (2015a) *Belliella kenyensis* sp. nov., isolated from an alkaline lake. Int J Syst Evol Microbiol 65:457–462

Akhwale JK, Göker M, Rohde M, Spröer C, Schumann P, Klenk H-P, Boga HI (2015b) *Streptomyces alkaliphilus* sp. nov., isolated from sediments of Lake Elmenteita in the Kenyan Rift Valley. Anton van Leeuwenhoek 107:1249–1259

Alazard D, Badillo C, Fardeau M-L, Cayol J-L, Thomas P, Roldan T, Tholozan J-L, Ollivier B (2007) *Tindallia texcoconensis* sp. nov., a new haloalkaliphilic bacterium isolated from lake Texcoco, Mexico. Extremophiles 11:33–39

Al-Sharawi SZRI, Abdelnasser SS, El-Shatoury EH, Gebreel HM, Eldiwany A (2013) A new low molecular mass alkaline cyclodextrin glucanotransferase from *Amphibacillus* sp. NRC-WN isolated from an Egyptian soda lake. Electron J Biotechnol 16:3–8

Amoozegar AM, Shahinpei A, Sepahy AA, Makhdoumi-Kakhki A, Seyedmahdi SS, Schumann P, Ventosa A (2014) *Pseudomonas salegens* sp. nov., a halophilic member of the genus *Pseudomonas* isolated from a wetland. Int J Syst Evol Microbiol 64:3565–3570

Andrei AŞ, Banciu HL, Oren A (2012) Living with salt: metabolic and phylogenetic diversity of archaea inhabiting saline ecosystems. FEMS Microbiol Lett 330:1–9

Antony CP, Kumaresan D, Ferrando L, Boden R, Moussard H, Scavino AF et al (2010) Active methylotrophs in the sediments of Lonar Lake, a saline and alkaline ecosystem formed by meteor impact. ISME J 4:1470–1480

Antony CP, Murrell JC, Shouche YS (2012a) Molecular diversity of methanogens and identification of *Methanolobus* sp. as active methylotrophic Archaea in Lonar crater lake sediments. FEMS Microbiol Ecol 81:43–51

Antony CP, Doronina NV, Boden R, Trotsenko YA, Shouche YS, Murrell JC (2012b) *Methylophaga lonarensis*, a novel moderately haloalkaliphilic methylotroph isolated from the soda lake sediments of a meteorite impact crater. Int J Syst Evol Microbiol 62:1613–1618

Antony CP, Kumaresan D, Hunger S, Drake HL, Murrell JC, Shouche YS (2013) Microbiology of Lonar Lake and other soda lakes. ISME J 7:468–476

Asao M, Jung DO, Achenbach LA, Madigan MT (2006) *Heliorestis convoluta* sp. nov., a coiled, alkaliphilic heliobacterium from the Wadi Natrun, Egypt. Extremophiles 10:403–410

Asao M, Takaichi S, Madigan MT (2007) *Thiocapsa imhoffii*, sp. nov., an alkaliphilic purple sulfur bacterium of the family Chromatiaceae from Soap Lake, Washington (USA). Arch Microbiol 188:665–675

Asao M, Pinkart HC, Madigan MT (2011) Diversity of extremophilic purple phototrophic bacteria in Soap Lake, a Central Washington (USA) Soda Lake. Environ Microbiol 13:2146–2157

Asao M, Takaichi S, Madigan MT (2012) Amino acid-assimilating phototrophic heliobacteria from soda lake environments: *Heliorestis acidaminivorans* sp. nov. and 'Candidatus *Heliomonas lunata*'. Extremophiles 16:585–595

Baesman SM, Stolz JF, Kulp TR, Oremland RS (2009) Enrichment and isolation of *Bacillus beveridgei* sp. nov., a facultative anaerobic haloalkaliphile from Mono Lake, California, that respires oxyanions of tellurium, selenium, and arsenic. Extremophiles 13:695–705

Banciu H, Sorokin DY (2013) Adaptation in haloalkaliphiles and natronophilic bacteria. In: Seckbach J, Oren A, Stan-Lotter H (eds) Polyextremophiles: life under multiple forms of stress. Springer, Dordrecht, pp 121–178

Banciu H, Sorokin DY, Galinski EA, Muyzer G, Kleerebezem R, Kuenen JG (2004) *Thialkalivibrio halophilus* sp. nov., a novel obligately chemolithoautotrophic, facultatively alkaliphilic, and extremely salt-tolerant, sulfur-oxidizing bacterium from a hypersaline alkaline lake. Extremophiles 8:325–334

Banciu HL, Sorokin DY, Tourova TP, Galinski EA, Muntyan MS, Kuenen JG et al (2008) Influence of salts and pH on growth and activity of a novel facultatively alkaliphilic, extremely salt-tolerant, obligately chemolithoautotrophic sulfur-oxidizing Gammaproteobacterium *Thioalkalibacter halophilus* gen. nov., sp. nov. from South-Western Siberian soda lakes. Extremophiles 12:391–404

Baxter BK, Mangalea MR, Willcox S, Sabet S, Nagoulat M-V, Griffith JD (2011) Haloviruses of Great Salt Lake: a model for understanding viral diversity. In: Ventosa A, Oren A, Ma Y (eds) Halophiles and hypersaline environments: current research and future trends. Springer, Tokyo, pp 173–190

Bhandari V, Gupta R (2014) Molecular signatures for the phylum (class) *Thermotogae* and a proposal for its division into three orders (*Thermotogales*, *Kosmotogales* ord. nov. and *Petrotogales* ord. nov.) containing four families (*Thermotogaceae*, *Fervidobacteriaceae* fam. nov., *Kosmotogaceae* fam. nov. and *Petrotogaceae* fam. nov.) and a new genus

Pseudothermotoga gen. nov. with five new combinations. Anton van Leeuwenhoek 105:143–168

Bivin DB, Stoeckenius W (1986) Photoactive retinal pigments in haloalkaliphilic bacteria. J Gen Microbiol 132:2167–2177

Blum JS, Burns BA, Buzzelli J, Stolz JF, Oremland RS (1998) *Bacillus arsenicoselenatis* sp. nov., and *Bacillus selenitireducens* sp. nov.: two haloalkaliphiles from Mono Lake, California that respire oxyanions of selenium and arsenic. Arch Microbiol 171:19–30

Blum JS, Han S, Lanoil B, Witte B, Saltikov C, Tabita FR, Langley S, Beveridge TJ, Jahnke L, Oremland RS (2009) Ecophysiology of "*Halarsenatibacter silvermanii*" strain SLAS-1 T, gen. nov., sp. nov., a facultative chemoautotrophic arsenate respirer from salt-saturated Searles Lake, California. Appl Environ Microbiol 75:1950–1960

Blum JS, Kulp TR, Han S, Lanoil B, Saltikov CW, Stolz JF et al (2012) *Desulfohalophilus alkaliarsenatis* gen. nov., sp. nov., an extremely halophilic sulfate- and arsenate-respiring bacterium from Searles Lake, California. Extremophiles 16:727–742

Boldareva EN, Bryantseva IA, Tsapin A, Nelson K, Sorokin DY, Tourova TP et al (2007) The new bacteriochlorophyll *a*-containing bacterium *Roseinatronobacter monicus* sp. nov. from the hypersaline soda Mono Lake (California, United States). Microbiology 76:82–92

Boldareva EN, Akimov VN, Boychenko VA, Stadnichuk IN, Moskalenko AA, Makhneva ZK et al (2008) *Rhodobaca barguzinensis* sp. nov., a new alkaliphilic purple nonsulfur bacterium isolated from a soda lake of the Barguzin Valley (Buryat Republic, Eastern Siberia). Microbiology 77:206–218

Boldareva EN, Tourova TP, Kolganova TV, Moskalenko AA, Makhneva ZK, Gorlenko VM (2009a) *Roseococcus suduntuyensis* sp. nov., a new aerobic bacteriochlorophyll *a*-containing bacterium isolated from a low-mineralized soda lake of Eastern Siberia. Microbiology 78:92–101

Boldareva EN, Moskalenko AA, Makhneva ZK, Tourova TP, Kolganova TV, Gorlenko VM (2009b) *Rubribacterium polymorphum* gen. nov., sp. nov., a novel alkaliphilic nonsulfur purple bacterium from an Eastern Siberian soda lake. Microbiology 78:732–740

Boltyanskaya YV, Antipov AN, Kolganova TV, Lysenko AM, Kostrikina NA, Zhilina TN (2004) *Halomonas campisalis*, an obligatorily alkaliphilic, nitrous oxide-reducing denitrifier with a molybdenum cofactor-lacking nitrate reductase. Microbiology 73:271–278

Boltyanskaya YV, Kevbrin VV, Lysenko AM, Kolganova TV, Turova TP, Osipov GA (2007) *Halomonas mongoliensis* sp. nov., and *Halomonas kenyensis* sp. nov., new haloalkaliphilic denitrifiers capable of N_2O reduction, isolated from soda lakes. Microbiology 76:739–747

Boone DR (2001) Genus IV. *Methanohalophilus* Paterek and Smith 1988, 122^{VP}. In: Boone DR, Castenholz RW, Garrity GM (eds) Bergey's manual of systematic bacteriology. The Archaea and the deeply branching and phototrophic bacteria, vol 1, 2nd edn. Springer, New York, pp 281–283

Boone DR, Baker CC (2001) Genus VI. *Methanosalsum* gen. nov. In: Boone DR, Castenholz RW, Garrity GM (eds) Bergey's manual of systematic bacteriology. The Archaea and the deeply branching and phototrophic bacteria, vol 1, 2nd edn. Springer, New York, pp 287–289

Borsodi AK, Khayer B, Szabo G, Marialigeti K, Schumann P, Toth EM (2008a) *Nesterenkonia pannonica* sp. nov., a novel bacterium isolated from a Hungarian soda lake. Unpublished. GenBank Acc. No. FM179648.1. http://www.ncbi.nlm.nih.gov/nuccore/FM179648.1 Accessed 21 May 2015

Borsodi AK, Marialigeti K, Szabo G, Palatinszky M, Pollák B, Kéki Z et al. (2008b) *Bacillus aurantiacus* sp nov., an alkaliphilic and moderately halophilic bacterium isolated from Hungarian soda lakes. Int J Syst Evol Microbiol 58:845–851

Borsodi AK, Pollak B, Keki Z, Rusznyák A, Kovács AL, Spröer C et al (2011) *Bacillus alkalisediminis* sp. nov., an alkaliphilic and moderately halophilic bacterium isolated from sediment of extremely shallow soda ponds. Int J Syst Evol Microbiol 61:1880–1886

Borsodi AK, Knab M, Czeibert K, Marialigeti K, Voeroes L, Somogyi B (2013) Planktonic bacterial community composition of an extremely shallow soda pond during a phytoplankton bloom revealed by cultivation and molecular cloning. Extremophiles 17:575–584

Bowers KJ, Wiegel J (2011) Temperature and pH optima of extremely halophilic archaea: a mini-review. Extremophiles 15:119–128

Bowers KJ, Mesbah NM, Wiegel J (2008) *Natranaerobius 'grantii'* and *Natranaerobius 'jonesii'*, spp. nov., two anaerobic halophilic alkaliphiles isolated from the Kenyan-Tanzanian Rift. Abst Gen Meet Am Soc Microbiol Boston, MA I-007

Bowers KJ, Mesbah NM, Wiegel J (2009) Biodiversity of polyextremophilic Bacteria: does combining the extremes of high salt, alkaline pH and elevated temperature approach a physicochemical boundary for life? Saline Syst 5:9–15

Brown SD, Begemann MB, Mormile MR, Wall JD, Han CS, Goodwin LA et al (2011) Complete genome sequence of the haloalkaliphilic, hydrogen-producing bacterium *Halanaerobium hydrogenoformans*. J Bacteriol 193:3682–3683

Brum JR, Steward GF (2010) Morphological characterization of viruses in the stratified water column of alkaline, hypersaline Mono Lake. Microb Ecol 60:636–643

Brum JR, Steward GF, Jiang SC, Jellison R (2005) Spatial and temporal variability of prokaryotes, viruses, and viral infections of prokaryotes in an alkaline, hypersaline lake. Aquat Microb Ecol 41:247–260

Bryantseva I, Gorlenko VM, Kompantseva EI, Imhoff JF, Süling J, Mityushina L (1999) *Thiorhodospira sibirica* gen. nov., sp. nov., a new alkaliphilic purple sulfur bacterium from a Siberian soda lake. Int J Syst Evol Microbiol 49:697–703

Bryantseva IA, Gorlenko VM, Kompantseva EI, Achenbach LA, Madigan MT (2000a) *Heliorestis daurensis*, gen. nov., sp. nov., an alkaliphilic rod-to-coiled-shaped phototrophic heliobacterium from a Siberian soda lake. Arch Microbiol 172:167–174

Bryantseva IA, Gorlenko VM, Kompantseva EI, Tourova TP, Kuznetsov BB, Osipov GA (2000b) Alkaliphilic heliobacterium *Heliorestis baculata* sp. nov. and emended description of the genus *Heliorestis*. Arch Microbiol 174:283–291

Bryantseva IA, Gorlenko VM, Kompantseva EI, Imhoff JF (2000c) *Thioalkalicoccus limnaeus* gen. nov., sp. nov., a new alkaliphilic purple sulfur bacterium with bacteriochlorophyll *b*. Int J Syst Evol Microbiol 50:2157–2163

Bryantseva IA, Tourova TP, Kovaleva OL, Kostrikina NA, Gorlenko VM (2010) *Ectothiorhodospira magna* sp. nov., a new large alkaliphilic purple sulfur bacterium. Microbiology 79:780–790

Bryantseva IA, Gaisin V, Gerlenko VM (2015) *Rhodobaculum claviforme* gen. nov., sp. nov., a new alkaliphilic nonsulfur purple bacterium. Microbiology 84:247–255

Canganella F, Wiegel J (2014) Anaerobic thermophiles. Life 4:77–104

Cao S-J, Qu J-H, Yang J-S, Sun Q, Yuan H-L (2008) *Halolactibacillus alkaliphilus* sp. nov., a moderately alkaliphilic and halophilic bacterium isolated from a soda lake in Inner Mongolia, China, and emended description of the genus *Halolactibacillus*. Int J Syst Evol Microbiol 58:2169–2173

Cao S-J, Ou J-H, Yuan H-L, Li B-Z (2010) *Salsuginibacillus halophilus* sp. nov., a halophilic bacterium isolated from a soda lake. Int J Syst Evol Microbiol 60:1339–1343

Carini S, Bano N, LeCleir G, Joye SB (2005) Aerobic methane oxidation and methanotroph community composition during seasonal stratification in Mono Lake, California (USA). Environ Microbiol 7:1127–1138

Carrasco IJ, Marquez MC, Xue Y, Ma Y, Cowan DA, Jones BE et al (2007a) *Salsuginibacillus kocurii* gen. nov., sp. nov., a moderately halophilic bacterium from soda-lake sediment. Int J Syst Evol Microbiol 57:2381–2386

Carrasco IJ, Marquez MC, Xue Y, Cowan DA, Jones BE, Grant WD et al (2007b) *Bacillus chagannoriensis* sp. nov., a moderate halophile from a soda lake in Inner Mongolia, China. Int J Syst Evol Microbiol 57:2084–2088

Casanova J (1986) East African rift stromatolites. In: Frostick LE, Renaut RW, Reid I, Tiercelin JJ (eds) Sedimentation in the African Rifts, Geological Society, London, Special Publications 25. Blackwell Scientific, Oxford, pp 201–210

Cherni NE, Solovieva ZV, Fedorova VD, Kondratieva EN (1969) Ultrastructure of cells of two species of purple sulfur bacteria. Mikrobiologiya 38:479–484 (in Russian)

Corral P, Gutiérrez MC, Castillo AM, Domínguez M, Lopalco P, Corcelli A et al (2013) *Natronococcus roseus* sp. nov., a haloalkaliphilic archaeon from a hypersaline lake. Int J Syst Evol Microbiol 63:104–108

Corral P, Corcelli A, Ventosa A (2015) *Halostagnicola bangensis* sp. nov., an alkaliphilic haloarchaeon from a soda lake. Int J Syst Evol Microbiol 65:754–759

Cui HL, Tohty D, Liu HC, Liu SJ, Oren A, Zhou PJ (2007) *Natronorubrum sulfidifaciens* sp. nov., an extremely haloalkaliphilic archaeon isolated from Aiding salt lake in Xin-Jiang, China. Int J Syst Evol Microbiol 57:738–740

Dadheech PK, Glöckner G, Casper P, Kotut K, Mazzoni CJ, Mbedi S et al (2013) Cyanobacterial diversity in the hot springs, pelagic and benthic habitats of a tropical soda lake. FEMS Microbiol Ecol 85:389–401

De Graaff M, Bijmans MF, Abbas B, Euverink GJ, Muyzer G, Janssen AJ (2011) Biological treatment of refinery spent caustics under halo-alkaline conditions. Bioresour Technol 102:7257–7264

Delgado O, Quillaguaman J, Bakhtiar S, Mattiasson B, Gessesse A, Hatti-Kaul R (2006) *Nesterenkonia aethiopica* sp. nov., an alkaliphilic, moderate halophile isolated from an Ethiopian soda lake. Int J Syst Evol Microbiol 56:1229–1232

Dimitriu PA, Shukla SK, Conradt J, Márquez MC, Ventosa A, Maglia A et al (2006) *Nitrincola lacisaponensis* gen. nov., sp. nov., a novel alkaliphilic bacterium isolated from an alkaline, saline lake. Int J Syst Evol Microbiol 55:2273–2278

Dimitriu PA, Pinkart HC, Peyton BM, Mormile MR (2008) Spatial and temporal patterns in the microbial diversity of a meromictic soda lake in Washington State. Appl Environ Microbiol 74:77–88

Dirmeier R, Keller M, Hafenbradl D, Braun FJ, Rachel R, Burggraf S, Stetter KO (1998) *Thermococcus acidaminovorans* sp. nov., a new hyperthermophilic alkalophilic archaeon growing on amino acids. Extremophiles 2:109–114

Doronina N, Darmaeva T, Trotsenko Y (2003a) *Methylophaga natronica* sp. nov., a new alkaliphilic and moderately halophilic, restricted-facultatively methylotrophic bacterium from Soda Lake of the Southern Transbaikal Region. Syst Appl Microbiol 26:382–389

Doronina NV, Darmaeva TD, Trotsenko YA (2003b) *Methylophaga alcalica* sp. nov., a novel alkaliphilic and moderately halophilic, obligately methylotrophic bacterium from an East Mongolian saline soda lake. Int J Syst Evol Microbiol 53:223–229

Duckworth AW, Grant WD, Jones BE, van Steenbergen R (1996) Phylogenetic diversity of soda lake alkaliphiles. FEMS Microbiol Ecol 9:181–191

Duckworth AW, Grant S, Grant WD, Jones BE, Meijer D (1998) *Dietzia natronolimnaios* sp. nov., a new member of the genus *Dietzia* isolated from an East African Soda Lake. Extremophiles 2:359–366

Duckworth AW, Grant WD, Jones BE, Meijer D, Marquez MC, Ventosa A (2000) *Halomonas magadii* sp. nov., a new member of the genus *Halomonas* isolated from a soda lake of the East African Rift Valley. Extremophiles 4:53–60

Eder W, Huber R (2002) New isolates and physiological properties of the *Aquificales* and description of *Thermocrinis albus* sp. nov. Extremophiles 6:309–318

Engle M, Li Y, Woese C, Wiegel J (1995) Isolation and characterization of a novel alkalitolerant thermophile, *Anaerobranca horikoshii* gen. nov., sp. nov. Int J Syst Bacteriol 45:454–461

Engle M, Li Y, Rainey F, DeBlois S, Mai V, Reichert A et al (1996) *Thermobrachium celere* gen. nov., sp. nov., a rapidly growing thermophilic, alkalitolerant, and proteolytic obligate anaerobe. Int J Syst Bacteriol 46:1025–1033

Falb M, Pfeiffer F, Palm P, Rodewald K, Hickmann V, Tittor J et al (2005) Living with two extremes: conclusions from the genome sequence of *Natronomonas pharaonis*. Genome Res 15:1336–1343

Fan HP, Xue YF, Ma Y, Ventosa A, Grant WD (2004) *Halorubrum tibetense* sp. nov., a novel haloalkaliphilic archaeon from Lake Zabuye in Tibet, China. Int J Syst Evol Microbiol 57:1137–1142

Feng J, Zhou P, Zhou YG, Liu S-J, Warren Rhodes K (2005) *Halorubrum alkaliphilum* sp. nov., a novel haloalkaliphile isolated from a soda lake in Xinjiang, China. Int J Syst Evol Microbiol 55:149–152

Fritze D (1996) *Bacillus haloalkaliphilus* sp. nov. Int J Syst Evol Microbiol 46:98–101

Gao MA, Wang L, Chen SF, Zhou YG, Liu HC (2010) *Salinicoccus kekensis* sp. nov., a novel alkaliphile and moderate halophile isolated from Keke Salt Lake in Qinghai, China. Anton van Leeuwenhoek 98:351–357

Garnova ES, Zhilina TN, Tourova TP, Lysenko AM (2003) *Anoxynatronum sibiricum* gen.nov., sp. nov. alkaliphilic saccharolytic anaerobe from cellulolytic community of Nizhnee Beloe (Transbaikal region). Extremophiles 7:213–220

Garnova ES, Zhilina TN, Tourova TP, Kostrikina NA, Zavarzin GA (2004) Anaerobic, alkaliphilic, saccharolytic bacterium *Alkalibacter saccharofermentans* gen. nov., sp. nov. from a soda lake in the Transbaikal region of Russia. Extremophiles 8:309–316

Ghai R, Pašić L, Fernandéz AB, Mizuno CM, McMahon KD, Papke RT et al (2011) New abundant microbial groups in aquatic hypersaline environments. Sci Rep 1:135. doi:10.1038/srep00135

Ghozlan H, Dong HL, Zhang GX (2006) Microbial diversity in sediments of saline Qinghai Lake, China: linking geochemical controls to microbial ecology. Microb Ecol 51:65–82

Godfroy A, Meunier JR, Guezennec J, Lesongeur F, Raguénès G, Rimbault A et al (1996) *Thermococcus fumicolans* sp. nov., a new hyperthermophilic archaeon isolated from a deep-sea hydrothermal vent in the North Fiji Basin. Int J Syst Bacteriol 46:1113–1119

Gorlenko VM, Bryantseva IA, Panteleeva EE, Tourova TP, Kolganova TV, Makhneva ZK et al (2004) *Ectothiorhodosinus mongolicum* gen. nov., sp. nov., a new purple bacterium from a soda lake in Mongolia. Microbiology 73:66–73

Gorlenko VM, Bryantseva IA, Rabold S, Tourova TP, Rubtsova D, Smirnova E et al (2009) *Ectothiorhodospira variabilis* sp. nov., an alkaliphilic and halophilic purple sulfur bacterium from soda lakes. Int J Syst Evol Microbiol 59:658–664

Gorlenko VM, Buryukhaev SP, Matyugina EB, Borzenko SV, Namsaraev ZB, Bryantseva IA et al (2010) Microbial communities of the stratified soda Lake Doroninskoe (Transbaikal region). Microbiology 79:390–401

Gorlenko VM, Bryantseva IA, Kalashnikov AM, Gaisin VA, Sukhacheva MV, Gruzdev DS, Kuznetsov BB (2014) Candidatus '*Chloroploca asiatica*' gen. nov., sp. nov., a new mesophilic filamentous anoxygenic phototrophic bacterium. Microbiology 83:838–848

Govender L, Naidoo L, Setati ME (2013) *Nesterenkonia suensis* sp. nov., a haloalkaliphilic actinobacterium isolated from a salt pan. Int J Syst Evol Microbiol 63:41–46

Grant WD (1992) The alkaline environment. In: Lederberg J (ed) Encyclopaedia of microbiology, vol 1. Academic Press, New York, pp 73–80

Grant WD (2003) Half a lifetime in soda lakes. In: Ventosa A (ed) Halophilic microorganisms. Springer, Heidelberg, pp 17–32

Grant WD (2006) Alkaline environments and biodiversity. In: Gerday C, Glansdorff N (eds) Extremophiles, encyclopedia of life support systems (EOLSS), developed under the auspices of the UNESCO. Eolss Publishers, Oxford, Available via http://www.eolss.net. Accessed 21 May 2015

Grant WD, Heaphy S (2010) Metagenomics and recovery of enzyme genes from alkaline saline environments. Environ Technol 31:1135–1143

Grant WD, Horikoshi K (1992) Alkaliphiles: ecology and biotechnological applications. In: Herbert RA, Sharpe RJ (eds) Molecular biology and biotechnology of extremophiles. Blackie, Glasgow, pp 143–162

Grant WD, Jones BE (2000) Alkaline environments. In: Lederberg J (ed) Encyclopaedia of microbiology, vol 1, 2nd edn. Academic, New York, pp 126–133

Grant WD, Mwatha WE (1989) Alkaliphiles: ecology and diversity. In: Hattori T (ed) Recent advances in microbial ecology. Japan Scientific Societies Press, Tokyo, pp 64–68

Grant WD, Ross HNM (1986) The ecology and taxonomy of halobacteria. FEMS Microbiol Rev 39:9–15

Grant WD, Sorokin DY (2011) Distribution and diversity of soda lake alkaliphiles. In: Horikoshi K,

Antranikian G, Bull AT, Robb FT, Stetter KO (eds) Extremophiles handbook. Springer, New York, pp 27–54

Grant WD, Tindall BJ (1980) The isolation of alkalophilic bacteria. In: Gould GW, Corry JEL (eds) Microbial growth and survival in extremes of environment. Academic, London, pp 27–36

Grant WD, Tindall BJ (1986) The alkaline saline environment. In: Herbert RA, Codd GA (eds) Microbes in extreme environments. Academic, London, pp 25–54

Grant WD, Mills AA, Schofield AK (1979) An alkalophilic species of *Ectothiorhodospira* from a Kenyan soda lake. J Gen Microbiol 110:137–142

Grant WD, Mwatha WE, Jones BE (1990) Alkaliphiles: ecology, diversity and applications. FEMS Microbiol Rev 75:255–270

Grant WD, Gemmell RT, McGenity TJ (1998) Halophiles. In: Horikoshi K, Grant WD (eds) Extremophiles: microbial life in extreme environments. Wiley-Liss, New York, pp 93–132

Grant S, Grant WD, Jones BE, Kato C, Li L (1999) Novel archaeal phylotypes from an East African alkaline saltern. Extremophiles 3:139–145

Grant S, Sorokin DY, Grant WD, Jones BE, Heaphy S (2004) Phylogenetic analysis of Wadi Natrun soda lake cellulose enrichment cultures and identification of cellulase genes from these cultures. Extremophiles 8:421–424

Grant WD, Pagaling E, Márquez MC, Gutiérrez MC, Cowan DA, Jones BE et al (2011) The hypersaline LAKES of Inner Mongolia: the MGAtech project. In: Ventosa A, Oren A, Ma Y (eds) Halophiles and hypersaline environments: current research and future trends. Springer, Tokyo, pp 65–109

Groth I, Schumann P, Rainey FA, Martin K, Schuetze B, Augsten K (1997) *Bogoriella caseilytica* gen. nov., sp. nov., a new alkaliphilic actinomycete from a soda lake in Africa. Int J Syst Bacteriol 47:788–794

Gupta M, Aggarwal S, Kumar N, Choudhury B (2015) Isolation and characterization of a protease-producing novel haloalkaliphilic bacterium *Halobiforma* sp. strain BNMIITR from Sambhar lake in Rajasthan, India. Ann Microbiol 65:677–686

Hassan HA, Rizk NMH, Hefnawy MA, Awad AM (2012) Isolation and characterization of a halophilic aromatic and chloroaromatic degrader from Wadi Natrun soda lake. Life Sci J 9:1565–1570

Hedlund BP, Cole JK, Williams JA, Hou W, Zhou E, Li W, Dong H (2012) A review of the microbiology of the Rehai geothermal field in Tengchong, Yunnan Province, China. GSF 3:273–288

Hezayen FF, Tindall BJ, Steinbüchel A, Rehm BHA (2002) Characterization of a novel halophilic archaeon, *Halobiforma haloterrestris* gen. nov., sp. nov., and transfer of *Natronobacterium nitratireducens* to *Halobiforma nitratireducens* comb. nov. Int J Syst Evol Microbiol 52:2271–2280

Hirsch P, Hoffman B (1989) *Dichotomicrobium thermohalophilum*, gen. nov., spec. nov., budding prosthecate bacteria from the solar lake (Sinai) and some related strains. Syst Appl Microbiol 11:291–301

Hoeft SE, Switzer BJ, Stolz JF, Tabita FR, Witte B, King GM et al (2007) *Alkalilimnicola ehrlichii* sp. nov., a novel arsenite-oxidizing, haloalkaliphilic gammaproteobacterium capable of chemoautotrophic or heterotrophic growth with nitrate or oxygen as the electron acceptor. Int J Syst Evol Microbiol 57:504–512

Hoover RB, Pikuta EV, Bej AK, Marsic D, Whitman W, Tang J et al (2003) *Spirochaeta americana* sp. nov., a new haloalkaliphilic, obligately anaerobic spirochaete isolated from soda Mono Lake in California. Int J Syst Evol Microbiol 53:815–821

Horikoshi K (2006) Alkaliphiles: genetic properties and application of enzymes. Kodansha\Springer, Tokyo \Berlin

Horikoshi K (2011) Enzymes isolated from alkaliphiles. In: Horikoshi K, Antranikian G, Bull AT, Robb FT, Stetter KO (eds) Extremophiles handbook. Springer, New York, pp 163–182

Hu LF, Pan HL, Xue YF, Cowan DA, Jones BE, Grant WD et al (2008) *Halorubrum luteum* sp. nov., isolated from Lake Chagannor, Inner Mongolia, China. Int J Syst Evol Microbiol 58:1705–1708

Huber R, Eder W (2006) Aquificales. In: Dworkin et al (eds) Prokaryotes. Springer, New York, pp 925–938

Huber R, Stetter KO (1992) The order *Thermotogales*. In: Balows A et al (eds) The prokaryotes. Springer, New York, pp 3809–3815

Humayoun SB, Bano N, Hollibaugh JT (2003) Depth distribution of microbial diversity in Mono Lake, a meromictic soda lake in California. Appl Environ Microbiol 69:1030–1042

Ibrahim ASS, El-Tayeb MA, Elbadawi YB, Al-Salamah AA (2011) Bioreduction of Cr (VI) by potent novel chromate resistant alkaliphilic *Bacillus* sp. strain KSUCr5 isolated from hypersaline soda lakes. Afr J Biotechnol 10:7207–7218

Ibrahim ASS, Al-Salamah AA, El-Tayeb MA, El-Badawi YB, Antranikian G (2012) A novel cyclodextrin glycosyltransferase from alkaliphilic *Amphibacillus* sp. NPST-10: purification and properties. Int J Mol Sci 13:10505–10522

Imhoff JF, Süling J (1996) The phylogenetic relationship among *Ectothiorhodospiraceae*: a reevaluation of their taxonomy on the basis of 16S rDNA analyses. Arch Microbiol 165:106–113

Imhoff JF, Trüper HG (1977) *Ectothiorhodospira halochloris* sp. nov., a new extremely halophilic phototrophic bacterium containing bacteriochlorophyll *b*. Arch Microbiol 114:115–121

Imhoff JF, Haswa F, Trüper HG (1978) Isolation of extremely halophilic phototrophic bacteria from the alkaline Wadi Natrun, Egypt. Arch Hydrobiol 84:381–388

Imhoff JF, Sahl HG, Soliman GSH, Trüper HG (1979) The Wadi Natrun chemical composition and microbial mass developments in alkaline brines of eutrophic desert lakes. Geomicrobiol J 1:219–234

Imhoff JF, Tindall BJ, Grant WD, Trüper HG (1981) *Ectothiorhodospira vacuolata* sp. nov., a new phototrophic bacterium from soda lakes. Arch Microbiol 130:238–242

Isachenko BL (1951) Selected works, vol 2. Academy of Sciences of the Union of Soviet Socialist Republics, Moscow

Ishikawa M, Tanasupawat S, Nakajima K, Kanamori H, Ishizaki S, Kodama K et al (2009) *Alkalibacterium thalassium* sp. nov., *Alkalibacterium pelagium* sp. nov., *Alkalibacterium putridalgicola* sp. nov. and *Alkalibacterium kapii* sp. nov., slightly halophilic and alkaliphilic marine lactic acid bacteria isolated from marine organisms and salted foods collected in Japan and Thailand. Int J Syst Evol Microbiol 59:215–1226

Ito S (2011) Alkaline enzymes in current detergency. In: Horikoshi K, Antranikian G, Bull AT, Robb FT, Stetter KO (eds) Extremophiles handbook. Springer, New York, pp 229–254

Itoh T, Yamaguchi T, Zhou P, Takashina T (2005) *Natronolimnobius baerhuensis* gen. nov., sp. nov. and *Natronolimnobius innermongolicus* sp. nov., novel haloalkaliphilic archaea isolated from soda lakes in Inner Mongolia, China. Extremophiles 9:111–116

Jacquet S, Zhong X, Parvathi A, Sriram A, Ram P (2013) First description of a cyanophage infecting the cyanobacterium *Arthrospira platensis* (*Spirulina*). J Appl Phycol 25:195–203

Jannasch HW (1957) Die bakterielle Rotfärbung der Salzseen des Wadi Natrun (Ägypten). Arch Hydrobiol 53:425–433

Jan-Roblero J, Magos X, Fernandez L, Hernandez-Rodriguez C, Le Borgne S (2004) Phylogenetic analysis of bacterial populations in waters of the former Texcoco Lake, Mexico. Can J Microbiol 50:1049–1059

Janssen AJ, Lens PN, Stams AJ, Plugge CM, Sorokin DY, Muyzer G, Buisman CJ (2009) Application of bacteria involved in the biological sulfur cycle for paper mill effluent purification. Sci Total Environ 407:1333–1343

Jeon CO, Lim JM, Lee JM (2005) Reclassification of *Bacillus haloalkaliphilus* Fritze 1996 as *Alkalibacillus haloalkaliphilus* gen. nov., comb. nov. and the description of *Alkalibacillus salilacus* sp. nov., a novel halophilic bacterium isolated from a salt lake in China. Int J Syst Evol Microbiol 55:1891–1896

Jiang S, Steward G, Jellison R, Chu W, Choi S (2004) Abundance, distribution, and diversity of viruses in alkaline, hypersaline Mono Lake, California. Microb Ecol 47:9–17

Jiang F, Cao S-J, Li Z-H, Fan H, Li HF, Lui WJ, Yuan HL (2012) *Salisediminibacterium halotolerans* gen. nov., sp. nov., a halophilic bacterium from soda lake sediment. Int J Syst Evol Microbiol 62:2127–2132

Jones BE, Grant WD (2000) Microbial diversity and ecology of alkaline environments. In: Seckbach J (ed) Journeys to diverse microbial worlds. Kluwer, Dordrecht, pp 177–190

Jones BE, Grant WD (2002) Haloalkaliphilic microorganisms. US Patent 6420147B1

Jones BE, Grant WD, Collins NC, Mwatha WE (1994) Alkaliphiles: diversity and identification. In: Priest FG, Ramos-Cormenzana A, Tindall BJ (eds) Bacterial diversity and systematics. Plenum Press, New York, pp 195–229

Jones BE, Grant WD, Collins NC (1995) Gram-positive alkaliphilic microorganisms. US Patent 5401657

Jones BE, Herweijer MA, Danson MJ, Hough DW, Thompson CR (1997) Alkaliphilic and thermophilic microorganism and enzymes obtained therefrom. International patent application WO 97/10342 [Also published as granted US Patent US6432689B1 (13 Aug 2002) and European Patent EP0850307B1 (9 Nov 2005)]

Jones BE, Grant WD, Collins NC (1998a) Gram-negative alkaliphilic microorganisms. US Patent 5719047

Jones BE, Grant WD, Duckworth AW, Owenson GG (1998b) Microbial diversity of soda lakes. Extremophiles 2:191–200

Jones BE, Grant WD, Duckworth AW, Schumann P, Weiss N, Stackebrandt E (2005) *Cellulomonas bogoriensis* sp. nov., an alkaliphilic cellulomonad. Int J Syst Evol Microbiol 55:1711–1714

Joshi AA, Kanekar PP, Kelkar AS, Sarnaik SS, Shouche Y, Wani A (2007) Moderately halophilic, alkalitolerant *Halomonas campisalis* MCM B-365 from Lonar Lake, India. J Basic Microbiol 47:213–221

Joshi AA, Kanekar PP, Kelkar AS, Shouche YS, Vani AA, Borgave SB, Sarnaik SS (2008) Cultivable bacterial diversity of alkaline Lonar Lake, India. Microb Ecol 55:163–172

Kaggwa MN, Gruber M, Odour SO, Schagerl M (2013a) A detailed time series assessment of the diet of Lesser Flamingos–further explanation for their itinerant behaviour. Hydrobiologia 110:83–93

Kaggwa MN, Burian A, Odour SO, Schagerl M (2013b) Ecomorphological variability of *Arthrospira fusiformis* (Cyanoprocaryota) in African saline lakes. Microbiol Open 2:881–891

Kalyuzhnaya M, Khmelenina V, Eshinimaev B (2001) Taxonomic characterization of new alkaliphilic and alkalitolerant methanotrophs from soda lakes of the Southeastern Transbaikal region and description of *Methylomicrobium buryatense* sp.nov. Syst Appl Microbiol 24:166–176

Kalyuzhnaya MG, Khmelenina V, Eshinimaev B, Sorokin DY, Fuse H, Lidstom M, Trotsenko Y (2008) Classification of halo(alkali)philic and halo(alkali)tolerant methanotrophs provisionally assigned to the genera *Methylomicrobium* and *Methylobacter* and emended description of the genus *Methylomicrobium*. Int J Syst Evol Microbiol 58:591–596

Kamekura M, Dyall-Smith M, Upasani V, Ventosa A, Kates M (1997) Diversity of alkaliphilic halobacteria:

proposals for transfer of *Natronobacterium vacuolatum*, *Natronobacterium magadii*, and *Natronobacterium pharaonis* to *Halorubrum*, *Natrialba*, and *Natronomonas* gen nov., respectively, as *Halorubrum vacuolatum* comb. nov., *Natrialba magadii* comb. nov. and *Natronomonas pharaonis* comb. nov., respectively. Int J Syst Bacteriol 47:853–857

Kanai H, Kobayashi T, Aono R, Kudo T (1995) *Natronococcus amylolyticus*, sp. nov., a haloalkaliphilic archaeon. Int J Syst Bacteriol 45:762–766

Keller M, Braun FJ, Dirmeier R, Hafenbradl D, Burggraf S, Stetter KO (1995) *Thermococcus alcaliphilus* sp. nov., a new hyperthermophilic archaeum growing on polysulfide at alkaline pH. Arch Microbiol 164:390–395

Kevbrin VV, Lysenko AM, Zhilina TN (1997) Physiology of the alkaliphilic methanogen Z-7936, a new strain of *Methanosalsus zhilinaeae* isolated from Lake Magadi. Microbiology 66:261–266

Kevbrin VV, Zhilina TN, Rainey FA, Zavarzin GA (1998) *Tindallia magadii* gen. nov., sp. nov.: an alkaliphilic anaerobic ammonifier from soda lake deposits. Curr Microbiol 37:94–100

Kevbrin V, Boltyanskaya Y, Zhilina T, Kolganova T, Lavrentjeva E, Kuznetsov B (2013) *Proteinivorax tanatarense* gen. nov., sp. nov., an anaerobic, haloalkaliphilic, proteolytic bacterium isolated from a decaying algal bloom, and proposal of *Proteinivoraceae* fam. nov. Extremophiles 17:747–756

Khmelenina VN, Eshinimaev BT, Kalyuzhnaya MG, Trotsenko YA (2000) Potential activity of methane and ammonium oxidation by methanotrophic communities from the soda lakes of Southern Transbaikal. Microbiology 69:460–465

Kompantseva EI, Bryantseva IA, Komova AV, Namsaraev BB (2007) The structure of phototrophic communities of soda lakes of the southeastern Transbaikal Region. Microbiology 7:211–219

Kompantseva EI, Komova AV, Kostrikina NA (2010) *Rhodovulum steppense* sp. nov., an obligately haloalkaliphilic purple nonsulfur bacterium widespread in saline soda lakes of Central Asia. Int J Syst Evol Microbiol 60:1210–1214

Kompantseva EI, Komova AV, Novikov AA, Kostrikina NA (2012) *Rhodovulum tesquicola* sp. nov., a haloalkaliphilic purple non-sulfur bacterium from brackish steppe soda lakes. Int J Syst Evol Microbiol 62:2962–2966

Kovaleva OL, Tourova TP, Muyzer G, Kolganova TV, Sorokin DY (2011) Diversity of RuBisCO and ATP citrate lyase genes in soda lake sediments. FEMS Microbiol 75:37–47

Krammer M, Velimirov B, Fischer U, Farnleitner AH, Herzig A, Kirschner AKT (2008) Growth response of soda lake bacterial communities to simulated rainfall. Microb Ecol 55:194–211

Krienitz L, Kotut K (2010) Fluctuating algal food populations and the occurrence of Lesser Flamingo (*Phoeniconaias minor*) in three Kenyan Rift Valley lakes. J Phycol 46:1088–1096

Krienitz L, Ballot A, Kotut K, Wiegand C, Pülz S, Metcalf JS et al (2003) Contribution of hot spring cyanobacteria to the mysterious deaths of Lesser Flamingos at Lake Bogoria, Kenya. FEMS Microbiol Ecol 43:141–148

Kumar V, Satyanarayana T (2013) Thermoalkaliphilic microbes. In: Seckbach J, Oren A, Stan-Lotter H (eds) Polyextremophiles: life under multiple forms of stress. Springer, Dordrecht, pp 273–296

Kumar PA, Srinivas TNR, Madhu S, Manorama R, Shivaji S (2010a) *Indibacter alkaliphilus* gen. nov., sp. nov., an alkaliphilic bacterium isolated from a haloalkaline lake. Int J Syst Evol Microbiol 60:721–726

Kumar PA, Srinivas TNR, Kumar PP, Madhu S, Shivaji S (2010b) *Nitritalea halalkaliphila* gen. nov., sp. nov., an alkaliphilic bacterium of the family 'Cyclobacteriaceae', phylum Bacteroidetes. Int J Syst Evol Microbiol 60:2320–2325

Kumar PA, Srinivas TNR, Madhu S, Sravan R, Singh S, Naqvi SWA et al (2012) *Cecembia lonarensis* gen. nov., sp. nov., a haloalkalitolerant bacterium of the family Cyclobacteriaceae, isolated from a haloalkaline lake and emended descriptions of the genera *Indibacter, Nitritalea* and *Belliella*. Int J Syst Evol Microbiol 62:2252–2258

Lefèvre CT, Frankel RB, Posfai M, Prozorov T, Bazylinski DA (2011) Isolation of obligately alkaliphilic magnetotactic bacteria from extremely alkaline environments. Environ Microbiol 13:2342–2350

Li Y, Mandelco L, Wiegel J (1993) Isolation and characterization of a moderately thermophilic anaerobic alkaliphile, *Clostridium paradoxum* sp. nov. Int J Syst Bacteriol 43:450–460

Li Y, Engle M, Weiss N, Mandelco L, Wiegel J (1994) *Clostridium thermoalcaliphilum* sp. nov., an anaerobic and thermotolerant facultative alkaliphile. Int J Syst Bacteriol 44:111–118

Lin JL, Radajewski S, Eshinimaev BT, Trotsenko YA, McDonald IR, Murrell JC (2004) Molecular diversity of methanotrophs in Transbaikal soda lake sediments and identification of potentially active populations by stable isotope probing. Environ Microbiol 6:1049–1060

Lin JL, Joye SB, Scholte JCM, Schafer H, McDonald IR, Murrell JC (2005) Analysis of methane monooxygenase genes in Mono Lake suggests that increased methane oxidation activity may correlate with a change in methanotroph community structure. Appl Environ Microbiol 71:6458–6462

Liu YP, Boone DR, Choy C (1990) *Methanohalophilus oregonense* sp. nov., a methylotrophic methanogen from an alkaline saline aquifer. Int J Syst Bacteriol 40:111–116

Liu YP, Wang YX, Li YX, Feng FY, Liu HR, Wang J (2012) *Mongoliicoccus roseus* gen. nov., sp. nov., an

alkaliphilic bacterium isolated from a haloalkaline lake. Int J Syst Evol Microbiol 62:2206–2212
Lopez-Garcia P, Kazmierczak J, Benzerara K, Kempe S, Guyot F, Moreira D (2005) Bacterial diversity and carbonate precipitation in the giant microbialites from the highly alkaline Lake Van, Turkey. Extremophiles 9:263–274
Ma Y, Zhang W, Xue Y, Zhou P, Ventosa A, Grant WD (2004a) Bacterial diversity of the Inner Mongolian Baer Soda Lake as revealed by 16S rRNA gene sequence analyses. Extremophiles 8:45–51
Ma Y, Xue Y, Grant WD, Collins NC, Duckworth AW, van Steenbergen RP, Jones BE (2004b) *Alkalimonas amylolytica* gen. nov., sp. nov., and *Alkalimonas delamerensis* gen. nov., sp. nov., novel alkaliphilic bacteria from soda lakes in China and East Africa. Extremophiles 8:193–200
Marquez MC, Carrasco IJ, de la Haba RR, Jones BE, Grant WD, Ventosa A (2011) *Bacillus locisalis* sp. nov., a new haloalkaliphilic species from hypersaline and alkaline lakes of China, Kenya and Tanzania. Syst Appl Microbiol 34:424–428
Martins RF, Davids W, Abu Al-Soud W, Levander F, Radstrom P, Hatti-Kaul R (2001) Starch-hydrolyzing bacteria from Ethiopian soda lakes. Extremophiles 5:135–144
Mathrani IM, Boone DR, Mah RA, Fox GE, Lau PP (1988) Methanohalophilus zhilinae sp. nov., an alkaliphilic, halophilic, methylotrophic methanogen. Int J Syst Bacteriol 38:139–142
McCall J (2010) Lake Bogoria, Kenya: hot and warm springs, geysers and Holocene stromatolites. Earth Sci Rev 103:71–79
Meijer D, van Steenbergen R, Jones BE (1996) Isolation and characterization of a novel alkaliphilic thermophilic bacterium from the soda lake environment belonging to the order *Thermotogales*. In: Conference abstracts “Thermophiles ‘96”, international conference on “The Biology, Ecology and Biotechnology of Thermophilic Microorganisms”, Abstr. #P1.A.4 University of Georgia, Athens, GA, USA, Sept 1996
Melack JM, Kilham P (1974) Photosynthetic rates of phytoplankton in East African soda lakes. Limnol Oceanogr 19:743–755
Mesbah NM, Wiegel J (2008) Life at extreme limits: the anaerobic halophilic alkalithermophiles. Ann NY Acad Sci 1125:44–57
Mesbah NM, Wiegel J (2009) *Natronovirga wadinatrunensis* gen. nov., sp. nov. and *Natranaerobius trueperi* sp. nov., halophilic, alkalithermophilic microorganisms from soda lakes of the Wadi An Natrun, Egypt. Int J Syst Evol Microbiol 59:2042–2048
Mesbah NM, Wiegel J (2011) Halophiles exposed concomitantly to multiple stressors: adaptive mechanisms of halophilic alkalithermophiles. In: Ventosa A, Oren A, Ma Y (eds) Halophiles and hypersaline environments. Springer, Berlin, pp 249–273
Mesbah NM, Wiegel J (2012) Life under multiple extreme conditions: diversity and physiology of the halophilic alkalithermophiles. Appl Environ Microbiol 78:4074–4082
Mesbah NM, Abou-El-Ela SH, Wiegel J (2007a) Novel and unexpected prokaryotic diversity in water and sediments of the alkaline, hypersaline lakes of the Wadi an Natrun, Egypt. Microb Ecol 54:598–617
Mesbah NM, Hedrick DB, Peacock AD, Rohde M, Wiegel J (2007b) *Natranaerobius thermophilus* gen. nov., sp. nov., a halophilic, alkalithermophilic bacterium from soda lakes of the Wadi An Natrun, Egypt, and proposal of *Natranaerobiaceae* fam. nov. and *Natranaerobiales* ord. nov. Int J Syst Evol Microbiol 57:2507–2512
Milford AD, Achenbach LA, Jung DO, Madigan MT (2000) *Rhodobaca bogoriensis* gen. nov. and sp. nov., an alkaliphilic purple nonsulfur bacterium from African Rift Valley soda lakes. Arch Microbiol 174:18–27
Mormile MR, Romine MF, Garcia MT, Ventosa A, Bailey TJ, Peyton BM (1999) *Halomonas campisalis* sp. nov., a denitrifying, moderately haloalkaliphilic bacterium. Syst Appl Microbiol 22:551–558
Muyzer G, Sorokin DY, Mavromatis K, Lapidus A, Foster B, Sun H et al (2011) Complete genome sequence of *Thioalkalivibrio* sp. K90mix. Stand Genomic Sci 5:341–355
Mwatha WE (1991) Microbial ecology of Kenyan soda lakes. PhD thesis, University of Leicester, UK
Mwatha WE, Grant WD (1993) *Natronobacterium vacuolata* sp.nov., a haloalkaliphilic archaeon isolated from Lake Magadi, Kenya. Int J Syst Bacteriol 43:401–404
Mwirichia R, Muigai AW, Tindall BJ, Boga HI, Stackebrandt E (2010a) Isolation and characterisation of bacteria from the haloalkaline Lake Elmenteita, Kenya. Extremophiles 14:339–348
Mwirichia R, Cousin S, Muigai AW, Boga HI, Stackebrandt E (2010b) Archaeal diversity in the haloalkaline Lake Elmenteita in Kenya. Curr Microbiol 60:47–52
Mwirichia R, Cousin S, Muigai AW, Boga HI, Stackebrandt E (2011) Bacterial diversity in the haloalkaline Lake Elmenteita, Kenya. Curr Microbiol 62:209–221
Namsaraev Z, Akimov V, Tsapin A, Barinova E, Nealson K, Gorlenko V (2009) *Marinospirillum celere* sp. nov., a novel haloalkaliphilic, helical bacterium isolated from Mono Lake. Int J Syst Evol Microbiol 59:2329–2332
Narasingarao P, Podell S, Ugalde JA, Brochier-Armanet-C, Emerson JB, Brock JJ et al (2011) De novo metagenomic assembly reveals abundant novel major lineage of Archaea in hypersaline microbial communities. ISME J 6:181–193
Ntougias S, Russell NJ (2001) *Alkalibacterium olivoapovliticus* gen. nov., sp. nov., a new obligately alkaliphilic bacterium isolated from edible-olive wash-waters. Int J Syst Evol Microbiol 51:1161–1170
Nuyanzina-Boldareva EN, Gorlenko VM (2014) *Roseibacula alcaliphilum* gen. nov., sp. nov., a new alkaliphilic aerobic anoxygenic phototrophic bacterium from a meromictic soda lake Doroninskoe (East Siberia, Russia). Microbiology 83:381–390

Oduor OS, Schagerl M (2007) Phytoplankton primary productivity characteristics in response to photosynthetically active radiation in three Kenyan Rift valley saline-alkaline lakes. J Plankton Res 29:1041–1050

Oremland RS, Marsh L, Des Marais DJ (1982) Methanogenesis in Big Soda Lake, Nevada: an alkaline, moderately hypersaline desert lake. Appl Environ Microbiol 43:462–468

Oremland RS, Stolz JF, Hollibaugh JT (2004) The microbial arsenic cycle in Mono Lake, California. FEMS Microbiol Ecol 48:15–27

Oremland RS, Kulp TR, Blum JS, Hoeft SE, Baesman S, Miller LG et al (2005) A microbial arsenic cycle in a salt-saturated, extreme environment. Science 308:1305–130

Oremland RS, Kulp TR, Blum JS, Hoeft SE, Baesman S, Miller LG, Stolz JF (2009) A microbial arsenic cycle in a salt-saturated, extreme environment. Science 308:1305–1308

Oren A (2011) Ecology of halophiles. In: Horikoshi K, Antranikian G, Bull AT, Robb FT, Stetter KO (eds) Extremophiles handbook. Springer, New York, pp 334–361

Oren A (2013) Two centuries of microbiological research in the Wadi Natrun, Egypt: a model system for the study of the ecology, physiology, and taxonomy of haloalkaliphilic microorganisms. In: Seckbach J, Oren A, Stan-Lotter H (eds) Polyextremophiles: life under multiple forms of stress. Springer, Dordrecht, pp 273–296

Osanjo GO, Muthike EW, Tsuma L et al (2009) A salt lake extremophile, *Paracoccus bogoriensis* sp. nov., efficiently produces xanthophyll carotenoids. Afr J Microbiol 3:426–433

Pagaling E (2007) Microbial diversity of Chinese Lakes. PhD thesis, University of Leicester, UK. Available via https://lra.le.ac.uk/handle/2381/7662. Accessed 21 May 2015

Pagaling E, Haigh R, Grant W, Cowan DA, Jones BE, Ma Y et al (2007) Sequence analysis of an archaeal virus isolated from a hypersaline lake in Inner Mongolia, China. BMC Genomics 8:410

Pagaling E, Wang H, Venables M, Wallace A, Grant WD, Cowan DA et al (2009) Microbial biogeography of six salt lakes in Inner Mongolia and one salt lake in Argentina. Appl Environ Microbiol 75:5750–5760

Pagaling E, Grant WD, Cowan DA, Jones BE, Ma Y, Ventosa A, Heaphy S (2012) Bacterial and archaeal diversity in two hot spring microbial mats from the geothermal region of Tengchong, China. Extremophiles 16:607–618

Peduzzi P, Gruber M, Gruber M, Schagerl M (2014) The virus's tooth: cyanophages affect an African flamingo population in a bottom-up cascade. ISME J 8:1346–1351

Pikuta EV, Zhilina TN, Zavarzin GA, Kostrikina NA, Osipov GA, Rainey FA (1998) *Desulfonatronum lacustre* gen. nov., sp. nov.: a new alkaliphilic sulfate-reducing bacterium utilizing ethanol. Microbiology 67:105–113

Pikuta EV, Hoover RB, Bej AK, Marsic D, Detkova EN, Whitman WB et al (2003a) *Tindallia californiensis* sp. nov., a new anaerobic, haloalkaliphilic, spore-forming acetogen isolated from Mono Lake in California. Extremophiles 4:327–334

Pikuta EV, Hoover RB, Bej AK, Marsic D, Whitman WB, Cleland D et al (2003b) *Desulfonatronum thiodismutans* sp. nov., a novel alkaliphilic, sulfate-reducing bacterium capable of lithoautotrophic growth. Int J Syst Evol Microbiol 53:1327–32

Pikuta EV, Itoh T, Krader P, Tang J, Whitman WB, Hoover RB (2006) *Anaerovirgula multivorans* gen. nov., sp. nov., a novel spore-forming, alkaliphilic anaerobe isolated from Owens Lake, California, USA. Int J Syst Evol Microbiol 56:2623–2629

Pikuta EV, Hoover RB, Bej AK, Marsic D, Whitman WB, Krader P (2009) *Spirochaeta dissipatitropha* sp. nov., an alkaliphilic, obligately anaerobic bacterium, and emended description of the genus *Spirochaeta* Ehrenberg 1835. Int J Syst Evol Microbiol 59:1466–5026

Pinkart HC, Storrie-Lombardi MC (2007) Diversity, evolution, and horizontal gene transfer (HGT) in soda lakes. In: Hoover RB, Levin GV, Rozanov AY, Davies PCW (eds) Instruments, methods, and missions for astrobiology, Proceedings of SPIE 6694, Article Number: 66940Y

Poser A, Lohmayer R, Vogt C, Knoeller K, Planer-Friedrich B, Sorokin D et al (2013) Disproportionation of elemental sulfur by haloalkaliphilic bacteria from soda lakes. Extremophiles 17:1003–1012

Prowe SG, Antranikian G (2001) *Anaerobranca gottschalkii* sp. nov., a novel thermoalkaliphilic bacterium that grows anaerobically at high pH and temperature. Int J Syst Evol Microbiol 51:457–465

Ramana CV, Srinivas A, Subhash Y, Tushar L, Mukherjee T, Kiran PU et al (2013) *Salinicoccus halitifaciens* sp. nov., a novel bacterium participating in halite formation. Anton van Leeuwenhoek 103:885–898

Reddy SV, Aspana S, Tushar DL, Sasikala C, Ramana CV (2013) *Spirochaeta sphaeroplastigenens* sp. nov., a halo-alkaliphilic, obligately anaerobic spirochaete isolated from soda lake Lonar. Int J Syst Evol Microbiol 63:2223–2228

Reddy SV, Thirumala M, Farooq M, Sasikala C, Ramana CV (2015a) *Bacillus lonarensis* sp. nov., an alkalitolerant bacterium isolated from a soda lake. Arch Microbiol 197:27–34

Reddy SV, Mothe T, Farooq M (2015b) *Bacillus caseinilyticus* sp. nov., a novel alkali and thermo tolerant bacterium isolated from a soda lake. Int J Syst Evol Microbiol 65:2441–2446. doi:10.1099/ijs.0.000275

Reddy SV, Mothe T, Farooq M (2015c) *Salisediminibacterium haloalkalitolerans* sp. nov., isolated from Lonar soda lake, India, and a proposal for reclassification of *Bacillus locisalis* as *Salisediminibacterium locisalis* comb. nov., and the emended description of the genus *Salisediminibacterium* and of the species

Salisediminibacterium halotolerans. Arch Microbiol 197:553–560

Reddy SV, Mothe T, Farooq M (2015d) *Streptomyces alkalithermotolerans* sp. nov., a novel alkaliphilic and thermotolerant actinomycete isolated from a soda lake. Anton van Leeuwenhoek 107:337–344

Rees HC, Grant WD, Jones BE, Heaphy S (2004) Diversity of Kenyan soda lake alkaliphiles assessed by molecular methods. Extremophiles 8:63–71

Renaut RW, Jones B, Tiercelin J-J (1998) Rapid in situ silicification of microbes at Loburu hot springs, Lake Bogoria, Kenya Rift Valley. Sedimentology 45:1083–1103

Renaut RW, Bernhart Owen R, Jones B, Tiercelins J-J, Tarits C, Ego JK, Konhauser KO (2013) Impact of lake-level changes on the formation of thermogene travertine in continental rifts: evidence from Lake Bogoria, Kenya Rift Valley. Sedimentology 60:428–468

Romano I, Lama L, Nicolaus B, Gambacorta A, Giordana A (2005a) *Bacillus saliphilus* sp. nov., isolated from mineral pool in Campania, Italy. Int J Syst Evol Microbiol 55:159–163

Romano I, Giordano A, Lama L, Nicolaus B, Gambacorta A (2005b) *Halomonas campaniensis* sp. nov., a haloalkaliphilic bacterium isolated from a mineral pool of Campania Region, Italy. Syst Appl Microbiol 28:610–618

Ruiz-Romero E, de los Angeles Coutino-Coutino M, Valenzuela-Encinas C, Patricia Lopez-Ramirez M, Marsch R, Dendooven L (2013a) *Texcoconibacillus texcoconensis* gen. nov., sp. nov., alkalophilic and halotolerant bacteria isolated from soil of the former lake Texcoco (Mexico). Int J Syst Evol Microbiol 63:3336–3341

Ruiz-Romero E, Valenzuela-Encinas C, Lopez-Ramirez MP, Coutino-Coutino MD, Marsch R, Dendooven L (2013b) *Natronorubrum texcoconense* sp. nov., a haloalkaliphilic archaeon isolated from soil of the former lake Texcoco (Mexico). Arch Microbiol 195:145–151

Ruiz-Romero E, Sánchez-López KB, Coutiño-Coutiño MDLA et al (2013c) *Natronobacterium texcoconense* sp. nov., a haloalkaliphilic archaeon isolated from soil of the former lake Texcoco (Mexico). Int J Syst Evol Microbiol 63:4163–4166

Rusznyak A, Szabo G, Pollak B, Vagany V, Palatinszky M (2007) Diversity of reed (*Phragmites australis*) stem biofilm bacterial communities in two Hungarian soda lakes. Acta Microbiol Immunol Hung 54:339–352

Ryzhmanova Y, Nepomnyashchaya Y, Abashina T, Ariskina E, Troshina O, Vainshtein M et al (2013) New sulfate-reducing bacteria isolated from Buryatian alkaline brackish lakes: description of *Desulfonatronum buryatense* sp. nov. Extremophiles 17:851–859

Sahay H, Mahfooz S, Singh AK, Singh S, Kaushik R, Saxena AK, Arora DK (2012) Exploration and characterization of agriculturally and industrially important haloalkaliphilic bacteria from environmental samples of hypersaline Sambhar Lake, India. World J Microbiol Biotechol 28:3207–3217

Scholten JCM, Joye SB, Hollibaugh JT, Murrell JC (2005) Molecular analysis of the sulfate reducing and archaeal community in a meromictic soda lake (Mono Lake, California) by targeting 16S rRNA, mcrA, apsA, and dsrAB genes. Microb Ecol 50:29–39

Schumann P, Cui XL, Stackebrandt E, Kroppenstedt RM, Xu LH, Jiang CL (2004) *Jonesia quinghaiensis* sp. nov., a new member of the suborder *Micrococcineae*. Int J Syst Evol Microbiol 54:2181–2184

Schweinfurth G, Lewin L (1898) Beiträge zur Topographie und Geochemie des ägyptischen Natron-Thals. Erde 33:1–25

Shapovalova AA, Khijniak TV, Tourova TP, Sorokin DY (2009) *Halomonas chromatireducens* sp. nov., a new denitrifying facultatively haloalkaliphilic bacterium from solonchak soil capable of aerobic chromate reduction. Microbiology 78:102–111

Sickenberger E (1892) Briefe aus Egypten. I. Wady Atrun Das Natronthal. Chemiker-Zeitung 16:1645–1646, 1691

Soliman HSG, Trüper HG (1982) *Halobacterium pharaonis* sp. nov., a new extremely halophilic bacterium with an extremely low magnesium requirement. Zbl Bakt Hyg I Abt Orig C 3:318–329

Song J, Zhai Y, Zhang C, Gerritsen J, Wang H, Chen X, Li Y, Zhao B, Zhao B, Ruan Z (2015) *Romboutsia sedimentorum* sp. nov., isolated from an alkaline-saline lake sediment and emended description of the genus *Romboutsia*. Int J Syst Evol Microbiol 65:1193–1198

Sorokin DY (1998) Occurrence of nitrification in extremely alkaline natural habitats. Microbiology 67:404–408

Sorokin DY, Kuenen JG (2005a) Alkaliphilic chemolithotrophs from soda lakes. FEMS Microbiol Ecol 52:287–295

Sorokin DY, Kuenen JG (2005b) Haloalkaliphilic sulphur-oxidizing bacteria in soda lakes. FEMS Microbiol Rev 29:685–702

Sorokin DY, Muyzer G (2010a) Haloalkaliphilic spore-forming sulfidogens from soda lake sediments and description of *Desulfitispora alkaliphila* gen. nov., sp. nov. Extremophiles 14:313–320

Sorokin DY, Muyzer G (2010b) *Desulfurispira natronophila* gen. nov. sp. nov.: an obligately anaerobic dissimilatory sulphur-reducing bacterium from soda lakes. Extremophiles 14:349–355

Sorokin DY, Muyzer G, Brinkhoff T, Kuenen JG, Jetten MSM (1998) Isolation and characterization of a novel facultatively alkaliphilic *Nitrobacter* species, *N. alkalicus* sp. nov. Arch Microbiol 170:345–352

Sorokin DY, Jones BE, Kuenen JG (2000a) An obligate methylotrophic, methane-oxidizing *Methylomicrobium* species from a highly alkaline environment. Extremophiles 4:145–155

Sorokin DY, Tourova TP, Kuznetsov BB, Bryantseva IA, Gorlenko VM (2000b) *Roseinatronobacter thiooxidans* gen. nov., sp. nov., a new alkaliphilic aerobic bacteriochlorophyll *a* - containing bacterium isolated from a soda lake. Microbiology 69:75–82

Sorokin DY, Tourova TP, Schmid M, Wagner M, Koops HP, Kuenen JG et al (2001a) Isolation and properties of obligately chemolithoautotrophic and extremely alkali-tolerant ammonia-oxidizing bacteria from Mongolian soda lakes. Arch Microbiol 176:170–177

Sorokin DY, Lysenko AM, Mityushina LL, Tourova TP, Jones BE, Rainey F et al (2001b) *Thioalkalimicrobium aerophilum* gen. nov., sp. nov. and *Thioalkalimicrobium sibiricum*. sp. nov., and *Thioalkalivibrio versutus* gen. nov., sp. nov., *Thioalkalivibrio nitratis* sp. nov. and *Thioalkalivibrio denitrificans* sp. nov., novel obligately alkaliphilic and obligately chemolithoautotrophic sulfur-oxidizing bacteria from soda lakes. Int J Syst Evol Microbiol 51:565–580

Sorokin D, Tourova T, Schmid MC, Wagner M, Koops H-P, Kuenen GJ, Jetten M (2001c) Isolation and properties of obligately chemolithoautotrophic and extremely alkali-tolerant ammonia-oxidizing bacteria from Mongolian soda lakes. Arch Microbiol 176:170–177

Sorokin DY, Tourova TP, Lysenko AM, Mityushina LL, Kuenen JG (2002a) *Thioalkalivibrio thiocyanooxidans* sp. nov. and *Thioalkalivibrio paradoxus* sp. nov., novel alkaliphilic, obligately autotrophic, sulfur-oxidizing bacteria from the soda lakes able to grow with thiocyanate. Int J Syst Evol Microbiol 52:657–664

Sorokin DY, Gorlenko VM, Tourova TP, Tsapin AI, Nealson KH, Kuenen GJ et al (2002b) *Thioalkalimicrobium cyclicum* sp. nov. and *Thioalkalivibrio jannaschii* sp. nov., new species of alkaliphilic, obligately chemolithoautotrophic sulphur-oxidizing bacteria from a hypersaline alkaline Mono Lake (California). Int J Syst Evol Microbiol 52:913–920

Sorokin DY, Tourova TP, Kolganova TV, Sjollema KA, Kuenen JG (2002c) *Thioalkalispira microaerophila* gen. nov., sp. nov., a novel lithoautotrophic, sulfur-oxidizing bacterium from a soda lake. Int J Syst Evol Microbiol 52:2175–2182

Sorokin DY, Tourova TP, Sjollema KA, Kuenen JG (2003) *Thialkalivibrio nitratireducens* sp. nov., a nitrate-reducing member of an autotrophic denitrifying consortium from a soda lake. Int J Syst Evol Microbiol 53:1779–1783

Sorokin DY, Tourova TP, Antipov AN, Muyzer G, Kuenen JG (2004a) Anaerobic growth of the haloalkaliphilic denitrifying sulphur-oxidising bacterium *Thialkalivibrio thiocyanodenitrificans* sp. nov. with thiocyanate. Microbiology 150:2435–2442

Sorokin DY, Gorlenko VM, Namsaraev BB, Namsaraev Z, Lysenko A, Eshinimaev B et al (2004b) Prokaryotic communities of the north-eastern Mongolian soda lakes. Hydrobiologia 522:235–248

Sorokin DY, Banciu H, Robertson LA, Kuenen JG (2006) Haloalkaliphilic sulfur-oxidizing bacteria. In: Dworkin M, Falkow S, Rosenberg E, Schleifer K-H, Stackebrandt E (eds) The prokaryotes: a handbook on the biology of bacteria, vol 2, 3rd edn. Springer, New York, pp 969–984

Sorokin DY, Foti M, Tindall BJ, Muyzer G (2007a) *Desulfurispirillum alkaliphilum* gen. nov., sp. nov., a novel obligately anaerobic sulfur- and dissimilatory nitrate-reducing bacterium from a full-scale sulfide-removing bioreactor. Extremophiles 11:363–370

Sorokin DY, van Pelt S, Tourova TP, Takaichi S, Muyzer G (2007b) Acetonitrile degradation under haloalkaline conditions by *Natronocella acetinitrilica* gen. nov., sp. nov. Microbiology 153 1157–1164

Sorokin DY, Foti M, Pinkart HC, Muyzer G (2007c) Sulfur-oxidizing bacteria in Soap Lake (Washington State), a meromictic, haloalkaline lake with an unprecedented high sulfide content. Appl Environ Microbiol 73:451–455

Sorokin DY, Trotsenko YA, Doronina NV, Tourova TP, Galinski EA, Kolganova TV et al (2007d) *Methylohalomonas lacus* gen. nov., sp. nov. and *Methylonatrum kenyense* gen. nov., sp. nov., methylotrophic gammaproteobacteria from hypersaline lakes. Int J Syst Evol Microbiol 57:2762–2769

Sorokin DY, Tourova TP, Bezsoudnova EY, Pol A, Muyzer G (2007e) Denitrification in a binary culture and thiocyanate metabolism in *Thiohalophilus thiocyanoxidans* gen. nov. sp. nov.: a moderately halophilic chemolithoautotrophic sulfur-oxidizing *Gammaproteobacterium* from hypersaline lakes. Arch Microbiol 187:441–450

Sorokin DY, Tourova TP, Henstra AM, Stams AJM, Galinski EA, Muyzer G (2008a) Sulfidogenesis at extremely haloalkaline conditions by *Desulfonatronospira thiodismutans* gen. nov., sp. nov., and *Desulfonatronospira delicata* sp. nov.: a novel lineage of *Deltaproteobacteria* from hypersaline soda lakes. Microbiology 154:1444–1453

Sorokin DY, Tourova TP, Mussmann M, Muyzer G (2008b) *Dethiobacter alkaliphilus* gen. nov., sp. nov., and *Desulfurivibrio alkaliphilus* gen. nov., sp. nov. - two novel representatives of reductive sulfur cycle from soda lakes. Extremophiles 12:431–439

Sorokin ID, Kravchenko IK, Doroshenko EV, Boulygina ES, Zadorina EV, Tourova TP et al (2008c) Haloalkaliphilic diazotrophs in soda solonchak soils. FEMS Microbiol Ecol 65:425–433

Sorokin ID, Zadorina EV, Kravchenko IK, Boulygina ES, Tourova TP, Sorokin DY (2008d) *Natronobacillus azotifigens* gen. nov., sp. nov., an anaerobic diazotrophic haloalkaliphile from soda-rich habitats. Extremophiles 12:819–827

Sorokin DY, van den Bosch PLF, Janssen AJH, Muyzer G (2008e) Microbiological analysis of the population of extremely haloalkaliphilic sulfur-oxidizing bacteria dominating in lab-scale sulphide-removing bioreactors. Appl Microbiol Biotechnol 80:965–975

Sorokin ID, Kravchenko IK, Tourova TP, Kolganova TV, Boulygina ES, Sorokin DY (2008f) *Bacillus alkalidiazotrophicus* sp. nov., a diazotrophic, low salt concentration-tolerant alkaliphile from Mongolian soda soil. Int J Syst Evol Microbiol 58:2459–2464

Sorokin DY, van Pelt S, Tourova TP, Evtushenko LI (2009) *Nitriliruptor alkaliphilus* gen. nov., sp. nov., a deep lineage haloalkaliphilic actinobacterium from soda lakes capable of growth on aliphatic nitriles and proposal of *Nitriliruptoraceae* fam. nov. and *Nitriliruptorales* ord. nov. Int J Syst Evol Microbiol 59:248–253

Sorokin DY, Detkova EN, Muyzer G (2010a) Propionate and butyrate dependent bacterial sulfate reduction at extremely haloalkaline conditions and description of *Desulfobotulus alkaliphilus* sp. nov. Extremophiles 14:71–77

Sorokin DY, Kovaleva OL, Tourova TP, Kuenen JG, Muyzer G (2010b) Aerobic carboxydotrophy at extremely haloalkaline conditions in *Alkalispirillum*/*Alkalilimnicola* strains isolated from soda lakes. Microbiology 156:819–827

Sorokin DY, Tourova TP, Detkova EN, Kolganova TV, Galinski EA, Muyzer G (2011a) Culturable diversity of lithotrophic haloalkaliphilic sulfate-reducing bacteria in soda lakes and the description of *Desulfonatronum thioautotrophicum* sp. nov., *Desulfonatronum thiosulfatophilum* sp. nov., *Desulfonatronovibrio thiodismutans* sp. nov., and *Desulfonatronovibrio magnus* sp. nov. Extremophiles 15:391–401

Sorokin DY, Panteleeva AN, Tourova TP, Kaparullina EN, Muyzer G (2011b) *Natronoflexus pectinivorans* gen. nov., sp. nov., an obligately anaerobic and alkaliphilic fermentative member of *Bacteroidetes* from soda lakes. Extremophiles 15:691–696

Sorokin DY, Detkova EN, Muyzer G (2011c) Sulfur-dependent respiration under extremely haloalkaline terms in the description of soda lake 'acetogens' and *Natroniella sulfidigena* sp. nov. FEMS Microbiol Lett 319:88–95

Sorokin DY, Tourova TP, Sukhacheva MV, Mardanov AV, Ravin NV (2012a) Bacterial chitin utilisation at extremely haloalkaline conditions. Extremophiles 16:883–894

Sorokin DY, Muntyan MS, Panteleeva AN, Muyzer G (2012b) *Thioalkalivibrio sulfidiphilus* sp. nov., a haloalkaliphilic, sulfur-oxidizing gammaproteobacterium from alkaline habitats. Int J Syst Evol Microbiol 62:1884–1889

Sorokin DY, Zacharova EE, Pimenov NV, Tourova TP, Panteleeva AN, Muyzer G (2012c) Sulfidogenesis in hypersaline chloride-sulfate lakes of Kulunda Steppe (Altai, Russia). FEMS Microbiol Ecol 79:445–453

Sorokin DY, Panteleeva AN, Tourova TP, Muyzer G (2012d) Haloalkaliphilic heterotrophic sulfate-reducing bacteria from soda lakes and description of *Desulfonatronobacter acidivorans* gen. nov., sp. nov., and *Desulfobulbus alkaliphilus* sp. nov. Int J Syst Evol Microbiol 62:2107–2113

Sorokin DY, Tourova TP, Panteleeva AN, Kaparullina E, Muyzer G (2012e) Anaerobic utilization of pectinous substrates at extremely haloalkaline conditions by *Natranaerovirga pectinivora* gen. nov., sp. nov., and *Natranaerovirga hydrolytica* sp. nov., isolated from hypersaline soda lakes. Extremophiles 16:307–315

Sorokin DY, Tourova TP, Sukhacheva MV, Muyzer G (2012f) *Desulfuribacillus alkaliarsenatis* gen. nov., sp. nov., a deep lineage, obligately anaerobic, dissimilatory sulfur and arsenate reducing, haloalkaliphilic representative of the order *Bacillales* from soda lakes. Extremophiles 16:597–605

Sorokin DY, Tourova TP, Abbas B, Suhacheva MV, Muyzer G (2012g) *Desulfonatronovibrio halophilus* sp. nov., a novel moderately halophilic sulfate-reducing bacterium from hypersaline chloride-sulfate lakes in Central Asia. Extremophiles 16:411–417

Sorokin DY, Tourova TP, Muyzer G (2013) Isolation and characterization of two novel alkalitolerant sulfidogens from a Thiopaq bioreactor, *Desulfonatronum alkalitolerans* sp. nov., and *Sulfurospirillum alkalitolerans* sp. nov. Extremophiles 17:535–543

Sorokin DY, Berben T, Melton EM, Overmars L, Vavourakis CD, Muyzer G (2014a) Microbial diversity and biogeochemical cycling in soda lakes. Extremophiles 18:791–809

Sorokin DY, Gumerov VM, Rakitin AL, Beletsky AV, Damste JS, Mardanov AV, Ravin NV (2014b) Genome analysis of *Chitinivibrio alkaliphilus* gen. nov., sp. nov., a novel extremely haloalkaliphilic anaerobic chitinolytic bacterium from the candidate phylum TG3. Environ Microbiol 16:1549–1565

Sorokin DY, Abbas B, Tourova TP, Bumazhkin BK, Kolganova TV, Muyzer G (2014c) Sulfate-dependent acetate oxidation at extremely natron-alkaline conditions by syntrophic associations from hypersaline soda lakes. Microbiology 160:723–732

Sorokin DY, Banciu H, Muyzer G (2015a) Functional microbiology of soda lakes. Curr Opin Microbiol 25:88–96

Sorokin DY, Abbas B, Geleijnse M, Pimenov NV, Sukhacheva MV, van Loosdrecht MCM (2015a) Methanogenesis at extremely haloalkaline conditions in soda lakes of Kulunda Steppe (Altai, Russia). FEMS Microbiol Ecol 91. doi:10.1093/femsec/fiv016

Sorokin DY, Abbas B, Geleijnse M, Pimenov NV, Sukhacheva MV, van Loosdrecht MCM (2015c) Methanogenesis at extremely haloalkaline conditions in soda lakes of Kulunda Steppe (Altai, Russia). FEMS Microbiol Ecol 91. doi:10.1093/femsec/fiv016

Srinivas A, Rahul K, Sasikala C, Subhash Y, Ramaprasad EVV, Ramana CV (2012) *Georgenia satyanarayanai* sp. nov., an alkaliphilic and thermotolerant amylase-producing actinobacterium isolated from a soda lake. Int J Syst Evol Microbiol 62:2405–2409

Srinivas TN, Aditya S, Bhumika V, Kumar PA (2014) *Lunatimonas lonarensis* gen. nov., sp. nov., a haloalkaline bacterium of the family *Cyclobacteriaceae* with nitrate reducing activity. Syst Appl Microbiol 37:10–16

Stöhr R, Waberski A, Völker H, Tindall BJ, Thomm M (2001) *Hydrogenothermus marinus* gen. nov., sp. nov., a novel thermophilic hydrogen-oxidizing bacterium, recognition of *Calderobacterium hydrogenophilum* as a member of the genus *Hydrogenobacter* and proposal of the reclassification of *Hydrogenobacter acidophilus* as

Hydrogenobaculum acidophilum gen. nov., comb. nov., in the phylum '*Hydrogenobacter/Aquifex*'. Int J Syst Evol Microbiol 51:1853–1862

Sultanpuram VR, Lodha TD, Chintalapati VR, Chintalapati S (2013) *Cohaesibacter haloalkalitolerans* sp. nov., isolated from a soda lake, and emended description of the genus *Cohaesibacter*. Int J Syst Evol Microbiol 63:4271–4276

Surakasi VP, Wani AA, Shouche YS, Ranade DR (2007) Phylogenetic analysis of methanogenic enrichment cultures obtained from Lonar Lake in India: isolation of *Methanocalculus* sp. and *Methanoculleus* sp. Microb Ecol 54:697–704

Surakasi VP, Antony CP, Sharma S, Patole MS, Shouche YS (2010) Temporal bacterial diversity and detection of putative methanotrophs in surface mats of Lonar crater lake. J Basic Microbiol 50:465–47

Suttle CA (2007) Marine viruses - major players in the global ecosystem. Nature Rev 5:801–812

Suzuki S, Kuenen JG, Schipper K, van der Velde S, Ishii S, Wu S et al (2014) Physiological and genomic features of highly alkaliphilic hydrogen-utilizing *Betaproteobacteria* from a continental serpentinizing site. Nat Commun 5:3900. doi:101038/ncomms4900

Svetlitshnyi V, Rainey F, Wiegel J (1996) *Thermosyntropha lipolytica* gen. nov., sp. nov., a lipolytic, anaerobic, alkalitolerant, thermophilic bacterium utilizing short- and long-chain fatty acids in syntrophic coculture with a methanogenic archaeum. Int J Syst Bacteriol 46:1131–1137

Talling JF, Wood RB, Prosser MV, Baxter RM (1973) The upper limit of photosynthetic productivity by phytoplankton: evidence from Ethiopian soda lakes. Freshw Biol 3:53–76

Thakker CD, Ranade DR (2002) Alkalophilic *Methanosarcina* isolated from Lonar Lake. Curr Sci 82:455–458

Thompson CR (1998) Alpha-amylase and pullulanase from '*Thermopallium natronophilum*'. PhD thesis, University of Bath, UK. http://search.proquest.com/docview/899791476. Accessed 22 May 2015

Tian S-P, Wang Y-X, Hu B, Zhang X-X, Xiao W, Chen Y et al (2010) *Litoribacter ruber* gen. nov., sp. nov., an alkaliphilic, halotolerant bacterium isolated from a soda lake sediment. Int J Syst Evol Microbiol 60:2996–3001

Tindall BJ, Mills AA, Grant WD (1980) An alkalophilic and halophilic bacterium with a low magnesium requirement from a Kenyan soda lake. J Gen Microbiol 116:252–260

Tindall BJ, Ross HNM, Grant WD (1984) *Natronobacterium* gen. nov. and *Natronococcus* gen. nov., two new genera of haloalkaliphilic archaebacteria. Syst Appl Microbiol 5:41–57

Tourova TP, Garnova ES, Zhilina TN (1999) Phylogenetic diversity of alkaliphilic anaerobic saccharolytic bacteria isolated from soda lakes. Microbiology 68:615–622

Tourova TP, Spiridonova EM, Berg IA, Slobodova NV, Boulygina S, Sorokin DY (2007) Phylogeny and evolution of the family *Ectothiorhodospiraceae* based on comparison of 16S rRNA, cbbL and nifH genes. Int J Syst Evol Microbiol 57:2387–2398

Tourova TP, Kovaleva OL, Bumazhkin BK, Patutina EO, Kuznetso BB, Bryantseva IA et al (2011) Application of ribulose-1,5-bisphosphate carboxylase/oxygenase genes as molecular markers for assessment of the diversity of autotrophic microbial communities inhabiting the upper sediment horizons of the saline and soda lakes of the Kulunda Steppe. Microbiology 80:812–825

Tourova TP, Slobodova NV, Bumazhkin BK, Kolganova TV, Muyzer G, Sorokin DY (2013) Analysis of community composition of sulfur-oxidizing bacteria in hypersaline and soda lakes using soxB as a functional molecular marker. FEMS Microbiol Ecol 84:280–289

Tourova TP, Slobodova NV, Bumazhkin BK, Sukhacheva MV, Sorokin DY (2014) Diversity of diazotrophs in the sediments of saline and soda lakes analyzed with the use of the *nifH* gene as a molecular marker. Microbiology 83:634–647

Valenzuela-Encinas C, Neria-Gonzalez I, Alcantara-Hernandez RJ, Estrada-Alvarado I, Zavala-Diaz de la Serna FJ, Dendooven L et al (2009) Changes in the bacterial populations of the highly alkaline saline soil of the former lake Texcoco (Mexico) following flooding. Extremophiles 13:609–621

van de Vossenberg JL, Ubbink-Kok T, Elferink MG, Driessen AJ, Konings WN (1995) Ion permeability of the cytoplasmic membrane limits the maximum growth temperature of bacteria and archaea. Mol Microbiol 18:925–932

van den Bosch PL, van Beusekom OC, Buisman CJ, Janssen AJ (2007) Sulfide oxidation at halo-alkaline conditions in a fedbatch bioreactor. Biotechnol Bioeng 97:1053–1063

Vargas VA, Delgado OD, Hatti-Kaul R, Mattiasson B (2004) Lipase-producing microorganisms from a Kenyan alkaline soda lake. Biotechnol Lett 26:81–86

Vargas VA, Delgado OD, Hatti-Kaul R, Mattiasson B (2005) *Bacillus bogoriensis* sp. nov., a novel alkaliphilic, halotolerant bacterium isolated from a Kenyan soda lake. Int J Syst Evol Microbiol 55:899–902

Wang S, Yang Q, Liu Z-H, Sun L, Wei D, Zhang J-Z et al (2011) *Halomonas alkalitolerans* sp. nov., a novel moderately halophilic bacterium isolated from doda meadow saline soil in Daqing, China. J Microbiol 49:24–28

Wang G, Wu H, Zhang X, Zhang H, Yang X, Tian X et al (2013) *Aliidiomarina sanyensis* sp. nov., a hexabromocyclododecane assimilating bacterium from the pool of *Spirulina platensis* cultivation, Sanya, China. Anton van Leeuwenhoek 104:309–314

Wang S, Sun L, Wei D, Zhou B, Zhang J, Gu X et al (2014) *Bacillus daqingensis* sp. nov., a halophilic,

alkaliphilic bacterium isolated from saline-sodic soil in Daqing, China. J Microbiol 52:548–53
Wani AA, Surakasi VP, Siddharth J, Raghavan RG, Patole MS, Ranade D et al (2006) Molecular analyses of microbial diversity associated with the Lonar soda lake in India: an impact crater in a basalt area. Res Microbiol 15:928–937
Whittaker RH, Likens GE (1973) Primary productivity: the biosphere and man. Hum Biol 1:357–369
Wiegel J (1998) Anaerobic alkalithermophiles, a novel group of extremophiles. Extremophiles 2:257–267
Wiegel J (2000) Ecology and diversity of anaerobic alkali-thermophiles. In: Bell CR, Brylinsky M, Johnson-Green P (eds) Microbial biosystems: new frontiers. Proceedings of the 8th international symposium on microbial ecology, Atlantic Canada Society for Microbial Ecology, Halifax, Canada, p 673–680
Wiegel J, Kevbrin VV (2004) Alkalithermophiles. Biochem Soc Trans 32:193–198
Williams RAD, Da Costa MS (1992) The genus *Thermus* and related organisms. In: Balows A et al (eds) The prokaryotes. Springer, New York, pp 3745–3753
Worakit S, Boone DR, Mah RA, Abdel-Samie ME, El-Halwagi MM (1986) *Methanobacterium alcaliphilum* sp. nov., an H_2-utilizing methanogen that grows at high pH values. Int J Syst Bacteriol 36:380–382
Wu X-Y, Zheng G, Zhang W-Wu XX-W, Wu M, Zhu X-F (2010) *Amphibacillus jilinensis* sp. nov., a facultatively anaerobic, alkaliphilic bacillus from a soda lake. Int J Syst Evol Microbiol 60:2540–2543
Xin H, Itoh T, Zhou P, Suzuki K, Nakase T (2001) *Natronobacterium nitratireducens* sp. nov., a aloalkaliphilic [sic] archaeon isolated from a soda lake in China. Int J Syst Evol Microbiol 51:1825–1829
Xu Y, Zhou P, Tian X (1999) Characterization of two novel haloalkaliphilic archaea *Natronorubrum bangense* gen. nov., sp. nov. and *Natronorubrum tibetense* gen. nov., sp. nov. Int J Syst Bacteriol 49:261–266
Xu Y, Wang ZX, Xue Y, Zhou PJ, Ma YH, Ventosa A, Grant WD (2001) *Natrialba hulunbeirensis* sp. nov, and *Natrialba chahannoensis* sp. nov., novel haloalkaliphilic archaea from soda lakes in Inner Mongolia Autonomous Region, China. Int J Syst Evol Microbiol 51:693–698
Xue Y, Fan H, Ventosa A, Grant WD, Jones BE, Cowan DA, Ma Y (2005) *Halalkalicoccus tibetensis* gen. nov., sp. nov., representing a novel genus of haloalkaliphilic archaea. Int Syst Evol Microbiol 55:2501–2505
Xue Y, Zhang X, Zhou C, Zhao Y, Cowan DA, Heaphy S et al (2006) *Caldalkalibacillus thermarum* gen. nov., sp. nov., a novel alkalithermophilic bacterium from a hot spring in China. Int J Syst Evol Microbiol 56:1217–1221
Yakimov MM, Giuliano L, Chernikova TN, Gentile G, Abraham WR, Lünsdorf H et al (2001) *Alcalilimnicola halodurans* gen. nov., sp. nov., an alkaliphilic, moderately halophilic and extremely halotolerant bacterium, isolated from sediments of soda-depositing Lake Natron, East Africa Rift Valley. Int J Syst Evol Microbiol 51:2133–2143
Yang CX, Liu YP, Bao QH, Feng FY, Liu HR, Zhang XJ et al (2012) *Mongoliitalea lutea* gen. nov., sp. nov., an alkaliphilic, halotolerant bacterium isolated from a haloalkaline lake. Int J Syst Evol Microbiol 62:647–653
Yutin N, Galperin MY (2013) A genomic update on clostridial phylogeny: gram-negative spore formers and other misplaced clostridia. Environ Microbiol 15:2631–2641
Zakharyuk AG, Kozyreva LP, Khijniak TV, Namsaraev BB, Shcherbakova VA (2015) *Desulfonatronum zhilinae* sp. nov., a novel haloalkaliphilic sulfate-reducing bacterium from soda Lake Alginskoe, Trans-Baikal Region, Russia. Extremophiles 19:673–680
Zavarzin GA, Zhilina TN (2000) Anaerobic chemotrophic alkaliphiles. In: Seckbach J (ed) Journey to diverse microbial worlds. Kluwer, Dordrecht, pp 191–208
Zavarzin GA, Zhilina TN, Kevbrin VV (1999) The alkaliphilic microbial community and its functional diversity. Microbiology 68:503–521
Zavarzina DG, Kolganova TV, Boulygina ES, Kostrikina NA, Tourova TP, Zavarzin GA (2006) *Geoalkalibacter ferrihydriticus* gen. nov., sp. nov., the first alkaliphilic representative of the family *Geobacteraceae*, isolated from a soda lake. Microbiology 75:673–682
Zavarzina DG, Tourova TP, Kolganova TV, Boulygina ES, Zhilina TN (2009) Description of *Anaerobacillus alkalilacustre* gen. nov., sp. nov.-strictly anaerobic diazotrophic bacillus isolated from soda lakes and transfer of *Bacillus arseniciselenatis, Bacillus macyae*, and *Bacillus alkalidiazotrophicus* to *Anaerobacillus* as the new combinations *A. arseniciselenatis* comb. nov., *A. macyae* comb. nov., and A. *alkalidiazotrophicus* comb. nov. Microbiology 78:723–731
Zavarzina DG, Zhilina TN, Kuznetsov BB, Kolganova TV, Osipov GA, Kotelev MS et al (2013) *Natranaerobaculum magadiensse* gen. nov., sp. nov., an anaerobic, alkalithermophilic bacterium from soda lake sediment. Int J Syst Evol Microbiol 63:4456–4461
Zhai L, Liao T, Xue Y, Ma Y (2012) *Bacillus daliensis* sp. nov., an alkaliphilic, gram-positive bacterium isolated from a soda lake. Int J Syst Evol Microbiol 62:949–953
Zhai L, Ma Y, Xue Y, Ma Y (2014) *Bacillus alkalicola* sp. nov., an alkaliphilic, gram-positive bacterium isolated from Zhabuye Lake in Tibet, China. Curr Microbiol 69:311–316
Zhang W, Xue Y, Ma Y, Grant WD, Ventosa A, Zhou P (2002a) *Marinospirillum alcaliphilum* sp. nov., a new alkaliphilic helical bacterium from Haoji soda lake in Inner Mongolia Autonomous Region of China. Extremophiles 6:33–38
Zhang WZ, Xue YF, Ma Y, Zhou PJ, Ventosa A, Grant WD (2002b) *Salinicoccus alkaliphilus* sp. nov., a novel alkaliphile and moderate halophile from Baer

Soda Lake in Inner Mongolia Autonomous Region, China. Int J Syst Evol Microbiol 52:789–793

Zhang T, Datta S, Eichler J, Ivanova N, Axen SD, Kerfeld CA, Rubin E (2011) Identification of a haloalkaliphilic and thermostable cellulose with improved ionic liquid tolerance. Green Chem 13:2083–2099

Zhao B, Chen S (2012) *Alkalitalea saponilacus* gen. nov., sp. nov., an obligately anaerobic, alkaliphilic, xylanolytic bacterium from a meromictic soda lake. Int J Syst Evol Microbiol 62:2618–2623

Zhao D-P, Zhang W-Z, Xue Y, Ma Y (2004) *Amphibacillus haojiensis* sp. nov., a novel alkaliphilic and slightly halophilic bacterium from Haoji Soda Lake in Inner Mongolia Autonomous Region, China. Acta Microbiol Sin 44:720–723

Zhao W, Weber C, Zhang CL, Romanek CS, King GM, Mills G et al (2006) *Thermalkalibacillus uzonensis* gen. nov., sp. nov., a novel aerobic alkali-tolerant thermophilic bacterium isolated from a hot spring in Uzon Caldera, Kamchatka. Extremophiles 10:337–345

Zhao W, Zhang CL, Romanek CS, Wiegel J (2008) Description of *Caldalkalibacillus uzonensis* sp. nov., and emended description of the genus *Caldalkalibacillus*. Int J Syst Evol Microbiol 58:1106–1108

Zhilina TN, Zavarzin GA (1994) Alkaliphilic anaerobic community at pH-10. Curr Microbiol 29:109–112

Zhilina TN, Zavarzin GA, Rainey F, Kevbrin VV, Kostrikina NA, Lysenko AM (1996a) *Spirochaeta alkalica* sp. nov, *Spirochaeta africana* sp. nov, and *Spirochaeta asiatica* sp. nov, alkaliphilic anaerobes from the Continental soda lakes in Central Asia and the East African Rift. Int J Syst Evol Microbiol 46:305–312

Zhilina TN, Zavarzin GA, Detkova EN, Rainey FA (1996b) *Natroniella acetigena* gen nov. sp. nov., an extremely haloalkaliphilic, homoacetic bacterium: a new member of *Haloanaerobiales*. Curr Microbiol 32:320–326

Zhilina TN, Zavarzin GA, Rainey FA, Pikuta EN, Osipou GA, Kostrikina NA (1997) *Desulfonatronovibrio hydrogenovorans* gen. nov., sp. nov., an alkaliphilic, sulphate-reducing bacterium. Int J Syst Bacteriol 47:144–149

Zhilina TN, Detkova EN, Rainey FA, Osipov GA, Lysenko AM, Kostrikina NA et al (1998) *Natronoincola histidinovorans* gen. nov., sp. nov., a new alkaliphilic acetogenic anaerobe. Curr Microbiol 37:177–185

Zhilina TN, Garnova ES, Tourova TP, Kostrikina NA, Zavarzin GA (2001a) *Amphibacillus fermentum* sp. nov. and *Amphibacillus tropicus* sp. nov., new alkaliphilic, facultatively anaerobic, saccharolytic bacilli from Lake Magadi. Microbiology 70:711–722

Zhilina TN, Garnova ES, Tourova TP, Kostrikina NA, Zavarzin GA (2001b) *Halonatronum saccharophilum* gen. nov. sp. nov.: a new haloalkaliphilic bacterium of the order *Haloanaerobiales* from Lake Magadi. Microbiology 70:64–72

Zhilina TN, Appel R, Probian C, Brossa EL, Harder J, Widdel F et al (2004) *Alkaliflexus imshenetskii* gen. nov., sp. nov., a new alkaliphilic gliding carbohydrate-fermenting bacterium with propionate formation from a soda lake. Arch Microbiol 182:244–253

Zhilina TN, Zavarzina DG, Kolganova TV, Tourova TP, Zavarzin GA (2005a) '*Candidatus Contubernalis alkalaceticum*', an obligately syntrophic alkaliphilic bacterium capable of anaerobic acetate oxidation in a coculture with *Desulfonatronum cooperativum*. Microbiology 74:695–703

Zhilina TN, Kevbrin VV, Tourova TP, Lysenko AM, Kostrikina NA, Zavarzin GA (2005a) *Clostridium alkalicellum* sp. nov., an obligately alkaliphilic cellulolytic bacterium from a soda lake in the Baikal region. Microbiology 74:557–566

Zhilina TN, Zavarzina DG, Kuever J, Lysenko AM, Zavarzin GA (2005b) *Desulfonatronum cooperativum* sp. nov., a novel hydrogenotrophic alkaliphilic, sulfate-reducing bacterium, from a syntrophic culture growing on acetate. Int J Syst Evol Microbiol 55:1001–1006

Zhilina TN, Zavarzina DG, Osipov GA, Kostrikina NA, Tourova TP (2009a) *Natronincola ferrireducens* sp. nov., and *Natronincola peptidovorans* sp. nov., new anaerobic alkaliphilic peptolytic iron-reducing bacteria isolated from soda lakes. Microbiology 78:455–467

Zhilina TN, Zavarzina DG, Kolganova TV, Lysenko AM, Tourova T (2009b) *Alkaliphilus peptidofermentans* sp. nov., a new alkaliphilic bacterial soda lake isolate capable of peptide fermentation and Fe(III) reduction. Microbiology 78:445–454

Zhilina TN, Zavarzina DG, Panteleeva AN, Osipov GA, Kostrikina NA, Tourova TP et al (2012) *Fuchsiella alkaliacetigena* gen. nov., sp. nov., an alkaliphilic, lithoautotrophic homoacetogen from a soda lake. Int J Syst Evol Microbiol 62:1666–1673

Zhilina TN, Zavarzina DG, Kevbrin VV, Kolganova TV (2013) *Methanocalculus natronophilus* sp. nov., a new alkaliphilic hydrogenotrophic methanogenic archaeon from a soda lake, and proposal of the new family *Methanocalculaceae*. Microbiology 82:698–706

Zhilina TN, Zavarzina DG, Detkova EN, Patutina EO, Kuznetsov BB (2015) *Fuchsiella ferrireducens* sp. nov., a new haloalkaliphilic, lithoautotrophic homoacetogen capable of iron reduction, and emendation of the description of the genus *Fuchsiella*. Int J Syst Evol Microbiol. doi:10.1099/ijs.0.000278 [Epub ahead of print]

Zhu S (2008) Use of ionic liquids for the efficient utilization of lignocellulosic materials. J Chem Technol Biotechnol 83:777–779

Tiny and Tough: Microphytes of East African Soda Lakes

6

Lothar Krienitz and Michael Schagerl

Abstract

Soda lakes of the East African Rift System and their connected hot springs offer environments that promote growth of extremophiles. We provide an overview of the cyanobacterial and algal flora and characterise the taxa observed in these habitats. East African soda lakes (EASL) are listed amongst the most productive ecosystems worldwide. The main driving force of these systems is phytoplankton; extremely high light attenuation minimises growth of phytobenthos, and larger plants occur only sporadically along shores. Some EASL are home to *Arthrospira fusiformis*, formerly called *Spirulina platensis*, which is the main food of Lesser Flamingos. This species and other Cyanobacteria are, however, also discussed as a possible cause for sporadic flamingo die-offs.

We summarise the somewhat confused taxonomy of *Arthrospira* and *Spirulina* and also show that—contrary to common opinion—*Arthrospira* blooms are not persistent. Under unfavourable growth conditions, *Arthrospira* is replaced by other taxa such as the prasinophyte *Picocystis salinarum* at increased ion concentrations. Besides Cyanobacteria, some of them diazotrophic, the algal community comprises mainly diatoms, coccoid green algae and flagellates of various algal groups. These extreme biotopes are usually treated as systems of low diversity, but, very recently, studies applying molecular tools revealed an unexpected hidden diversity. We demonstrate shifts of the phytoplankton community in these saline alkaline waters and discuss approaches to interpret this phenomenon based on biotic interactions and abiotic factors. Special attention is given to effects of changing salinity on the food web.

L. Krienitz (✉)
Department of Experimental Limnology, Leibniz-Institute of Freshwater Ecology and Inland Fisheries, 16775 Stechlin, Germany
e-mail: krie@igb-berlin.de

M. Schagerl
Department of Limnology and Bio-Oceanography, University of Vienna, Althanstrasse 14, 1090 Vienna, Austria
e-mail: michael.schagerl@univie.ac.at

M. Schagerl (ed.), *Soda Lakes of East Africa*, DOI 10.1007/978-3-319-28622-8_6

6.1 Introduction

Around 45 % of the global net productivity originates from aquatic ecosystems (Field et al. 1998), which is achieved by microphytes (Cyanobacteria and algae). These small photoautotrophs form the base of most aquatic food webs including those of the East African soda lakes (EASL). In contrast to other salt lakes, the salinity of soda lakes is mainly defined by high amounts of sodium, carbonate and hydrogen-carbonate, which is the reason for the high pH. The special environment promotes growth of alkaliphilic, halo- and thermotolerant organisms (Grant et al. 1990). Some lakes show strong gradients in salinity, for example, from hot springs with lower salinity or diffuse groundwater inflows along the shores towards the lake's centre with higher salinity. This interesting phenomenon has mostly been neglected. Most EASL are shallow and prone to water level changes, which is accompanied with large variations in salinity. These shifts in chemistry influence algal communities and other organisms.

It is commonly assumed that EASL are *Arthrospira*-lakes, but the reality is different: only a handful of lakes actually host this organism in high densities, such as Lake Chitu (Ethiopia), Lake Bogoria (Kenya) and Lake Big Momela (Tanzania). In other soda lakes, this taxon occurs in lower densities and is associated with other Cyanobacteria such as various species of *Anabaenopsis* or *Cyanospira* (Krienitz et al. 2016). Moreover, the cyanobacterial blooms are not persistent and collapse from time to time. Major confusion exists about the naming of the most prominent cyanobacterium. It is sometimes still called *Spirulina platensis* (Gomont) Geitler, but the correct identification is *Arthrospira fusiformis* (Voronikhin) Komárek et Lund. This taxon is found in Africa and Eurasia, whereas another species, *Arthrospira maxima* Setchell et Gardner, with similar morphology is known from America. Both species are exploited commercially as a food additive.

A first impression of microphytes in the soda lakes was provided by the zoologist Penelope Jenkin, who joined the famous Percy Sladen Expedition in 1929. In her letter to the editor of the journal 'Nature', Jenkin reported on the 'microphagous' feeding of flamingos at Lake Nakuru on the bloom-forming 'blue-green alga, *Spirulina* sp., in the plankton' (loc. cit. Jenkin 1929, p. 574). Furthermore, she assumed that diatoms provide another food source of flamingos. Jenkin handed over her algal samples to Florence Rich for more detailed investigations. Rich (1932) studied phytoplankton of the three Kenyan soda lakes Nakuru, Elmentaita and Sonachi (sometimes referred to as Green Crater Lake) and found that the dominating population of *Arthrospira* was interspersed with other Cyanobacteria such as *Anabaenopsis* and diatoms. In 1932–1933, the Mission Scientifique de l'Omo, headed by Jeannel and Arambourg and accompanied by the hydrobiologist Pierre-Alfred Chappuis, again visited lakes of the East African Rift System. Chappuis handed over samples to the phycologist Hans Bachmann, who amongst other findings provided a detailed description of the plankton community of Lake Elmentaita, showing once again the invasion of *Anabaenopsis* into the dominating *Arthrospira* populations (Bachmann 1939). Later on, scientists occasionally collected samples and confirmed earlier findings that soda lakes are characterised by only few taxa of microphytes establishing extremely high biomass (summarised in Wood and Talling 1988). In contrast to current opinion, molecular sequence analyses of environmental samples from five soda lakes of Ethiopia (Lanzén et al. 2013) and one sample from Kenya (Luo et al. 2013) revealed a much higher diversity of pro- and eukaryotes than was previously expected. This calls for further detailed studies on the microbial community.

In some EASL, the cyanobacterial and algal biomass forms the basis of a unique direct interaction between microphytes as primary producers and warm-blooded primary consumer birds, the Lesser Flamingos. This sensitive interaction has equally thrilled microbiologists, phycologists, ornithologists and ecologists until today. Ground-breaking research on the food web and energy flows of Lake Nakuru was conducted by Ekkehard Vareschi and co-workers, making this ecosystem the best-studied soda lake in the East African Rift System (Vareschi 1978, 1982; Vareschi and Jacobs 1985). Since around 20 years, recurring mass mortalities of flamingos have been reported (Ndetei and Muhandiki 2005). Besides bird diseases (Kock et al. 1999; Sileo et al. 1979), intoxification with pesticides or heavy metals (Greichus et al. 1978; Kairu 1996), also cyanotoxins have been forwarded as potential causes for the die-offs (Ballot et al. 2004, 2005; Metcalf et al. 2013). Recently, Peduzzi et al. (2014) proved that crashes of *Arthrospira* blooms are caused by cyanophages, which also affect the flamingo population via cascading effects: the breakdowns seem to trigger flamingo movements from one lake to another (Kaggwa et al. 2013; Kihwele et al. 2014).

6.2 Lake Systems and Their Algal Communities

Phytoplankton of **Ethiopian soda lakes** was investigated by Wood and Talling (1988), Kebede and Willén (1998) and Kebede (2002) and interpreted in a context of salinity. A clear pattern was observed with lower species number at higher salinity (see Chap. 12). Interestingly, some lakes formerly treated as characteristic *Arthrospira* systems now show different communities because of drastic changes in chemistry caused by human activities such as soda ash abstraction in the case of Lake Abijata (Kebede and Willén 1996) or eutrophication and seismic explosion experiments in the area of Lake Arenguade (Girma et al. 2012), one of the small crater lakes around the city of Debre Zeyit. In this lake, a small-celled ecotype of *A. fusiformis* was observed (pers. observation in autumn 2014), which was originally described from Lake Oloidien, Kenya (Krienitz et al. 2013b). In the crater lake Chitu, *A. fusiformis* is still persistent (Ogato and Kifle 2014, Fig. 6.1). Located just north of Lake Chitu, Lake Shala is the largest (~330 km^2) and deepest crater lake (~260 m) of Ethiopia and envisaged as a water reservoir for drying Lake Abijata. Information on its limnology and phytoplankton, however, is almost lacking. Kebede (2002) reported high abundances of halophilic diatoms such as *Chaetoceros* and *Thalassiosira* for this lake. Lake Basaka (Metahara), whose surface area has increased considerably in the past 40 years (Dinka 2012; Dinka et al. 2014), showed a community comparable to Lake Shala (Kebede and Willén 1998). Diatom assemblages of EASL including also some water bodies from the Afar Depression have been studied by Gasse et al. (1983). This thorough survey revealed five groups of diatoms related to water chemistry, one of them representing saline, alkaline environments.

Most phycological research on EASL has been conducted in the **Kenyan lakes** Bogoria, Elmentaita, Nakuru and Sonachi. Detailed studies focusing on nutrients and phytoplankton dynamics were already carried out in the 1970s and 1980s (Hecky and Kilham 1973; Milbrink 1977; Melack and Kilham 1974; Melack 1982; Tuite 1981; Vareschi 1982). The water bodies were low in species number and indicated temporal shifts towards taxa other than *Arthrospira fusiformis* such as *Anabaenopsis arnoldii* Aptekar, *Synechococcus* spp., *Monoraphidium minutum* (Nägeli) Komárková-Legnerová and the diatoms *Anomoeoneis sphaerophora* Pfitzer, *Navicula elkab* Müller and *Nitzschia frustulum* (Kützing) Grunow. The limnology of crater lake Sonachi was reconstructed for the last 175 years by Verschuren et al. (1999). The community shifted from a dominance of *Arthrospira fusiformis* to Chroococcales (Njugana 1988), after which *Arthrospira* prevailed again (Ballot et al. 2005). Harper et al. (2003) conducted a biodiversity study of Lake Bogoria including phytoplankton. Ballot et al. (2004) included a description of the phytoplankton community composition in lakes Bogoria, Elmentaita and Nakuru. Schagerl and Oduor (2008) and Schagerl et al. (2015) studied the influence of environmental factors on phytoplankton community composition. Ballot et al. (2005), Kaggwa et al. (2012),

Fig. 6.1 Impressions of flamingo lakes. (**a**, **c**) The small Ethiopian crater lake Chitu is home to thousands of Lesser Flamingos (November 2014); (**b**) *Arthrospira fusiformis* sampled from Lake Chitu; (**d**) Lake Bogoria (Kenya, November 2011)

Krienitz et al. (2003), Krienitz and Kotut (2010) and Schagerl and Oduor (2008) examined algae communities and reported periodic dominance and paucity amongst the different phytoplankton groups especially at extreme salinity levels. The genus *Cyanospira* was originally described by Florenzano et al. (1985) from Lake Magadi. Lake Magadi is dominated by *Arthrospira* and *Synechococcus*, but the chlorophyte *Picocystis salinarum* Lewin was also recorded there (Krienitz et al. 2012). Lake Turkana is almost neglected in phycological research, although it covers about 7500 km^2 and is the biggest lake in Kenya (Kolding 1992; the northernmost part is located in Ethiopia). Källqvist et al. (1988) found generally low phytoplankton amounts here (<10 μg L^{-1} chlorophyll-a); net plankton samples showed mostly *Botryococcus braunii* Kützing and *Microcystis aeruginosa* Kützing in the open water, reflecting the increased turbidity of Lake Turkana. Around the shoreline of Ferguson's Gulf, *Anabaenopsis arnoldii* was abundant, which indicated limitation of nitrogen supply.

Scarce information on algal communities in **Tanzania** is available. Melack and Kilham (1974) studied pelagic primary productivity in lakes of northern Tanzania (Big Momela, Reshitani, Magad), which were dominated by *Arthrospira fusiformis* and *Glaucospira laxissima* (G.S.West) Simic, Komárek and Dordevic. Lugomela et al. (2006) reported high abundances of *Arthrospira fusiformis* from Lake Big Momela and the small crater lake Embagai, interspersed with filamentous Cyanobacteria and pennate diatoms. In Lake Manyara, Lugomela et al. (2006) observed *Pseudanabaena*, *Oscillatoria* and *Aphanocapsa*. In this lake, the community shifted towards *Arthrospira fusiformis* in 2008 (Kihwele et al. 2014); this increase was highly related to Lesser Flamingo occurrence and nitrogen brought into the system by the birds.

6.3 Extreme Biotopes: The Hot Spring Tributaries

Hot springs and geysers are unique habitats in the catchment area of soda lakes around the world and considered to be inhabited by 'subcosmopolitan' taxa (species occurring throughout the world but only in 'appropriate' habitats; Padisák 2009). Interestingly, the geographic isolation of hot spring Cyanobacteria has led to evolutionary divergences on both the local and global scale (Castenholz 1996; Papke et al. 2003). In the East African Rift System, hot springs are located along the shorelines of lakes Bogoria (Fig. 6.2), Elmentaita and Magadi in Kenya, Manyara and Natron in Tanzania and Abijatta and Shala in Ethiopia (Fig. 6.2). Hot spring water is considerably lower in salinity compared to soda lakes; nevertheless, it shows marked differences in salinity: for example, hot springs at Elmentaita are subsaline (1.6–2.3‰), at Bogoria hyposaline (3.5–3.9‰) and at Magadi mesosaline (29–32‰). Exchange and contact of the spring organisms with the surrounding habitats are given by the streamlets of the springs discharging into the lake. Furthermore, flamingos frequently visit the outlets of hot springs for drinking and maintaining their plumage. Depending on the lake water level, hot spring areas can be covered by lake water, which enables a direct exchange of organisms and of the genetic pool. A question of interest is whether the cyanobacterial and algal flora of the various lakes and spring habitats differ or not, whereas the lakes are extreme in terms of increased salinity and pH, the hot springs show extreme temperatures.

The ecotones between hot springs and the lake are peculiar habitats and have been categorised into six major types by Owen et al. (2004). Based on phycological considerations, three groups are relevant: Type 1A = proximal hot/warm spring streamlets with water temperatures > 40 °C; Type 1B = hot/warm spring marshes of shallow

Fig. 6.2 Hot springs. (**a**, **b**) Lake Shala (2014); (**c**) hot water discharging into Lake Bogoria (2011); (**d**) hot springs and geysers along the shore of Bogoria (2008)

standing waters with temperatures ranging from 30 to 40 °C; and Type 5 = hypersaline lake littoral wetlands along the shorelines. The habitats are populated by microphytes that are adapted to the high disturbances in these transitional zones. Even minor changes in the lake water level and the lake-marginal groundwater can impact the 'behaviour' of the hot springs and adjacent habitats (Renaut et al. 2008, 2013), resulting in fluctuations of cyanobacterial populations (Krienitz et al. 2005). Owen et al. (2004) reported dissimilarities in the diatom flora at interface zones in the Bogoria area: in hot springs, the diversity was low, and common diatoms such as *Anomoeoneis sphaerophora* var. *guentheri* Müller, *Navicula cryptonella* Lange-Bertalot, *Craticula cuspidata* (Kützing) Mann and *Nitzschia invisitata* Hustedt were observed. The wetland habitats were populated by *Fragilaria brevistriata* Grunow, *Gomphonema parvulum* (Kützing) Kützing, *Navicula tenelloides* Hustedt, *Nitzschia communis* Rabenhorst, *N. latens* Hustedt, *N. sigma* (Kützing) W. Smith, *Rhopalodia gibberula* (Ehrenberg) Müller and *Stauroneis anceps* Ehrenberg. The hot spring diatom assemblages of Lake Baringo (freshwater) and Elmentaita were studied by Mpawenayo and Mathooko (2004), who reported a rich flora of 58 species in Elmentaita with Naviculaceae (22 taxa) and Epithemiaceae (13) being best represented. Similar to Bogoria hot springs, the dominant species were *Anomoeoneis sphaerophora* (38 %) and *Rhopalodia gibberula* (32 %) followed by *Aulacoseira granulata* (Ehrenberg) Simonsen (12 %).

Dadheech et al. (2013) studied biofilms in the hot springs and pelagic and littoral zones of Lake Bogoria by means of light microscopy, multilocus 454-amplicons sequencing and metagenomics. The flora was characterised by a high number of cyanobacterial taxa from different systematic groups. Compared to hot springs of other continents, most of the phylogenetic lineages occurred exclusively in the Bogoria hot springs, suggesting biogeographic speciation. The prevalent phylotypes were members of the Oscillatoriales mainly related to *Leptolyngbya*, *Spirulina*, *Oscillatoria*-like and *Planktothricoides*. Chroococcales were represented by different *Synechococcus* clades. Not a single phylotype of *Synechococcus* from Bogoria hot springs clustered with one of the lineages described in earlier studies from other continents. All lineages identified in the springs were not observed in the lake, showing that they were unable to survive under the strong alkaline conditions of soda lakes where *Arthrospira* and *Anabaenopsis* dominated. Despite the close habitat connection, cyanobacterial communities of the hot springs and the lake differed considerably. This reflects their restricted survival capability under particular extreme conditions. Surprisingly, a few phylotypes of *Arthrospira* and *Anabaenopsis* were detected in the hot springs, indicating their wide adaptability.

The transition zones could be very important for buffering future changes of the ecosystems. They harbour a rich diversity of cyanobacterial phylotypes, providing a genetic pool for evolutionary developments in the lake. According to Kirkwood et al. (2008), varying salinity promotes diversification, which might have a positive or negative impact on the organisms in the food webs. For instance, Lesser Flamingos are highly sensitive against changes in food quality (Krienitz and Kotut 2010) and mostly rely on high densities of *Arthrospira fusiformis* (see also Chaps. 10 and 13). Hot springs and transition zones can provide alternative food resources or establish producer populations containing inedible or even toxic organisms. Global climate-related changes along with disruption and alteration due to increasing anthropogenic disturbance also considerably affect these archaic habitats of Africa's soda lakes.

6.4 Cyanobacteria

6.4.1 *Arthrospira fusiformis*

Arthrospira fusiformis is a character species of soda lakes in Africa and Asia. It was mainly reported from the tropics (Komárek and Anagnostidis 2005), but is also found in temperate regions, e.g. Serbia (Fuzinato et al. 2010). Apart from their main occurrence in soda lakes, *Arthrospira* is capable of surviving in a wide range of conditions extending from freshwater (e.g. lakes Chad and Naivasha) to hypersaline water bodies (e.g. lakes Abbe and Katwe, Ciferri 1983; Dadheech et al. 2010, Krienitz and Schagerl unpubl. results). *A. fusiformis* thrives at salinities between 22 and 62‰ (Iltis 1968) and tolerates temperatures of > 35 °C (Vonshak and Tomaselli 2000). At laboratory conditions, Kebede (1997) studied growth of *A. fusiformis* strains over a range of salinities and different salts and found an inverse relationship, although growth was still possible at a salinity of about 90‰. It is an alkaliphilic species (Grant et al. 1990; Mesbah and Wiegel 2011) that thrives even at pH beyond 9.5 (Belkin and Boussiba 1991). *A. fusiformis* probably depends on elevated sodium concentrations to maintain pH homoeostasis at high pH (Schlesinger et al. 1996), because sodium is needed for the Na^+/H^+ antiport through the plasma membrane

(Krulwich et al. 2011; Pogoryelov et al. 2003) (see Chap. 12).

Strong saline alkaline conditions promote mass development of this taxon, which is well documented in lakes of East Africa such as lakes Bogoria (Hindák 1985; Harper et al. 2003; Kaggwa et al. 2013), Nakuru (Krienitz and Kotut 2010; Vareschi 1978, 1982), Simbi (Melack 1979a), Elmentaita (Ballot et al. 2004; Melack 1988; Schagerl and Oduor 2008), Sonachi (Ballot et al. 2005; Rich 1932), Oloidien (Krienitz et al. 2013b), Big Momela (Lugomela et al. 2006), Abijata (Kebede and Willén 1998) and Chitu (Kebede and Ahlgren 1996; Ogato and Kifle 2014). Remote sensing of chlorophyll-a as a proxy for phytoplankton biomass was applied for Lake Bogoria (Tebbs et al. 2013). This promising technique offered new insights into the horizontal distribution of the algal blooms and might also be applied in the future for studies focusing on phytoplankton development and crashes.

The biomass of *Arthrospira* reached maximal values of 200–1430 mg L^{-1} dry mass (Kaggwa et al. 2013; Tuite 1981; Vareschi 1978, 1982). An exceptional high primary production places the African soda lakes amongst the world's most productive ecosystems (Oduor and Schagerl 2007a and references therein). Using the diurnal changes in oxygen concentration of the water column, Melack and Kilham (1974) obtained gross primary productivity rates of 36 g m^{-2} O_2 $(12\ h)^{-1}$ for Lake Nakuru; a community rate of 57 g m^{-2} O_2 d^{-1} was recorded for Lake Arenguade (Ethiopia) by Talling et al. (1973), and similar rates were measured for Lake Simbi in Kenya (Melack 1979a). Maximum daily net productivities (12 h) for Kenyan lakes Bogoria and Elmentaita reached almost 20 g m^{-2} $(12\ h)^{-1}$ (Oduor and Schagerl 2007a); Nakuru showed even slightly higher rates (Oduor and Schagerl 2007a; Vareschi 1982).

Most of the studies focusing on EASL system productivity refer to *Arthrospira* as the main food source of Lesser Flamingos, which occur there in numbers between two million (Brown 1959) and three million (Mlingwa and Baker 2006). Other main consumers of *Arthrospira* are the fish *Alcolapia grahami* Boulenger; the rotifers *Brachionus dimidiatus* Bryce, *B. plicatilis* Müller and *Hexarthra jenkinae* De Beauchamps; the copepod *Lovenula africana* Daday and insects such as chironomids (Harper et al. 2003; Vareschi 1982). Populations of *Arthrospira*, however, turned out to be quite unstable and collapse in irregular intervals (Table 6.1). Peduzzi et al. (2014) proved virus infection as a major cause for *A. fusiformis* crashes in Kenyan lakes. *Arthrospira fusiformis* is a planktonic organism whose cells contain well-discernible aerotopes

Table 6.1 Crashes of *Arthrospira* in Kenyan soda lakes

Date of crash	Lake	Reference
1971	Elmentaita	Melack and Kilham (1974)
1973	Nakuru	Vareschi et al. (1981)
1973 (Mar–Sep)	Bogoria	Melack (1976)
1973, 1974	Elmentaita	Melack (1988)
1974	Nakuru	Vareschi (1982)
1974, 1976	Elmentaita	Tuite (1981)
1974 (Jan–Mar)	Nakuru	Tuite (2000)
2001 (Jun)	Nakuru	Ballot et al. (2004)
2001 (Oct)	Bogoria	Harper et al. (2003)
2002 (Sep)	Elmentaita	Ballot et al. (2004)
2004	Elmentaita	Schagerl and Oduor (2008)
2004 (Jul–Oct)	Bogoria	Schagerl and Oduor (2008)
2006	Bogoria	Krienitz and Kotut (2010)
2008 (Dec)–2009 (Apr) and 2009 (Sep–Oct)	Nakuru	Kaggwa et al. (2012)
2011	Elmentaita	Krienitz et al. (2013a)
2012	Nakuru	Krienitz et al. (2013a)
2013–2015	Bogoria	Krienitz et al. (2016)

regulating buoyancy in the water column. Coding genes of the aerotopes were recently elucidated by Miklaszewska et al. (2012). The non-heterocytous genus shows screw-like coiled trichomes which elongate by division (binary fission) of the cells; multiplication of trichomes involves transcellular fragmentation of the filaments (Tomaselli et al. 1981) leading to break-up of an intercalary, necridic cell. Hence, apices of a freshly fragmented filament show remains of the destroyed cell, appearing in the form of a mucilaginous sheath, whereby the cell itself is missing. Often, diameters of helices are narrowed towards the ends. Cross walls in the filament are clearly visible, with no or slight constrictions. Cells are cylindrical and shorter than wide; apical cells are rounded (Figs. 6.2 and 6.3).

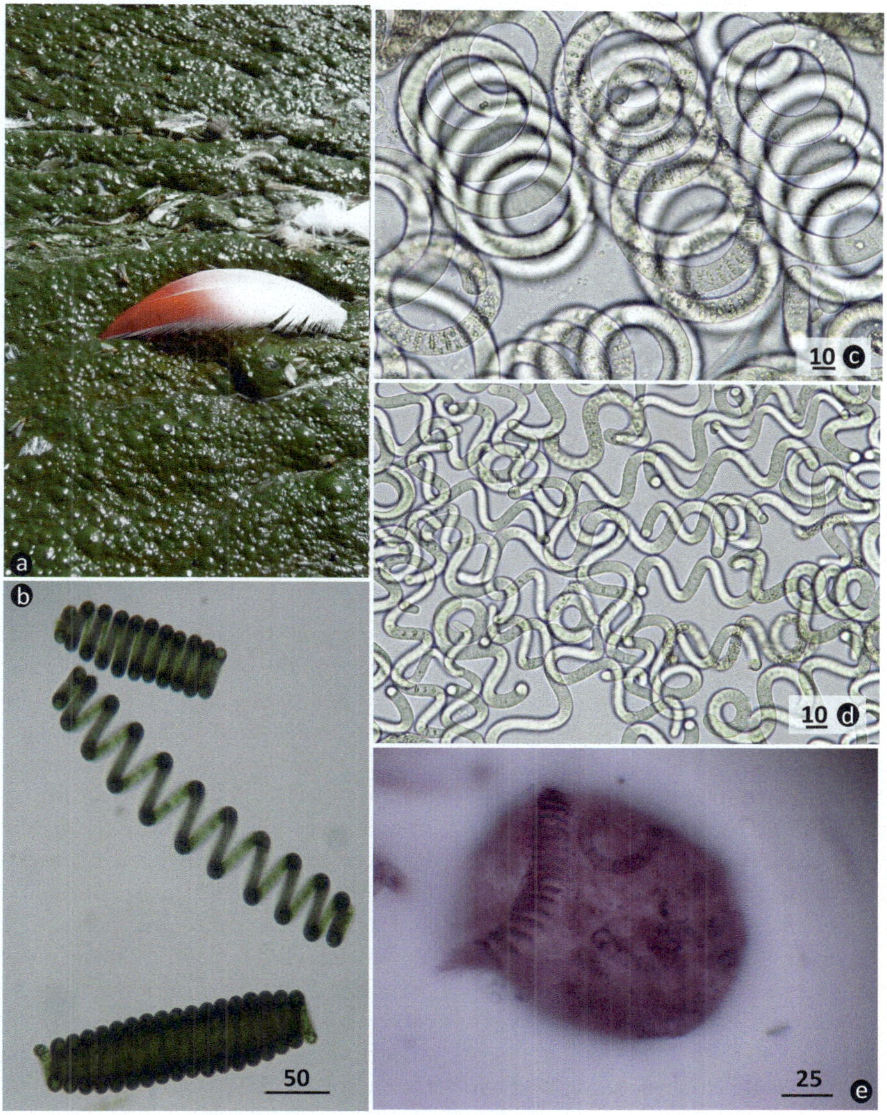

Fig. 6.3 Cyanobacteria 1. (**a**) Scums of Cyanobacteria on the lake shore; (**b–d**) *Arthrospira fusiformis* morphotypes; (**b**) densely and loosely coiled filaments; (**c**) detail of loosely coiled filaments; (**d**) small ecotype found in lakes of lower salinity such as Oloidien (Kenya) and Arenguade (Ethiopia); (**e**) a large ciliate with ingested *Arthrospira*. Scale bar in μm

The size and shape of trichomes differ considerably: trichome width (cell diameter) 3–15 μm, coil diameter 15–60 μm, pitch (distance between two coils) 0–80 μm. In a thorough sampling survey (16 months in weekly intervals) of lakes Bogoria and Nakuru, Kaggwa et al. (2013) focused on the morphology of *Arthrospira* and related it to environmental factors. Pulse-amplified modulated (PAM) fluorescence measurements revealed that large and loosely coiled filaments represent healthy conditions; this type dominated at elevated levels of soluble reactive phosphorus (SRP), wind speed (worse light supply because of increased turbulence), temperature and conductivity. The opposite was true for short and small helices. Large narrow-coiled filaments prevailed under grazing pressure, infections by cyanophages and elevated nitrate concentrations (Kaggwa et al. 2013). In lakes with lower salinity such as Lake Oloidien (Kenya), a small ecotype of *Arthrospira* was observed (Krienitz et al. 2013b; Fig. 6.3). The material from Oloidien exhibited filament diameters of 4–5 μm and coil diameters of 20–40 μm in nature. Under culture conditions, *Arthrospira* from Oloidien lost its small size within 30 days and reached larger dimensions comparable to those measured in field material of Lake Bogoria (filament diameter 12–15 μm, coil diameter 25–40 μm). This ecotype is also found in the Ethiopian Lake Arenguade (pers. observation).

The remarkable phenotypic differences of *Arthrospira* raised the question of hidden genetic diversity in this taxon. Phylogenetic analyses of the internal transcribed spacer (ITS) between 16S rRNA and 23S rRNA genes, along with partial sequences of the intergenic spacer (cpcBA-IGS) of the phycocyanin operon of several phenotypes isolated from EASL, revealed that they belong to the same genotype (Dadheech et al. 2010). The genus *Arthrospira* is considered to be phylogenetically homogeneous. Based on 16S rRNA sequences, it belongs to one joint lineage (Nelissen et al. 1994). ITS analyses showed two clades, each divided in two subclades IA, IB, IIA, and IIB (Baurain et al. 2002; Scheldeman et al. 1999). Members of clade I were assigned to the planktonic, aerotope-bearing morphospecies *Arthrospira maxima*, which prefers waters of tropical America, whereas clade II includes mainly strains of *A. fusiformis* from tropical Africa and Asia (Dadheech et al. 2010). These findings confirm the traditional split-up of the two taxa (Tomaselli 1997) and investigations based on total DNA restriction profiles (Viti et al. 1997). Comte et al. (2013) defined clade III in *Arthrospira* consisting of *A. jenneri* Stizenberger ex Gomont, *A. erdosensis* Tian et al. and a strain formerly designated as *Lyngbya laxespiralis* Skuja. Future phylogenetic studies on benthic types of *Arthrospira* may reveal more lineages of coiled filamentous Cyanobacteria placed within or outside the *Arthrospira* clade.

The taxonomy and nomenclature of *Arthrospira fusiformis* are highly complicated. This subject was discussed in detail by Komárek and Lund (1990) and Vonshak and Tomaselli (2000). Confusingly, several synonyms are still in use, most commonly *Spirulina platensis* and *Arthrospira platensis* (Nordstedt) Gomont. The problems have historical roots: within the genus *Spirulina* Turpin ex Gomont, *Spirulina jenneri* (Hassal) Kützing was described from a European stagnant water body and *Spirulina jenneri* f. *platensis* Nordstedt as a benthic variety from a pond in Uruguay (see Gomont 1892). Stizenberger (1852) established the new genus *Arthrospira*; the key characteristic distinguishing it from *Spirulina* was visible cross walls: *Spirulina* was considered to be unicellular, whereas *Arthrospira* showed clear septa in the filaments. Gardner (1917) described *Arthrospira maxima* from a saline pond in California. Geitler (1925), however, merged the two genera under the older name *Spirulina* and renamed *Spirulina jenneri* f. *platensis* into *Spirulina platensis*. In 1936, Voronichin (1934, cited in Fuzinato et al. 2010) described *Spirulina fusiformis* from Tanatar, a Siberian soda lake. The genera *Spirulina* and *Arthrospira* were re-established again by Castenholz (1989) and Desikachary (1959). There are strong arguments for keeping both genera *Arthrospira* and *Spirulina* based on phenotypic, genotypic and ecological perspectives.

Spirulina filaments are considerably smaller than those of *Arthrospira*. Major diacritic characteristics of the two genera are summarised by Anagnostidis and Komárek (1988) and Vonshak and Tomaselli (2000). Today, *Arthrospira* and *Spirulina* are considered as members of two different orders: *Arthrospira* is a member of the Oscillatoriales, and *Spirulina* belongs to Spirulinales (Komárek et al. 2014).

To make it more complicated, there is also a discrepancy in the typification of *Arthrospira*. According to the International Botanical Code of Nomenclature, the (lecto)type of the genus is the benthic *A. jenneri*, whereas according to the Bacteriological Code, the reference strain is *A. platensis* PCC 7345 isolated from a saline marsh (Vonshak and Tomaselli 2000).

Arthrospira fusiformis (in Africa and Asia) and *A. maxima* (in America) were utilised as food by indigenous human populations (Ciferri 1983). Later, biotechnological applications were established based on highly productive *Arthrospira* strains in mass cultures (Belay 2013; Sili et al. 2012; Vonshak and Richmond 1988). In commercially available *Arthrospira* products, the name *Spirulina* is still commonly used.

6.4.2 *Spirulina subsalsa* Oersted ex Gomont

S. subsalsa is a benthic species like other *Spirulina* taxa. *Spirulina* is an abundant member of cyanobacterial mats in hot springs (Hindák 2001; Dadheech et al. 2013, pers. observation) and found only sporadically in the plankton (Ballot et al. 2004), which may be due to drifting from spring rivulets discharging into the lakes.

Morphologically, *S. subsalsa* is characterised by fine, more or less long cylindrical filaments which are 1–2 μm wide and regularly tightly screw-like coiled; the coils are mostly joined to each other (Fig. 6.4a). Trichomes lack a mucilaginous sheath; the apical cell is rounded. Cross walls in the filaments are nearly invisible. The cell content is homogeneous with fine granules, without aerotopes, living solitary or in thalli of very different colours (bright or dirty blue green, blackish or reddish). Apart from the dominating taxon *S. subsalsa*, the tiny *S. subtilissima* Kützing is also found in EASL (Hindák 2001), differing by considerably finer filaments (maximum 1 μm) and wider helices with pitches of 1.5–3.0 μm.

6.4.3 *Anabaenopsis* and *Cyanospira*

In EASL, five different species frequently compete against the *Arthrospira* dominance: *Anabaenopsis abijatae* Kebede et Willén, *A. arnoldii*, *A. elenkinii* Miller, *Cyanospira capsulata* Florenzano et al. and *C. rippkae* Florenzano et al. (Figs. 6.4 and 6.5). The main diacritic criterium between *Anabaenopsis* and *Cyanospira* is the apoheterocytic development of akinetes by the latter taxon, i.e. the ability to transform all vegetative cells between two heterocytes into akinetes (Sili et al. 2011). Only a few molecular criteria are available to differentiate between the two genera (Ballot et al. 2008; Iteman et al. 2002; Komárek and Mareš 2012; Sili et al. 2011). Other diacritic characteristics of the species include the shape and size of the vegetative cells, the presence and shape of heterocytes and akinetes and the general organisation of the colonies, along with the production of mucilage (Krienitz et al. 2013a). The filaments of these nostocalean taxa have barrel-shaped cells forming widely arcuated or irregularly screw-like spirals united into large colonies embedded in thick mucilage. The intercalary heterocytes develop in pairs. Filaments disintegrate soon after new formation of heterocytes, bringing each heterocyte into a terminal position of the filament fragments. Akinetes are typically established in rows, which are longer in the genus *Cyanospira* containing tens of akinetes. The dense, botryoid colonies of *A. abijatae*, originally described from Lake Abijata (Ethiopia), have a very special appearance. They might be confused with *Microcystis* (see below). Huge mucilaginous colonies can be established by *C. capsulata*. They reach dimensions in the millimetre range,

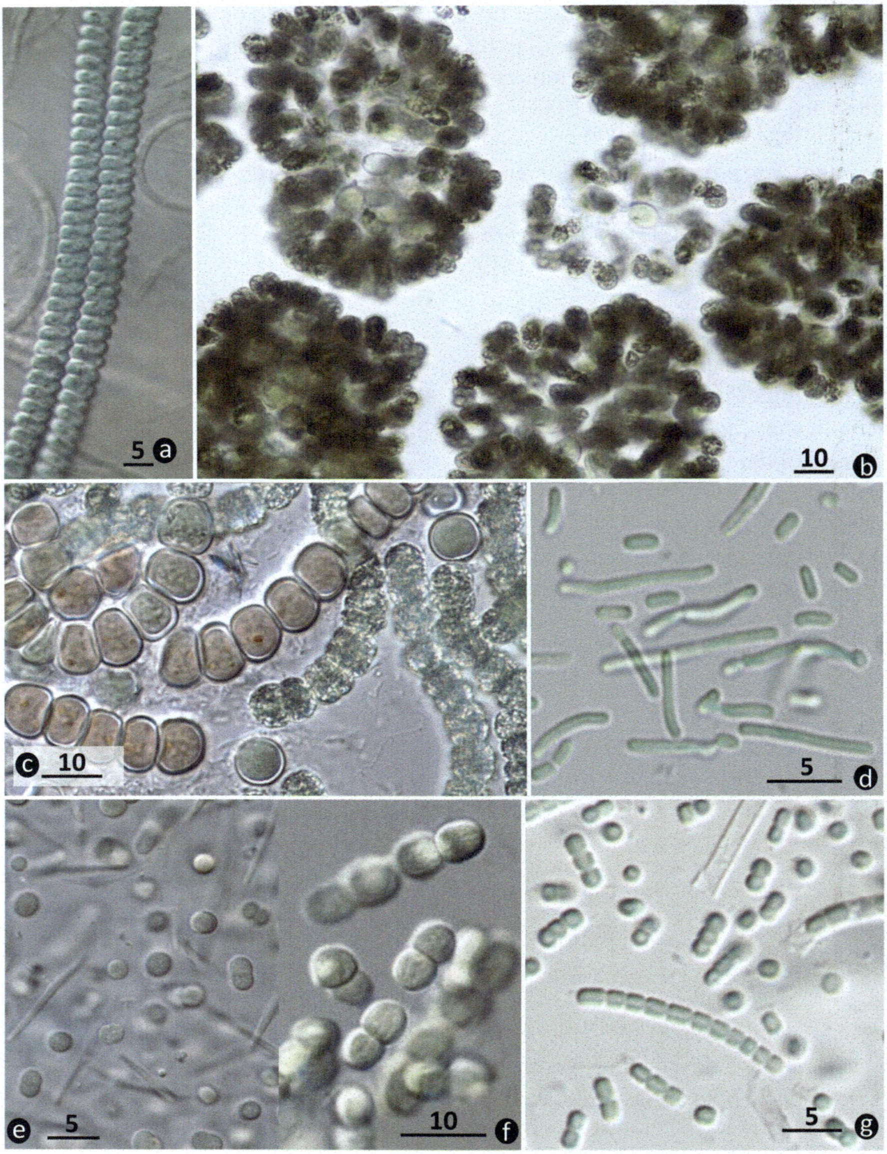

Fig. 6.4 Cyanobacteria 2. (**a**) *Spirulina subsalsa*; (**b**) *Anabaenopsis abijatae*; (**c**) *Cyanospira rippkae* (*brownish cells* are akinetes); (**d**) *Synechococcus* sp.; (**e**) *Myxobactron* sp. (*needle-like*, small cells) and *Synechocystis* cf. *salina* (± *spherical cells*); (**f**) *Cyanobacterium minervae*; (**g**) *Haloleptolyngbya alcalis*. Scale bar in µm

which is too large for ingestion by Lesser Flamingos.

Species of *Anabaenopsis* and *Cyanospira* prefer different salinity ranges (Krienitz et al. 2013a): the highest salinity between 30 and 40‰ supports growth of *C. capsulata* in Bogoria. *C. rippkae* is increasingly observed at a salinity range between 25 and 35‰ in Nakuru. *A. arnoldii* and *A. abijatae* prefer a salinity of 15 and 30‰ in Elmentaita and Nakuru, respectively, but disappears when salinity falls below 10‰. *A. elenkinii* growth is promoted at almost

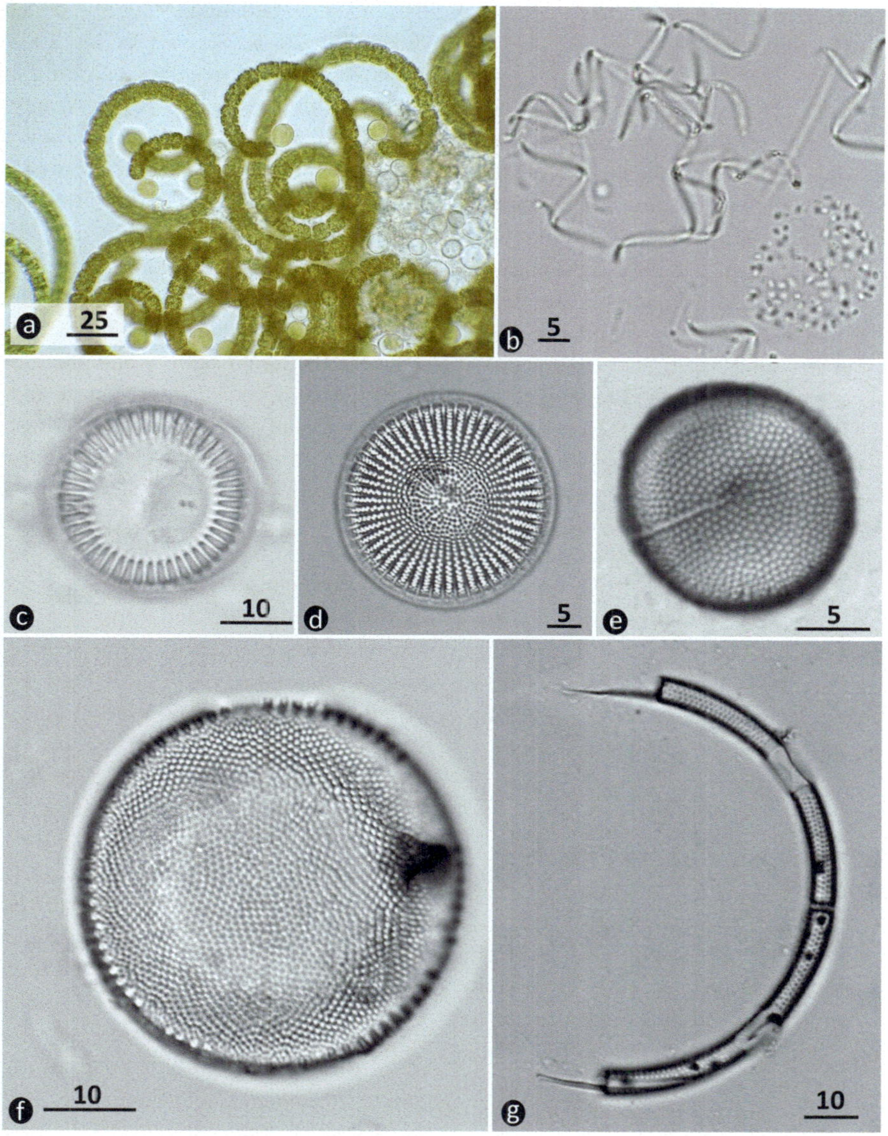

Fig. 6.5 Cyanobacteria 3 and diatoms 1. (**a**) *Anabaenopsis arnoldii*; (**b**) *Raphidiopsis curvata* and *Aphanothece clathrata*; (**c**) *Cyclotella meneghiniana* (central area tangentially undulating, two strutted processes visible as dots); (**d**) *Stephanodiscus* sp.; (**e**) *Thalassiosira* cf. *rudolfii*; (**f**) *Thalassiosira faurii*; (**g**) *Aulacoseira granulata* var. *angustissima* morphotype *curvata*. Scale bar in µm

freshwater conditions in Oloidien, but is displaced by *Arthrospira* when salinity values rise above 3‰ (Ballot et al. 2008; Krienitz et al. 2013a, b; Oduor and Schagerl 2007a, b; Schagerl and Oduor 2008). Interestingly, *A. elenkinii* is still able to survive at a salinity range from 5 to 15‰ in Elmenteita and Nakuru, in contrast to Lake Oloidien.

6.4.4 *Haloleptolyngbya alcalis* Dadheech et al.

This taxon was recently described from Lake Nakuru (Dadheech et al. 2012, Fig. 6.4g) and placed in the Leptolyngbyaceae (Komárek et al. 2014). The thallus is thin, pale to bright blue. Trichomes are solitary or densely entangled to form floating mats, tychoplanktonic or attached to the substrate. Filaments are long, straight or wavy, with rounded ends, 1.2–1.9 µm wide, and constricted at the cross walls. A colourless sheath envelops the trichomes. Cells are cylindrical, elongated or isodiametric, without aerotopes. This taxon represents a group of very tiny filamentous Cyanobacteria that are difficult to identify and that were formerly assigned to *Pseudanabaena* Lauterborn (e.g. Kaggwa et al. 2013); they may occur in phases of low *Arthrospira* abundance. Leptolyngbyaceae of the EASL probably have a higher diversity than previously expected and may act as an alternative food source for Lesser Flamingos.

6.4.5 *Raphidiopsis curvata* Fritsch et Rich

This nostocacean species occurs in EASL in phases of low salinity (<3‰). The thin (1–2 µm), spirally twisted filaments are up to 20 µm long and consist of only few cells (Fig. 6.5). The cross walls are very difficult to discern. The terminal cells are hyaline and pointed. Aerotopes are rarely visibly. Until now, no heterocytes have been observed.

6.4.6 *Microcystis* div. spec

The bloom- and scum-forming genus is found worldwide in inland waters and treated as one of the main cyanotoxin producers (Codd et al. 2005). High densities of *Microcystis* have been reported from Bogoria, Elmentaita and Nakuru and related to mass die-offs of Lesser Flamingos (Githaiga 2003; Ndetei and Muhandiki 2005; Stewart et al. 2008). There are, however, controversies regarding the occurrence of this taxon in EASL. As reported by Wood and Talling (1988), *Microcystis* is an important phytoplankton component of East African lakes, but restricted to salinities below 3‰. Kolding (1992) documented *Microcystis aeruginosa* as one of the dominant phytoplankton species in Lake Turkana in the 1980s, which at that time had a salinity of around 2.5‰. Krienitz and Kotut (2010) found salinity and alkalinity ranges in the Kenyan soda lakes Nakuru and Bogoria from 17–55 and 438–1760 meq L^{-1}, respectively. In a thorough survey with weekly sampling over 18 months, Jirsa et al. (2013) measured 23–96‰ salinity and 270–1560 meq L^{-1} alkalinity for both lakes. Active growth or even blooms of *Microcystis* can be ruled out under such saline alkaline conditions, and reports are probably due to errors in identification (e.g. confusion with *Anabaenopsis abijatae*, see Kotut and Krienitz 2011). Several detailed studies of the EASL did not mention *Microcystis*, e.g. the work on Lake Nakuru in the 1970s by Vareschi (1978); studies of Lake Bogoria from 1972–1978 and 2000–2003 by Harper et al. (2003); and the phytoplankton investigation of lakes Bogoria, Elmentaita and Nakuru by Oduor and Schagerl (2007a) and Schagerl and Oduor (2008) over the period 2003 to 2005 with 3-week sampling intervals. Molecular phylogenetic studies on Kenyan soda lakes recovered eight different phylotypes belonging to the genera *Anabaenopsis* (Nostocales), *Umezakia* (Nostocales), *Arthrospira* (Oscillatoriales) and *Chroococcidiopsis* (Chroococcales). However, no single phylotype belonging to the genus *Microcystis* was recovered (Dadheech et al. 2009). *Microcystis* might occur near lake inflows imported from freshwaters in the catchment area, but stable populations cannot establish in EASL.

Starting in 2010, some Kenyan soda lakes have received increased water inflow, resulting in a drastic increase in water level; the reasons for this are heavy rainfalls coupled with deforestation of the catchment area (see Chap. 1). The water level increase is directly related to a dilution of salts (from salinity of 50‰ in 2010 to below 4‰ in January 2015, Krienitz unpubl. results), which further impacts communities living in these ecosystems. Under such unusual situations, it is possible to observe *Microcystis* as ephemeral residents in the plankton of these lakes. Hecky and Kilham (1973) referred to such a phenomenon in which single samples contain atypical species due to unusual, temporary hydrological conditions.

Interestingly, Hammer et al. (1983) observed *Microcystis aeruginosa* over a wide range of salinities in Saskatchewan (Canada) lakes up to about 100‰. These systems are, however, dominated by sodium, magnesium and sulfate (Hammer 1978); in contrast to soda lakes, bicarbonate and carbonate play only a minor role. In culture experiments focusing on salt tolerance of *Microcystis aeruginosa*, Tonk et al. (2007) reported a collapse of growth if the cultures were exposed for several weeks to salt concentrations (NaCl) exceeding 10 mg L^{-1}. Apparently, *M. aeruginosa* can cope with highest sulfate concentrations (salinity of 100‰), is more sensitive against chloride (10‰), and restricted to carbonate systems lower than 3‰. This statement remains to be proved under defined culture conditions, also in combination with different pH.

6.4.7 *Synechococcus* and *Synechocystis* Sauvageau

Tiny solitary or clustering coccoid Cyanobacteria in EASL plankton (Fig. 6.4) were summarised under the names *Synechococcus* sp. and *Synechocystis* sp. (Ballot et al. 2004; Schagerl and Oduor 2008) or identified in Lake Sonachi as *Synechococcus bacillaris* Butcher (current name, *Cyanobium bacillare* (Butcher) Komárek, Kopecky et Cepák) and *Synechocystis aquatilis* Sauvageau (Melack et al. 1982). *Synechococcus* exhibits a rod-like cell shape measuring 1.0–3.0 × 0.5–1.5 µm; these cells are often arranged in short filaments. *Synechocystis* produces more or less spherical cells that are 1–3 µm in diameter. Both types propagate by binary fission. The taxonomy of these genera is highly complicated and far from clearly resolved. Molecular studies revealed that *Synechocystis* and *Synechococcus* phenotypes evolved in several different phylogenetic lines (Castenholz 2001; Korelusová et al. 2009; Choi and Noh 2009; Komárek 2010; Dvořák et al. 2014). We cannot provide a clear view of the actual taxonomic position of *Synechocystis* and *Synechococcus* originating from EASL, because further studies are needed to resolve this question. A 16S rRNA clone library established from Lake Elmentaita revealed a wide diversity of cyanobacterial lineages living in mud, sediment and water (Kachiuru 2009).

In the hot springs located at the shoreline of Lake Bogoria, the following species were detected by microscope: *Synechocystis salina* Wislouch, *Synechococcus sigmoideus* (Moore et Carter) Komárek and *S. bigranulatus* Skuja (Hindák 2001; Krienitz et al. 2003). Sequencing of field samples yielded *Synechococcus* lineages that do not resemble to any currently described clade (Dadheech et al. 2013). Hence, the species delineation discussed above needs to be proved.

6.4.8 Other Coccoid Cyanobacteria

Cyanobacterium minervae (Copeland) Komárek, Kopecký and Cepák established dense populations in the northern part of Lake Natron where the lake water covers several hot springs. Cells of this taxon occur solitary or in short pseudofilaments (Fig. 6.4f) and are 5–9 × 3–4 µm in size; their chromoplasma is striated (pers. observation).

In Lake Magadi, an interesting plankton community dominated by three coccoid taxa was observed in January 2010 at a salinity of 80‰ (pers. observation, Fig. 6.4): *Aphanothece* Nägeli sp. with cylindrical cells, rounded ends, cells 6–14 × 3–5 µm with mucilaginous cover, solitary or with few cells arranged in pseudofilaments, never in large colonies. *Synechocystis* cf. *salina* with spherical to elongated cells, 5–6 μm wide, after division live as two closely adjoining, hemispherical cells. *Myxobaktron* Schmidle sp. has fusiform cells, tapered or rounded ends, 15–40 × 2–3 μm, solitary or, after division, in short filaments, cell content hyaline, containing small droplets but no aerotopes (Fig. 6.4e). Type material of *M. salinum* Walsby et Hindák was discovered in Solar Lake at Sinai (Hindák 1992; Walsby et al. 1983) and showed broader cells (up to 6–8 μm) and clearly visible aerotopes. Further, *Euhalothece* sp. was isolated from Lake Magadi and the provisional name *Euhalothece natronophila* Garcia-Pichel given (Mikhodyuk et al. 2008).

Aphanothece clathrata W. et G.S. West (Fig. 6.5b) was observed in EASL at low salinity (<2‰).

6.4.9 Possible Poisoning of Lesser Flamingos Induced by Cyanobacteria

Since the mid-1990s, an increased mortality of Lesser Flamingos has been observed at the EASL. In 1993, 1995, 1999, 2001, 2006 and 2008, mass mortality involving more than 10,000 birds in each episode were reported (Ndetei and Muhandiki 2005). The causes of these die-offs are far from being completely understood but appear to be related to habitat degradation (see Chaps. 10 and 13). A combination of factors such as infectious diseases (Kock et al. 1999; Sileo et al. 1979) and poisoning by pesticides and heavy metals (Greichus et al. 1978; Kairu 1996) are subjects of ongoing speculation. The die-offs could also be linked to changes in food quantity (Sileo et al. 1979; Ton 2007) and quality (Motelin et al. 2000; Ndetei and Muhandiki 2005), although Kaggwa et al. (2013) found no relation between mass mortality and food quality of Lesser Flamingos. Malnutrition resulting from a decline in the *Arthrospira* food concentration is thought to have caused the birds to suffer immunodeficiency, leading to infection by microbacteriosis (Sileo et al. 1979).

According to literature surveys, almost all of the main Cyanobacteria inhabiting the lakes have the potential to produce toxins (Carmichael et al. 2001; Codd et al. 2005). The flamingos are also confronted with toxic Cyanobacteria, which occasionally become dominant and displace the previously healthy and nutritious populations of *Arthrospira* (Ballot et al. 2004; Codd et al. 2003; Koenig 2006; Kotut et al. 2006). Although several studies have revealed the high nutritional and pharmaceutical value as well as the non-toxic nature of this cyanobacterium (Belay et al. 1993; Ciferri 1983; Fox 1996), a few reports have commented on its potential toxicity. Iwasa et al. (2002) reported on the hepatotoxic symptoms of a patient after treatment with *Arthrospira*-based tablets. Recently, neurotoxic effects of beta-N-methylamino-L-alanin (BMAA) from *Arthrospira*-based dietary supplements or pharmaceuticals were reported (Mazokopakis et al. 2008; Papapetropoulos 2007). Out of a total of 40 *Arthrospira* strains isolated from Kenyan waters, two potentially toxic strains containing low concentrations of microcystins and anatoxin-a have been identified (Ballot et al. 2004, 2005). Nonetheless, a survey of three other *Arthrospira* strains (two from a sewage pond in Mozambique, one from Lake Nakuru, Kenya) did not reveal toxicity under test conditions (Mussagy et al. 2006). Mouse bioassays using crude extracts of monospecific *Arthrospira* blooms from the soda lake Big Momela (Tanzania) revealed the potential toxicity of these natural cyanobacterial populations to Lesser Flamingos (Lugomela et al. 2006). Further evidence pointing to the potential exposure of flamingos to cyanotoxins was provided by the detection of aminotransferase domains for cyanotoxin production in sediments of the EASL (Dadheech et al. 2009). Metcalf et al. (2006, 2013) detected microcystins and anatoxin-a as well as BMAA in extracts of feathers of Lesser Flamingos. The authors hypothesised that under natural conditions the toxins are sequestered in the feathers. Due to the growing environmental stress, however, the sequestration mechanism in the weakened flamingos may not be effective enough to protect the birds against cyanotoxicosis.

There are also contrary results in which no cyanotoxins were detected even during sudden deaths of flamingos: Straubinger-Gansberger et al. (2014) screened plankton samples of lakes Nakuru and Bogoria in weekly intervals for cyanotoxins, and although three smaller die-offs were recognised in this period, no toxins could be detected, even by applying different methods. Also, the tissues of 20 flamingo carcasses yielded no cyanotoxins. All these findings (see Chaps. 10 and 13) underline the urgent need for further studies to clarify the contribution of toxic Cyanobacteria to flamingo die-offs in the EASL.

6.5 Diatoms

Diatoms were investigated thoroughly in the EASL during the 1960s–1980s (Gasse et al. 1983; Gasse 1986; Hecky and Kilham 1973). Diatoms are used as bioindicators of environmental conditions; this feature is also applied in paleaolimnology and paleoclimate reconstructions (Lamb 2001; Verschuren et al. 1999) and provides insights into historic lake basins (Richardson and Richardson 1972; Roubeix et al. 2014). Especially at low biomass of planktonic Cyanobacteria, benthic diatoms might become important in the food webs of soda lakes (Tuite 1981), but the net primary productivity of benthic diatoms in EASL is usually one to two orders of magnitude less than that of *Arthrospira*, and the carrying capacity of the habitat with diatoms is lower in the same order (Tuite 2000). Lesser Flamingos also consume benthic diatoms in the case of low *Arthrospira* availability (Kummsa and Bekele 2014).

Salinity considerably influences diatom abundance (Gasse et al. 1983; Hecky and Kilham 1973). In their survey of Kenyan and Tanzanian soda lakes, Hecky and Kilham (1973) found only six diatom species predominating. *Nitzschia frustulum* and *Anomoeoneis sphaerophora* were dominant in most benthic samples; others were confined to a lower salinity range (Figs. 6.5 and 6.6). Further taxa such as *Rhopalodia gibberula* never reached high densities, but occurred in nearly every lake. Other common benthic diatoms are *Craticula elkab* originally described from Egyptian soda pans, *Craticula halophila* (Grunow) Mann and *Nitzschia sigma* (Kützing) W. Smith (Gasse 1986; Melack 1982, 1988; Schagerl and Oduor 2008).

Not only benthic, but also planktonic diatoms–mainly comprising of Centrales—are found in the EASL. *Cyclotella meneghiniana* Kützing was observed at lower salinity (Hecky and Kilham 1973, whereas *Thalassiosira rudolfii* (Bachmann) Hasle occurred at intermediate alkalinities (Hecky and Kilham 1973). *T. rudolfii* was described from Lake Turkana (formerly called Lake Rudolf) and seems to be restricted to lakes of East and Central Africa such as the Tanzanian lakes Big Momela and El Kekhotoito (Hasle 1978) or the Ethiopian lakes Shala, Metahara and Langano (Kebede and Willén 1998), the latter being a freshwater lake. The smaller *Thalassiosira faurii* (Gasse) Hasle was originally described from the Afar region (Hasle 1978) and seems to be restricted to African lakes. Although *T. faurii* is used for paleoclimate reconstructions, its autecology is still unresolved: laboratory studies by Roubeix et al. (2014) showed that net growth stopped at conductivities beyond 2 mS cm^{-1}, but this species is found in the natural habitat also at higher ion concentrations up to 13.5 mS cm^{-1} (Chalie and Gasse 2002; Hasle 1978).

6.6 Eustigmatophyceae

Microchloropsis salina (Hibberd) Fawley et al. (syn. *Monallantus salina* Bourrelly, *Nannochloropsis salina*) was found in high abundance in Lake Sonachi (Melack et al. 1982). The ovoid cells of this picoplanktonic taxon are about

Fig. 6.6 Diatoms 2. (**a**) *Epithemia* cf. *sorex*; (**b**, **c**) *Epithemia* cf. *adnata*; (**b**) girdle view; (**c**) valve view; (**d**) *Rhopalodia* cf. *gibberula*; (**e**) *Anomoeoneis sphaerophora*; (**f**) *Craticula halophila*. Scale bar in µm

2 × 3 µm. This species is usually treated as a marine organism and might be interesting for industrial exploitation because of its high content of fatty acids (Mourente et al. 1990). Furthermore, *Nannochloropsis limnetica* Krienitz et al. was detected in Lake Nakuru in a phase

of near freshwater conditions (see Chaps. 10 and 13).

6.7 Coccoid Green Algae (Chloro- and Prasinophytes)

The most common green alga in the soda lakes is the picoplanktonic prasinophyte *Picocystis salinarum* (Fig. 6.7e). Because of its small size (2 μm or even less), this organism is hard to recognise and probably often overlooked; moreover, identification in field samples is difficult. Because of its key position as a highly productive primary producer in food webs of saline habitats, much more attention should be paid to its distribution, which is probably widespread. Recently, this alga was found to be abundant in lakes Bogoria, Magadi and Nakuru (Krienitz and Kotut 2010; Krienitz et al. 2013a).

The vegetative cells of *Picocystis* are mostly spherical, 1.5–2.0 μm in diameter; the two to four autospore-bearing mother cells do not exceed 2.5 μm. Some of the cells are lobated, their shape somewhat resembling pine pollen (Lewin et al. 2000) or the contour of the 'Mickey Mouse head' (Lewin, pers. communication): two lateral lobes contain the chloroplast and a third central lobe contains the nucleus. *Picocystis* is characterised by an exclusive set of pigments with an unusual carotenoid pattern (Lewin et al. 2000; Roesler et al. 2002). *Picocystis* is able to outcompete *Arthrospira* in phases of high salinity. In Bogoria, first mass developments of *Picocystis* were observed in February 2006, when salinity reached 55‰. In Lake Nakuru, mass developments were observed for the first time in late 2009 (Schagerl et al. 2015) and early 2010 (Krienitz and Kotut 2010) at salinities > 50‰. *Arthrospira* was completely suppressed during these periods. For Lesser Flamingos, the dominance of *Picocystis* has fatal consequences. At only about 2 μm in diameter, the cells are too small to be retained by the filter lamellae in the bills of these birds (mesh size about 50 μm).

In the nanoplankton of the EASL, coccoid chlorophytes of the families Selenastraceae, Scenedesmaceae, Oocystaceae and Chlorellaceae are known to be ubiquitous (Schagerl and Oduor 2008; Vareschi 1982). These chlorophytes probably vegetate at sediment surfaces in the soda lakes and re-establish populations under suitable growth conditions. Other sources of green algae are catchment areas, from where they are discharged into the lake.

Ferroni et al. (2007) conducted salinity experiments with a crescent-shaped strain (*Kirchneriella* sp. AW15) isolated from the freshwater Lake Awasa (Ethiopia). In batch cultures, the alga was grown under different NaCl concentrations between 0.4 and 19.4‰. The strain tolerated high salinity levels without destruction of the photosynthetic apparatus. The crescent-shaped morphotype (Fig. 6.7c) is always present in the soda lakes in phases of low-performing *Arthrospira*. This phenotype harbours several phylotypes assigned to different genera such as *Kirchneriella*, *Monoraphidium*, *Raphidocelis*, *Selenastrum* and *Tetranephris*. We assume that members of all these genera are able to survive in the soda lakes. A clone culture isolated from the plankton of Lake Nakuru was assigned to *Tetranephris brasiliensis* Leite et Bicudo based on molecular phylogenetic analyses (Krienitz et al. 2011). The marine species *Picochlorum atomus* (Butcher) Henley et al. was observed in high abundance in Lake Sonachi in 1980 (Melack et al. 1982).

Mass development of *Atractomorpha* Hoffman sp. was observed in a northern cove of Lake Magadi in January 2010 (Fig. 6.7d; pers. observation). *Atractomorpha* is a rare member of the green algal order Sphaeropleales and characterised by fusiform, multinucleate vegetative cells (Hoffman 1983). The apices of the cells are needle shaped and elongated; cell length exceeds 100 μm. Asexual reproduction takes place via zoospores or aplanospores (not observed in this sample). Sexual reproduction involves anisogamy or oogamy; zygotes have a

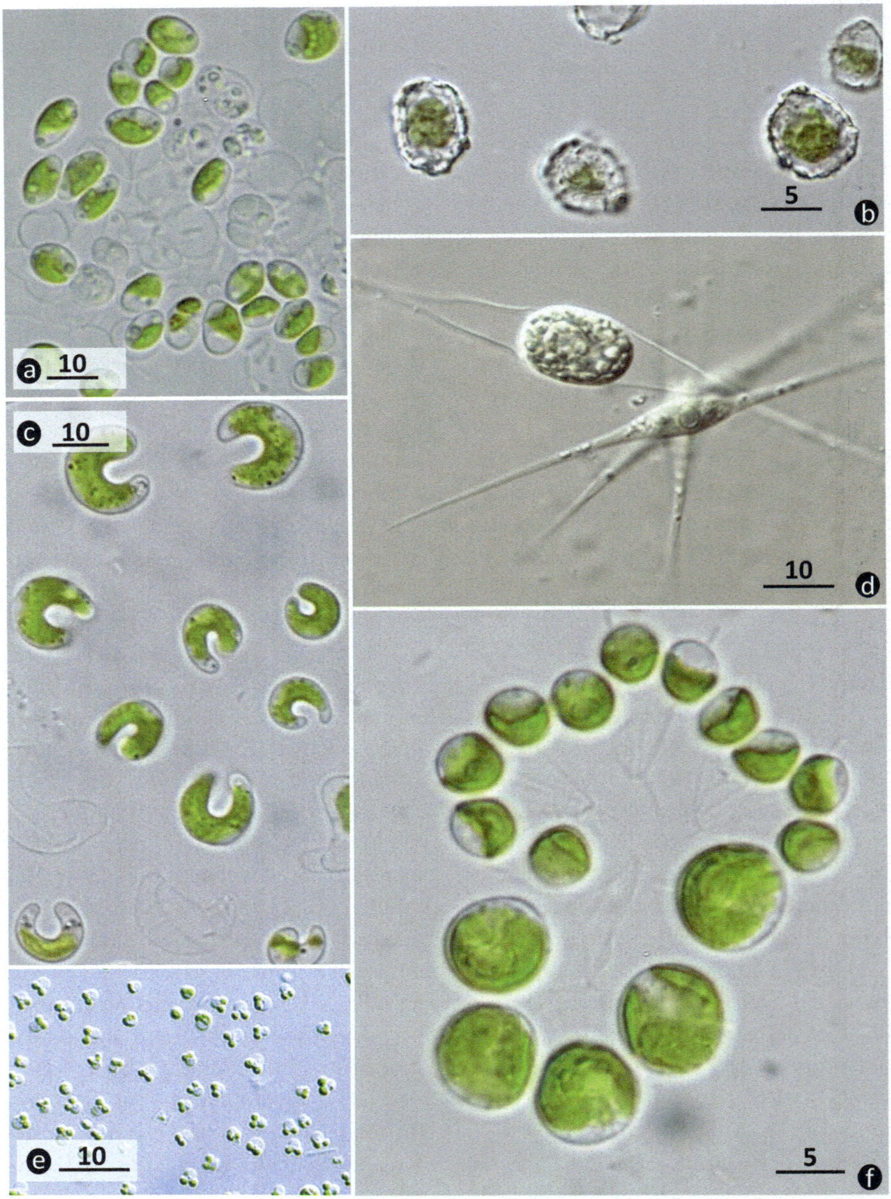

Fig. 6.7 The green line. (**a**) *Oocystis* sp.; (**b**) *Wislouchiella* cf. *planctonica*; (**c**) *Raphidocelis* sp.; (**d**) *Atractomorpha* sp.; (**e**) *Picocystis salinarum*; (**f**) *Masaia oloidia*. Scale bar in μm

prominent wall and orange to red content. The material observed in Lake Magadi was related to the type species *Atractomorpha echinata* Hoffman (known from soil), but no clear designation to species level was possible because of missing developmental stages and the different ecology of *A. echinata*.

6.8 Phytoflagellates

6.8.1 Chlorophyta

We assume that the diversity of green flagellates in the EASL is very high in phases of low salinity: an 18S rRNA clone library set-up from a sample collected in Lake Nakuru in a phase in which the salinity was 9.7‰ in November 2011 provided a first insight into the hidden diversity (Luo et al. 2013). The library contained phylotypes of green flagellates of *Gungnir* (a clade near *Dunaliella*), relatives of *Chlamydomonas* and chloroflagellates equipped with a lorica related to *Pteromonas* and *Wislouchiella*.

6.8.2 Cryptophyta

These small-celled auto- and mixotrophic flagellates were frequently observed in Lake Nakuru (Schagerl and Oduor 2008; Vareschi 1982) and provisionally placed in the genera *Cryptomonas* or *Rhodomonas* (Fig. 6.8a). The closest match of the 18S rRNA gene clone library from Nakuru collected in November 2011 is *Guillardia theta* Hill et Wetherbee, a cryptophyte known from coastal waters of North America (Hill and Wetherbee 1990; Luo et al. 2013).

6.8.3 Euglenophyta

This group is usually found in only low abundance (Fig. 6.8f to i). Kaggwa et al. (2013) reported a mean contribution of 6 % in lakes Nakuru and Bogoria. Only during conditions of low *Arthrospira* biomass might euglenoids reach larger numbers together with cryptomonads (Dadheech et al. 2012; Luo et al. 2013). Euglenoids were also found in Lake Manyara, Tanzania (Kihwele et al. 2014).

6.8.4 Dinophyta

This group was sporadically detected, but might have been misinterpreted because dinoflagellates are most prevalent in marine systems (Fig. 6.8). In their study of hidden diversity in Lake Nakuru, Luo et al. (2013) found a small dinoflagellate in their sample from November 2011 in the 18S rRNA gene clone library; the closest match was the heterotrophic dinoflagellate *Cryptoperidiniopsis broadyii* Steidinger et al., which is known from estuaries of North America (Steidinger et al. 2006).

6.9 Seasonal Succession of Phytoplankton

In temperate systems, seasonal plankton succession is highly influenced by allochthonous factors defining mixing and stratification of water bodies and their consequences (Sommer et al. 1986, 2012). In lower latitudes, annual variation of insolation is much less pronounced, and therefore tropical systems are expected to show no regular patterns over the year (Lewis 1996; see Chap. 12). Because EASL are special ecosystems, they host extremophiles, and diversity is thus further reduced. Although the systems are often treated as *Spirulina*/*Arthrospira* lakes, an increasing number of studies demonstrated the high instability of the producer populations (Krienitz and Kotut 2010; Krienitz et al. 2013a; Oduor and Schagerl 2007a, b; Schagerl and Oduor 2008; Schagerl et al. 2015; Tuite 1981; Vareschi 1978). Occasionally, blooms of the dominant *Arthrospira* are replaced by coccoid Cyanobacteria, green algae, heterocytous Cyanobacteria such as *Anabaenopsis*, diatoms and/or cryptomonads-a condition that could last for several months (Vareschi 1982; Melack 1979b).

Melack (1988) listed eight possible causes of the abrupt switch from one phytoplankton assemblage to another in lakes Elmentaita and Nakuru. These included changes in salinity, nutrient concentration and water level, outcompeting of filamentous Cyanobacteria by coccoid or flagellated plankton and attacks by cyanophages or autolysis. Peduzzi et al (2014) proved viral infection as one major cause for *Arthrospira* crashes (Fig. 6.9). Schagerl et al. (2015) linked the abiotic factors and biotic interactions potentially influencing the community patterns and concluded that *Arthrospira*'s response to rapid environmental changes is limited. During rapid changes, *Arthrospira* is outcompeted by other

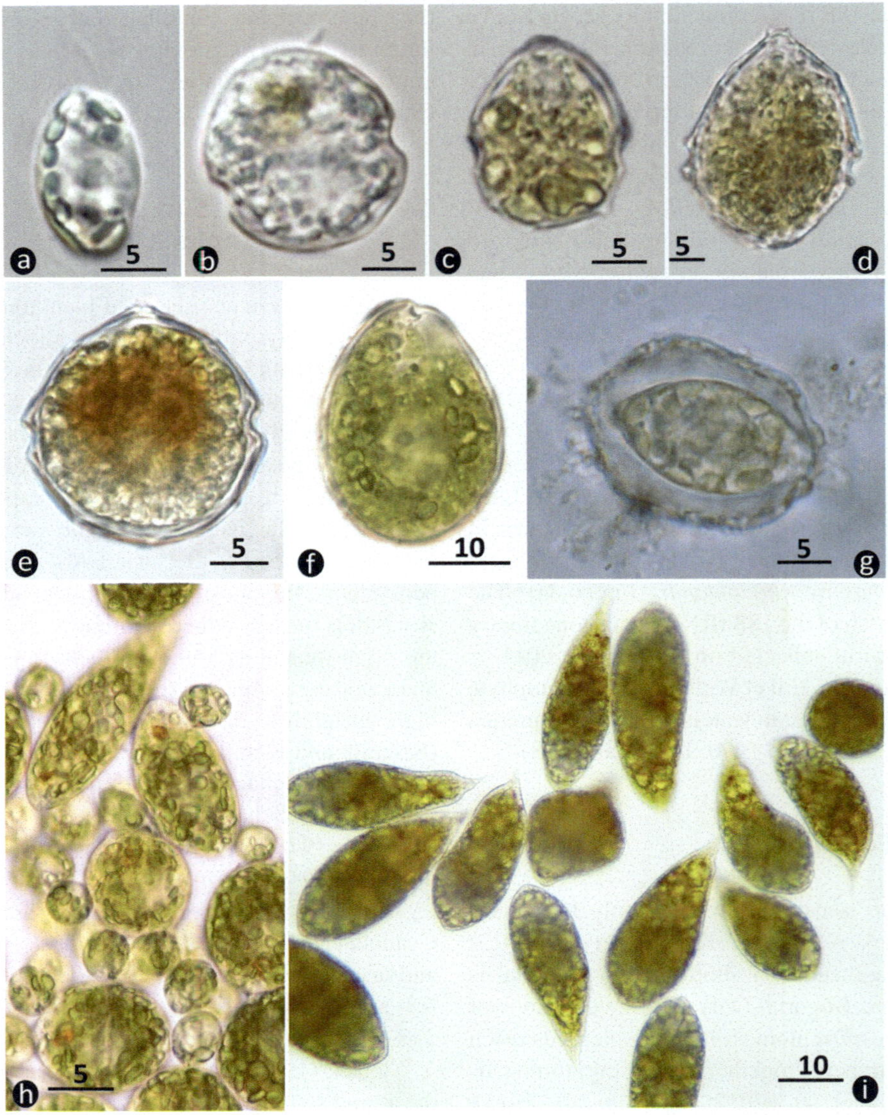

Fig. 6.8 Phytoflagellates. (**a**) Cryptophyta; (**b–e**) Dinophyta; (**f–i**) Euglenophyta; (**f**) *Euglena texta* var. *salina*; (**g**) *Strombomonas* sp.; (**h** and **i**) *Euglena* cf. *caudata* (*elongated cells*), *Euglena* sp. (*spherical cells*). Scale bar in μm

taxa, resulting in a more diverse phytoplankton community. Light penetration, consumers of *Arthrospira* and cyanophages attacking this taxon significantly influenced *Arthrospira* abundance in Lake Nakuru. Nevertheless, the key factors are still under discussion, and a model needs to be developed that considers multiple factors along with possible interactions that may increase or diminish the effects of one factor against another (see Chap. 12). Because of high fluctuations in water level, Lake Nakuru provides an ideal system for studying community changes caused by varying ion concentrations. Between January and October 2009, salinity increased from around 20‰ to more than 60‰, and the phytoplankton community changed from a dominance of flagellates to *Picocystis salinarum* and *Haloleptolyngbya alcalis* (listed in Kaggwa

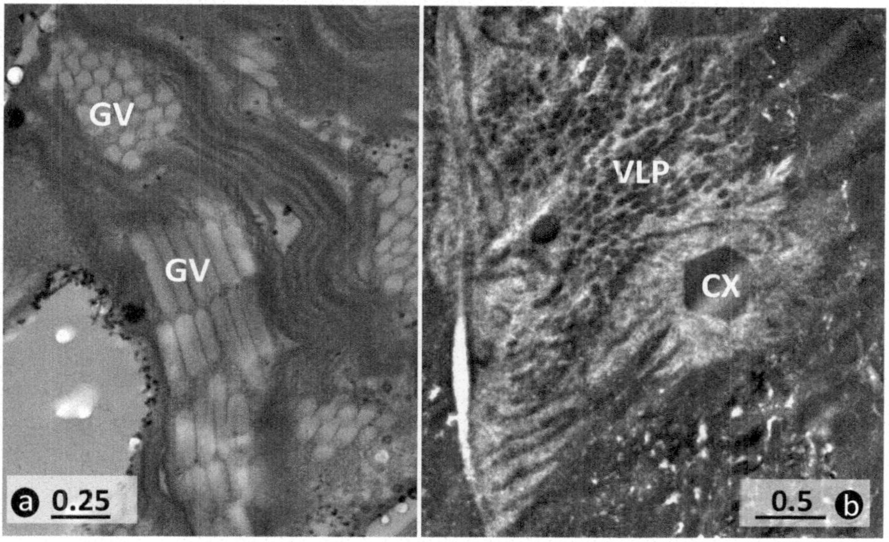

Fig. 6.9 Cyanophage infection of *Arthrospira*. (**a**) Healthy cytoplasm; (**b**) = infected part. GV gas vesicles in longitudinal and cross section, VLP virus-like particles, CX carboxysome. Scale bar in μm

et al. 2013 as cf. *Pseudanabaena acicularis* (Nygaard) Anagnostidis). In November 2010, a low water level continued to promote high concentration of salts (51‰), which facilitated *Picocystis salinarum* dominance (Krienitz et al. 2012). A contrary situation was recognised in November 2011, when high water levels after heavy rains resulted in a salinity decrease down to 9‰. During this phase, Cyanobacteria were outcompeted by algae preferring lower salinity. These algae were probably discharged into the lake from connected rivers and an outlet of the Nakuru oxidation sewage treatment plant. Furthermore, the sewage pond hosts an extraordinary high diversity of relatives of the spherical green alga *Chlorella*, which could act as inoculum of the lake's algal flora (Krienitz et al. 2012). Luo et al. (2013) observed a unique phytoplankton species composition characterised by a high diversity of monadoid and coccoid green algae. In the period 2013–2015, the water level increased again, and the shoreline of Lake Nakuru was flooded (see also Chaps. 1 and 12). The lake's salinity in this time dropped to extremely low levels between 4.6 and 3.6‰ (pers. observation). In February 2015, no flamingos were observed at Lake Nakuru due to lacking food algae. Our personal observations revealed that other important habitats of Lesser Flamingos in Kenya also showed considerably higher water levels accompanied by reduced salinity and disappearance of main food algae. At Lake Bogoria (salinity 10–20‰), only a few thousand flamingos sustained by the reduced feeding on *Anabaenopsis*. At Lake Elmentaita (salinity 3.8–4.6‰), the numbers of flamingos dropped below 1000. On this lake, flamingo numbers already started to drop in 2010, when the only food resource consisted of pennate diatoms. At Lake Oloidien, due to massive growth of *Arthrospira*, the flamingo numbers increased between 2006 and 2010 to about 200,000 but dropped to zero by 2015 when food algal populations collapsed under reduced salinity of 1.0‰. The question remains open how long the soda lake ecosystems will need to re-establish the characteristic food web driven by *Arthrospira* and its main feeder, the Lesser Flamingo.

6.10 Concluding Remarks

Although some EASL are highly threatened, only a few are protected by law. As endorheic systems, they are moreover highly sensitive to pollution and human impacts. Many of the lakes experience rapid changes in water level, and the accompanying shifts in chemistry may in turn influence the biota. Surprisingly, no regular monitoring system has yet been established along with accessible databases, which would greatly assist our understanding of their chemistry, physics and biology. For most of the EASL, only snapshot studies have been conducted, providing merely fragmentary insight into both short- and long-term changes. How do compartments of the food web interact? Although EASL are amongst the most productive systems, almost no information exists about interactions between photoautotrophs and consumers, the microbial loop or energy fluxes.

The recent insight into hidden diversity strongly suggests the discovery of new species, which, however, will require the use of highly sophisticated molecular tools for confirmation. This calls for intensified international cooperation. Some species might be interesting for commercial purposes because extremophiles have developed strategies to survive under harsh conditions. This monetary interest, however, can also be a major drawback: some East African countries might significantly restrict cooperation in light of potential biopiracy issues. Here, partner countries are requested to pave the way for basic research—only an intense exchange of knowledge will enable us to answer the many open questions and ultimately contribute to the conservation efforts in EASL.

References

Anagnostidis K, Komárek J (1988) Modern approach to the classification system of cyanophytes. 3—Oscillatoriales. Arch Hydrobiol Suppl 80 (Algol Stud 50–53):327–472

Bachmann H (1939) Mission scientifique de l'Omo. Beiträge zur Kenntnis des Phytoplanktons ostafrikanischer Seen. Z Hydrologie 8:119–140

Ballot A, Krienitz L, Kotut K, Wiegand C, Metcalf JS, Codd GA, Pflugmacher S (2004) Cyanobacteria and cyanobacterial toxins in three alkaline Rift Valley lakes of Kenya—Lakes Bogoria, Nakuru and Elmenteita. J Plankton Res 26:925–935

Ballot A, Krienitz L, Kotut K, Wiegand C, Pflugmacher S (2005) Cyanobacteria and cyanobacterial toxins in the alkaline crater lakes Sonachi and Simbi, Kenya. Harmful Algae 4:139–150

Ballot A, Dadheech PK, Haande S, Krienitz L (2008) Morphological and phylogenetic analysis of *Anabaenopsis abijatae* and *Anabaenopsis elenkinii* (Nostocales, Cyanobacteria) from tropical inland water bodies. Microb Ecol 55:608–618

Baurain D, Renquin L, Grubisic S, Scheldeman P, Belay A, Wilmotte A (2002) Remarkable conservation of internally transcribed spacer sequences of *Arthrospira* ("*Spirulina*") (Cyanophyceae, Cyanobacteria) strains from four continents and of recent and 30-year-old dried samples from Africa. J Phycol 38:384–393

Belay A (2013) Biology and industrial production of *Arthrospira* (*Spirulina*). In: Richmond A, Hu Q (eds) Handbook of microalgal culture: applied phycology and biotechnology, 2nd edn. Wiley, Hoboken, p 736

Belay A, Ota Y, Miyakawa K, Shimamatsu H (1993) Current knowledge on potential health benefits of *Spirulina*. J Appl Phycol 5:235–241

Belkin S, Boussiba S (1991) Resistance of *Spirulina platensis* to ammonia at high pH values. Plant Cell Physiol 32:953–958

Brown LH (1959) The mystery of the flamingos. Country Life Ltd, London

Carmichael WW, Azevedo SMFO, An JS, Molica JR, Jochimsen EM, Lau S, Rinehart KL, Shaw GR, Eaglesham GK (2001) Human fatalities from cyanobacteria: chemical and biological evidence for cyanotoxins. Environ Health Perspect 109:663–668

Castenholz RW (1989) Subsection III, order Oscillatoriales. In: Staley JT, Bryant MP, Pfennig N, Holt JG (eds) Bergey's manual of systematic bacteriology 3. Williams and Wilkins, Baltimore, pp 1771–1780

Castenholz RW (1996) Endemism and biodiversity of thermophilic cyanobacteria. Nova Hedwig 112:33–47

Castenholz RW (2001) Oxygenic photosynthetic bacteria. In: Boone DR, Castenholz RW (eds) Bergey's manual of systematic bacteriology, vol 1, 2nd edn. Springer, New York, pp 473–600

Chalie F, Gasse F (2002) Late Glacial- Holocene diatom record of water chemistry and lake level change from the tropical East African Rift Lake Abiyata (Ethiopia). Palaeogeogr Palaeoclimatol Palaeoecol 187:259–283

Choi DH, Noh JH (2009) Phylogenetic diversity of *Synechococcus* strains isolated from the East China Sea and the East See. FEMS Microbiol Ecol 69:439–448

Ciferri O (1983) *Spirulina*, the edible microorganism. Microbiol Rev 47:551–578

Codd GA, Metcalf JS, Morrison LF, Krienitz L, Ballot A, Pflugmacher S, Wiegand C, Kotut K (2003) Susceptibility of flamingos to cyanobacterial toxins via feeding. Vet Rec 152:722–723

Codd GA, Morrison LF, Metcalf JS (2005) Cyanobacterial toxins: risk management for health protection. Toxicol Appl Pharmacol 203:264–272

Comte K, Coursin T, Carre-Mlouka A (2013) A new genotype in the genus *Arthrospira* (Oscillatoriales,

Cyanobacteria) revealed by a mosaic-like structure of the 16S-23SrRNA intergenic spacer region in strain PCC 9901. Phycologia 52:333–337

Dadheech PK, Krienitz L, Kotut K, Ballot A, Casper P (2009) Molecular detection of uncultured cyanobacteria and aminotransferase domains for cyanotoxin production in sediments. FEMS Microbiol Ecol 68:340–350

Dadheech PK, Ballot A, Casper P, Kotut K, Novelo E, Lemma B, Pröschold T, Krienitz L (2010) Phylogenetic relationship and divergence among planktonic strains of *Arthrospira* (Oscillatoriales, Cyanobacteria) of African, Asian and American origin deduced by 16S-23S ITS and phycocyanin operon sequences. Phycologia 49:361–372

Dadheech PK, Mahoud H, Kotut K, Krienitz L (2012) *Haloleptolyngbya alcalis* gen. et sp. nov., a new filamentous cyanobacterium from the soda lake Nakuru, Kenya. Hydrobiologia 691:269–283

Dadheech PK, Glöckner G, Casper P, Kotut K, Mazzoni CJ, Mbedi S, Krienitz L (2013) Cyanobacterial diversity in the hot spring, pelagic and benthic habitats of a tropical soda lake. FEMS Microbiol Ecol 85:389–401

Desikachary TV (1959) Cyanophyta. monographs on algae 2. Indian Council of Agricultural Research, New Delhi, p 686

Dinka MO (2012) Analysing the extent (size and shape) of Lake Basaka expansion (Main Ethiopian Rift Valley) using remote sensing and GIS. Lakes Reserv Res Manag 17:131–141

Dinka MO, Loiskandl W, Ndambuki JM (2014) Hydrologic modelling for Lake Basaka: development and application of a conceptual water budget model. Environ Monit Assess 186:5363–5379

Dvořák P, Casamatta DA, Poulíčková A, Hašler P, Ondřej V, Sanges R (2014) *Synechococcus*: 3 billion years of global dominance. Mol Ecol 23:5538–5551

Ferroni L, Baldisserotto C, Pantaleoni L, Billi P, Fasulo MP, Pancaldi S (2007) High salinity alters chloroplast morpho-physiology in a freshwater *Kirchneriella* species (Selenastraceae) from Ethiopian Lake Awasa. Am J Bot 94:1972–1983

Field CB, Behrenfeld MJ, Randerson JT, Falkowski P (1998) Primary production of the biosphere: integrating terrestrial and oceanic components. Science 281(5374):237–240

Florenzano G, Sili C, Pelosi E, Vincenzini M (1985) *Cyanospira rippkae* and *Cyanospira capsulata* (gen. nov. and spp. nov.): new filamentous heterocystous cyanobacteria from Magadi lake (Kenya). Arch Microbiol 140:301–306

Fox RD (1996) *Spirulina*: production and potential. EDISUD, Aix-en-Provence, 232 pp

Fuzinato S, Fodora A, Subakov-Simic G (2010) *Arthrospira fusiformis* (Voronichin) Komárek et Lund (Cyanoprokaryota)—a new species for Europe. Algol Stud 134:17–24

Gardner NL (1917) New Pacific coast marine algae I. University of California Publications in Botany 6:377–416

Gasse F (1986) East African diatoms. Taxonomy, ecological distribution. Bibl Diatomol 11:1–202 + 44 pl

Gasse F, Talling JF, Kilham P (1983) Diatom assemblages in East Africa: classification, distribution and ecology. Rev Hydrobiol Trop 16:3–34

Geitler L (1925) Cyanophyceae. In: Pascher A (ed) Die Süßwasser-Flora Deutschlands, Österreichs und der Schweiz, 12. Gustav Fischer, Jena, p 450

Girma MB, Kifle D, Jebessa H (2012) Deep underwater seismic explosion experiments and their possible ecological impact—the case of Lake Arenguade-Central Ethiopian highlands. Limnologica 42:212–219

Githaiga JM (2003) Ecological Factors Determining utilization patterns and interlake movements of lesser flamingos (*Phoeniconaias minor* Geoffrey) in Kenyan alkaline lakes. Dissertation. University of Nairobi, p 228

Gomont M (1892) Monographie des Oscillariées (Nostocacées Homocystées). Deuxième partie—Lyngbyées Annales des Sciences Naturelles Botanique Série 7:91–264

Grant WD, Mwatha WE, Jones BE (1990) Alkaliphiles: ecology, diversity and applications. FEMS Microbiol Lett 75:255–269

Greichus YA, Greichus A, Amman BB, Hopcraft J (1978) Insecticides, polychlorinated biphenyls and metals in African lake ecosystems. III. Lake Nakuru, Kenya. Bull Environ Contam Toxicol 19:454–461

Hammer UT (1978) The Saline lakes of Saskatchewan III. Chemical characterization. Internationale Revue der gesamten Hydrobiologie und Hydrographie 63:311–335

Hammer UT, Shamess J, Haynes RC (1983) The distribution and the abundance of algae in saline lakes of Saskatchewan, Canada. Hydrobiologia 105:1–26

Harper DM, Childress RB, Harper MM, Boar RR, Hickley PH, Mills SC, Otieno N, Drane T, Vareschi E, Nasirwa O, Mwatha WE, Darlington JPEC, Escuté-Gasulla X (2003) Aquatic biodiversity and saline lakes: Lake Bogoria National Reserve, Kenya. Hydrobiologia 500:259–276

Hasle GR (1978) Some freshwater and brackish water species of diatom genus *Thalassiosira* Cleve. Phycologia 17:263–292

Hecky RE, Kilham P (1973) Diatoms in alkaline, saline lakes: ecology and geochemical implications. Limnol Oceanogr 18:53–71

Hill DRA, Wetherbee R (1990) *Guillardia theta* gen. et sp. nov. (Cryptophyceae). Can J Bot 68:1873–1876

Hindák F (1985) Morphology of trichomes in *Spirulina fusiformis* Voronichin from Lake Bogoria, Kenya. Arch Hydrobiol Suppl 71 (Algol Stud 38/39):201–218

Hindák F (1992) Several interesting planktic cyanophytes. Algol Stud 66:1–15

Hindák F (2001) Thermal microorganisms from a hot spring on the coast of Lake Bogoria. Kenya. Nova Hedwig 123:77–93

Hoffman LR (1983) *Atractomorpha echinata* gen. et sp. nov., a new anisogamous member of the Sphaeropleaceae (Chlorophyceae). J Phycol 19:76–86

Iltis A (1968) Tolerance de salinite de *Spirulina platensis* (Gom.). Geitl., (Cyanophyta) dans les mares natronees du Kanem (Tchad). Cahiers O.R.S.T.O.M—Série Hydrobiologie 2:119–125

Iteman I, Rippka R, Tandeau de Marsac N, Herdman M (2002) rDNA analyses of planktonic heterocystous cyanobacteria, including members of the genera *Anabaenopsis* and *Cyanospira*. Microbiology 148:481–496

Iwasa M, Yamamoto M, Tanaka Y, Kaito M, Adach Y (2002) *Spirulina*-associated hepatotoxicity. Am J Gastroenterol 97:3212–3213

Jenkin PM (1929) Biology of lakes in Kenya. Nature 124:574

Jirsa F, Gruber M, Stojanovic A, Odour SO, Mader D, Körner W, Schagerl M (2013) Major and trace element geochemistry of Lake Bogoria and Lake Nakuru, Kenya, during extreme drought. Chem Erde-Geochem 73:275–282

Kachiuru RM (2009) Microbial diversity of Lake Elmenteita, Kenya. Dissertation, University of Agriculture and Technology

Kaggwa MN, Gruber M, Oduor SO, Schagerl M (2012) A detailed time series assessment of the diet of Lesser Flamingos: further explanation for their itinerant behaviour. Hydrobiologia 710:83–93

Kaggwa MN, Burian A, Oduor SO, Schagerl M (2013) Ecomorphological variability of *Arthrospira fusiformis* (Cyanoprokaryota) in African soda lakes. Microbiol Open 2:881–891

Kairu JK (1996) Heavy metals residues in birds of Lake Nakuru, Kenya. Afr J Ecol 34:397–400

Källqvist T, Lien L, Liti D (1988) Lake Turkana Limnological Study 1985–1988. NIVA Norwegian Institute for Water Research report No. 0-85313, Oslo, p 98

Kebede E (1997) Response of *Spirulina platensis* (= *Arthrospira fusiformis*) from Lake Chitu, Ethiopia, to salinity stress from sodium salts. J Appl Phycol 9:551–558

Kebede E (2002) Phytoplankton distribution in lakes of the Ethiopian Rift Valley. In: Tudorancea C, Taylor WD (eds) Ethiopian Rift Valley Lakes. Backhuys Publishers, Leiden, pp 61–93

Kebede E, Ahlgren G (1996) Optimum growth conditions and light utilization efficiency of *Spirulina platensis* (= *Arthrospira fusiformis*)(Cyanophyta) from lake Chitu, Ethiopia. Hydrobiologia 332:99–109

Kebede E, Willén E (1996) *Anabaenopsis abijatae*, a new cyanophyte from Lake Abijata, an alkaline, saline lake in the Ethiopian Rift Valley. Algol Stud 80:1–8

Kebede E, Willén E (1998) Phytoplankton in a salinity-alkalinity series of lakes in the Ethiopian Rift valley. Algol Stud 89:63–96

Kihwele ES, Lugomela C, Howell KM (2014) Temporal changes in the Lesser Flamingos population (*Phoenicopterus minor*) in relation to phytoplankton abundance in Lake Manyara, Tanzania. Open J Ecol 4:145–161

Kirkwood AE, Buchheim JA, Buchheim MA, Henley WJ (2008) Cyanobacterial diversity and halotolerance in a variable hypersaline environment. Microbial Ecol 55:453–465

Kock ND, Kock RA, Wambua J, Kamau GJ, Mohan K (1999) *Mycobacterium avium*-related epizootic in free ranging Lesser Flamingos in Kenya. J Wildl Dis 35:297–300

Koenig R (2006) The pink death: Die-offs of the Lesser Flamingo raise concern. Science 313:1724–1725

Kolding J (1992) A summary of Lake Turkana—an ever changing environment. Mitt Internat Verein Limnol 23:25–35

Komárek J (2010) Recent changes (2008) in cyanobacteria taxonomy based on a combination of molecular background with phenotypic and ecological consequences (genus and species concept). Hydrobiologia 639:245–259

Komárek J, Anagnostidis K (2005) Cyanoprokaryota 2. Oscillatoriales. In: Büdel B, Gärtner G, Krienitz L, Schagerl M (eds) Süßwasserflora von Mitteleuropa 19/2. Spektrum Akademischer Verlag, Heidelberg, p 759

Komárek J, Lund JWG (1990) What is "*Spirulina platensis*" in fact? Algol Stud 58:1–13

Komárek J, Mareš J (2012) An update to modern taxonomy (2011) of freshwater planktic heterocytous cyanobacteria. Hydrobiologia 698:327–351

Komárek J, Kaštovský J, Mareš J, Johansen JR (2014) Taxonomic classification of cyanoprokaryotes (cyanobacterial genera) 2014, using a polyphasic approach. Preslia 86:295–335

Korelusová J, Kaštovský J, Komárek J (2009) Diversity of the cyanobacterial genus *Synechocystis* Sauvegeau and *Geminocystis* genus nova. J Phycol 45:928–937

Kotut K, Ballot A, Krienitz L (2006) Toxic cyanobacteria and their toxins in standing waters of Kenya: implications for water resource use. J Water Health 4:233–245

Kotut K, Krienitz L (2011) Does the potentially toxic cyanobacterium *Microcystis* exist in the soda lakes of East Africa? Hydrobiologia 664:219–225

Krienitz L, Ballot A, Casper P, Kotut K, Wiegand C, Pflugmacher S (2005) Cyanobacteria in hot springs of East Africa and their potential toxicity. Algol Stud 117:297–306

Krienitz L, Ballot A, Kotut K, Wiegand C, Pütz S, Metcalf JS, Codd GA, Pflugmacher S (2003) Contribution of hot spring cyanobacteria to the mysterious deaths of Lesser Flamingos at Lake Bogoria, Kenya. FEMS Microbiol Ecol 1437:141–148

Krienitz L, Bock C, Nozaki H, Wolf M (2011) SSU rRNA gene phylogeny of morphospecies affiliated to the bioassay alga "*Selenastrum capricornutum*" recovered the polyphyletic origin of crescent-shaped Chlorophyta. J Phycol 47:880–893

Krienitz L, Bock C, Kotut K, Luo W (2012) *Picocystis salinarum* (Chlorophyta) in saline lakes and hot springs of East Africa. Phycologia 51:22–32

Krienitz L, Dadheech PK, Kotut K (2013a) Mass developments of a small sized ecotype of *Arthrospira fusiformis* in Lake Oloidien, Kenya, a new feeding ground for Lesser Flamingos in East Africa. Fottea 13:215–225

Krienitz L, Dadheech PK, Kotut K (2013b) Mass developments of the cyanobacteria *Anabaenopsis* and *Cyanospira* (Nostocales) in the soda lakes of Kenya: ecological and systematic implications. Hydrobiologia 703:79–93

Krienitz L, Kotut K (2010) Fluctuating algal food populations and the occurrence of Lesser Flamingos (*Phoeniconaias minor*) in three Kenyan Rift Valley lakes. J Phycol 46:1088–1096

Krienitz L, Krienitz D, Dadheech PK, Hübener T, Kotut K, Luo W, Teubner K, Versfeld WD (2016) Algal food for Lesser Flamingos: a stocktaking. Hydrobiologia 775:21–50

Krulwich TA, Sachs G, Padan E (2011) Molecular aspects of bacterial pH sensing and homeostasis. Nat Rev Microbiol 9:330–343

Kummsa T, Bekele A (2014) Feeding ecology of Lesser Flamingos (*Phoeniconaias minor*) in Abijata-Shalla Lakes National Park (ASLNP) with special reference to lakes Abijata and Chitu, Ethiopia. Asian J Biol Sci 7:57–67

Lamb HF (2001) Multi-Proxy Records of Holocene climate and vegetation change from Ethiopian crater lakes. Biol Environ Proc Roy Irish Acad 101B:35–46

Lanzén A, Simachew A, Gessesse A, Chmolowska D, Jonassen I, Øvreås L (2013) Surprising prokaryotic and eukaryotic diversity, community structure and biogeography of Ethiopian soda lakes. PLoS One 8, e72577

Lewin RA, Krienitz L, Goericke R, Takeda H, Hepperle D (2000) *Picocystis salinarum* gen.et sp. nov. (Chlorophyta)—a new picoplanktonic green alga. Phycologia 39:560–65

Lewis WM Jr (1996) Tropical lakes: how latitude makes a difference. In: Schiemer F, Boland KT (eds) Perspectives in tropical limnology. SPB Academic Publishing, Amsterdam, pp 43–64

Lugomela C, Pratap HB, Mgaya YD (2006) Cyanobacteria blooms—a possible cause of mass mortality of Lesser Flamingos in Lake Manyara and Lake Big Momela, Tanzania. Harmful Algae 5:534–541

Luo W, Kotut K, Krienitz L (2013) Hidden diversity of eukaryotic plankton in the soda lake Nakuru, Kenya, during a phase of low salinity revealed by a SSU rRNA gene clone library. Hydrobiologia 702:95–103

Mazokopakis EE, Karefilakis CM, Tsartsalis AN, Milkas AN, Ganotakis ES (2008) Acute rhabdomyolysis caused by *Spirulina* (*Arthrospira platensis*). Phytomedicine 15:525–527

Melack JM (1976) Limnology and dynamics of phytoplankton in equatorial African lakes. PhD Thesis, Duke University, Durham, NC, 453p

Melack JM (1979a) Photosynthesis and growth of *Spirulina platensis* (Cyanophyta) in an equatorial lake (Lake Simbi, Kenya). Limnol Oceanogr 24:753–760

Melack JM (1979b) Temporal variability of phytoplankton in tropical lakes. Oecologia 44:1–7

Melack JM (1982) Photosynthetic activity and respiration in an equatorial African soda lake. Freshw Biol 12:381–400

Melack JM (1988) Primary producer dynamics associated with evaporative concentration in a shallow, equatorial soda lake (Lake Elmenteita, Kenya). Hydrobiologia 158:1–14

Melack JM, Kilham P (1974) Photosynthetic rates of phytoplankton in East African alkaline, saline lakes. Limnol Oceanogr 19:743–755

Melack JM, Kilham P, Fisher TR (1982) Responses of phytoplankton to experimental fertilization with ammonium and phosphate in an African Soda Lake. Oecologia 52:321–326

Mesbah NM, Wiegel J (2011) Halophiles exposed concomitantly to multiple stressors: adaptive mechanisms of halophilic alkalithermophiles. In: Ventosa A, Oren A, Ma Y (eds) Halophiles and hypersaline environments. Springer, Heidelberg, pp 249–273

Metcalf JS, Morrison LF, Krienitz L, Ballot A, Krause E, Kotut K, Pütz S, Wiegand C, Pflugmacher S, Codd GA (2006) Analysis of cyanotoxins anatoxin-a and microcystins in Lesser Flamingo feathers. Toxicol Environ Chem 88:159–167

Metcalf JS, Banack SA, Kotut K, Krienitz L, Codd GA (2013) Amino acid neurotoxins in feathers of the Lesser Flamingo, *Phoeniconaias minor*. Chemosphere 90:835–839

Mikhodyuk OS, Gerasimenko LM, Akimov VN, Ivanovsky RN, Zavarzin GA (2008) Ecophysiology and polymorphism of the unicellular extremely natronophilic cyanobacterium *Euhalothece* sp. Z-M001 from Lake Magadi. Microbiology 77:717–725

Miklaszewska M, Waleron M, Morin N, Calusinska M, Wilmotte A, De Marsac NT, Rippka R, Waleron K (2012) Elucidation of the gas vesicle gene clusters in cyanobacteria of the genus *Arthrospira* (Oscillatoriales, Cyanophyta) and correlation with ITS phylogeny. Eur J Phycol 47:233–244

Milbrink G (1977) On the limnology of two alkaline lakes (Nakuru and Naivasha) in the East Rift Valley System in Kenya. Int Rev Ges Hydrobiol Hydrograph 62:1–17

Mlingwa C, Baker N (2006) Lesser Flamingo *Phoenicopterus minor* counts in Tanzanian soda lakes: implications for conservation. In: Boere GC, Galbraith CA, Stroud DA (eds) Waterbirds around the world. The Stationery Office, Edinburgh, pp 230–233

Motelin G, Thampy R, Doros D (2000) An ecotoxicological study of the potential roles of metals, pesticides and algal toxins on the 1993/5 Lesser Flamingo mass die-offs in Lake Bogoria and Nakuru, Kenya; and the health status of the same species of birds in the Rift Valley Lakes during the 1990s. Proc East Afr Environ Forum 2000:11–12

Mourente G, Lubián LM, Odriozola JM (1990) Total fatty acid composition as a taxonomic index of some marine microalgae used as food in marine aquaculture. Hydrobiologia 203:147–154

Mpawenayo B, Mathooko JM (2004) Diatom assemblages in the hotsprings associated with Lakes Elmenteita and Baringo in Kenya. Afr J Ecol 42:363–367

Mussagy A, Annadotter H, Cronberg G (2006) An experimental study of toxin production in *Arthrospira fusiformis* (Cyanophyceae) isolated from African waters. Toxicon 48:1027–1034

Ndetei R, Muhandiki VS (2005) Mortalities of lesser flamingos in Kenyan Rift Valley saline lakes and the implications for sustainable management of the lakes. Lakes Reserv Res Manag 10:51–8

Nelissen B, Wilmotte A, Neefs JM, De Wachter R (1994) Phylogenetic relationships among filamentous helical cyanobacteria investigated on the basis of 16S ribosomal-RNA gene sequence analysis. Syst Appl Microbiol 17:206–210

Njugana SG (1988) Nutrient–phytoplankton relationships in a tropical meromictic soda lake. Hydrobiologia 158:15–28

Oduor SO, Schagerl M (2007a) Phytoplankton photosynthetic characteristics in three Kenyan Rift Valley saline-alkaline lakes. J Plankton Res 29:1041–1050

Oduor SO, Schagerl M (2007b) Temporal trends of ion contents and nutrients in three Kenyan Rift Valley saline-alkaline lakes and their influence on phytoplankton biomass. Hydrobiologia 584:59–68

Ogato T, Kifle D (2014) Morphological variability of *Arthrospira* (*Spirulina*) *fusiformis* (Cyanophyta) in relation to environmental variables in the tropical soda lake Chitu, Ethiopia. Hydrobiologia 738:21–33

Owen RB, Renaut RW, Hover VC, Ashley GM, Muasya AM (2004) Swamps, springs and diatoms: wetlands of semi-arid Bogoria-Baringo Rift, Kenya. Hydrobiologia 518:59–78

Padisák J (2009) The psychogeography of freshwater algae. In: Likens G (ed) Encyclopedia of inland waters, vol I. Elsevier, Oxford, pp 219–223

Papapetropoulos S (2007) Is there a role for naturally occurring cyanobacterial toxins in neurodegeneration? The beta-N-methylamino-L-alanine (BMAA) paradigm. Neurochem Int 50:998–1003

Papke RT, Ramsing NB, Bateson MM, Ward DM (2003) Geographical isolation in hot spring cyanobacteria. Environ Microbiol 5:650–659

Peduzzi P, Gruber M, Gruber M, Schagerl M (2014) The virus's tooth—cyanophages affect the flamingo population of an African soda lake in a bottom up cascade. ISME J 8:1346–1351

Pogoryelov D, Sudhir PR, Kovács L, Gombos Z, Brown I, Garab G (2003) Sodium dependency of the photosynthetic electron transport in the alkaliphilic cyanobacterium *Arthrospira platensis*. J Bioenerg Biomembr 35:427–437

Renaut RW, Owen RB, Ego JK (2008) Recent changes in geyser activity at Loburu, Lake Bogoria, Kenya Rift Valley. GOSA Trans 10:4–7

Renaut RW, Owen RB, Jones B, Tiercelins J-J, Tarits C, Ego JK, Konhauser KO (2013) Impact of lake-level changes on the formation of thermogene travertine in continental rifts: evidence from Lake Bogoria, Kenya Rift Valley. Sedimentology 60:428–468

Rich F (1932) Reports on the Percy Sladen Expedition to some Rift Valley Lakes in Kenya in 1929.—IV. Phytoplankton from the Rift Valley Lakes in Kenya. Annu Mag Nat Hist Ser 10 10:233–262

Richardson JL, Richardson AE (1972) History of an African Rift Lake and its climatic implications. Ecol Monogr 42:499–534

Roesler CS, Culbertson CW, Etheridge SM, Goericke R, Kiene RP, Miller LG, Oremland RS (2002) Distribution, production, and ecophysiology of *Picocystis* strain ML in Mono Lake, California. Limnol Oceanogr 47:440–452

Roubeix V, Chalié F, Gasse F (2014) The diatom *Thalassiosira faurii* (Gasse) Hasle in the Ziway–Shala lakes (Ethiopia) and implications for paleoclimatic reconstructions: case study of the Glacial–Holocene transition in East Africa. Palaeogeogr Palaeoclim Palaeoecol 402:104–112

Schagerl M, Oduor SO (2008) Phytoplankton community relationship to environmental variables in three Kenyan Rift Valley saline-alkaline lakes. Mar Freshw Res 59:125–136

Schagerl M, Burian A, Gruber-Dorninger M, Oduor SO, Kaggwa MN (2015) Phytoplankton succession of African soda lakes with a special focus on *Arthrospira fusiformis*. Fottea 15:245–257

Scheldeman P, Baurain D, Bouhy R, Scott M, Mühling M, Whitton BA, Belay A, Wilmotte A (1999) *Arthrospira* ('*Spirulina*') strains from four continents are resolved into only two clusters, based on amplified ribosomal DNA restriction analysis of the internally transcribed spacer. FEMS Microbiol Lett 172:213–222

Schlesinger P, Belkin S, Boussiba S (1996) Sodium deprivation under alkaline conditions causes rapid death of the filamentous cyanobacterium *Spirulina platensis*. J Phycol 32:608–613

Sileo L, Grootenhuise JG, Tuite GH, Hopcraft HD (1979) Microbacteriosis in the lesser flamingo of Lake Nakuru, Kenya. J Wildl Dis 15:387–390

Sili C, Mascalchi C, Ventura S (2011) Evolutionary differentiation of the sister cyanobacterial genera *Cyanospira* Florenzano, Sili, Pelosi et Vincenzini and *Anabaenopsis* (Woloszynska) Miller in response to extreme life conditions. Fottea 11:107–117

Sili C, Torzillo G, Vonshak A (2012) *Arthrospira* (*Spirulina*). In: Whitton BA (ed) Ecology of Cyanobacteria II: their diversity in space and time. Springer, Dordrecht, pp 677–705

Sommer U, Gliwicz ZM, Lampert W, Duncan A (1986) The PEG-model of seasonal succession of planktonic events in fresh waters. Arch Hydrobiol 106:433–471

Sommer U, Adrian R, De Senerpont DL, Elser JJ, Gaedke U, Ibelings B, Jeppesen E, Lürling M, Molinero JC, Mooij WM, van Donk E, Winder M (2012) Beyond the plankton ecology group (PEG) model: mechanisms driving plankton succession. Annu Rev Ecol Evol Syst 43:429–448

Steidinger KA, Landsberg JH, Mason PL, Vogelbein WK, Tester PA, Litaker RW (2006) *Cryptoperidiniopsis broadyi* gen. et sp. nov. (Dinophyceae), a small lightly armoured dinoflagellate in the Pfiesteriaceae. J Phycol 42:951–961

Stewart I, Seawright AA, Shaw GR (2008) Cyanobacterial poisoning in livestock, wild mammals and birds—an overview. In: Kenneth HH (ed) Cyanobacterial harmful algal blooms: state of the science and research needs. Springer, Heidelberg, pp 613–638

Stizenberger E (1852) *Spirulina* and *Arthrospira* (nov. gen.). Hedwigia 1:32–41

Straubinger-Gansberger N, Lawton L, Gruber M, Kaggwa MN, Oduor SO, Schagerl M (2014) Cyanobacterial toxins in East African Rift Valley lakes—do they really cause flamingo mass mortality? Wildl Biol 20:185–189

Talling JF, Wood RB, Prosser V, Baxter RM (1973) The upper limit of photosynthetic productivity by phytoplankton: evidence from Ethiopian soda lakes. Freshw Biol 3:53–76

Tebbs EJ, Remedios JJ, Harper DM (2013) Remote sensing of chlorophyll-a as a measure of cyanobacterial biomass in Lake Bogoria, a hypertrophic, saline–alkaline, flamingo lake, using Landsat ETM+. Remote Sens Environ 135:92–106

Tomaselli L (1997) Morphology, ultrastructure and taxonomy of *Arthrospira* (*Spirulina*) *maxima* and *Arthrospira* (*Spirulina*) *platensis*. In: Vonshak A (ed) *Spirulina platensis* (*Arthrospira*): physiology, cell-biology and biotechnology. Taylor and Francis, London, pp 1–15

Tomaselli L, Giovanetti L, Sacchi A, Margheri MC (1981) On the mechanism of trichome breakage in *Spirulina platensis* and *S. maxima*. Ann Microbiol 31:27–31

Ton T (2007) Pink starvation: Flamingos in Africa malnourished. http://en.epochtimes.com/news/7-1-23/50812.html. Accessed 18 Jan 2015

Tonk L, Bosch K, Visser PM, Huisman J (2007) Salt tolerance of the harmful cyanobacterium *Microcystis aeruginosa*. Aquatic Microbial Ecol 46:117–123

Tuite CH (1981) Standing crop densities and distribution of *Spirulina* and benthic diatoms in East African alkaline saline lakes. Freshw Biol 11:345–360

Tuite CH (2000) The distribution and density of Lesser Flamingos in East Africa in relation to food availability and productivity. Waterbirds (Spec Publ) 23:52–63

Vareschi E (1978) The ecology of Lake Nakuru (Kenya) I. Abundance and feeding of the Lesser Flamingo. Oecologia 32:11–35

Vareschi E (1982) The ecology of Lake Nakuru (Kenya) III. Abiotic factors and primary production. Oecologia 55:81–101

Vareschi E, Jacobs J (1985) The ecology of Lake Nakuru VI. Synopsis of production and energy flow. Oecologia 65:412–424

Vareschi E, Melack JM, Kilham P (1981) Saline waters. In: Symoens JJ, Burgis MJ, Gaudet JJ (eds) The ecology and utilization of African inland waters. UNEP Reports and Proceeding Series No. 1, Nairobi, pp 93–102

Verschuren D, Cocquyt C, Tibby J, Roberts CN, Leavitt PR (1999) Long-term dynamics of algal and invertebrate communities in a small, fluctuating tropical soda lake. Limnol Oceanogr 44:1216–1231

Viti C, Ventura S, Lotti E, Capolino E, Tomaselli L, Giovannetti L (1997) Genotypic diversity and typing of cyanobacterial strains of the genus *Arthrospira* by very sensitive total DNA restriction profile analysis. Res Microbiol 148:605–611

Vonshak A, Richmond A (1988) Mass production of the blue-green alga *Spirulina*: an overview. Biomass 15:233–247

Vonshak A, Tomaselli L (2000) *Arthrospira* (*Spirulina*): systematics and eco-physiology. In: Whitton BA, Potts M (eds) The ecology of cyanobacteria. Kluwer, Dordrecht, pp 505–522

Walsby AE, Van Rijn J, Cohan Y (1983) The biology of a new gas-vacuolate cyanobacterium, *Dactylococcopsis salina* sp. nov., in Solar Lake. Proc R Soc Lond B 217:417–447

Wood RB, Talling JF (1988) Chemical and algal relationships in a salinity series of Ethiopian inland waters. Hydrobiologia 158:29–67

The Protozoa of Soda Lakes in East Africa 7

Andrew W. Yasindi and William D. Taylor

Abstract

Protozoa, mainly ciliates, have been studied in East African soda lakes (EASL) since the 1980s. A few studies have noted the presence of flagellates, including *Chlamydomonas*, *Ceratium* and chilomonids. Amoebae have been even less studied, although *Biomyxa* and *Platamoeba* have been observed. The abundance, biomass and production of ciliates in the EASL are higher than in tropical freshwater, subtropical and temperate lakes. This is attributed to the high phytoplankton and bacterial abundance in these lakes. Accordingly, the ciliate fauna is dominated by herbivorous and bacterivorous taxa. Estimates of growth rates for ciliates in EASL range from 0.18 to 4.75 day^{-1}. High water temperatures allow for these fast growth rates, which together with high biomass lead to high secondary production. Some of this production can reach higher trophic organisms such as zooplankton and fish, while some is consumed by predators within the microbial food web.

7.1 Introduction

Soda lakes are highly alkaline-saline aquatic ecosystems confined to semiarid and arid areas worldwide (Antony et al. 2013). They are often remote from human activity and therefore less studied than freshwater lakes (Jones and Grant 1999). As a result, they have received limited scientific investigation (Antony et al. 2013; Williams 1981). They are even less studied when areas of research such as the ecology of free-living protozoa are considered (Finlay et al. 1987). Nonetheless, some of the best-studied soda lakes are located in the East African Rift System (EARS; Fig. 7.1).

A.W. Yasindi (✉)
Department of Biological Sciences, Egerton University, PO Box 536-20115, Egerton, Kenya
e-mail: ayasindi@yahoo.com

W.D. Taylor
Department of Biology and Department of Earth and Environmental Sciences, University of Waterloo, Waterloo, ON, Canada N2L 3G1
e-mail: wdtaylor@uwaterloo.ca

M. Schagerl (ed.), *Soda Lakes of East Africa*, DOI 10.1007/978-3-319-28622-8_7

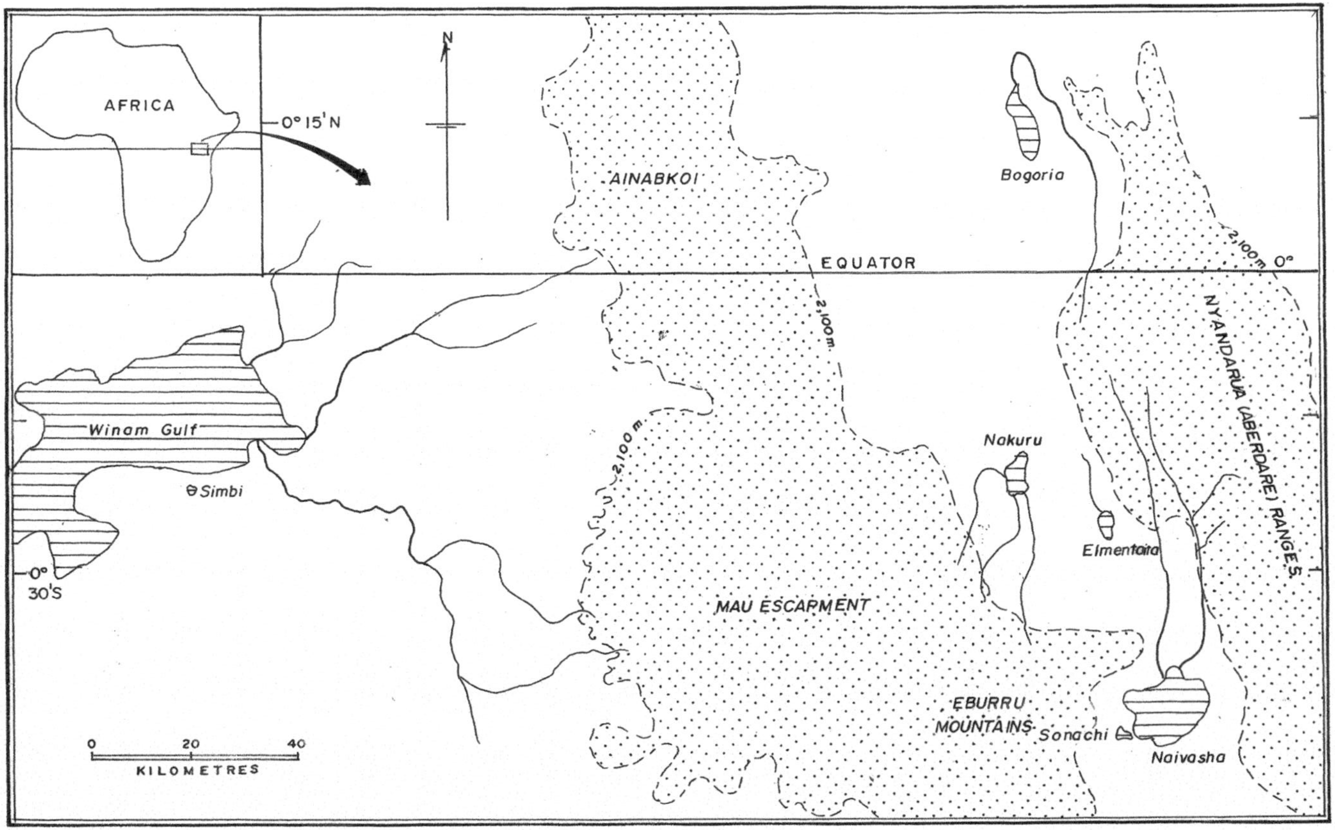

Fig. 7.1 A Map showing some of the soda lakes of East Africa (from Melack 1981). The Winam Gulf of Lake Victoria and Lake Naivasha are, however, freshwater. Although alkaline-saline, Lake Simbi is not located in the EARS

For a long time, studies of planktonic communities in tropical lakes rarely included the protozoa except to mention their presence. For example, Jenkin (1936) reported finding ciliates in Lake Elmenteita, while Kilham (1981) observed *Paramecium*-like organisms in unpreserved water samples from Lake Sonachi. Therefore, data on protozoa in soda lakes are still fragmentary. Despite the few studies, accumulating empirical evidence suggests that protozoa, or at least ciliates, may be extremely abundant and diverse in soda lakes (Finlay et al. 1987; Lanzén et al. 2013; Yasindi et al. 2002, 2007).

Studies from temperate freshwater lakes (e.g. Esteban et al. 2012; Taylor and Heynen 1987) and the sea (e.g. Gifford 1991; Worden et al. 2015) emphasize the importance of protozoa in the functioning of aquatic systems. Protozoa feed predominantly on pico- and nanoplankton organisms (Xu and Cronberg 2010), and they, in turn, are consumed by metazooplankton (Xu and Cronberg 2010; Wickham and Gilbert 1993), by fish larvae and as well by some adult fish (Jack and Gilbert 1997; Lair et al. 1994; Mageed et al. 2002). Therefore, protozoa and other microplanktonic organisms are commonly implicated as trophic links between picoplankton organisms and metazoan consumers (Gifford 1991; Xu and Cronberg 2010). Because some protozoan prey, mainly bacteria, other pico- and nanoplanktonic organisms and even viruses (Hennemuth et al. 2008) are too small to be effectively grazed by metazoa, protozoa repackage these organisms into protozoan biomass, which is large enough to be effectively consumed by larger organisms (Adrian and Frost 1993; Gifford 1991; Weisse and Müller 1998). Protozoa also recycle nutrients for phytoplankton in aquatic systems (Guildford and Taylor 2011), and some contribute directly to primary production via phototrophic endosymbionts (Esteban et al. 2010; Finlay et al. 1988; Sanders 1991). These diverse and ecologically important functions of protozoa that have been demonstrated in marine and temperate freshwaters are still unclear in tropical lakes in general and soda lakes in particular. The purpose of the following review is to take stock of protozoan ecology in East African soda lakes (EASL). These lakes are among the most productive aquatic systems in the world (Antony et al. 2013; Melack 1981; Talling et al. 1973; Vareschi et al. 1981) and exhibit simple food webs (Vareschi and Jacobs 1985) that include extremely productive microbial populations (Finlay et al. 1987; Kilham 1981; Lanzén et al. 2013; Yasindi et al. 2002). Our review is biased towards ciliates because these protozoans are the most studied in EASL (Fig. 7.2). We will, however, include the sparse information available on amoeboid and flagellated protists.

7.2 Composition of Protozoan Communities in Soda Lakes

The composition of protozoa in soda lakes is variable both within and among lakes. Although most taxa encountered are similar to ones described from freshwater habitats (Table 7.1), a few are unique. Due to the high salinity of soda lakes, they are thought to support only species adapted to these conditions (Colburn 1988; Finlay et al. 1987), so we expect species from soda lakes to be unique. Nonetheless, we are unaware of studies attempting to determine whether species from soda lakes can be reared in freshwater and vice versa. The high abundance and biomass exhibited by ciliates (Yasindi et al. 2002) in these lakes show that some are well adapted to these hostile environments. Their normal size and shape, their ability to divide and conjugate and their normal contractile vacuole activity as observed in *Cyclidium* and *Prorodon* (Finlay et al. 1987) support this. The contractile vacuoles of ciliates in soda lakes may expel sodium, enabling them to survive in lakes of high salt concentration (Hampton and Schwartz 1976).

Yasindi et al. (2002) recorded 33 species of ciliates in Lake Nakuru, while Yasindi et al. (2007) identified 49 species belonging to 14 orders in 17 East African lakes of salinity ranging from freshwater to saline (soda) conditions. Of the 14 orders, ciliates of Scuticociliatida and Oligotrichida were the most abundant, with the

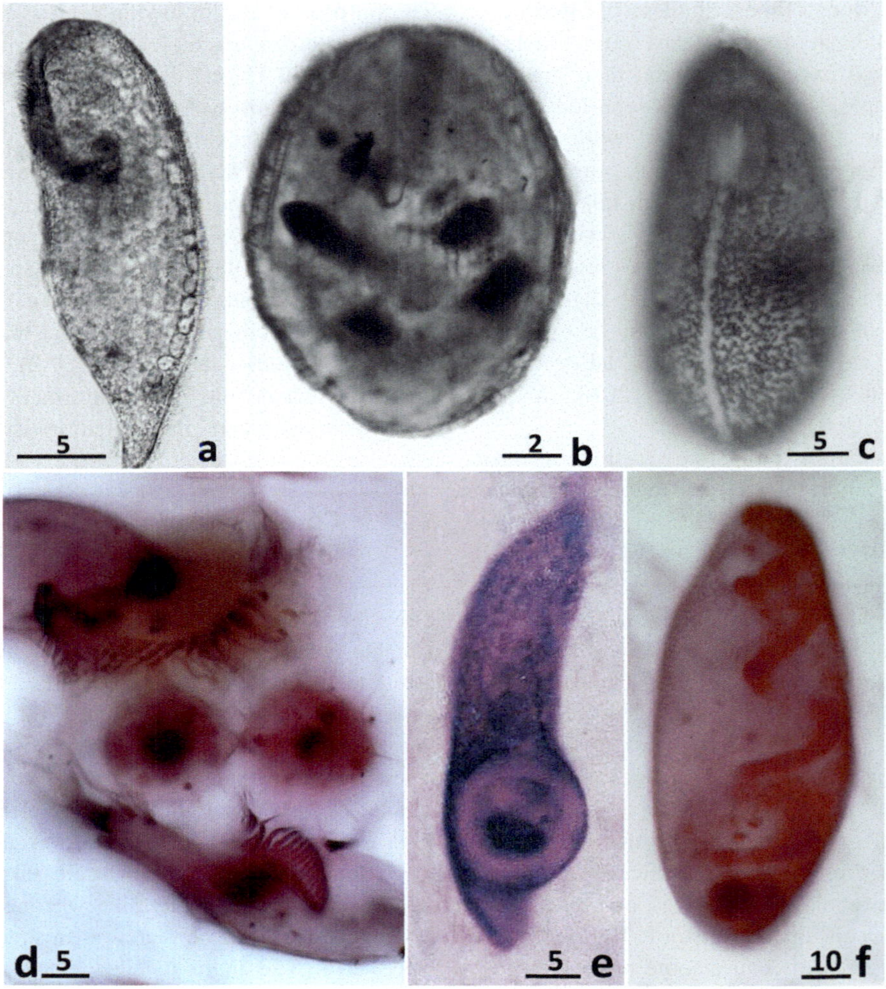

Fig. 7.2 Protists from Kenyan soda lakes. (a) *Condylostoma elephantinus*, (b) *Holophrya* sp., (c) *Frontonia* sp., (d) carnivorous *Sphaerophrya* sp. dividing (centre) while feeding on two *Euplotes* ciliates at top and bottom attached to the tentacles, (e) *Acineria* containing ingested flagellates, (f) *Frontonia* sp. with an ingested *Arthrospira fusiformis*. Scale bar refers to size in μm

contribution of the former increasing with trophy and salinity and the latter decreasing. In contrast to observations in subtropical lakes, where scuticociliates and haptorids co-dominate eutrophic freshwater lakes (Beaver and Crisman 1982), haptorids are generally a small component of the fauna in eutrophic EASL (Yasindi et al. 2007). The haptorids *Askenasia* (Blochmann) and *Mesodinium* (Stein), which are common in temperate lakes (Pace et al. 1990; Sonntag et al. 2011; Taylor and Heynen 1987), have not been found in EASL, although they are thought to be widely distributed (Laybourn-Parry 1992). The low abundance and diversity of haptorids probably reflect a low abundance of flagellates on which many small haptorids feed (Stensdotter-Blomberg 1998; Yasindi et al. 2007). The large haptorid *Dileptus* (Dujardin), which consumes other ciliates and rotifers, contributes significantly to the ciliate biomass of Lake Bogoria (Ong'ondo et al. 2013). This suggests that bacterivorous flagellates play a lesser role in East African soda lakes than in

Table 7.1 Some common ciliated protozoa and their sizes (volumes) in East African soda lakes (from Yasindi 2001)

Order	Genus	Volume (μm^3)
Oligotrichida	*Strobilidium*	$2.59 \times 10^3 \pm 1.38 \times 10^3$
	Halteria	$2.62 \times 10^3 \pm 1.56 \times 10^3$
Scuticociliatida	*Cyclidium* spp.	$3.40 \times 10^3 \pm 2.28 \times 10^3$
	Pleuronema	$1.35 \times 10^3 \pm 1.02 \times 10^3$
	Cristigera	$2.32 \times 10^3 \pm 2.9 \times 10^3$
	Cinetochilum	3.18×10^3
	Uronema	4.39×10^3
Peritrichida	*Vorticella* 1	1.47×10^4
	Vorticella 2	$3.16 \times 10^3 \pm 1.35 \times 10^3$
Heterotrichida	*Linostoma*	$2.87 \times 10^5 \pm 2.24 \times 10^5$
	Spirostomum	$3.25 \times 10^5 \pm 7.27 \times 10^4$
	Stentor	$2.13 \times 10^5 \pm 5.01 \times 10^4$
Stichotrichida	*Stichotricha*	$1.26 \times 10^4 \pm 1.8 \times 10^5$
	Stylonychia	7.56×10^3
	Uroleptus	$6.72 \times 10^3 \pm 1.36 \times 10^3$
Haptorida	*Phialina* 1	$1.76 \times 10^4 \pm 6.76 \times 10^3$
	Phialina 2	$2.09 \times 10^3 \pm 1.7 \times 10^3$
	Dileptus	$6.14 \times 10^4 \pm 4 \times 10^3$
	Paradileptus	$4.12 \times 10^6 \pm 2.08 \times 10^6$
	Monodinium	$1.45 \times 10^4 \pm 7.83 \times 10^3$
	Lagynophrya	$1.05 \times 10^4 \pm 1.17 \times 10^3$
	Spathidium	2.8×10^4
	Trachelius	4.2×10^4
Prostomatida	*Holophrya×*	$1.32 \times 10^5 \pm 4.74 \times 10^4$
	Urotricha	$9.60 \times 10^3 \pm 5.24 \times 10^3$
	Metacystis	$7.78 \times 10^4 \pm 3.33 \times 10^4$
Pleurostomatida	*Acineria*	$1.15 \times 10^4 \pm 6.45 \times 10^3$
	Litonotus	$1.38 \times 10^4 \pm 8.93 \times 10^3$
Hypotrichida	*Aspidisca*	$3.08 \times 10^3 \pm 8.59 \times 10^2$
	Euplotes	$2.58 \times 10^4 \pm 1.98 \times 10^4$
Cyrtophorida	*Dysteria*	$7.12 \times 10^2 \pm 1.13 \times 10^2$
	Chlamydodon	$9.18 \times 10^3 \pm 4.01 \times 10^3$
Peniculida	*Frontonia*	$1.98 \times 10^5 \pm 1.11 \times 10^5$
Suctoria	*Sphaerophrya*	$4.16 \times 10^3 \pm 1.57 \times 10^3$
	Periacineta	1.51×10^4
Tintinnida	*Tintinnidium*	$6.37 \times 10^3 \pm 5.45 \times 10^3$
Armophorida	Armophorid	$3.55 \times 10^4 \pm 5.48 \times 10^3$
	Metopus	9.09×10^3
	Caenomorpha	2.07×10^4

Error ranges are standard deviations

temperate freshwater lakes (Zinabu and Taylor 1997). Nonetheless, the role of nanoflagellates and their predators in African soda lakes deserves further attention because specialized techniques suitable to quantify heterotrophic forms have not been applied.

The occurrence of *Halteria* in soda lakes as well as freshwater lakes of East Africa is consistent with the observation that *Halteria* spp. are distributed worldwide across the trophic spectrum of lakes (Laybourn-Parry 1992; Xu and Cronberg 2010). In contrast, the oligotrichs

Strombidium and *Strobilidium*, which are ubiquitous in freshwater and marine plankton, are not abundant in EASL (Yasindi et al. 2007). A single specimen of *Strombidium* was observed in Lake Nakuru (Yasindi et al. 2002) although this genus is present in saline and hypersaline lakes in Antarctica (Parker et al. 1982).

The peritrich genus *Vorticella* is widely found in Kenyan and Ethiopian Rift Valley lakes, including lakes Sonachi, Oloidien and Abijata (Yasindi et al. 2007). *Vorticella* and other peritrichs are usually abundant in mesotrophic and eutrophic lakes (Laybourn-Parry 1992; Xu and Cronberg 2010) containing abundant filamentous Cyanobacteria or diatoms, to which they are often attached, and that are rich in Bacteria (Yasindi et al. 2007). Therefore, soda lakes with high bacterial and cyanobacterial abundance may be suitable habitats for *Vorticella*. *Cyclidium* is particularly abundant in Kenyan soda lakes, although it is occasionally outnumbered by other small ciliates in the soda lakes Elmenteita, Sonachi and Bogoria. In Elmenteita, *Cyclidium* was outnumbered by a small cyrtophorid, *Dysteria* (Huxley), in 1998 and another scuticociliate, *Pleuronema* (Dujardin), in 1999 (Yasindi and Taylor 2006) *Cyclidium* was abundant in Lake Nakuru (Finlay et al. 1987; Yasindi et al. 2002) in earlier studies and in 1998, but it was absent from this lake in 1999, when *Stylonychia* was dominant (Yasindi and Taylor 2006). A heterotrich, *Linostomella*, and a peniculid, *Frontonia*, were abundant in the plankton of Lake Sonachi in 1998, but were absent in 1999. *Frontonia*, usually rare in the plankton of temperate lakes, was common in EASL (Yasindi et al. 2007). *Linostomella* also occurred in the plankton of another soda lake, Oloidien, where it contributed significantly to the biomass. Another heterotrich, *Spirostomum*, was recorded in the hypolimnion of Lake Sonachi (Yasindi et al. 2007), similar to observations in subtropical lakes (Beaver and Crisman 1989a) and temperate lakes (Bark 1981). The large heterotrich *Condylostoma* was abundant in Lake Nakuru (Ong'ondo et al. 2013; Yasindi et al. 2002, 2007). The presence of *Tintinidium* (Kent) in Lake Elmenteita as well as in freshwater lakes Naivasha and Crescent supports the suggestion that this genus spans the whole salinity spectrum from freshwater to soda lakes (Yasindi et al. 2007). It was reported earlier in Lake Tanganyika (Hecky and Kling 1981) and Lake Albert (Green 1994).

7.3 Diversity of Protozoa

Data on protozoan diversity in tropical lakes in general and soda lakes in particular are still fragmentary. Finlay et al. (1987) identified 21 ciliate species on two sampling occasions in Lake Nakuru in 1985 and 6 species of ciliates in Lake Simbi on a single sampling occasion in 1984. They also observed two species of cryptomonad and several colourless bodonid flagellates in Lake Nakuru. Other protozoa they observed included a dinoflagellate, *Chlamydomonas* sp., a heliozoan and three different naked Amoebae including possibly *Biomyxa* and *Platamoeba* or *Vanella*. Yasindi et al. (2002), in their study of ciliates in Lake Nakuru from June 1993 to January 1994, found an average Shannon Diversity of 1.31. Diversity was negatively correlated with conductivity. In another study, Yasindi (2001) recorded ciliate diversity ranging from 0.3 to 1.15 with mean of 0.94 in soda lakes Solai, Bogoria, Simbi, Nakuru and Sonachi. Solai had the lowest species diversity of 0.3, whereas Elmenteita had the highest. The mean species diversity of 0.94 for soda lakes was only slightly lower than the mean for freshwater lakes (1.05) but might reflect the exclusion of some species (Vareschi and Jacobs 1985; Williams et al. 1990). However, the tolerances for salinity among ciliates inhabiting soda lakes are not known. A weak negative relationship between the Shannon Diversity index and trophic state in East African lakes contradicts the positive relationship recorded in subtropical (Florida) lakes (Beaver and Crisman 1989b), where the highest diversity and number of species were found in hypereutrophic lakes. Beaver and Crisman (1989b) attributed the low species diversity of ciliates in oligotrophic lakes to reduced phytoplankton abundance and diversity.

According to Mathes and Arndt (1994), as lake trophy increases, the role of protozoa as food for zooplankton also increases, indicating the importance of the microbial food web in eutrophic and hypereutrophic systems.

7.4 Abundance of Protozoa in Soda Lakes

A number of studies have estimated the abundance of ciliated protozoa in EASL (Table 7.2). In a study that covered 17 East African lakes of different salinity and trophy, Yasindi and Taylor (2005) recorded a range in ciliate abundance from 2 to 1220 ciliates mL^{-1}. In contrast, in temperate (Quebec) lakes, abundance ranged from 3.3 to 21.6 ciliates mL^{-1} (Pace 1986), and values in subtropical (Florida) lakes ranged from 10.8 to 155.5 ciliates mL^{-1} (Beaver and Crisman 1982). Yasindi and Taylor (2005) found that ciliate abundance increased with increasing trophy as occurs in temperate (Pace 1986) and subtropical lakes (Beaver and Crisman 1982) but also increased with salinity, so that soda lakes had the highest mean abundance (Fig. 7.3). That study found that tropical lakes generally have higher ciliate abundance than either subtropical or temperate lakes, a phenomenon that was attributed to higher chl-*a* concentrations in tropical lakes. When regression models predicting abundance and biomass of ciliates from chl-*a* were compared for tropical East African lakes, subtropical and temperate lakes (Fig. 7.4a), the slopes were not significantly different (Yasindi et al. 2007). Therefore, temperate, subtropical and tropical lakes have similar ciliate abundance and biomass per unit chl-*a*. The soda lakes Bogoria, Sonachi and Elmenteita supported a very high abundance of ciliates, probably due to high phytoplankton biomass and bacterial abundances of $> 10^7\ mL^{-1}$ (Finlay et al. 1987; Yasindi et al. 2002). Apparently, ciliate cell sizes decrease with increase in chl-*a* concentration, probably as a response to increasing importance of bacterivory and the generally smaller size of bacterivores. This is congruent with the finding that most ciliates in EASL are bacterivorous (Yasindi and Taylor 2006) and agrees with the assertion that despite the presence of some large species, eutrophic soda lakes have many species which are bacterivores (Laybourn-Parry 1992). The salinity of EASL is due to sodium carbonate, which likely affects the protozoa in them. Ong'ondo et al. (2013) attributed the smaller biovolume of *Frontonia* in Lake Bogoria relative to Nakuru to a difference in salinity. Similarly, Lake Bogoria had fewer species or morphospecies (17) compared to 26 in Lake Nakuru. This could also be due to

Table 7.2 The range, means and standard deviation of abundance and biomass of protozoa in East African soda lakes

Lake	Abundance Range	Mean	Biomass Range	Mean	Reference
Ciliates					
Nakuru		675,851			Finlay et al. (1987)
Nakuru	1.0×10^4 -1.4×10^7	1.2×10^6 $\pm 3.1 \times 10^5$	$16.9–5.0 \times 10^5$	1.8×10^4 $\pm 7.1 \times 10^5$	Yasindi et al. (2002)
Bogoria	14.4–72.6	43.5 ± 8.2	69.6–369.9	220 ± 150	Yasindi and Taylor (2005)
Elmenteita	29.4–3575.3	640.3 ± 578	20.9–4520.0	971 ± 763	Yasindi and Taylor (2005)
Oloidien	4.7–8.8	6.9 ± 1.2	74.5–54.9	30 ± 12.6	Yasindi and Taylor (2005)
Sonachi	57.5–567.5	377.9 ± 70	347.1–3319.3	1900 ± 1344	Yasindi and Taylor (2005)
Simbi	0.1–6.1	2.9 ± 1.2	0.1–36.1	12.0 ± 8.5	Yasindi and Taylor (2005)
Simbi		400			Finlay et al. (1987)
Flagellates					
Nakuru		8.4×10^4			Finlay et al. (1987)
Elmenteita	1.0×10^3				Kilham (1981)
Simbi	$1.4–7.5 \times 10^3$				Finlay et al. (1987)

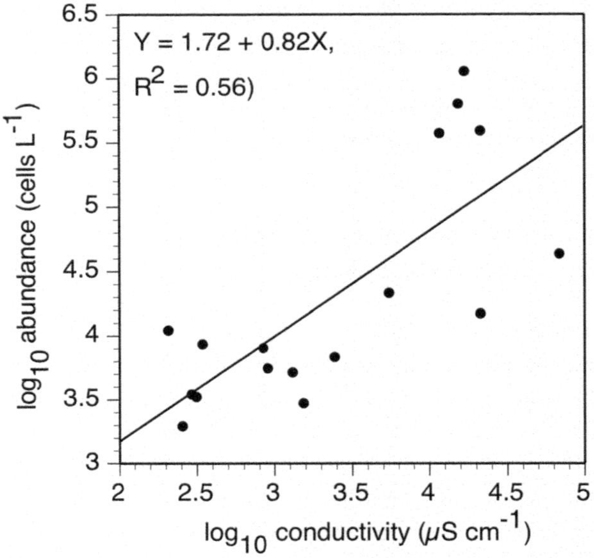

Fig. 7.3 The relationship between Log_{10} ciliate abundance and Log_{10} conductivity in East African lakes (from Yasindi and Taylor 2005)

the two-fold higher cyanobacterial biovolume in Lake Bogoria, which reflects lower availability of edible algae to ciliates. Finlay et al. (1987) attributed the higher composition and abundance of ciliated protozoa in Lake Nakuru versus Lake Simbi to a lesser abundance of flagellates and Bacteria in the latter.

Although most ciliates in EASL are heterotrophic, mostly bacterivorous, there are also many herbivores and predators (Yasindi and Taylor 2006). Note, however, that most soda lakes had $> 50\ \%$ of the total ciliate production from bacterivorous ciliates (Yasindi and Taylor 2006). This suggests that ciliate bacterivory is an important link between bacterioplankton production and metazoan zooplankton grazers in EASL. In all these lakes, herbivorous and bacterivorous ciliates together accounted for at least 75 % of the ciliate population production, some of which is consumed by metazoan zooplankton (Yasindi and Taylor 2006) and fish (Milbrink 1977). Therefore, ciliates may form an important linkage through which the pico- and nanoplankton productions from the microbial loop are transferred to the classic food webs in EASL.

Protozoa other than ciliates have been less studied in EASL. A few studies, however, indicate that flagellates may be abundant in soda lakes (Table 7.2).

7.5 Ciliate Sizes in Soda Lakes

The biomass of protozoa is estimated from their volume, which is often compared among species of different shapes as equivalent spherical diameter (ESD). In Lake Nakuru, the ESD of ciliates ranged from 11 to 279 μm, while the mean ± SE

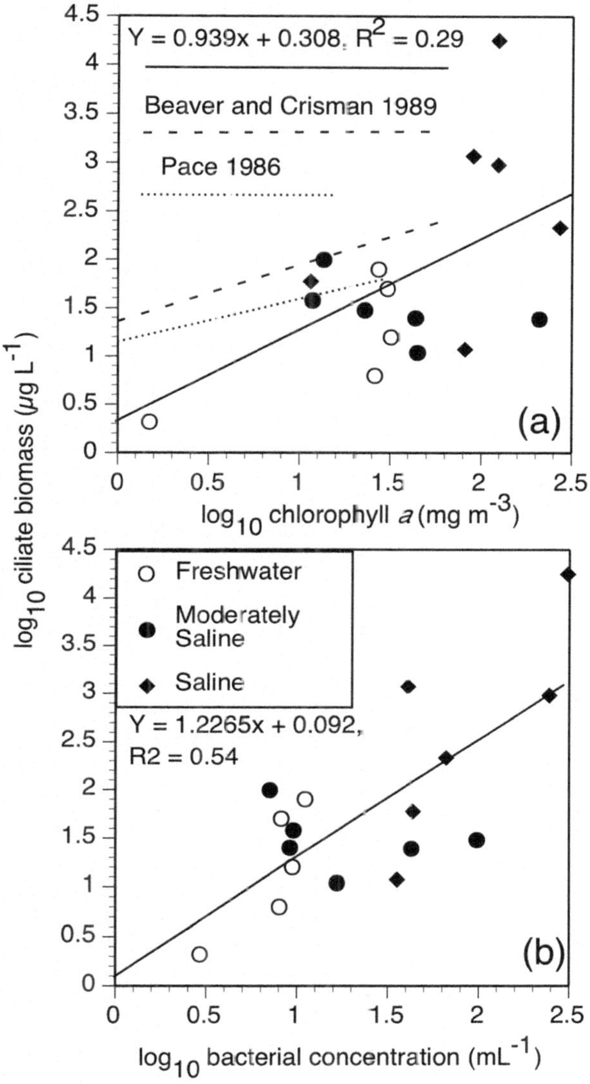

Fig. 7.4 The relationship between (**a**) Log_{10} ciliate biomass and Log_{10} chlorophyll *a* concentration in East African lakes compared to subtropical and temperate lakes, (**b**) Log_{10} ciliate biomass and Log_{10} bacterial concentration in East African lakes (from Yasindi et al. 2007)

for various taxa ranged from 15.1 ± 0.5 μm to 184.3 ± 6.7 μm (Yasindi et al. 2002). Ciliates with ESD of 11–40 μm made up 75.7 % of the total abundance, while large ciliates with ESD > 50 μm comprised 7.5 % of abundance (Fig. 7.5) and included species in the genera *Prorodon*, *Frontonia* and *Condylostoma*. The overall mean ESD for ciliates in this soda lake was 43 ± 2 μm (Yasindi et al. 2002). The mean ESD of ciliates in EASL of varying salinity and trophy ranged from 10.3 ± 0.2 to 193 ± 35 μm. *Strobilidium* was the smallest ciliate, with a mean ESD of 7.1 μm, while *Paradileptus elephantinus* was the largest, with an ESD of 227 μm. Ciliate biovolumes ranged from 10^2 to 10^6 $μm^3$ (Yasindi and Taylor 2005). Ciliates within the ESD size class 11–20 were the most abundant in most of the soda lakes (Fig. 7.5). Ciliates in the > 50 μm ESD size class, though fewer in abundance, dominated the biomass in the soda lakes. The 20–30-μm ESD size class included ciliates in the genera *Cyclidium* (Scuticociliatida); *Dysteria*, *Halteria*, *Aspidisca* and *Urotricha* and *Monodinium* (Prostomatida);

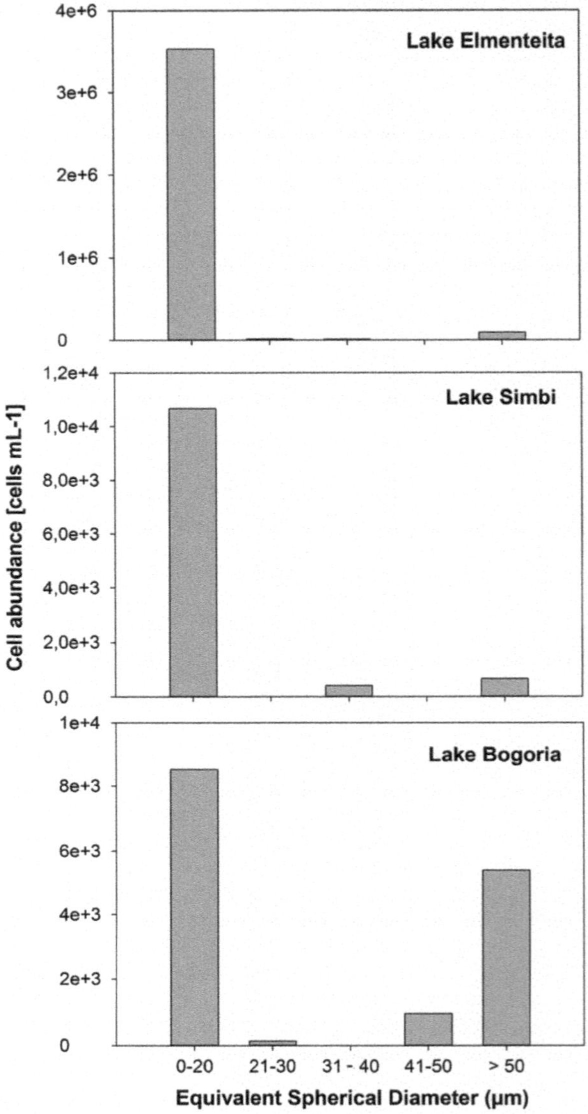

Fig. 7.5 The abundance as a function of equivalent spherical diameter (ESD) of ciliates in Lakes Simbi, Elmenteita and Bogoria, Kenya (Yasindi 2001)

Lagynophrya (Haptorida); *Stylonychia* and *Stichotrichia* (Stichotrichida); and *Tintinidium* (Tintinida). The > 50-μm ESD size class was made up of large ciliates in the genera *Linostoma* and *Spirostomum* (Heterotrichida), *Holophrya* (Prostomatida), *Frontonia* (Peniculida) and *Paradileptus* (Haptorida). This size class dominated the biomass in most lakes.

7.6 Ciliate Biomass in East African Soda Lakes

Ciliate biomass is variable within and among lakes. For example, Finlay et al. (1987) estimated the biomass of ciliates in Lake Nakuru to be about 0.24 and 1.2 g m^{-3} on 30 June 1985 and 5 July 1985, respectively (Table 7.2). In their study,

Yasindi et al. (2002) found that ciliate biomass in Lake Nakuru was dominated by large ciliates, mainly *Frontonia* and *Condylostoma*. *Cyclidium*, which dominated the abundance, contributed only 0.84 % of the total biomass.

In East African Rift system lakes of different salinity, Yasindi and Taylor (2005) found that biomass increased with trophy and salinity (Fig. 7.4a). The highest biomass was recorded in soda lakes Nakuru, Abijata, Bogoria, Elmenteita and Sonachi, with mean biomasses (± standard deviation) of 31 ± 68, 60 ± 10.5, 220 ± 150, 971 ± 763 and 1900 ± 1344 μg C L^{-1}, respectively.

While the high biomass in EASL is attributed to high chl-*a* (Laybourn-Parry 1992), Bogoria, Nakuru, Elmenteita and Sonachi also harbour abundant flamingos and other waterbirds (Lanzén et al. 2013; Yasindi et al. 2007). The faeces of these birds likely elevate the bacterial concentration relative to chl-*a* (Finlay et al. 1987). An empirical study of bacterial abundance in Ethiopian lakes found that the relationship between bacterial abundance and chl-*a* was steeper for soda lakes than for freshwater ones (Zinabu and Taylor 1997). A particularly high bacterial abundance in soda lakes may explain why ciliate biomass is more correlated with bacterial abundance than with chl-*a* (Fig. 7.4 a, b) and why bacterivores are particularly abundant. In general, the EASL are more variable in their ciliate biomass than subtropical and temperate lakes, even removing the covariation between ciliate biomass and chl-*a* (Yasindi et al. 2007). The coefficient of determination for ciliate biomass versus chl-*a* is 0.95 for temperate lakes (Pace 1986), whereas in EASL it is 0.29, reflecting that the latter are much more variable in terms of their physical, chemical and biological characteristics. It could also reflect that fewer samples, and sometimes only single samples, were used to characterize a lake rather than means of monthly sampling (Yasindi et al. 2007).

In general, the biomass range for ciliates from 1.9 to 1900.0 μg C L^{-1} is greater than observed in both temperate lakes (12.3 to 56.3 μg C L^{-1}; Pace 1986) and subtropical (9.3 to 126.0 μg C L^{-1}; Beaver and Crisman 1982) lakes. As the biomass increases with increasing salinity and trophy, soda lakes have the highest ciliate biomass among the EASL.

7.7 Distribution of Protozoa in Soda Lakes

Lakes are heterogeneous environments, varying temporally, vertically and horizontally (Laybourn-Parry et al. 1990b). Such differences influence variation in the abundance and distribution of organisms in lakes. Many studies link the composition and abundance of bacteria (Fortin et al. 2000; Gustavson et al. 2000), phytoplankton (Hyenstrand et al. 2000; Saros and Fritz 2000), zooplankton (Giebelhausen and Lampert 2001; Hall and Burns 2001; Sommer 1989) and ciliated protozoa (Krashevska et al. 2014; Pace 1986; Pace and Orcutt 1981) to physical, chemical and biological characteristics in temperate lakes. Studies in temperate waters also attribute seasonal maxima in ciliate abundance and biomass to the presence of food (Pace and Orcutt 1981), absence of predators (Colburn 1988; Madoni 1990; Sonntag et al. 2011; Xu and Cronberg 2010) and lack of competition for resources (Reynolds 1989; Sonntag et al. 2011). Most authors, however, agree that food quality and quantity is the most important factor determining ciliate distribution in aquatic systems (Esteban et al. 2012; Hwang and Heath 1997; Madoni 1990; Müller et al. 1991; Roberts et al. 2000; Shukla and Gupta 2001). This conclusion is in agreement with the higher diversity, abundance and biomass of ciliates observed in epilimnia (Fenchel et al. 1990; Roberts et al. 2000; Taylor and Heynen 1987) and at oxyclines or chemoclines (Fenchel et al. 1990; Finlay and Esteban 2009; Madoni 1990; Roberts et al. 2000) of stratified aquatic systems. It is also based on observations that ciliate abundance and biomass increase with increasing trophic status of lakes and are significantly correlated to their food items (Beaver and Crisman 1982,

1989b; Pace 1986; Pace and Orcutt 1981; Sonntag et al. 2011; Weisse et al. 2001).

Taylor and Sanders (2010) suggested that most protozoa can live within a broad range of physical and chemical conditions, but that temperature, oxygen, light and environmental contaminants are among the key factors affecting their distribution. Light is important in the nutrition of autotrophic and mixotrophic protozoa, although UV light is also detrimental to protozoa (Sommaruga et al. 1999; Sonntag et al. 2011). There is accumulating literature with regard to the effects of environmental contaminants on the distribution of protozoa in natural water systems (Taylor and Sanders 2010). The amount and type of organic pollutants in aquatic habitats may influence the community structure and biomass of the inhabiting protozoa (Krashevska et al. 2014). Thus, various types of protozoans can be used to classify the environmental state of aquatic habitats, and some ciliates are regarded as biological indicators of organic pollution in freshwater (Bick 1972; Cairns et al. 1972; Foissner and Berger 1996; Krashevska et al. 2014).

Oxygen and its influence on protozoa have been well investigated (e.g. Finlay 1981; Finlay and Esteban 2009; Wetzel 1983), and these studies suggest that the occurrence of microaerophilic ciliates in the plankton is related to thermal stratification, deoxygenation and accumulation of reducing compounds in the hypolimnion and sediment-water interface (Goulder 1980 and references therein). One such ciliate is *Spirostomum*, a large and well-known microaerophilic ciliate that lives in low-oxygen conditions in the sediments (Goulder 1971) but sometimes migrates upwards into oxygenated waters including that of Lake Sonachi (see next section). It has also been argued that large ciliates track prey that migrate upwards during stratification or that they move away from increasing concentrations of reduced compounds in the hypolimnion but not necessarily towards an oxygen source (Finlay 1981; Finlay et al. 1983; Guhl and Finlay 1993). The migration of these ciliates from the benthos to the water column of productive lakes is linked to seasonal changes in oxygen availability (Finlay 1981). Other large ciliates such as *Loxodes* respire anaerobically, utilizing nitrate as the electron acceptor. *Loxodes* also aggregate at areas of low oxygen, which they detect using unique organelles, the Muller's vesicles (Fenchel and Finlay 1991).

According to Fenchel (1969), true anaerobic ciliates or "sulphide ciliates" live in reducing sediments and are intolerant of oxygen, lack mitochondria and cytochrome oxidase, but use hydrogenosomes (Finlay 1981), microbodies that reduce pyruvate (Fenchel and Finlay 1991) to generate energy. Although ciliate predators may be absent in anaerobic hypolimnia (Guhl and Finlay 1993), the anaerobic ciliates are always low in number probably due to limited niches or fewer food sources provided by anoxic habitats (Finlay and Esteban 1998; Guhl et al. 1996; Massana and Pedrós-Alió 1994).

7.7.1 Vertical Distribution of Ciliate Biomass in East African Soda Lakes

Vertical zonation of ciliates is commonly observed in lakes (Fenchel 1969; Gasol et al. 1992; Massana and Pedrós-Alió 1994; Müller et al. 1991; Sonntag et al. 2011) and is related to stratification of the water column. Some EARS lakes > 5 m deep stratify vertically: for example, Lakes Sonachi and Simbi are both volcanic crater lakes and are usually chemically and thermally stratified (Yasindi 2001). Lake Sonachi was stratified from September to December 1998 (Fig. 7.6) with the thermocline and chemocline between 1 and 5 m. Hypolimnetic water samples had no oxygen (Fig. 7.6c) during September and October and smelled of hydrogen sulphide. Lake Simbi was also stratified (Fig. 7.7a, b) during sampling in October 1999 with the thermocline and chemocline at 1 m (Fig. 7.7a) and the oxycline between 5 and 10 m (Fig. 7.7b).

Vertical heterogeneity in the distribution of ciliate community composition and biomass was observed in Lakes Sonachi and Simbi (Yasindi 2001). Lake Sonachi had the highest ciliate biomass in areal units (6244 mg m^{-2}) compared to 171 mg m^{-2} in Lake Simbi. In

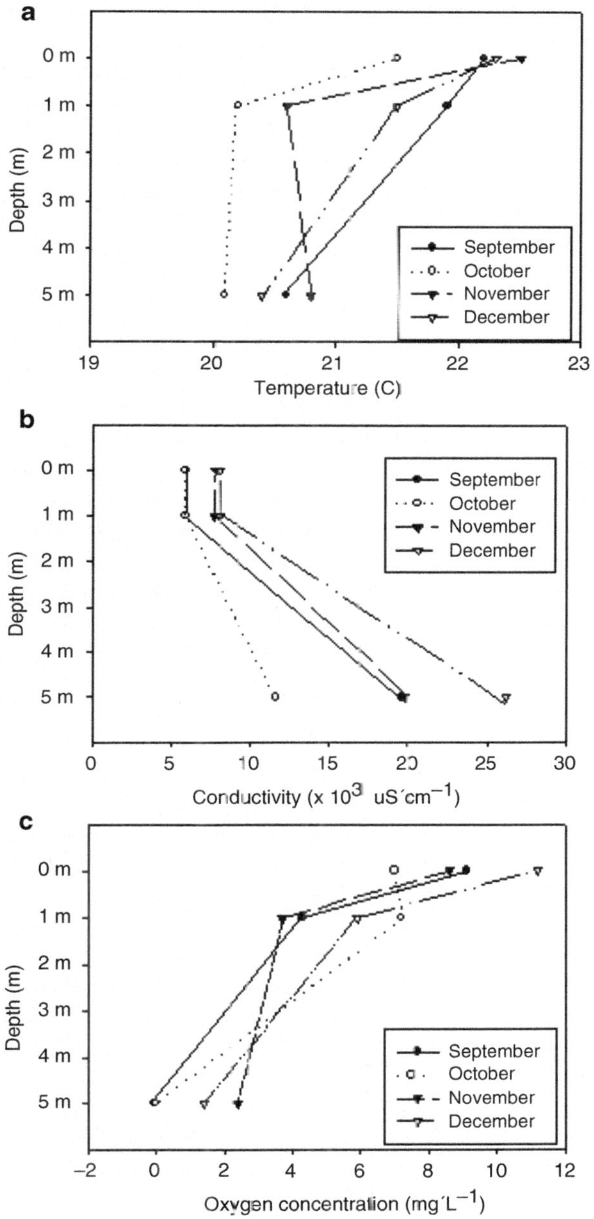

Fig. 7.6 Temperature (**a**), conductivity (**b**) and oxygen (**c**) profiles in Lake Sonachi in 1998 (Yasindi 2001)

Lake Sonachi (and also Lake Oloidien), the biomass was greatest at 1 m, but in Lake Simbi it was greatest at 5 m (Fig. 7.8).

During stratification in Lake Sonachi, many ciliate species including *Frontonia*, *Acineria*, *Monodinium*, *Holophrya*, *Spirostomum* and *Metacystis* were found in the oxygenated epilimnetic zone, while *Cyclidium*, *Pleuronema* and *Cristigera* formed dense populations at the oxycline. A few ciliate species with low abundance occurred in the anoxic hypolimnion. These included *Metacystis*, *Caenomorpha*, an unknown armophorid and sometimes *Cyclidium*.

In holomictic lakes (e.g. shallow Lakes Elmenteita, Nakuru and Solai) and unstratified deep lakes, ciliate zonation was absent and ciliate species composition was nearly homogeneous throughout the water column (Yasindi 2001).

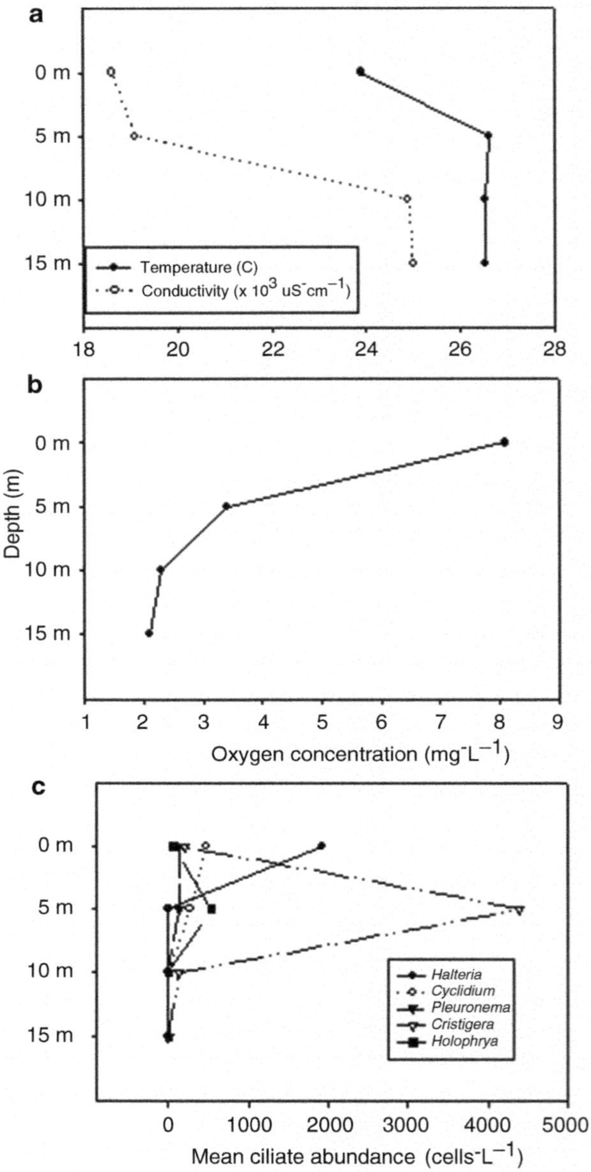

Fig. 7.7 The temperature and conductivity (**a**) and oxygen (**b**) profiles and the variation in abundance of *Halteria*, *Cyclidium*, *Pleuronema*, *Cristigera* and *Holophrya* with depth sampled (**c**) in Lake Simbi in October, 1999 (Yasindi 2001)

The ciliate fauna in the epilimnia of stratified lakes (Sonachi and Simbi) was similar to that in shallow, holomictic lakes or lakes that experienced daily superficial stratification (Lakes Elmenteita and Nakuru). Shallow lakes, by virtue of their mixing, have uniform temperature, chemical parameters and food throughout the water column. The high diversity of ciliates in the epilimnia of stratified lakes has been attributed to food availability (Beaver and Crisman 1990; Iriberri et al. 1995; Sonntag et al. 2011; Witek 1998). However, the high ciliate biomass at oxyclines (Guhl and Finlay 1993) and chemoclines (Madoni 1990; Roberts et al. 2000) may be due to the absence of predators (Bloem and Bär-Gilissen 1989; Guhl et al. 1996; Zöllner et al. 2003) in addition to increased availability of food, usually Bacteria (Fenchel et al. 1990). In the EASL, the oxyclines and chemoclines of stratified lakes had a higher

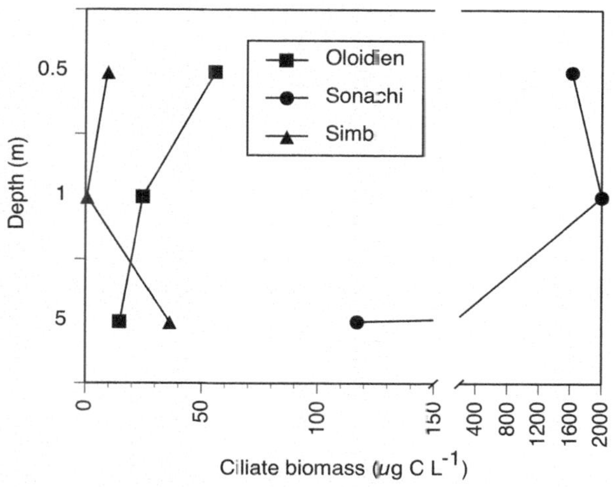

Fig. 7.8 The variation of mean ciliate biomass (µg C L^{-1}) with depth in Lakes Oloidien, Simbi and Sonachi (Yasindi 2001)

biomass but lower diversity of ciliates (Yasindi 2001). A similar observation was made in Lake Fryxell, Antarctica (Roberts et al. 2000). The higher biomass of Bacteria at oxyclines and chemoclines (Roberts et al. 2000) is due to higher concentrations of inorganic nutrients and DOC, which diffuse upwards from the monimolimnion of meromictic lakes (Finlay and Esteban 1998; Priscu et al. 1989). The Bacteria in these zones are larger and have higher productivity (Fenchel et al. 1990). Massana and Pedrós-Alió (1994) found that anaerobic Bacteria were larger than aerobic ones in Lake Cisó, Spain. Therefore, Bacteria and their grazers (heterotrophic nanoflagellates and ciliates) have biomass maxima at oxyclines (Bloem and Bär-Gilissen 1989).

The high density of Bacteria near the oxycline in Lake Sonachi may have attracted *Cyclidium*, *Halteria*, *Aspidisca* and *Cristigera* as the main bacterivores. Yasindi (2001) found *Spirostomum* in the epilimnion during September and October 1998 in Lake Sonachi, when the lake was stratified. This is because microaerophilic ciliates such as *Spirostomum* migrate into the epilimnion during stratification as discussed above. Other hypolimnetic ciliates such as *Caenomorpha*, an unidentified armophorid and *Metacystis* were either microaerophilic or anaerobic and tolerated anoxia and reducing conditions (Fenchel 1969). Similar zonations were noted in Esthwaite (Bark 1981; Goulder 1974) and Lake Cisó, Spain (Gasol et al. 1992; Massana and Pedrós-Alió 1994). The hypolimnetic ciliates *Spirostomum* and *Metopus* found in EASL have also been recorded in the hypolimnia of temperate lakes (Finlay 1981; Goulder 1971) and subtropical lakes (Beaver and Crisman 1989a; Goulder 1971). According to Yasindi (2001) some of these ciliates are confined to the hypolimnion because of their unique characteristics. For example, pigmented *Stentor* are sensitive to strong light when oxygen is available (Giese 1973) and are usually absent from the epilimnia of lakes. *Metacystis* dwells in anoxic zones, where it feeds on sulfur Bacteria (Curds 1982), and its occurrence and sometimes maximum abundance are coincident with high hydrogen sulphide (Guhl et al. 1996), such as observed in the hypolimnion of Lake Sonachi (MacIntyre and Melack 1982; Yasindi 2001).

Lake Simbi usually has a chemocline at 2–6 m but sometimes at 11–15 m (Melack 1979). Ochumba and Kibaara (1988) observed a chemocline at 4 m and an oxycline at 2.8 m. Yasindi (2001), however, recorded an oxycline at 5 m (Fig. 7.7b). Melack (1979) recorded high abundance of the cyanobacterium *Arthrospira fusiformis* within 1 m of the surface, but the biomass declined below 10 m. This agrees with the observations of Ochumba and Kibaara (1988), who found most phytoplankton biomass in the upper 5 m in Lake Simbi. An oxycline near 5 m (Fig. 7.7b) may have accounted for the ciliate biomass maximum at that depth in Lake

Simbi. Although *Cyclidium* and *Halteria* have higher biomass at 0.5 m, *Cristigera* was more abundant at the oxycline (Fig. 7.6c).

7.7.2 Temporal Distribution

Stratification and development of anoxia in lakes may be linked to seasonal changes in weather. While the strong seasonality in temperate lakes reflects major changes in temperature and solar radiation, seasonal changes in tropical systems are not clearly defined because solar radiation and temperature are more constant throughout the year. While ciliate biomass maxima occur during early spring and late summer in temperate lakes (Gates 1984; Xu and Cronberg 2010), ciliate biomass is highest during the onset of rains in tropical lakes, when they mix. Wind and decreased heat input are major determinants of the mixing cycle in tropical lakes and related to the seasonal rainfall cycle in East Africa (Talling and Lamoalle 1998). The rainy season in the EARS occurs in April to July (long rains) and is followed by a dry spell in August to September. The season of short rains follows from October to November, whereas December to March is a dry season during which winds from the northeast cause mixing in Kenyan Rift Valley lakes. The northerly winds are stronger during the drier months of December to March, decreasing in June/July (Vareschi 1982), but the mixing of lakes in May to August is due to heavy southeast trade winds (Beadle 1981; Vareschi 1982).

Seasonal cycles of planktonic ciliates in temperate lakes (Beaver and Crisman 1990; Pace and Orcutt 1981; Xu and Cronberg 2010) are attributed mainly to the availability of suitable food resources (Beaver and Crisman 1989a). Nonetheless, the seasonal dynamics of planktonic ciliates remain poorly understood in EARS although tropical lakes are known to show seasonal as well as diurnal changes (Hecky and Fee 1981) in phytoplankton (Hart and Hart 1977; Hecky and Fee 1981; Kebede and Belay 1994; Talling 1986; Wood and Talling 1988), zooplankton (Martens and Tudorancea 1991; Mavuti 1990; Mengestou and Fernando 1991; Nogrady 1983; Twombly 1983) and bacterioplankton (Robarts 1988; Zinabu and Taylor 1989). Only a few studies, such as those of Barbieri and Orlando (1989), Hecky and Kling (1981) and Lewis (1985), have shown that abundance and biomass of some ciliates vary seasonally in tropical lakes. A few data exist for EASL (Fig. 7.9; see also Yasindi et al. (2002).

7.8 Growth Rates and Productivity of Ciliated Protozoa

Most studies of protozoan growth rates in the field have been undertaken in either temperate lakes (see Jack and Gilbert 1997 for review) or in

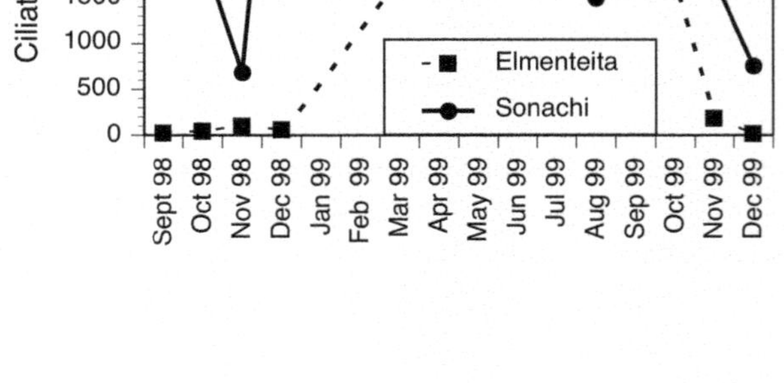

Fig. 7.9 Ciliate biomass (µg C L^{-1}) sampled at two depths (0.5 and 1 m) and averaged in two lakes from September to December, 1998, and between August and December, 1999 (from Yasindi 2001)

marine ecosystems (e.g. Verity 1986). The only information from EARS lakes is for ciliates (Yasindi 2001; Yasindi and Taylor 2006). In these studies, three polyethylene carboys containing 4 L of whole lake water and three carboys with same amount of lake water sieved through a 40-μm-mesh net to remove zooplankton (>40 μm in size) were incubated in situ for 18 h. After the incubations, the rates of change (C) of ciliate cell populations were calculated as follows: $C = (\ln N_t - N_0)/t$, where N_t and N_0 are the final and initial abundances of ciliate taxa and t is the time elapsed (Yasindi and Taylor 2006). EASL where ciliate growth rates were estimated include Lakes Nakuru (Yasindi et al. 2002) and Bogoria, Elmenteita, Sonachi and Simbi (Yasindi and Taylor 2006). The estimated rates of change for individual ciliate taxa were from −4.30 day^{-1} to 4.78 day^{-1} in unscreened lake water (with zooplankton) but from −2.47 day^{-1} to 5.57 day^{-1} in water samples where zooplankton had been removed using a 40-μm net (Table 7.3). The growth rates of abundant taxa during October 1999 varied from 0.18 day^{-1} for *Pleuronema* in Lake Elmenteita to 4.78 day^{-1} for *Cyclidium* in Lake Simbi (Table 7.3). Some of the in situ ciliate growth rate estimates in these studies are higher than previously reported in marine (Banse 1982; Montagnes 1996) and temperate freshwaters (e.g. Carrick et al. 1992; Jack and Gilbert 1997). They are higher than 0.3–2.2 day^{-1} estimated for marine oligotrichs (Montagnes 1996), −0.2–0.87 day^{-1} in Lake Søbygärd, Denmark (Hansen 2000), and 0.79–0.87 day^{-1} in Storrs Pond, Vermont (Wickham and Gilbert 1993), and for three marine ciliates (Strom and Morello 1998). They are, however, plausible based on the relationship between cell size and maximum growth rate given by Fenchel (1969).

Ciliates vary widely in their maximum growth rates, even among species of the same volume at the same temperature (Finlay 1977; Taylor and Shuter 1981; Müller and Geller 1993). Their growth rates in the field, however, likely depend on the quality and quantity of food most of the

Table 7.3 Abundance of ciliates in whole lake water at the beginning of the incubation

Lake	Date	Ciliate	Mean counted	r_{wlw}	$r_{<\ 40\ \mu m}$	Predation > 40 μm
Bogoria	Sept/99	*Dileptus*	67	3.91	2.86*	−1.05
	Sept/99	*Cristigera*	67	4.42*	1.23	−3.19
	Sept/99	*Cyclidium*	9800	−1.10	1.85*	2.95
Elmenteita	Oct/98	*Dysteria*	346	−0.84	0.92*	1.76
	Oct/98	*Cyclidium*	79	−1.12	−0.27	0.85*
	Oct/98	*Aspidisca*	214	−4.21	3.27*	7.48
	Nov/98	*Halteria*	85	−3.47	0.51*	3.98
	Nov/98	*Dysteria*	220	−1.11	1.76*	2.87
	Nov/98	*Aspidisca*	90	−4.30	−2.47	1.83*
	Dec/98	*Phialina*	53	−2.11	−1.39	0.72*
	Dec/98	*Halteria*	85	−0.19	−0.36	−0.17
	Dec/98	*Dysteria*	388	0.23*	−0.35	−0.58
	Oct/99	*Pleuronema*	232	−0.08	0.18*	0.26
Simbi	Oct/99	*Dysteria*	333	2.63*	−1.63	−4.26
	Oct/99	*Cyclidium*	667	4.78*	5.57	0.79
	Oct/99	*Halteria*	667	2.90	2.55*	−0.35
Sonachi	Oct/99	*Cyclidium*	13,800	2.19*	0.76	−1.43
	Oct/99	*Aspidisca*	1200	1.59*	−0.28	−1.87
	Oct/99	*Acineria*	3333	4.29	4.06*	−0.23
	Oct/99	*Dysteria*	4400	3.48	2.74*	−0.74

r_{wlw} is the exponential rate of increase (day^{-1}) in unscreened whole lake water while $r < 40$ μm is the exponential rate of increase in water screened through a 40-μm-mesh net. Only ciliates that had a total count of > 100 in each experiment were considered. The number of ciliates counted represents the mean of three replicates (Yasindi 2001) time (Gifford 1991; Taylor 1978). Yasindi (2001) was the first to estimate growth rates of ciliates in tropical waters of Africa. Some taxa demonstrated high growth rates in some experiments. Nonetheless, compared to growth rate estimates in temperate lakes (Carrick et al. 1992; Hansen 2000; Taylor and Johannsson 1991), subtropical marine systems (Gilron and Lynn 1989) and Lake Qaroun, Egypt (Mageed et al. 2002), ciliates in soda lakes actually may be slow growers when the high ambient temperatures are taken into account. Yasindi (2001) concluded that incubations in the carboys might have underestimated ciliate growth rates in the soda lakes. Some ciliates may not multiply in screened lake water (Furnas 1982). It is possible that removing grazers disrupts the recycling of nutrients required to sustain growth within a closed container (Eppley et al. 1971).

Another factor that likely reduces growth rates in screened treatments is predation. Fewer ciliates in control treatments with zooplankton than in screened treatments where crustacean zooplankton have been removed may be caused by releasing predatory ciliates from predation by zooplankton. This hypothesis will be explored in the next section. Therefore, Yasindi (2001) attributed instances in which there was no effect or a negative effect on ciliate abundance of screening treatments to predation within the ciliate community. A similar observation was made in Lake Qaroun, Egypt, where a higher growth rate (2.15 day^{-1}) was recorded in the control (without zooplankton and fish) than in the presence of zooplankton (1.18 day^{-1}) and fish (*Mugil cephalus*) 0.63 day^{-1} (Mageed et al. 2002).

The measured predation rates of zooplankton > 40 μm on ciliates was up to 0.79 day^{-1} (Yasindi 2001). Most crustacean zooplankton are omnivorous and feed on both phytoplankton and microzooplankton (Branstrator et al. 1996), including ciliates (Gifford 1991). Based on biochemical composition and size, ciliates are higher-quality food than phytoplankton to copepods (Hansen 2000) and other zooplankton (Jack and Gilbert 1993; Stoecker and Capuzzo 1990). Ciliates have a high fat and protein content and are assimilated more efficiently than phytoplankton (Conover 1982). Ciliates store energy as glycogen, lipid droplets and polyunsaturated fatty acids (Holz and Conner 1987). Also the C:N ratios of ciliates are about 4.0, close to pure protein (3.0), compared to the average ratio of 6.0 for phytoplankton (Banse 1974). Moreover, copepods select ciliates over phytoplankton (Fessenden and Cowles 1994). Burian et al. (2013) made similar observations; they found that the rotifers *Brachionus dimidiatus* and *B. plicatilis* preferred ciliates to the more abundant Cyanobacteria in Lake Nakuru.

Therefore, ciliates probably constitute an important food item for zooplankton in soda lakes. Zooplankton predation may be an important factor affecting ciliate abundance and the rate of population increase in these lakes, as observed in temperate (Jack and Gilbert 1997) and subtropical lakes (Havens and Beaver 1997).

The first estimates of production by ciliates in African lakes were based on growth rates in temperate waters (Yasindi et al. 2002). Later values based on field estimates of growth rates used the rates in Yasindi (2001) and Yasindi and Taylor (2006). This yielded estimates of ciliate production for EASL that were higher than earlier values for temperate lakes (e.g. Carrick et al. 1992; Taylor and Johannsson 1991) and marine environments (Lynn et al. 1991; Montagnes et al. 1988). The higher production in EARS lakes may be attributed to high growth rates resulting from abundant food and high temperatures (Yasindi et al. 2002). Production is also a function of biomass. Thus, as in temperate systems, large ciliates had high production due to their large biomass and despite lower growth rates (Yasindi and Taylor 2006). Production by small-sized ($< 10^3$ $μm^3$) bacterivorous ciliates was low in most of the soda lakes studied due to their small biomass, despite their high population growth rates (Yasindi and Taylor 2006). They, however, contributed 95 % of the total production by ciliates in Lake Nakuru. This corroborated earlier studies in this lake, where ingestion rates by bacterivorous ciliates were relatively high (Finlay et al. 1987).

7.9 Predaceous Ciliates

In his study on in situ growth rates of ciliates in EASL, Yasindi (2001) observed negative rates of population change for ciliates in some > 40-μm-screened water samples. This negative change in population growth was attributed to an increase in predaceous ciliates in the absence of zooplankton predators. Though predation among ciliates has been reported in freshwater (Taylor and Johannsson 1991; Zingel 1999) and marine systems (Paffenhöfer 1998), it has been overlooked in most plankton studies, mainly because of lack of information on the feeding biology of planktonic ciliates (Taylor and Johannsson 1991). However, it is well studied in the laboratory in ciliates such as *Didinium* (Hewett 1988), *Favella* (Stoecker and Evans 1985) and *Stylonychia* (Wiackowski and Staronska 1999).

In some EARS lakes, such as Lake Sonachi, screening lake water always decreased the population growth rate of ciliates (Yasindi 2001). This observation suggests that predaceous ciliates may potentially reduce the rate of population change when they are released from predation by screening. Yasindi (2001) noted that the growth rates of predaceous ciliates such as *Phialina* tended to be higher in the screened treatment without zooplankton, while growth rates of herbivores and bacterivores decreased.

To assess the predatory effect of carnivorous ciliates on their prey ciliates, the production of the prey was compared to the observed demands of their predators. In October 1999, in Lake Sonachi, *Monodinium* preyed on *Cyclidium* and *Dysteria*, which had production of 1.2×10^8 μg C L^{-1} day^{-1} and 1.9×10^5 μg C L^{-1} day^{-1}, respectively. Assuming *Monodinium* was the only source of mortality and had a growth efficiency of 50 % (Stoecker and Evans 1985), the production of the two ciliates consumed by *Monodinium* would account for the production of 1.2×10^8 μg L^{-1} day^{-1}. The observed growth rate of *Monodinium* was 2.45 day^{-1} and its biomass was 1.5×10^8 μg C L^{-1} day^{-1}, which gave a potential production of 3.7×10^8 μg C L^{-1} day^{-1} (Table 7.4). The conclusion was that *Monodinium* could consume the production of these two ciliates and effectively control their populations. Therefore, it was not surprising that > 40-μm screening enhanced predation of ciliates in Lake Sonachi (Table 7.3). A similar scenario was noted in Lake Bogoria, where the production of *Cyclidium* and *Cristigera* was effectively controlled by the predaceous *Dileptus* (Table 7.4). This predation by predaceous ciliates diminishes the importance of ciliates and the microbial loop (Sherr et al. 1986) to higher trophic levels. Predaceous ciliates lengthen food webs and thereby reduce the efficiency of transfer of energy to higher trophic level organisms such as flamingos, pelicans and fish.

7.10 Food Webs

The role of protozoa in the food webs of EARS lakes started emerging when Milbrink (1977)

Table 7.4 The biomass, production and growth rates of four ciliates in three different guilds in Lake Sonachi in October, 1999, and Bogoria in September, 1999 (Yasindi 2001)

Lake—taxon	Guild	Growth rate (r day^{-1})	Biomass (C pg L^{-1})	Production (C pg L^{-1} day^{-1})
Sonachi				
Cyclidium	Bacterivore	0.757	1.6E + 08	1.2E + 08
Acineria	Omnivore	4.291	4.7E + 06	1.9E + 07
Monodinium	Carnivore	2.194	1.5E + 08	3.6E + 08
Dysteria	Bacterivore	3.479	6.9E + 04	1.9E + 05
Bogoria				
Cyclidium	Bacterivore	0.749	5.9E + 06	1.1E + 07
Cristigera	Bacterivore	1.231	8.7E + 04	1.1E + 05
Dileptus	Carnivore	2.860	1.9E + 07	5.4E + 07

observed peritrich ciliates in the stomach contents of the cichlid fish *Alcolapia grahami* in Lake Nakuru. Finlay et al. (1987) identified small suspension-feeding ciliates, which increased with increasing bacterial numbers, and a variety of ciliates feeding on larger algal particles and cyanobacterial filaments. They also recorded some carnivorous ciliates such as *Spathidium*, which fed on other protozoa in Lake Nakuru, and others like *Holophrya* that fed on rotifers in Lake Simbi. These studies, therefore, indicate that protozoa occupy several different trophic levels in the food webs of EASL. Since then, other studies have confirmed these ecological roles (Burian et al. 2013, 2014; Ong'ondo et al. 2013; Yasindi et al. 2002, 2007; Yasindi and Taylor 2006). By observing ciliates using epifluorescent microscopy and quantitative protargol-stained samples (Montagnes and Lynn 1993), the contents of the food vacuoles of ciliates in Lake Nakuru were observed (Burian et al. 2014; Ong'ondo et al. 2013; Yasindi et al. 2002). They confirmed that several ciliates including *Cyclidium*, *Halteria*, *Dysteria*, *Aspidisca* and *Vorticella* ingested bacteria-sized fluorescently labelled latex beads and were considered bacterivorous (Yasindi 2001). From QPS-stained samples, *Frontonia* were observed to contain *Arthrospira* coils and flagellates, while *Prorodon* and *Chlamydodon* contained flagellates and diatoms and were, therefore, herbivores. The suctorian ciliate *Sphaerophrya* and *Frontonia* and *Condylostoma* were observed feeding on flagellates and other ciliates, while *Condylostoma* also fed on rotifers and their eggs. These ciliates, together with *Didinium*, *Monodinium*, *Dileptus*, *Paradileptus*, *Litonotus*, *Spathidium* and *Trachelius* among others in EASL, are predators (Yasindi and Taylor 2006). Yasindi and Taylor (2006) also observed some ciliates such as *Halteria* and *Stichotricha* with sequestered chloroplasts or zoochlorellae in their cytoplasm and classified them as mixotrophic protozoa. Yasindi and Taylor (2006) estimated the biomass and percentage contribution of these feeding guilds to the total ciliate biomass in the soda lakes (Table 7.5).

Large ciliates (>100 μm in length) such as *Condylostoma* and *Frontonia* feed on both phytoplankton and ciliates and are omnivorous. Animals in aquatic systems require flexible feeding behaviours (Adrian and Frost 1993), so omnivory may be advantageous to these ciliates during times of food limitation as it is to other aquatic animals and may contribute to the structural stability of planktonic communities in the soda lakes as in temperate lakes (Sprules and Bowerman 1988). As seen from Table 7.5, predaceous ciliates contributed a lot of biomass and, as discussed above, they may effectively control other ciliate populations (Yasindi and Taylor 2006).

These data clearly demonstrate that ciliates in EASL contribute significantly to secondary production. Nonetheless, the pathways of carbon flow through the ciliate community in these lakes are complex because they feed at different trophic levels and are consumed by a host of predators depending largely on their size.

Table 7.5 The biomass (C μg L^{-1}) and production (C μg L^{-1} day^{-1}) (in brackets) of each feeding guild to the total ciliate biomass in East African soda lakes (Yasindi 2001; Yasindi and Taylor 2006)

Lake	Bacterivores	Herbivores	Predators	Total
Bogoria	18 (34)	1187 (599)	16 (45)	1221 (678)
Nakuru	28 (46)	2 (2)	1 (1)	31 (49)
Oloidien	13 (11)	4 (5)	15 (4)	32 (20)
Elmenteita	509 (230)	8 (7)	1630 (403)	2147 (640)
Sonachi	452 (901)	1843 (1081)	117 253)	2412 (2235)
Simbi	7 (20)	0 (0)	23 (16)	31 (36)
Abijata	18 (41)	40 (77)	1 (1)	59 (119)

7.11 Importance of Protozoa in East African Soda Lakes

EASL have higher abundance and biomass but a lower diversity of ciliates than subtropical and temperate lakes. The higher ciliate abundance and biomass are mainly attributed to higher chl-*a* and bacterial abundance in soda lakes. In contrast, the low protozoan diversity is likely due to the unusual chemical composition of the water and several other unfavourable conditions in soda lakes. The difference in the concentration of salts has been shown to cause differences in the size of ciliates in soda lakes (Ong'ondo et al. 2013).

Most ciliates in EASL are bacterivores and herbivores; these two trophic groups account for 75 % of ciliate production (Yasindi and Taylor 2005). This production supports zooplankton (Finlay et al. 1987; Yasindi and Taylor 2005), which are in turn consumed by fish (Milbrink 1977) and possibly flamingos. Protozoa and ciliates in particular form an important link through which production from the microbial loop may be channelled to the classical food chains in EASL.

7.12 Future Work

Most protozoan work in EASL has involved pelagic protozoa (Burian et al. 2013; Ong'ondo et al. 2013; Yasindi et al. 2002; Yasindi and Taylor 2006), yet those living in and on sediments remain unknown. The population dynamics and productivity of these species need to be explored. Some studies in temperate aquatic ecosystems have shown reduced biological diversity in the pelagic zones of extreme environments such as inland hypersaline habitats (Galotti et al. 2014). In such hostile environments, many microorganisms including ciliates survive in the sediments as inactive cysts, spores or other dormant stages that wait for development of favourable growth conditions (Esteban and Finlay 2003). The dormant stages are hidden "seed banks" of microbial species in those environments. Experimental manipulations of water and sediment samples have revealed the role of microbial seed banks as repositories of microbial diversity in natural environments (Esteban and Finlay 2003; Finlay et al. 1996). Microbial seed banks in general and protozoan seed banks in particular remain unexplored fields in EASL, and their study may reveal hidden microbial diversity.

It is now evident that, despite their hypereutrophic status, EASL also have protozoa with endosymbiotic algae or chloroplasts (Yasindi et al. 2002; Yasindi and Taylor 2006). The significance of phototrophic symbionts to these species should be studied.

While some data on ciliates in EASL are accumulating, data on heterotrophic and mixotrophic flagellates are still fragmentary (e.g. Finlay et al. 1987), whereas data on sarcodines are almost totally missing. These two groups of protozoa definitely need investigations with a view to understanding their ecological roles. We need capacity building to increase expertise in aquatic ecology in this part of the world where these unique lakes are abundant and important.

References

Adrian R, Frost TM (1993) Omnivory in cyclopoid copepods: comparisons of algae and invertebrates as food for three differently sized species. J Plankton Res 15:643–658

Antony CP, Kumaresan D, Hunger S, Drake HL, Murrell C, Shouche YS (2013) Microbiology of Lonar Lake and other soda lakes. Int J Syst Evol Microbiol 7:468–476

Banse K (1974) On the interpretation of data for the carbon-to-nitrogen ratio of phytoplankton. Limnol Oceanogr 19:695–699

Banse K (1982) Cell volume, maximal growth rates of unicellular algae and ciliates, and the role of ciliates in the marine pelagial. Limnol Oceanogr 27:1059–1071

Barbieri SM, Orlando MJL (1989) Ecological studies on the planktonic protozoa of a eutrophic reservoir (Rio Grande Reservoir—Brazil). Hydrobiologia 183:1–10

Bark AW (1981) The temporal and spatial distribution of planktonic and benthic protozoan communities in a small productive lake. Hydrobiologia 85:239–255

Beadle LC (1981) The inland waters of tropical Africa. An introduction to tropical limnology, 2nd edn. Longman, London

Beaver JR, Crisman TL (1982) The trophic response of ciliated protozoans in freshwater lakes. Limnol Oceanogr 22:246–253

Beaver JR, Crisman TL (1989a) The role of ciliated protozoa in pelagic freshwater ecosystems. Microbiol Ecol 17:111–136

Beaver JR, Crisman TL (1989b) Analysis of the community structure of planktonic ciliated protozoa relative to trophic state in Florida lakes. Hydrobiologia 174:177–184

Beaver JR, Crisman TL (1990) Seasonality of planktonic ciliated protozoa in 20 subtropical Florida lakes of varying trophic state. Hydrobiologia 190:127–135

Bick H (1972) Ciliated protozoa. An illustrated guide to the species used as biological indicators in freshwater biology. World Health Organization, Geneva

Bloem J, Bär-Gilissen M-JB (1989) Bacterial activity and protozoan grazing potential in a stratified lake. Limnol Oceanogr 34:297–309

Branstrator DK, Lehman JT, Ndawula LM (1996) Zooplankton dynamics in Lake Victoria. In: Johnson TC, Odada EO (eds) The limnology, climatology and paleoclimatology of the past African lakes. Gordon and Breach, Toronto, pp 337–355

Burian A, Kaggwa M, Schagerl M, Yasindi AW (2013) Microzooplankton feeding behaviour: grazing on the microbial and the classical food web of African soda lakes. Hydrobiologia 710:61–72

Burian A, Kainz MJ, Schagerl M, Yasindi AW (2014) Species-specific separation of lake plankton reveals divergent food assimilation patterns in rotifers. Freshw Biol 59:1257–1265

Cairns J Jr, Lanza GR, Parker BC (1972) Pollution related structural and functional changes in aquatic communities with emphasis on freshwater algae and protozoa. Proc Acad Nat Sci USA 124:79–127

Carrick HJ, Fahnenstiel GL, Taylor WD (1992) Growth and production of planktonic protozoa in Lake Michigan: in situ versus in vitro comparisons and importance to food web dynamics. Limnol Oceanogr 37:1221–1235

Colburn EA (1988) Factors influencing species diversity in saline waters of Death Valley, USA. Hydrobiologia 158:215–226

Conover RJ Jr (1982) Interrelations between microplankton and other plankton organisms. Ann Inst Oceanogr 59(S):31–46

Curds CR (1982) British and other freshwater ciliated protozoa. Part I. Ciliophora: Kinetofragminophora. Keys and notes for the identification of the free-living genera. Cambridge University Press, Cambridge

Eppley RW, Carlucci AF, Holm HO, Kiefer D, McCarthy JJ, Venrick EL, Williams PM (1971) Phytoplankton growth and composition in shipboard cultures supplied with nitrate, ammonia or urea as the nitrogen source. Limnol Oceanogr 16:741–751

Esteban GF, Finlay BJ (2003) Cryptic freshwater ciliates in a hypersaline lagoon. Protist 154:411–418

Esteban GF, Fenchel T, Finlay BJ (2010) Mixotrophy in ciliates. Protist 161:621–641

Esteban GF, Finlay BJ, Clarke KJ (2012) Priest Pot in the English Lake district: a showcase of microbial diversity. Freshw Biol 57:321–330

Fenchel T (1969) The ecology of the marine microbenthos. IV. Structure and function of the benthic ecosystem, its chemical and physical factors and the microfauna communities with special reference to the ciliated protozoa. Ophelia 6:1–182

Fenchel T, Finlay BJ (1991) The biology of free-living anaerobic ciliates. Eur J Protistol 26:201–215

Fenchel T, Kristensen LD, Rasmussen L (1990) Water column anoxia: vertical zonation of planktonic protozoa. Mar Ecol Prog Ser 62:1–10

Fessenden L, Cowles TJ (1994) Copepod predation on phagotrophic ciliates in Oregon coastal waters. Mar Ecol Prog Ser 107:103–111

Finlay BJ (1977) The dependence of reproductive rate on cell size and temperature in freshwater ciliated protozoa. Oecologia 30:75–81

Finlay BJ (1981) Oxygen availability and seasonal migration of ciliated protozoa in a productive lake. J Gen Microbiol 123:173–178

Finlay BJ, Esteban GF (1998) Planktonic ciliate species diversity as an integral component of ecosystem function in a freshwater pond. Protist 149:155–165

Finlay BJ, Esteban GF (2009) Oxygen sensing drives predictable migrations in a microbial community. Environ Microbiol 11:81–85

Finlay BJ, Span ASW, Harman JMP (1983) Nitrate respiration in primitive eukaryotes. Nature 303:333–336

Finlay BJ, Curds CR, Bamforth SS, Bafort JM (1987) Ciliated protozoa and other microorganisms from two African soda lakes (Lake Nakuru and Lake Simbi, Kenya). Arch Protistenk 133:81–91

Finlay BJ, Clarke KJ, Cowling AJ, Hindle RM, Rogerson A, Berninger U-G (1988) On the abundance and distribution of protozoa and their food in a productive freshwater pond. Eur J Protistol 23:205–217

Finlay BJ, Esteban GF, Fenchel T (1996) Global diversity and body size. Nature 383:132–133

Foissner W, Berger H (1996) A user-friendly guide to the ciliates (Protozoa, Ciliophora) commonly used by hydrobiologists as bioindicators in rivers, lakes, and waste waters, with notes on their ecology. Freshw Biol 35:375–482

Fortin D, Roy M, Rioux J, Thibault P (2000) Occurrence of sulfate-reducing bacteria under a wide range of physico-chemical conditions in Au and Cu-Zn mine tailings. Microbiol Ecol 33:197–208

Furnas MJ (1982) An evaluation of two diffusion culture techniques for estimating phytoplankton growth rates in situ. Mar Biol 70:63–72

Galotti A, Finlay BJ, Jiménez-Gómez F, Guerrero F, Esteban G (2014) Most ciliated protozoa in extreme environments are cryptic in the 'seed-bank'. Aquat Microb Ecol 72:187–193

Gasol JM, Peters F, Guerrero R, Pedrós-Alió R (1992) Community structure in Lake Ciso: Biomass allocation to trophic groups and differing patterns of seasonal succession in the meta- and epilimnion. Arch Hydrobiol 123:275–303

Gates MA (1984) Quantitative importance of ciliates in the planktonic biomass of lake ecosystems. Hydrobiologia 108:233–238

Giebelhausen B, Lampert W (2001) Temperature reaction norms of *Daphnia magna*: the effect of food concentration. Freshw Biol 46:281–289

Giese AC (ed) (1973) Blepharisma. The biology of a light-sensitive protozoa. Stanford University Press, Stanford, CA

Gifford DJ (1991) The protozoa-metazoan trophic link in pelagic ecosystems. J Protozool 38:81–86

Gilron GL, Lynn DH (1989) Estimates of in situ population growth rates of four tintinnine ciliate species near Kingston Harbour, Jamaica. Estuar Coast Shelf Sci 29:1–10

Goulder R (1971) The effects of saprobic conditions on some ciliated protozoa in the benthos and hypolimnion of a eutrophic pond. Freshw Biol 1:307–318

Goulder R (1974) The seasonal and spatial distribution of some benthic ciliated protozoa in Esthwaite water. Freshw Biol 4:127–147

Goulder R (1980) The ecology of two species of primitive ciliated protozoa commonly found in standing freshwaters (*Loxodes magnus* Stokes and *L. striatus* Penard). Hydrobiologia 72:131–158

Green J (1994) The temperate-tropical gradient of planktonic Protozoa and Rotifera. Hydrobiologia 272:13–26

Guhl BE, Finlay BJ (1993) Anaerobic predatory ciliates track seasonal migrations of planktonic photosynthetic bacteria. FEMS Microbiol Lett 107:313–316

Guhl BE, Finlay BJ, Schink B (1996) Comparison of ciliate communities in the anoxic hypolimnia of three lakes: general features and the influence of lake characteristics. J Plankton Res 18:335–353

Guildford SJ, Taylor WD (2011) Evidence supporting the importance of nutrient regeneration by micro- and nanograzers for phytoplankton photosynthesis in Lake Malawi/Nyasa. J Great Lakes Res 37:54–60

Gustavson K, Garde K, Wängberg S, Selmer J (2000) Influence of UV-B radiation on bacterial activity in coastal waters. J Plankton Res 22:1501–1511

Hall CJ, Burns CW (2001) Effects of salinity and temperature on survival and reproduction of *Boeckella hamata* (Copepoda: Calanoida) from a periodically brackish lake. J Plankton Res 23:97–104

Hampton JR, Schwartz RL (1976) Contractile vacuole function in *Pseudocohnilembus persalinus:* responses to variations in ion and total solute concentration. Comp Biochem Physiol 55:1–4

Hansen AM (2000) Response of ciliates and *Cryptomonas* to the spring cohort of a cyclopoid copepod in a shallow hypereutrophic lake. J Plankton Res 22:185–203

Hart RC, Hart R (1977) The seasonal cycles of phytoplankton in Lake Sibaya: a preliminary investigation. Arch Hydrobiol 80:85–107

Havens KE, Beaver JR (1997) Consumer vs. resource control of ciliate protozoa in a copepod-dominated subtropical lake. Arch Hydrobiol 140:491–511

Hecky RE, Fee EJ (1981) Primary production and rates of algal growth in Lake Tanganyika. Limnol Oceanogr 26:532–547

Hecky RE, Kling HJ (1981) The phytoplankton and protozooplankton of the euphotic zone of Lake Tanganyika: species composition, biomass, chlorophyll content and spatio-temporal distribution. Limnol Oceanogr 26:548–564

Hennemuth W, Rhoads LS, Eichelberger H, Watanabe M, VanBell KM, Ke L, Kim H, Nguyen G, Jonas JD, Veith D, Van Bell CT (2008) Ingestion and inactivation of bacteriophages by *Tetrahymena*. J Eukaryot Microbiol 55:44–50

Hewett SW (1988) Predation by *Didinium nasutum*: effects of predator and prey size. Ecology 69:135–145

Holz GG Jr, Conner RL (1987) The composition, metabolism, and roles of lipids in *Tetrahymena*. In: Elliot AM (ed) The biology of *Tetrahymena*. Dowden, Hutchinson and Ross Inc, Stroudburg, PA, pp 99–122

Hwang S, Heath RT (1997) The distribution of protozoa across a trophic gradient, factors controlling their abundance and importance in the food web. J Plankton Res 19:491–518

Hyenstrand P, Rydin E, Gunnerhed M (2000) Response of pelagic cyanobacteria to iron additions-enclosure experiments from Lake Erken. J Plankton Res 22:1113–1126

Iriberri J, Ayo B, Santamaria E, Barcina I, Egea L (1995) Influence of bacterial density and water temperature on the grazing activity of two freshwater ciliates. Freshw Biol 33:223–231

Jack J, Gilbert JJ (1993) Susceptibilities of different-sized ciliates to direct suppression by small and large cladocerans. Freshw Biol 29:19–29

Jack J, Gilbert JJ (1997) Effects of metazoan predators on ciliates in freshwater plankton communities. J Eukaryot Microbiol 44:194–199

Jenkin PN (1936) Reports on the Percy Sladen Expedition to some Rift Valley Lakes in Kenya in 1929.- VII. Summary of the ecological results, with special reference to the alkaline lakes. Annu Mag Nat Hist Ser 10 18:161–181

Jones BE, Grant WD (1999) Microbial diversity and ecology of the Soda Lakes of East Africa. In: Proceedings of the 8th international symposium on microbial ecology. http://Plato.acadiau.ca/isme/Symposium22/jones. Accessed Dez 2014

Kebede E, Belay A (1994) Species composition and phytoplankton biomass in a tropical African lake (Lake Awassa, Ethiopia). Hydrobiologia 288:13–32

Kilham P (1981) Pelagic bacteria: extreme abundances in African saline lakes. Naturwiss 67:380–381

Krashevska V, Sandmann D, Maraun M, Scheu S (2014) Moderate changes in nutrient input alter tropical microbial and protist communities and belowground linkages. Int J Syst Evol Microbiol 8:1126–1134

Lair N, Leveille J-C, Reyes-Marchant P, Taleb H (1994) The feeding of a larval fish, *Lebistes reticulatus*, on ciliates and rotifers. Mar Microb Food Webs 8:337–346

Lanzén A, Simachew A, Gessesse A, Chmolowska D, Jonassen I, Øvre L (2013) Surprising prokaryotic and eukaryotic diversity, community structure and biogeography of Ethiopian soda lakes. PLOSOne 8, e72577

Laybourn-Parry J (1992) Protozoan plankton ecology. Chapman and Hall, London

Laybourn-Parry J, Olver J, Rees S (1990) The hypolimnetic protozoan plankton of a eutrophic lake. Hydrobiologia 203:111–119

Lewis WM Jr (1985) Protozoan abundances in the plankton of two tropical lakes. Arch Hydrobiol 104:337–343

Lynn DH, Roff JC, Hopcroft RR (1991) Annual abundance of aloricate ciliates in tropical neritic waters off Kingston, Jamaica. Mar Biol 110:437–448

MacIntyre S, Melack JM (1982) Meromixis in an equatorial African soda lake. Limnol Oceanogr 27:595–609

Madoni P (1990) The ciliated protozoa of the monomictic Lake Kinneret (Israel): species composition and distribution during stratification. Hydrobiologia 190:111–120

Mageed AA, Mohammadein A, Desouky M (2002) Importance of protozoa as food to zooplankton and some fish species in Lake Qaroun, Egypt. J Egypt Aquat Biol Fish 6:59–74

Martens K, Tudorancea C (1991) Seasonality and spatial distribution of the ostracods of Lake Zwai, Ethiopia (Crustacea: Ostracoda). Freshw Biol 25:233–241

Massana R, Pedrós-Alió C (1994) Role of anaerobic ciliates in planktonic food webs: abundance, feeding, and impact on bacteria in the field. Appl Environ Microbiol 60:1325–1334

Mathes J, Arndt H (1994) Biomass and composition of protozooplankton in relation to lake trophy in north German lakes. Mar Microb Food Webs 8:357–375

Mavuti KM (1990) Ecology and role of zooplankton in the fishery of Lake Naivasha. Hydrobiologia 208:131–140

Melack JM (1979) Photosynthesis and growth of *Spirulina platensis* (Cyanophyta) in an equatorial lake (Lake Simbi, Kenya). Limnol Oceanogr 24:753–760

Melack JM (1981) Photosynthetic activity of phytoplankton in tropical African soda lakes. Hydrobiologia 81:71–85

Mengestou S, Fernando CH (1991) Seasonality and abundance of the dominant crustacean zooplankton in a tropical Rift Valley lake, Awassa, Ethiopia. Hydrobiologia 226:137–152

Milbrink G (1977) On the limnology of two alkaline lakes (Nakuru and Naivasha) in the East Rift Valley system in Kenya. Int Rev Ges Hydrobiol 62:1–17

Montagnes DJS (1996) Growth responses of planktonic ciliates in the genera *Strombidium* and *Strobilidium*. Mar Ecol Prog Ser 130:241–254

Montagnes DJS, Lynn DH (1993) A quantitative protargol stain (QPS) for ciliates and other protists. In: Kemp PF, Sherr BF, Sherr EB, Cole JJ (eds) Handbook of methods in aquatic microbial ecology. Lewis Publishers, Boca Raton, pp 229–240

Montagnes DJS, Lynn DH, Roff JC, Taylor WD (1988) The annual cycle of heterotrophic planktonic ciliates in the waters surrounding the Isles of Shoals, Gulf of Maine: an assessment of their trophic role. Mar Biol 99:21–30

Müller H, Geller W (1993) Maximum growth rates of aquatic ciliated protozoa: the dependence on body size and temperature reconsidered. Arch Hydrobiol 126:315–327

Müller H, Schone A, Pinto-Coelho RM, Schweizer A, Weisse T (1991) Seasonal succession of ciliates in Lake Constance. Microbiol Ecol 21:119–138

Nogrady T (1983) Succession of planktonic rotifer population in some lakes of the Eastern Rift Valley, Kenya. Hydrobiologia 98:45–54

Ochumba PBO, Kibaara DI (1988) An instance of thermal instability in Lake Simbi, Kenya. Hydrobiologia 158:247–252

Ong'ondo GO, Yasindi AW, Oduor SO, Jost S, Schagerl M, Sonntag P, Boenigk J (2013) Ecology and community structure of ciliated protists in two alkaline-saline Rift Valley lakes in Kenya with special emphasis on *Frontonia*. J Plankton Res 35:759–771

Pace ML (1986) An empirical analysis of zooplankton community size structure across lake trophic gradients. Limnol Oceanogr 31:45–55

Pace ML, Orcutt JD (1981) The relative importance of protozoans, rotifers, and crustaceans in a freshwater zooplankton community. Limnol Oceanogr 25:822–830

Pace ML, McManus GB, Findlay SEG (1990) Planktonic community structure determines the fate of bacterial production in a temperate lake. Limnol Oceanogr 35:795–808

Paffenhöfer GA (1998) Heterotrophic protozoa and small metazoa: feeding rates and prey-consumer interactions. J Plankton Res 20:121–133

Parker BC, Simmons GM, Seaburg KG (1982) Comparative ecology of plankton communities in seven Antarctic oasis lakes. J Plankton Res 4:271–286

Priscu JC, Vincent WF, Howard-Williams C (1989) Inorganic nitrogen uptake and regeneration in perennially ice-covered Lake Fryxell and Vanda, Antarctica. J Plankton Res 11:335–351

Reynolds CS (1989) Physical determinants of phytoplankton succession. In: Sommer U (ed) Plankton ecology, succession in plankton communities. Springer, London, pp 9–56

Robarts RD (1988) Heterotrophic bacterial activity and primary production in a hypertrophic African lake. Hydrobiologia 148:97–107
Roberts E, Laybourn-Parry J, Mcnight DM, Novarino G (2000) Stratification and dynamics of microbial loop communities in lake Fryxell, Antarctica. Freshw Biol 44:649–661
Sanders RW (1991) Mixotrophic protists in marine and freshwater ecosystems. J Protozool 38:76–81
Saros JE, Fritz SC (2000) Changes in the growth rates of saline-lake diatoms in response to variation in salinity, brine type and nitrogen form. J Plankton Res 22:1071–1083
Sherr EB, Sherr BF, Fallon RD, Newell SY (1986) Small aloricate ciliates as a major component of the marine heterotrophic nanoplankton. Limnol Oceanogr 31:177–183
Shukla U, Gupta PK (2001) Assemblage of ciliated protozoan community in a polluted and non-polluted environment in a tropical lake of central Himalaya: Lake Naini Tal, India. J Plankton Res 23:571–584
Sommaruga R, Sattler B, Oberleiter A (1999) An in situ enclosure experiment to test the solar UVB impact on plankton in a high-altitude mountain lake. II. Effects on the microbial food web. J Plankton Res 21:859–876
Sommer U (ed) (1989) Plankton ecology: succession in plankton communities. Springer, London
Sonntag B, Summerer M, Sommaruga R (2011) Factors involved in the distribution pattern of ciliates in the water column of a transparent alpine lake. J Plankton Res 33:535–540
Sprules WG, Bowerman JE (1988) Omnivory and food chain length in zooplankton food webs. Ecology 69:418–426
Stensdotter-Blomberg U (1998) Factors controlling pelagic populations of ciliates and heliozoans- late summer investigations in an acidic lake before and after liming. J Plankton Res 20:423–442
Stoecker DK, Capuzzo JM (1990) Predation on protozoa: its importance to zooplankton. J Plankton Res 12:891–908
Stoecker DK, Evans GT (1985) Effects of protozoan herbivory and carnivory in a microplankton food web. Mar Ecol Prog Ser 25:159–167
Strom S, Morello T (1998) Comparative growth rates and yields of ciliates and heterotrophic dinoflagellates. J Plankton Res 20:571–584
Talling JF (1986) The seasonality of phytoplankton in African lakes. Hydrobiologia 138:139–160
Talling JF, Lamoalle J (1998) Ecological dynamics of tropical inland waters. Cambridge University Press, Cambridge
Talling JF, Wood RB, Prosser MV, Baxter RM (1973) The upper limit of photosynthetic productivity by phytoplankton: evidence from Ethiopian soda lakes. Freshw Biol 3:53–76
Taylor WD (1978) Growth responses of ciliate protozoa to the abundance of their bacterial prey. Microb Ecol 4:207–214
Taylor WD, Heynen ML (1987) Seasonal and vertical distribution of ciliophora in Lake Ontario. Can J Fish Aquat Sci 44:2185–2191
Taylor WD, Johannsson EO (1991) A comparison of estimates of productivity and consumption by zooplankton for planktonic ciliates in Lake Ontario. J Plankton Res 13:363–372
Taylor WD, Sanders RW (2010) Chapter 3. Protozoa. In: Thorp JH, Covich AP (eds) The ecology and classification of North American freshwater invertebrates, 3rd edn. Academic, London, pp 43–96
Taylor WD, Shuter BJ (1981) Body size, genome size, and intrinsic rate of increase in ciliated protozoa. Am Nat 118:160–172
Twombly S (1983) Seasonal and short term fluctuations in zooplankton abundance in tropical Lake Malawi. Limnol Oceanogr 28:1214–1224
Vareschi E (1982) The ecology of Lake Nakuru (Kenya) III. Abiotic factors and primary production. Oecologia 55:81–101
Vareschi E, Jacobs J (1985) The ecology of Lake Nakuru (Kenya). VI. Synopsis of production and energy flow. Oecologia 65:412–424
Vareschi E, Melack JM, Kilham P (1981) Saline waters. In: Symoens JJ, Burgis MJ, Gaudet JJ (eds) The ecology and utilization of African inland waters, UNEP Reports and Proceedings Series No. 1, Nairobi, pp 93–102
Verity PG (1986) Growth rates of natural tintinnid populations in Narragansett Bay. Mar Ecol Prog Ser 29:117–126
Weisse T, Müller H (1998) Planktonic protozoa and the microbial food web in Lake Constance. Arch Hydrobiol Spec Issues Adv Limnol 53:223–254
Weisse T, Karstens N, Meyer VCL (2001) Niche separation in common prostome freshwater ciliates: the effect of food and temperature. Aquat Microb Ecol 26:167–179
Wetzel RG (1983) Limnology, 2nd edn. Saunders College Publishing, Philadelphia
Wiackowski K, Staronska A (1999) The effect of predator and prey density on the induced defence of a ciliate. Funct Ecol 13:59–65
Wickham SA, Gilbert JJ (1993) The comparative importance of competition and predation by *Daphnia* on ciliated protists. Arch Hydrobiol 126:289–313
Williams WD (1981) Inland salt lakes: an introduction. Hydrobiologia 81:1–14
Williams WD, Boulton AJ, Taaffe RG (1990) Salinity as a determinant of salt lake fauna: a question of scale. Hydrobiologia 197:257–266
Witek M (1998) Annual changes of abundance and biomass of planktonic ciliates in the Gdalsk basin, southern Baltic. Int Rev Hydrobiol 83:163–182
Wood RB, Talling JF (1988) Chemical and algal relationships in a salinity series of Ethiopian inland waters. Hydrobiologia 158:29–67
Worden AZ, Follows MJ, Giovannoni SJ, Wilken S, Zimmerman AE, Keeling PJ (2015) Rethinking the marine carbon cycle: factoring in the lifestyles of microbes. Science 347:735

Xu R, Cronberg G (2010) Planktonic ciliates in Western Basin of Lake Ringsjön, Sweden: community structure, seasonal dynamics and long-term changes. Protistology 6:173–187

Yasindi AW (2001) The ecology of planktonic ciliates in tropical lakes of East Africa. Dissertation, University of Waterloo, Ontario

Yasindi AW, Taylor WD (2005) The abundance, biomass and composition of pelagic ciliates in East African lakes of different salinity and trophy. In: Proceedings of the 11th World Lake conference, vol II, Nairobi, Kenya, pp 374–378

Yasindi AW, Taylor WD (2006) The trophic position of planktonic ciliate populations in the food webs of some East African lakes. Afr J Aquat Sci 31:53–62

Yasindi AW, Lynn DH, Taylor WD (2002) Ciliated protozoa in Lake Nakuru, a shallow alkaline-saline lake in Kenya: seasonal variation, estimated production, and role in the food web. Arch Hydrobiol 154:311–325

Yasindi AW, Taylor WD, Lynn DH (2007) The community composition and biomass of pelagic ciliated protozoa in East African lakes. Afr J Aquat Sci 32:175–183

Zinabu G-M, Taylor WD (1989) Seasonality and spatial variation in abundance, biomass and activity of heterotrophic bacterioplankton in relation to some biotic and abiotic variables in an Ethiopian rift-valley lake (Awassa). Freshw Biol 22:355–368

Zinabu G-M, Taylor WD (1997) Bacteria-chlorophyll relationships in Ethiopian lakes of varying salinity: are soda lakes different? J Plankton Res 19:647–654

Zingel P (1999) Pelagic ciliated protozoa in a shallow eutrophic lake: community structure and seasonal dynamics. Arch Hydrobiol 146:495–511

Zöllner E, Santer B, Boersma M, Hoppe H-G, Jürgens K (2003) Cascading predation effects of *Daphnia* and copepods on microbial food web components. Freshw Biol 48:2174–2193

Invertebrates of East African Soda Lakes 8

Seyoum Mengistou

Abstract

Soda lakes are among the world's most productive natural ecosystems, but the fauna consists of only a few specialized groups which can attain high biomass, seasonally or throughout the year. The invertebrate fauna is generally poor in species diversity. Protozoa are restricted to a few ciliates dominated by *Condylostoma* and *Frontonia* spp. The zooplankton biomass is dominated by a calanoid, *Paradiaptomus africanus*, with a maximum value of 1.17 g m^{-3} reported in Lake Nakuru, Kenya. Rotifers dominate in terms of species diversity, with six species/lake being the highest recorded number at six. Cladocerans are rare and include some euryhaline forms also common in other Rift Valley lakes, such as *Alona* sp., *Macrothrix triseralis*, *Moina* and *Ceriodaphnia* spp. Rotifera are represented by a few euryhaline and halobiontic species such as *Brachionus dimidiatus*, *B. plicatilis* and *Hexarthra jenkinae*. At times, *B. dimidiatus* can attain extremely high abundances (800 million m^{-3}). The major food sources for zooplankton grazers are the large cyanoprokaryote *Arthrospira fusiformis*, Bacteria and detritus. Larger rotifers such as *B. plicatilis* obtain 48 % of their diet from fragments of *A. fusiformis*, whereas the smaller *B. dimidiatus* remove particles < 2 μm. Assimilation of this cyanoprokaryote was low. Despite this food constraint, secondary production of rotifers can exceed that of calanoids by a factor of 100x in soda lakes and over 600x in freshwater lakes. The macroinvertebrate community of soda lakes is dominated by insects of the order Heteroptera, Family Corixidae (water boatmen), genera *Micronecta* and *Sigara* and Family Notonectidae (backswimmers), genus *Anisops*. Other taxa include nematodes, oligochaetes, chironomids (non-biting

S. Mengistou (✉)
Department of Zoological Sciences, Fisheries and Aquatic Sciences Stream, College of Natural Sciences, Addis Ababa University, King George VI St, Addis Ababa 1000, Ethiopia
e-mail: seyoumeng@gmail.com

M. Schagerl (ed.), *Soda Lakes of East Africa*, DOI 10.1007/978-3-319-28622-8_8

midges), culicine mosquitoes, an anostracan (*Branchinella spinosa*) and ostracods (mussel shrimp). The chironomid community consists of halobiontic forms such as *Microchironomus deribae*, *Kiefferulus disparilis* and *Tanytarsus minutipalpis*. *M. deribae* can increase to enormous numbers seasonally and become a nuisance for lakeshore residents around soda lakes. In Lake Nakuru, Kenya, a high secondary production of 120 mg dry mass m^{-2} day^{-1} was recorded for *M. deribae*, but this is lower than the highest secondary production value of 182 mg dry mass m^{-2} day^{-1} for chironomids reported from the saline Australian Lake Werowrap. Secondary production can be high seasonally in soda lakes, but the biomass turnover rate (*P*/*B* ratio) of zooplankton and macroinvertebrates is low, probably because of the seasonal food limitations (e.g. *Anisops* is predatory on calanoids and fish) or the poor food quality of detritus and decomposing cyanoprokaryotes in the system. Soda lakes have simple food chains and are ideal as natural experimental models to study food web dynamics and energy flow in lakes.

8.1 Soda Lakes in East Africa

Saline lakes are restricted in their global distribution compared with the more common freshwater lakes (Hammer 1986). Soda lakes represent another set of saline lakes. They have a worldwide distribution but are mainly confined to (sub)tropical latitudes in continental interiors. Due to their hostile nature, they are often remote from the main centres of human activity. This is perhaps one reason why few studies have been conducted on these soda lakes.

Saline and soda lakes are more abundant in the Eastern Rift than in any other African ecoregion. Streams that feed the lakes flow over highly alkaline volcanic rocks, bringing Natron (a naturally occurring salt consisting of sodium carbonate and sodium bicarbonate) into the lakes' waters. Many of these lakes are endorheic, and high ambient temperature in the Rift Valley increases the evaporation rate, thereby enhancing the water's alkalinity by raising the concentrations of Na^{+}, HCO_3^{-} and CO_3^{2-}. Whereas the pH of natural freshwater lakes often ranges between 6 and 8, that of the soda lakes ranges from 9 to 12. Many of the soda lakes fluctuate in size and change in water salinity with dry and wet periods (Fig. 8.1). Soda lakes vary in shape from broad, shallow pans to narrow, deep depressions.

Alkaline-saline lakes, or soda lakes, in tropical Africa are inhabited by specialized biota that can tolerate high water temperature (e.g. 40 °C in Lake Bogoria, Kenya, and even as high as 50 °C in Lake Natron, Tanzania; Hughes and Hughes 1992), high salinity and low oxygen solubility (Hecky and Kilham 1973; Talling 1992). Only a few specialized groups of zooplankton and macroinvertebrates are able to survive in soda lakes. Most of these organisms also possess highly developed strategies of diapause and dispersal that enable them to persist in arid landscapes where scarce aquatic habitats are hydrologically unstable and of marginal quality.

Due to the presence of high solar radiation, moderate temperature, unlimited carbon dioxide reserves in the form of carbonate-bicarbonate salts and other essential nutrients such as phosphate, the East African soda lakes (EASL) have high rates of photosynthesis (Grant 2006; Melack and Kilham 1974; Oduor and Schagerl 2007; Talling et al. 1973). Primary production rates of >10 g C $m^{-2}day^{-1}$ have been recorded, making these among the most productive aquatic

Fig. 8.1 Lake Eyasi (Tanzania) in 2010 almost completely dried up (photo: Michael Schagerl)

environments in terms of biomass anywhere in the world. Indeed, the highest primary productivity for a natural habitat was earlier recorded from some of these lakes in Ethiopia (Talling et al. 1973). This remarkable photoautotrophic primary production is presumably the driving force behind all biological processes in what is essentially a closed environment. Cyanobacteria (notably *Arthrospira*) species and heterotrophic Bacteria are the most abundant primary producers in these habitats (Grant 2006; Mwatha 1991).

Salinity is variable and high in the soda lakes. Total salts vary from about 50‰ in the northern lakes (Bogoria, Nakuru, Elmenteita, Sonachi) to saturation in parts of the Magadi-Natron basin in the south. The high alkaline and high salt concentrations of soda lakes seem to be too extreme for most forms of life to exist, but these lakes support a diverse group of microorganisms. These microorganisms, generally termed extremophiles, are adapted to the extremes of alkaline pH and high salt concentration (Grant 2006). In the highly productive soda lakes, abundant organic matter (Jirsa et al. 2013) supports a huge assemblage of heterotrophic Bacteria, probably involved in organic matter decomposition and nutrient recycling (Grant 2006; Mutanga et al. 2000). In the process of oxidizing organic matter, these heterotrophic microbes consume the available oxygen. As a result, in some of these lakes (e.g. Lake Chitu and Lake Arenguade, Ethiopia), no measureable dissolved oxygen is detected in areas deeper than 35 cm and no fish are present.

The dominant cyanoprokaryote *Arthrospira fusiformis* has evolutionarily developed a mechanism of adaptation to the hyperosmotic environment in soda lakes (see Schagerl and Burian, Chap. 12). How have the invertebrates coped with the same problem? Benthic invertebrates have several mechanisms to adapt to saline and alkaline lakes. Larvae have a lime gland to remove carbonate and bicarbonate ions from the blood and excrete them (see Schagerl and Burian, Chap. 12).

With an abundance of food and a paucity of competitors, soda lake invertebrate populations grow to astounding densities. It is estimated that 37 million brine flies can be found on one linear mile of shore at the Great Salt Lake (USA). It is not unusual to see soda lake shores covered with immense numbers of corixids on which huge flocks of birds converge and feast. The unconsumed invertebrate biomass decomposes on the lake shores, giving off an offensive and obnoxious smell to the surrounding area, which depreciates the aesthetic and recreational value of soda lakes.

The algal and invertebrate communities of soda lakes are highly sensitive to environmental change, and some taxa can become locally extirpated following salinity fluctuations well within known tolerance ranges. Several observations indicate that soda lake communities are regulated by a variety of environmental factors, among which salinity may be important but not necessarily decisive for survival. Other factors may include nutrient or food availability, temporal variation in dissolved oxygen, water-column transparency and the stability of various substrata available to benthic organisms (McClanahan et al. 1996; Melack 1976; Vareschi and Vareschi 1984).

The main saline or alkaline soda lakes are Bogoria (42 km^2), Nakuru (49 km^2), Elmenteita (19 km^2) and Magadi (105 km^2) in Kenya; Natron (900 km^2), Manyara (470 km^2) and Eyasi (1050 km^2) in Tanzania; and some lakes in the Ethiopian Rift Valley such as Lakes Arenguade, Abijata, Shala and Chitu. There are also other small lakes (<20 km^2) on the valley floor, including Lakes Solai, Sonachi and Oloidien in Kenya and Lakes Lelu and Momera in Tanzania. These lakes are salty and some have high concentrations of fluoride and sodium carbonate salts. For a general overview, please refer to the Appendix.

8.1.1 The Ethiopian Soda Lakes

Many lakes in Ethiopia tend to concentrate salts due to evaporative concentration, hot springs and saline intrusions. However, the major lakes that concentrate carbonate-bicarbonate salts, mainly of the sodium type, are Arenguade, Abijata, Shala and Chitu (see Table 8.3). In Lake Abijata, the lake water has been withdrawn for industrial extraction of soda ash since 1985 with an annual production of 8500 tons. Ayenew (2002) documented that the water level of Lake Abijata has receded drastically since then, exposing dry bed of trona and posing a threat to the aquatic and bird life (Fig. 8.2). Lake Shala, one of the deepest lakes in Africa ($Z_{max} = 266$ m), has a rich reserve of carbonate-bicarbonate salts, despite the inflow of many small freshwater hot springs from its caldera rim (Fig. 8.3). Lake Chitu, a small lake near Lake Shala, has higher salinity and alkalinity than the other Ethiopian lakes and supports high populations of heterotrophic Bacteria, *Arthrospira fusiformis* and flamingos. The hypersaline lakes in the Afar Depression close to the Red Sea are brine solutions of NaCl and other salts such as K, Ca, Mg and SO_4; they accumulate very little carbonate salts. The geographical location, salinity and

Fig. 8.2 Lake Abijata (Ethiopia) in 2011—receding lake water and evaporative deposit of trona (photo: Steve Omondi)

Fig. 8.3 Lake Shala (Ethiopia)—fresh hot springs flowing into the lake (photo: Seyoum Mengistou)

carbonate-bicarbonate concentration of the four soda lakes in Ethiopia are given in the Appendix.

8.1.2 The Kenyan Soda Lakes

The major soda lakes in Kenya include Bogoria, Elmenteita, Oloidien (low in salinity with around 4), the well-studied Lake Nakuru and L. Sonachi and Lake Magadi, where industrial extraction of trona for the commercial production of soda ash has been taking place since 1924. The northern soda lakes include Lakes Logipi, Elmenteita, Nakuru and Bogoria, and some of which have lake water temperatures at about 40 °C (Hughes and Hughes 1992).

Lake Magadi It is at the southern end in the Kenya Rift Valley, and during the dry season, it is 80 % covered by trona and is well known for its wading birds, including flamingos. Lake Magadi is a saline, alkaline lake, approximately 100 km^2 in size. It is an example of a "saline pan", with some areas being up to 40 m thick with salt.

Lake Bogoria This is a moderately sized soda lake (area 30 km^2) with high salinity (conductivity 40–80 mS cm^{-1}), alkalinity (1500 meq L^{-1}) and pH 10.3. Hot fumaroles and springs fringe the lake shore and the deposited soda layers are evident from a long distance. The lake has a narrow and extended littoral zone, which is devoid of vegetation and harbours one of the largest Lesser Flamingo populations in the Rift Valley (Mutanga et al. 2000). Lake Bogoria is rich in concentrated carbonate-bicarbonate mixture, and the *Arthrospira* that thrives abundantly is restricted to certain parts of the lake, where large flocks of the Lesser Flamingo concentrate. The chironomid *Paratendipes* sp. dominates the invertebrate community; sometimes a mass emergence of Ephydridae can be observed (Figs. 8.4 and 8.5). Rotifers include *Brachionus dimidiatus* (Fig. 8.6), *B. plicatilis* and *Hexarthra jenkinae* (Burian et al. 2013). Ong'Ondo et al. (2013) also reported many ciliate species such as *Cyclidium*, *Rimaleptus*, *Halophyra*, *Acineria* and *Frontonia* spp., with the latter forming the dominant group.

Lake Nakuru It typically covers between 35 km^2 and 49 km^2 but occasionally dries out completely due to unknown reasons; this endorheic lake with a mean depth of 2.3 m is small for its large drainage area of 1760 km^2 (International Lake Environment Committee 2001) and as a result has high external loading of nutrients and sediment. The three major rivers feeding the lake are Njoro, Makalia and Nderit, and it also receives water from several alkaline springs. The lake was voted as a Ramsar site in 1990. Scientists estimated that the flamingo

Fig. 8.4 Mass emergence of Ephydridae at Lake Bogoria, 2010 (photo: Michael Schagerl)

Fig. 8.5 Exuviae of Ephydridae at shore of Lake Bogoria, 2010 (photo: Michael Schagerl)

population at Nakuru consumes about 250,000 kg of algae per hectare of surface area per year (International Lake Environment Committee 2001). The invertebrate community here has been comparatively well studied and is discussed under other sections.

Lake Sonachi It is a small (0.14 km^2), shallow (Z_{max} 4.25 m in 1993), alkaline-saline crater lake located 1884 m above sea level in the semiarid Rift Valley of central Kenya. Its water level is maintained against a strong moisture deficit by subsurface flow from nearby Lake Naivasha. The salt budget of Lake Sonachi is governed by rainfall and evaporation at the surface and dissolution of sedimentary evaporates by percolating groundwater. As in other African soda lakes, the ionic composition of lake water is dominated by sodium and bicarbonate, whereas pH values are very high (9.0–10.3; Melack 1976). Surface-

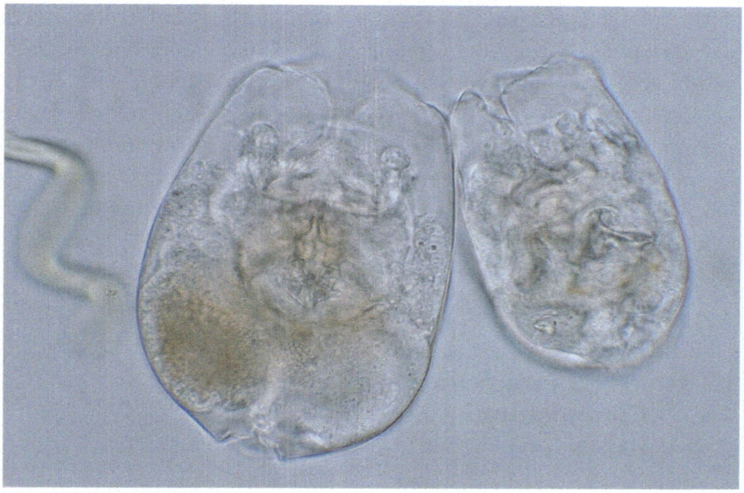

Fig. 8.6 *Brachionus dimidiatus* from Lake Nakuru (photo: Michael Schagerl)

water conductivity (K_{25}) since 1929 ranged from 3000 to 11,550 μS cm^{-1}, and years with good data coverage show seasonal fluctuations on the order of 1000–2000 μS cm^{-1} (Verschuren et al. 1999). The lake is rich in planktonic Cyanobacteria including *Arthrospira fusiformis*, whereas planktonic diatoms and green algae are scarce.

Lake Oloidien It is less saline and alkaline than the other Kenyan lakes (conductivity less than 5 mS cm^{-1}) and is situated at the southern tip of Lake Naivasha, with which it shares euryhaline zooplankton forms typically found in freshwater tropical lakes. These include the cyclopoid *Thermocyclops oblongatus*, the cladocerans *Ceriodaphnia cornuta* and *Diaphanosoma excisum* and the rotifers *Brachionus angularis* and *Hexarthra mira* (Uku and Mavuti 1994). The collected benthic macroinvertebrates included forms also present in other soda lakes such as oligochaetes (Tubificidae) and chironomids (Clark et al. 1989).

8.1.3 Tanzanian Soda Lakes

In the Tanzanian Rift Valley, the largest saline and soda lakes are Lakes Eyasi, Manyara and Natron, but many small saline alkaline lakes also exist (see appendix).

Lake Eyasi It is a seasonal shallow endorheic soda lake on the floor of the Great Rift Valley at the base of the Serengeti Plateau in the Crater Highlands of Tanzania. It is usually dry, but may occasionally flood up to a depth of 1 m (Hughes and Hughes 1992). The Sibiti River occasionally flows into the lake but its waters usually evaporate before reaching the dry lakebed. Even though permanent springs lie along the lakeshores, their waters evaporate quickly, leaving huge flakes of salt on the lake shore.

Lake Manyara It is a shallow lake (max depth 3.7 m) in the Natron-Manyara-Balangida branch of the Great Rift Valley in Tanzania. It was said by Ernest Hemingway to be the "loveliest [lake] . . . in Africa", with its huge flocks of flamingos that rely on the abundant *Arthrospira* and extensive dry out of soda salts during lake recession (Matagi 2004).

Lake Natron It lies to the north on the border between Kenya and Tanzania. Seasonal streams drain into Lake Natron from the Ngorongoro Highlands south of the lake (Hughes and Hughes 1992). About 28 springs, most of them saline, also feed into the lake. Despite these inflows, most of the lake's water is derived from direct precipitation. Evaporation exceeds precipitation, and the maximum depth is only 2 m. A large portion of the lake's bed is covered by a salt crust that dissolves during the rainy season. Lake Natron serves as an important breeding site for the Lesser Flamingo populations that fly over from other Rift Valley lakes as far away as Kenya and Ethiopia (Mutanga et al. 2000). Temperatures in the lake can reach 50 °C and, depending on rainfall, a pH of 9–10.5 can be reached (almost as alkaline as ammonia).

8.2 Invertebrate Species Composition in Ethiopian Soda Lakes

8.2.1 Zooplankton

Copepods are the dominant zooplankton in soda lakes. Cladocera are rare and Rotifera are represented by a few euryhaline salt-tolerant taxa such as *Brachionus dimidiatus*, *B. plicatilis*, *B. calyciflorus* and *B. angularis*.

Of the Copepoda, the calanoids are prominent in soda lakes and sometimes the sole zooplankton in small soda lakes such as Lake Arenguade in Ethiopia (pers. observation). The species most often encountered is *Lovenula africana* (*Paradiaptomus africanus*). This large calanoid (max length 1.48 mm) subsists on the rich cyanoprokaryote food resource in soda lakes, especially *Arthrospira fusiformis*. Fragments of *Arthrospira* sp. were often visible in the gut of this calanoid in Lake Arenguade (Afeworki Ghebrai, pers. comm.), and other large protozoans such as the ciliates belonging to genera *Strombidium*, *Strobilidium* and *Halteria* were observed to ingest large pieces of this cyanoprokaryote in East African lakes (Yasindi and Taylor 2006; see Yasindi and Taylor, Chap. 7).

Cyclopoids are less represented in soda lakes and often belong to a few euryhaline genera also common in freshwater lakes. *Mesocyclops* has wide range of tolerance to conductivity up to 5000 μS cm^{-1}, whereas *Afrocyclops* and *Thermocyclops* spp. extend to even higher conductivities close to 10,000 μS cm^{-1} (LaBarbera and Kilham 1974). *Mesocyclops* has occasionally been observed in Lake Abijata along with *Afrocyclops gibsoni* and *Eucyclops serrulatus*. These cyclopoids were present in low numbers and only at certain times of the year, especially when the lake became diluted during the rainy season (June–Sept.). Matagi (2004) reported two cyclopoids from Lake Shala—*Lytocyclops gibsoni* and *Cyclops agiloides*. Matagi (2004) did not state the source of the finding, and the present author has doubts as to the authenticity of the records. The *Lytocyclops gibsoni* probably refers to *Afrocyclops gibsoni*, which this author collected only from Lake Abijata and not Shala. The identity of the species identified as *Cyclops agiloides* remains unclear. Equally, there are no reports of any cladoceran from Lake Shala, and *Diaphanosoma excisum* was never recovered from this lake, even though it is abundant in other freshwater lakes in Ethiopia. A copepod parasite of fish, *Argulus africanus*, was regularly observed in the freshwater influx areas of Lake Shala, where fish are often present (see Fig. 8.3).

Of particular interest was the recording of a harpacticoid copepod in Lake Shala in 1983 (pers. observation). This harpacticoid (*Nitocra lacustris*) was the dominant and sole zooplankton in the lake during the survey. Tudorancea and Taylor (2002) also reported recovering this harpacticoid from the benthos of Lakes Abijata and Shala.

Cladocera are rare in soda lakes, mainly because of the high osmotic pressure, which they cannot withstand. The osmoregulatory mechanisms that calanoids have evolved to withstand the osmotic stress in soda lakes have apparently not been well developed in Cladocera. The rare species encountered in soda lakes include the ubiquitous *Moina* spp., *Alona* spp. and *Macrothrix triseralis*. *Moina belli* was encountered only in Lake Shala.

Rotifera are the most diverse zooplankton in soda lakes. The species most often encountered are the euryhaline, halophilic forms which can tolerate high salinity and are represented by *Brachionus* species—*B. dimidiatus*, *B. urceolaris*, *B. plicatilis*, *B. angularis* and

B. calyciflorus. Other rotifers encountered include: *Lecane bulla*, *L. luna* and *Trichocerca tetractis* in Lake Shala and *Hexarthra* sp. and *Asplanchna brightwelli* in Lake Chitu. Apparently, rotifers in the soda lakes face dual problems: they have to withstand the high ionic concentration and imbalance and must also subsist on detrital and bacterial food resources. This is because the large cyanoprokaryotes in the soda lakes, such as *Arthrospira* spp., *Anabaenopsis* spp. and *Synechococcus* spp., cannot be ingested by the small rotifers. *Oocystis*, a small chlorophyte, became the dominant planktonic algae in the soda Lake Abijata in 1961 (Wood and Talling 1988). Rotifers can potentially rely on such small primary producers (besides Bacteria and detritus) whenever they bloom in soda lakes.

In general, the zooplankton community of soda lakes is poor and restricted to a few halophilic and alkalophilic species, which is also true for the phytoplankton and microbial communities studied in these lakes (e.g. Grant et al. 1990). The list of the common zooplankton species encountered in the Ethiopian soda lakes is presented in Table 8.1.

8.2.2 Benthic Invertebrates

The list of benthic invertebrate species documented from the Ethiopian soda lakes is presented in Table 8.2; Table 8.3 provides a summary of invertebrates occurring in the EASL. The species diversity of benthic macroinvertebrates in soda lakes is low, with maximum of 20 taxa, and restricted to a few specialized taxa that have evolved mechanisms to tolerate the extreme alkalinity and salinity. From the limited studies done on the invertebrate composition and distribution in soda lakes, the dominant forms are from the Arthropoda, Order Hemiptera (bugs), the Nematoda (roundworms), Oligochaeta (aquatic earthworms) and Chelicerata. Two families are well represented in the Hemiptera, including the Families Corixidae (water boatmen) and Notonectidae (backswimmers). The Diptera are highly represented by the family Chironomidae, subfamily Chironominae (non-biting midges), which can reach enormous densities during emergence. Harrison (2002) published a detailed study of the Chironomidae in the soda lakes of Ethiopia and his records are included in Table 8.2.

Table 8.1 List of the zooplankton in Ethiopian soda lakes

Taxon	L. Arenguade	L. Abijata	L. Shala	L. Chitu
Protozoa	*Strombidium*, *Strobilidium*			*Strombidium*
Cyclopoida (Copepoda)		*Mesocyclops* sp.* *Afrocyclops gibsoni** *Eucyclops serrulatus**		
Harpacticoida (Copepoda)			*Nitocra lacustris*	
Calanoida (Copepoda)	*Paradiaptomus africanus*	Nauplii		
Cladocera		*Alona rectangula*	*Moina belli*	
		*Macrothrix triseralis**		
Rotifera		*Brachionus dimidiatus*	*Lecane luna*	*B. calyciflorus**
		B. plicatilis	*L. bulla*	*B. plicatilis*
		B. urceolaris	*Trichocerca tetractis*	*B. urceolaris*
		B. angularis		*Hexarthra* sp.*

*Also found in other non-soda Ethiopian lakes

Table 8.2 Benthic invertebrate taxa documented from Ethiopian soda lakes

Taxon (common name)	L. Arenguade	L. Abijata	L. Shala	L. Chitu
Order Hemiptera (bugs) Fam. Corixidae (water boatmen)				
Micronecta jenkinae	+	++		
Sigara hieroglyphica	+	++		
Order Hemiptera Fam. Notonectidae (backswimmers)				
Anisops sp.				+
Class Crustacea				
Order Ostracoda (mussel shrimp)				
Limnocythere barosi	+	(44,922) ++	(14,836) ++	
Gomphocythere sp.			(40) +	
Darwinula stevensoni			(60) +	
Potamocypris mastigophora		+		
Order Diptera				
Fam. Chironomidae (midges)				
Subfamily Chironominae				
Kiefferulus dispersalis	+	(27,360) ++	(1) rare	+
Cladotanytarsus pseudomancus		(281) +	(1592) +	+
Microchironomus deribae		(10,644)	(412) +	+
M. lendlin		++	+	+
Tanytarsus minutipalpis		+	+	+
Chelicerata				
Hydracarina sp. (water mites)		+	+	
Nematoda		(10 species)	(3 species)	
Monohystera		(539) ++	(49) +	
Mesodorylaimus		(326) ++	(949) +	
Oligochaeta (aquatic earthworms)				
Tubificidae		(167) +	(1761) ++	

Relative occurrence recorded as + = common and ++ = very common
Numbers in brackets are densities in number m^{-2}

Nematode worms of the genera *Monohystera* and *Mesodorylaimus* are quite common in soda lakes, while the Oligochaeta, represented by the family Tubificidae, can reach high densities in the lake bottom. Other macroinvertebrates in soda lakes include the chelicerate *Hydracarina* (water mites) and crustaceans belonging to the class Ostracoda. Members of the genus *Limnocythere* can reach extremely high densities in soda lakes such as Lake Abijata (~ 45,000 m^{-2}).

Eyualem (2002) conducted an exhaustive study on the nematodes of the Ethiopian Rift Valley lakes, including the soda lakes Abijata and Shala. The genus *Monhystrella* of the family Nordidae dominated in soda lakes, represented by about four species—*M. parvellus*, *M. hoogewejisi*, *M. jacobsi* and *M. macrura* (Table 8.4). This is followed by the family Dorylaimidae with the genera *Dorylaimus* and *Mesodorylaimus*. *M. macrospiculum* is present in both soda lakes, whereas the other two species are lacking from Lake Abijata.

Hughes and Hughes (1992) reported the presence of the molluscs *Bellamya unicolor* and *Helicarion ruppellianum* in Lake Abijata, which were not collected from Lakes Shala and Chitu. Apparently, these molluscs tolerated the lower salinity of Lake Abijata but not the higher salinities of Lakes Shala and Chitu.

Table 8.3 Checklist of invertebrates in East African soda lakes (after various sources); number codes of the lakes are (1) Arenguade, (2) Abijata, (3) Shala, (4) Chitu, (5) Elmenteita, (6) Nakuru, (7) Bogoria, (8) Sonachi

Group	Genus species	Lake Code
Protozoa	*Strombidium* sp.	1,4
	Strobilidium sp.	1
	Condylostoma sp.	1,6
	Euplotes sp.	6
	Pleuronema sp.	6
	Lionatus sp.	6
	Frontonia sp.	6, 7
Copepoda		
Calanoida	*Paradiaptomus africanus*	1,6,7
Cyclopoida	*Mesocyclops* sp.	2
	Afrocyclops gibsoni	2,3
	Eucyclops serrulatus	2
Harpacticoida	*Nitocra lacustris*	3
Cladocera	*Alona* sp.	2,8
	Macrothrix triseralis	2
	Moina spp.	2,3,6,8
	Ceriodaphnia cornuta	12
	Diaphanosoma excisum	2, 7
Rotifera	*Brachionus dimidiatus*	2,6,8
	B. plicatilis	2,4,6
	B. urceolaris	2,4
	B. angularis	2
	Lecane luna	3
	L. bulla	3
	Trichocerca tetractis	3
	Hexarthra jenkinae	4,6
	Keratella tropica	6
	Filinia longiseta	6
Macroinvertebrates		
Order Hemiptera (bugs)		
Fam. Corixidae (water boatmen)	*Micronecta* sp.	1,2,6
Fam. Notonectidae (backswimmers)	*Sigara* sp.	1,2,6
	Anisops sp.	4
Order Coleoptera (beetles)	Dytiscidae	8
Dytiscidae (diving beetles)	Hydrophilidae	8
Class Crustacea		
Order Anostraca (fairy shrimp)	*Branchinella spinosa*	5
Order Ostracoda (mussel shrimp)	*Limnocythere barosi*	2,3
	Gomphocythere sp.	1,3
	Darwinula stevensoni	2,3
	Potamocypris mastigophora	2

(continued)

Table 8.3 (continued)

Group	Genus species	Lake Code
Order Diptera		
Fam. Chironomidae (midges)		
Subfamily Chironominae	*Kiefferulus dispersalis*	1,2,8
	Cladotanytarsus pseudomancus	2,3,8
	Microchironomus deribae	2,3,4,6,8
	M. lendlin *Tanytarsus minutipalpis*	2,3,4, 2,3,4,6
	Dicrotendipes septemmaculatus	8
	Cladotanytarsus pseudomancus	7, 8
	Paratendipes sp.	7, 8
	Microtendipes spp.	7,8
Subfamily Orthocladiinae		
	Smittia spp.	8
Chelicerata	*Hydracarina* sp. (water mites)	2,3
Nematoda		
Family Dorylaimidae	*Dorylaimus generi*	3
	Mesodorylaimus bainsi	3
	M. macrospiculum	2,3
Family Monhysteridae		
	Monhystrella parvellus	3
	M. hoogewejisi	2,3
	M. jacobsi	2,3
Family Aphelechoididae	*M. macrura*	3
Family Isolamidae	*Brenolobrilus graciloides*	2,3
Family Rhabdolaimidae	*Isolaminium africanum*	2,3
	Rhabdolaimus aequatorialis	3
Oligochaeta (aquatic earthworms)	Tubificidae	2,3,6

Table 8.4 List of nematodes recovered from Ethiopian soda lakes (from Eyualem 2002) (Relative occurrence code same as for Table 8.2)

Taxon	Lake Abijata	Lake Shala
Family Aphelechoididae		
Brenolobrilus graciloides	+	+
Family Dorylaimidae		
Dorylaimus generi		+
Mesodorylaimus bainsi		+
M. macrospiculum	+	+
Family Isolamidae		
Isolaminium africanum		+
Family Nordidae		
Monhystrella parvellus		+
M. hoogewejisi	+	+
M. jacobsi	+	+
M. macrura		+
Family Rhabdolaimidae		
Rhabdolaimus aequatorialis		+
Total number of species	4	10

The ostracod taxa in the Ethiopian Rift Valley appear to show distinct segregation in saline and nonsaline lakes. Martens (2002) recorded that the freshwater Lakes Ziway, Langano and Awasa harboured the new subspecies *Limnocythere thomasi thomasi* (Ziway), *L. thomasi langanoensis* (Langano) and *L. thomasi awasensis* (Awasa), while the saline soda lakes Abijata and Shala harboured the subspecies *Limnocythere borisi borisi* (Abijata) and *L. borisi shalaensis* (Shala). Other ostracods present in the soda lakes were also represented in the freshwater lakes, such as *Potamocypris mastigophora*, *Gomphocythere angulata* and *Darwinula stevensoni*, the latter present almost in all Ethiopian Rift Valley lakes.

8.3 Invertebrates in Kenyan Soda Lakes

The main invertebrate taxa recorded from the Kenyan soda lakes (Table 8.5) closely resemble those from the Ethiopian soda lakes. Lake Nakuru was studied more extensively and more data are available for this lake (Burian et al. 2013, 2014; Matagi 2004; Oyoo-Okoth et al. 2011; Vareschi and Jacobs 1984, 1985).

The most abundant macroinvertebrates in the Kenyan soda lakes are the chironomids. In Lake Bogoria, chironomid dipterans have been recorded by Mutanga et al. (2000). In Lake Nakuru, *Culicinae* sp., the chironomid *Tanypus*, oligochaetes and *Microchironomus deribae* form the dominant species (Vareschi and Jacobs 1985). In Lake Elmenteita the anostracan *Branchinella spinosa* and the hemipterans *Micronecta* spp. were reported by Vareschi and Jacobs (1985). Four heteropterans, one notonectid and three corixids were identified in Lake Nakuru.

In Lake Nakuru, unsurprisingly, the same invertebrate species are present as those recorded from Ethiopian soda lakes. Two zooplankton groups, with a total of nine species, were reported by Matagi (2004) and Oyoo-Okoth et al. (2011). Rotifers were represented by *Brachionus dimidiatus*, *B. plicatilis*, *B. calyciflorus*, *Keratella*

Table 8.5 List of invertebrates reported from Lake Nakuru (Kenya)

	Taxon/species
Protozoa Ciliates	*Condylostoma* spp.
	Euplotes spp.
	Pleuronema spp.
	Lionatus spp.
	Frontonia spp.
Copepoda, Calanoida	*Paradiaptomus africanus*
Rotifera	*Brachionus dimidiatus*
	B. plicatilis
	B. calyciflorus
	Keratella tropica
	Filinia longiseta
	Hexarthra jenkinae
Heteroptera Notonectidae	*Anisops varia*
Corixidae	*Micronecta scutellaris*
	M. jenkinae
	Sigara hieroglyphica kilimanjaronis
Chironomidae	*Microchironomus deribae*
	Tanytarsus horni (=*T. minutipalpis*)
	Total number of taxa =17

tropica and *Filinia longiseta* (Burian et al. 2014), while ciliates were represented by *Condylostoma* spp., *Euplotes* spp., *Lionatus* spp. and *Pleuronema* spp. (Yasindi and Taylor 2006). In all the zooplankton samples, *B. dimidiatus* dominated, with a peak abundance (80–100,000 $\times 10^3$ individuals m^{-3}) and highest carbon biomass, and constituting about 80 % of the zooplankton samples (Chemoiwa et al. 2014). *Condylostoma* spp. dominated among the ciliates, although the ciliates were overall less abundant than the rotifers (Burian et al. 2013, 2014).

Macroinvertebrates in Lake Nakuru consist of three species of water bug (*Micronecta jenkinae*, *M. scutellaris* and *Sigara hieroglyphica kilimanjaronis*), which is typical of shallow, saline lakes and alkaline-saline lakes (Matagi 2004). The zooplankton included a copepod (*Paradiaptomus africanus*), beetle larvae, ceratopogonids, chironomid larvae (*Microchironomus deribae* and *Tanytarsus* spp.) and three species of rotifers (*Brachionus dimidiatus*, *B. plicatilis*, *Hexarthra jenkinae*).

Lake Oloidien The zooplankton of Lake Oloidien consists of freshwater forms such as the cladocerans *Ceriodaphnia cornuta* and *Diaphanosoma excisum* and cyclopoid copepods such as *Thermocyclops oblongatus*. Rotifera was dominated by *Brachionus angularis* and *Hexarthra mira*, and the euryhaline forms of *Brachionus* (*B. dimidiatus and B. plicatilis*) are missing (Uku and Mavuti 1994). The chironomids, however, are also found in other soda lakes and include *Chironomus*, *Dicrotendipes*, *Cladotanytarsus* and *Nilodorus* species. Hemipterans include Notonectidae, Corixidae, Pleidae, Hebridae and Microvelidae. Molluscs of the family Physidae dominate the benthos in Lake Oloidien (Clark et al. 1989).

Lake Sonachi The zooplankton community of Lake Sonachi has a poor species assemblage which is similar to other northern Ethiopian and Kenyan soda lakes. Earlier, De Beauchamp (1932) recorded the rotifer *Brachionus dimidiatus* in 1929, and Beadle (1932) encountered the calanoid copepod *Paradiaptomus africanus* in 1931, but the zooplankton seems to have disappeared since then. The nearshore zoobenthos consisted of salt-tolerant chironomids (*Kiefferulus disparilis*, *Microtendipes* sp. and *Cladotanytarsus pseudomancus*), with a total density of 13,500 organisms m^2 (Clark et al. 1989). Other invertebrates reported from Lake Sonachi include coleopterans such as Dytiscidae (diving beetles) and Hydrophilidae along with hemipterans such as Notonectidae and Corixidae (Clark et al. 1989).

Verschuren et al. (1999) conducted a paleaolimnologial study on fossil invertebrate communities in Lake Sonachi and reported freshwater forms which do not exist in the lake at present. These included benthic cladoceran remains of *Bosmina longirostris*, *Euryalona orientalis*, *Alona pulchella*, *Chydorus sphaericus*, *C. parvus* and *Pseudochydorus globosus*. Planktonic cladocerans such as *Moina micrura* and *Alona rectangula* and one species of Orthocladiinae, eight of Chironomini and two of Tanytarsini chironomids were reported (Verschuren et al. 1999). The authors noted that the palaeofauna was dominated by two anoxia-tolerant halobiontic species, *Kiefferulus disparilis* and *Microtendipes* sp. *Tanytarsus minutipalpis* is also a halobiont but intolerant to anoxia.

The freshwater chironomids recorded with varying degrees of tolerance to the high salinities typical of African soda lakes include *Dicrotendipes septemmaculatus*, *Cladotanytarsus pseudomancus*, *Chironomus formosipennis*, *C. alluaudi*, *Nilodorum brevipalpis* and the Orthocladiinae *Smittia* (Verschuren et al. 1999).

Lake Bogoria It had no zooplankton larger than the size of protozoa and only a single chironomid species was recorded. The latter was tentatively designated as *Paratendipes* sp. and occurred in high densities throughout the lake, with an enormous wet weight biomass in the order of 3 tons and a daily emergence of 210 kg for the whole lake (Harper et al. 2003). This sustained several thousand individual avian predators, the most numerous being swifts and swallows and the most important for biodiversity conservation being the Cape Teal and Black-necked Grebe. Several insects were also recorded from the shores of Lake Bogoria including Coleoptera, particularly tenebrionids such as *Gonocephalum* spp., *Sepidium* spp., *Vietomorpha* spp. and *Rhytinota praelonga*, which scavenge nocturnally and shelter by day under stones. During the day, the lake margin was dominated by *Zophosis* spp. and the predatory cicindelid (Tiger Beetle) *Lophyra boreodilatata* (Horn). The lake edge scrub supported many cerambycids (Longhorn Beetles), including the large prionids *Tithoes confinis* Castelnau and the ubiquitous *Macrotoma palmata* (Fabricius), which are primary agents in breaking down dead wood. Tiger Beetles caught in light traps were *Prothyma methneri methneri* (Horn), *Cylindera rectangularis* (Klug) and *Myriochile vicina pseudovicina* (Mandl). In wetter periods, Scarabaeidae dominated light trap catches, particularly rutelids and melolonthids. Elaterids were common, including the 6–7 cm long *Tetralobus* spp. (Harper et al. 2003).

Lake Magadi It is located 610 m below sea level and has an area of 100 km^2 and a depth range from 1 to 5 m (Grant et al. 1990). The lake has no zooplankton or other aquatic invertebrates and is dominated by heterotrophic Bacteria. The alkaline lake niche limits diversity in microbial life due to high pH and high salinity. The large presence of trona causes the lake water to form a sodium carbonate brine and an extremely high fluoride concentration. Nonetheless, this hostile environment hosts extremophile Bacteria, the Magadi Tilapia *Alcolapia grahami* Boulenger and flamingos, herons, pelicans and spoonbills. These animals have developed unique physiological mechanisms to withstand the extreme salinity, alkalinity and fluoride content.

8.4 Invertebrates in Tanzanian Soda Lakes

Less information is available on the invertebrate fauna of the Tanzanian soda lakes. The invertebrate composition of some Tanzanian lakes has been discussed in relation to lake protection and conservation (Sarunday 1999). Because many endangered birds depend on the invertebrates as a food source, it is argued that conservation of the larger avifauna must be planned in conjunction with reducing the threats to the invertebrate fauna. In Africa as a whole, however, the protection of invertebrate diversity has been accorded less status and attention compared with the issue of fish diversity extinction by Nile Perch piscivory in Lake Victoria, for example.

8.5 Secondary Production of Invertebrates in Soda Lakes

8.5.1 Zooplankton Secondary Production

The only extensive study done on secondary production rates in soda lakes is the work of Vareschi and Jacobs (1984, 1985) in Lake Nakuru, Kenya. They studied production, consumption and energy flow of the dominant consumer organisms in the lake from 1972 to 1976. In general, Vareschi and Jacobs (1985) estimated secondary production rates for the zooplankton from field abundance data and laboratory experiments on growth and filtration rates. They did not include detrital food chains in their study, making it difficult to extrapolate the relative contribution of energy provided by live algal food versus the detrital component for the grazers in the food chain. The major food source believed to be grazed by zooplankton was the dominant cyanoprokaryote *Arthrospira fusiformis* (therein called *Spirulina platensis*) and an unknown fraction of food contributed by Bacteria and detritus. Besides the zooplankton, the other important grazers on *Arthrospira* were the Lesser Flamingo (*Phoeniconaias minor*) and the cichlid fish *Alcolapia grahami*. Of the zooplankton, the major grazers on *Arthrospira* were the calanoid *Paradiaptomus africanus* and the three dominant rotifers—*Brachionus dimidiatus*, *B. plicatilis* and *Hexarthra jenkinae*.

8.5.1.1 Calanoida: *Paradiaptomus africanus*

Laboratory cultures were used to obtain data on the stage duration and dry mass (DM) of this large calanoid (Vareschi and Jacobs 1985). Field data were used to determine the abundance of each stage. Egg production (adult) was estimated from the field egg ratio. The daily production of each stage, nauplii, copepodites and eggs (adults) was estimated by the discrete-step method (growth summation method) of Winberg et al. (1971). Daily secondary production of *P. africanus* was highly variable from year to year. Total production of all stages (eggs, nauplii and copepodites) was 80, 55 and 30 mg $m^{-3}day^{-1}$ DM during 1972, 1973 and 1974, respectively. This calanoid had a prodigious abundance and very high production in Lake Nakuru when compared with other calanoids in large, freshwater Lake Tana (Wondie and Mengistou 2006). *Thermodiaptomus galebi*, the dominant calanoid in Lake Tana, had a mean daily production of only 1.04 mg m^{-3} day^{-1} DM.

The mean biomass of *P. africanus* was about 1.81 g m^{-3} DM (Vareschi and Jacobs 1985); this high value apparently reflects the lack of competition by other zooplankton and surplus food sources in soda lakes, which result in enormous numbers of the calanoids, even in the presence of fish predators such as *Alcolapia grahami*.

Daily biomass turnover rates (P/B ratio) of the calanoids in Lakes Nakuru and Tana were similar at about 0.05 day^{-1}, or 5 % of the biomass turnover every day. This indicates that although the biomass and production of calanoids can be extremely high in soda lakes, the productivity or daily biomass turnover rate is not correspondingly high. This could be because of few predators or higher consumers to remove the high biomass in soda lakes or because the excess food does not contribute to biomass production beyond threshold levels in soda lakes. These are interesting research questions to follow up.

What does the calanoid *Paradiaptomus africanus* feed on? Vareschi and Jacobs (1985) tackled this question by conducting feeding (clearance rate) experiments on live *Arthrospira* in laboratory cultures and noting the algal mortality with time. They found that consumption rates were highest for copepodites, followed by adults and nauplii, and that this calanoid consumed only about 15–25 % of the net primary production, the rest being consumed by the Lesser Flamingo and fish. This food source is probably insufficient to support such high calanoid production, and the copepods apparently supplement their additional energy needs with live Bacteria and detritus. Literature data on calanoid secondary production from different types of aquatic ecosystems were compiled (Table 8.6), with an effort to show the

Table 8.6 Biomass (*B*), daily production (*P*) and daily *P*/*B* ratio of selected calanoids from different tropical aquatic ecosystems (sources included)

Lake/country	Calanoid species	*B* (mg m^{-3})	*P* (mg DM m^{-3} day^{-1})	*P*/*B* (day^{-1})	Reference
Nakuru/Kenya	*Paradiaptomus africanus*	1165	67.20	0.05	Vareshi and Jacobs (1984, 1985)
Tana/Ethiopia	*Thermodiaptomus galebi*	21	1.04	0.05	Wondie and Mengistou (2006)
Brazil Reservoir	*Notodiaptomus iheringi*		8.08 mean 13.67 highest value		Santos-Wisniewski and Rocha (2007)

performance of soda lakes versus other systems. It can be concluded that calanoid daily production is highest in soda lakes but the *P/B* ratio is not. Introducing consumers such as fish could increase the productivity of the system, but apparently very few fish are adapted to survive in the harsh soda lake environment. This is one of the dilemmas of soda lakes!

8.5.1.2 Rotifera

The dominant rotifer species in the East African soda lakes are invariably *Brachionus dimidiatus*, *B. calyciflorus*, *B. angularis*, *B. urceolaris*, *Keratella tropica*, *Hexarthra jenkinae* and *Filinia longiseta* (Appendix). Recent studies on Rotifera from East Africa have focused on their feeding behaviour and their role in the food webs using stable isotope analyses (Burian et al. 2013, 2014).

More intensive studies on secondary production estimates were done earlier for two very abundant rotifer species in Lake Nakuru by Vareschi and Jacobs (1984, 1985). The growth increment method (Winberg et al. 1971) was used to estimate production of rotifer eggs and juveniles separately. Egg production was determined according to the equation $P = DMe \times Ne \times De^{-1}$ (mg DW m^{-3} day^{-1}), where *DM*e is the dry mass of eggs in mg, *N*e is the number of eggs per m^3 and *D*e is the duration of egg development in days. The authors determined De in laboratory experiments and obtained Ne from field data. Production of juvenile rotifers was calculated from $Pj = DMj \times Nf \times Df^{-1}$ (mg m^{-3} day^{-1} DM), where *DM*j is the dry mass of juveniles, *N*f is the number of females per m^3 and *D*f is an estimated life span. Rotifer life span was computed as the half-life of the mortality (*d*) of rotifers (= *Ln*2/*d*) between two successive time intervals, *t*1 and *t*2, which was obtained from the difference between the birth rate (*b*) and rate of population change (*r*). *b* was calculated as $b = \ln(1 + Ne \times De^{-1} \times Nf^{-1})$ and *r* as $(\ln N_2 - \ln N_1)/t2{-}t1$. It is unclear whether the growth increment method over- or underestimated secondary production, but it has recently been shown that the recruitment method of estimating secondary production is more appropriate for small species with small difference in biomass between eggs and adults, such as rotifers (Rodriguez and Tundisi 2002; Wondie and Mengistou 2014).

Results for daily production rates of *B. dimidiatus* and *B. plicatilis* were highly variable between the five years of investigation of Vareschi and Jacobs (1985; Table 8.7), indicating the dynamic nature of rotifer secondary production, which fluctuates erratically depending on temporal and spatial variations of food, temperature and biotic conditions. Overall, the larger species, *B. plicatilis*, was 3× more productive during the five study years. In recent studies using isotopic labelling, Burian et al. (2014) showed that *B. plicatilis* obtained almost half (48 %) of its diet from fragments of *A. fusiformis*, whereas *B. dimidiatus* mainly ingested particles < 2 μm. In other tropical freshwater lakes, we observed that rotifer production often peaks after some time lag of decomposition of cyanoprokaryote blooms, and we attributed this to their grazing on detrital particles and Bacteria (Assefa and Mengistou 2011; Wondie and Mengistou 2014). The poor quality of this detrital food was suggested as a possible factor for the low secondary production of rotifers in these freshwater lakes.

When compared with freshwater lakes, rotifer secondary production is clearly higher in soda lakes, even comparable to, and at times exceeding, production by calanoids and other grazers in the system. Vareschi and Jacobs (1985) argued that the main reasons for the high rotifer production are the large egg sizes and the short development times of eggs and juveniles. The latter are features common to many tropical rotifers, and perhaps a more plausible explanation for the high secondary production of large rotifers in soda lakes is the conducive high temperature and excess cyanoprokaryote food supply.

In general, secondary production by rotifers in soda lakes is higher by two orders of magnitude when compared with values in freshwater lakes, even taking computational differences into consideration (see Table 8.7). For example, comparing the secondary production of the smaller *B. dimidiatus* in Lake Nakuru with that of other brachionids in Lake Tana and Lake Kuriftu (both

Table 8.7 Daily production and mean biomass of Rotifera in Lake Nakuru during 1972 and 1976 and some other tropical African lakes (sources included)

Species/year/lake	Mean biomass (B) (mg DW m^{-3})	Daily P (mg DW m^{-3} day^{-1})	P/B (day^{-1})	Reference
B. dimidiatus				
1972	65	34	0.53	
1973	230	90	0.40	
1974	232	76	0.44	
1975	346	133	0.42	
1976	585	210	0.48	Vareschi and Jacobs (1985)
B. plicatilis				
1973	58	45	0.70	
1974	1 592	674	0.49	
1975	1 010	316	0.43	
1976	425	118	0.65	Vareschi and Jacobs (1985)
Brachionus spp.				
Lake Tana/Eth 2005	1.09	0.34	0.03	Wondie and Mengistou (2014)
B. calyciflorus				
Lake Kuriftu/Eth 2009	0.63	0.18	0.29	Assefa and Mengistou (2011)
Filinia pejleri				
Brazilian reservoir 2002	0.11	0.04	0.42	Rodriguez and Tundisi (2002)
Keratella americana				
Brazilian reservoir 2002	0.27	0.08	0.30	Rodriguez and Tundisi (2002)

Ethiopia) indicates that daily production in the freshwater lakes is < 1.00 mg, whereas an average of 100 mg m^{-3} day^{-1} DM was documented for the former in Lake Nakuru over a period of five years. Some of the reasons for the low rotifer production in the Ethiopian lakes were the low primary production in Lakes Tana and Kuriftu, and the low quality of the detrital food available for rotifers during the productive post-rainy months in these lakes (Assefa and Mengistou 2011; Wondie and Mengistou 2014). Rodriguez and Tundisi (2002) did not discuss the reasons for the extremely low rotifer production they calculated in a Brazilian reservoir.

Perhaps the daily biomass turnover rate (P/B ratio) is more insightful to explain the high daily production of rotifers in soda lakes. Close to 50 % of the rotifer biomass was replaced every day according to the 5-year study done by Vareschi and Jacobs (1985). This implies that half of the biomass is replaced daily due to the synergistic effects of food and temperature and lack of important rotifer predators. This is incredibly high when compared to a turnover rate of just 3 % in Lake Tana, which was attributed to factors mentioned above, such as poor food quality and low water temperature. A higher daily turnover rate of 30 % was documented in the small shallow Lake Kuriftu in Ethiopia (Assefa and Mengistou 2011). This reflected high primary production and water temperature, even though the food quality was still poor. In a tropical Brazilian reservoir, Rodriguez and Tundisi (2002) calculated quite high daily biomass turnover rates, closer to the values obtained in soda lakes (40 % per day) and higher than the values recorded for the Ethiopian shallow lakes. In general, the factors governing rates of secondary production and productivity (P/B) of rotifers in soda lakes need to be investigated in greater detail in the future.

8.5.2 Secondary Production of Macroinvertebrates in Soda Lakes

In view of the low species diversity of macroinvertebrates in soda lakes, less attention was paid to secondary production in this group. Nevertheless, the high seasonal biomass of some benthic macroinvertebrates during emergence has been noted by ecologists and others for centuries, most notably because of the spectacular impressions formed when millions or billions of insects emerge from lakes and literally "conquer" the surrounding environment for a short period.

8.5.2.1 Hemiptera (Bugs)

The best example of an extensive study on secondary production of macroinvertebrates is that of Vareschi and Jacobs (1985). They considered two families and three species for their study on secondary production and consumption rates in Lake Nakuru (Kenya). The species studied belonged to the phylum Arthropoda, class Insecta, order Hemiptera (bugs), family Corixidae (water boatmen) and family Notonectidae (backswimmers). Two corixid species were considered—*Micronecta scutellaris* and *Sigara hieroglyphica kilimanjaronis*—along with *Anisops varia* of the Notonectidae. Biomass and daily production were estimated for each species based on culturing animals in aquaria and noting weight increment with age (Mavuti 1975). Egg production was estimated from the egg ratio and sex ratio (1:1 assumed). The results indicated that the three macroinvertebrates had a very low daily secondary production of < 1.00 mg m^{-3} day^{-1} DM (Table 8.8) and made up no more than 1 % of the total consumer production in the lake. The daily biomass turnover rates of all three species were very low—less than 4 % of the biomass was replaced every day, reflecting very slow growth rates for all macroinvertebrates. This is in contrast to some chironomids, which can replace 15 % of their biomass per day and achieve annual biomass turnover rates of over 50 times, whereas the heteropterans could achieve a maximum annual P/B ratio of only 15 times (Benke 1998).

The low production and biomass turnover rate of these macroinvertebrates probably reflect the seasonally limited food supply. The notonectid *Anisops* is predatory and feeds on calanoids and even fish, and the corixids feed on *Spirulina*, other zooplankton and algae, which are highly seasonal, as well as on detritus in the lake.

8.5.2.2 Chironomidae (Non-biting Midges)

The most common observation in soda lakes is the phenomenal emergence of millions of flies from the lake at certain times of the year. They are feast for birds and other predators and fortunately do not transmit disease or act as vectors themselves. Several studies on the secondary production of chironomids have treated

Table 8.8 Biomass, daily production and *P*/*B* ratio of macroinvertebrates in Lake Nakuru during 1972 and 1973 (after Vareschi and Jacobs 1985)

Family/species/year	Biomass (mg DM m^{-3})	Daily production (mg DM m^{-3} day^{-1})	*P*/*B* ratio (day^{-1})
Notonectidae, *Anisops*			
1972	24.9	0.29	0.012
1973	13.8	0.07	0.005
Corixidae, *Micronecta*			
1972	7.9	0.30	0.038
1973	6.3	0.14	0.022
Corixidae, *Sigara*			
1972	0.22	0.01	0.014
1973	1.05	0.02	0.019

non-soda lakes (e.g. Benke 1998) because of the ubiquitous distribution of chironomids in lakes of different water chemistry and trophic status.

The highest secondary production of chironomids, and indeed of any inland benthic community, was reported from the saline Australian Lake Werowrap by Paterson and Walker (1974): 182 mg m^{-2} day^{-1} DM. The chironomid biomass and production in soda lakes are lower, with a maximum value of 120 mg m^{-2} day^{-1} DM and a minimum value of 1 mg m^{-2} day^{-1} DM for the dominant chironomid *Microchironomus deribae* in Lake Nakuru (Vareschi and Jacobs 1985). The authors cultured the chironomid in the laboratory and estimated its production based on growth increments of stages and size-frequency field data. They reported that, in general, chironomid production in the soda Lake Nakuru was high compared with temperate lakes but still lower than the peak values reported in the saline Australian lake. In very saline but less alkaline lakes, brine flies are quite common and can reach incredible numbers seasonally. In the soda lakes, chironomids did not reach such numbers, even though the detrital organic food in the lake sediment was by no means limiting. Benke (1998) reported that chironomids can reach extremely high biomass turnover rates in snag habitats of rivers, even as high as 15 % daily. Vareschi and Jacobs (1985) recorded a mean of 8.4 % daily P/B ratio for the chironomid in Lake Nakuru. In general, high secondary production and biomass turnover rates of chironomids have been documented from saline and soda lakes. It is thus interesting to investigate what factor(s) cause such differences. Could the salt difference contribute to this because food levels are apparently high in both cases?

8.6 Conclusions

- Soda lakes are populated by a few, yet highly specialized group of invertebrates which can assume high seasonal biomass. The species diversity of the invertebrate fauna is generally poor.
- The zooplankton is dominated by Rotifera. Especially the species *Brachionus dimidiatus* can attain incredibly high biomass (800 million m^{-3}) at favourable times.
- The macroinvertebrate community is dominated by corixids (genera *Micronecta* and *Sigara*) and notonectids (genus *Anisops*). The chironomid *Microchironomus deribae* is very abundant in many soda lakes.
- Vertebrates such as birds either feed directly on the benthic macroinvertebrates (Greater Flamingo) or filter the *Arthrospira* biomass in soda lakes. The ecotourism potential of soda lakes is extremely high because of the spectacular numbers of birds at the lakes at any one time.
- The secondary production of chironomids in alkaline soda lakes is much lower than that of saline Australian lakes, indicating some limitation for invertebrate production in the former.
- Soda lakes have simple food chains and are ideal as natural experimental models to study food web dynamics and energy flow using stable isotopes.
- The inclination to extract biological and chemical resources from soda lakes should be discouraged at best or should be exercised with extreme caution at worst.

References

Assefa E, Mengistou S (2011) Seasonal variation of biomass and secondary production of *Thermocyclops* (Cyclopoida) and *Brachionus* (Rotifera) spp. in a shallow tropical Lake Kuriftu, Ethiopia. SINET Ethiop J Sci 34(2):73–88

Ayenew T (2002) Recent changes in the level of Lake Abijata, central main Ethiopian Rift. Hydrol Sci J 47:493–503

Beadle LC (1932) Scientific results of the Cambridge expedition to the East African lakes. 1930–31. 4. The waters of some East African lakes in relation to their fauna and flora. J Linn Soc Zool 38:157–211

Benke W (1998) Production dynamics of riverine chironomids: extremely high biomass turnover rates of primary consumers. Ecology 79:899–910

Burian A, Schagerl M, Yasindi A (2013) Microzooplankton feeding behavior: grazing on the microbial and the classical food web of African soda lakes. Hydrobiologia 710:61–72

Burian A, Kainz MJ, Schagerl M, Yasindi A (2014) The role of rotifers in the food web of the saline-alkaline, hypereutrophic Kenyan Rift Valley Lake Nakuru: a snapshot-analysis of micro- and mesoplankton communities. Freshw Biol 59:1257–1265

Chemoiwa EJ, Oyoo-Okoth E, Mugo-Bundi J, Njerga EW, Matary EC, Korir RT, Njugi CC (2014) Elemental ratio (C:N) and stable isotope composition of dominant rotifer species in a tropical eutrophic alkaline-saline Lake Nakuru (Kenya). Hydrobiologia. doi:10.1007/s10750-014-2123-y

Clark F, Beeby A, Kirry P (1989) A study of macro-invertebrates of lakes Naivasha, Oloidien and Sonachi, Kenya. Rev Hydriobiol Trop 22(1):21–33

De Beauchamp P (1932) XIX. Reports on the Percy Sladen expedition to some Rift Valley lakes in Kenya in 1929. III. Rotifieres des lacs de la valee du Rift. Annu Nat Hist Ser 10 9(50):158–165

Eyualem A (2002) Free-living aquatic nematodes of the Ethiopian Rift Valley. In: Tudorancea C, Taylor WD (eds) Ethiopian Rift valley lakes. Backhuys Publishers, Leiden, pp 143–156

Grant WD (2006) Alkaline environments and biodiversity in extremophiles. UNESCO/Ecles Publishers, Oxford

Grant WD, Mwatha WE, Jones BE (1990) Alkaliphiles: ecology, diversity and applications. FEMS Microbiol Rev 75:255–270

Hammer TU (1986) Saline Lake ecosystems of the world. Dr. W. Junk, Dordrecht

Harper DM, Brooks Childress R, Harper MM, Boar RR, Hickley P, Mills SC, Otieno N, Drane T, Vareschi E, Nasirwa O, Mwatha WE, Darlington JPEC, Gasulla X (2003) Aquatic biodiversity and saline lakes: Lake Bogoria National Reserve, Kenya. Hydrobiologia 500:259–276

Harrison AD (2002) A general view on the communities of the Chironomidae adults in the Ethiopian Rift Valley. In: Tudorancea C, Taylor WD (eds) Ethiopian Rift valley lakes. Backhuys Publishers, Leiden, pp 157–162

Hecky RE, Kilham P (1973) Diatoms in alkaline, saline lakes: ecology and geochemical implications. Limnol Oceanogr 18:53–71

Hughes RH, Hughes JS (1992) A directory of African wetlands. IUCN, Gland

Jirsa F, Gruber M, Stojanovic A, Odour SO, Mader D, Körner W, Schagerl M (2013) Major and trace element geochemistry of Lake Bogoria and Lake Nakuru, Kenya, during extreme drought. Chem Erde-Geochem 73:275–282

LaBarbera MC, Kilham P (1974) The chemical ecology of copepod distribution in the lakes of East and Central Africa. Limnol Oceanogr 19:459–465

Martens K (2002) Ostracoda in the Ziway-Shala-Awasa basin. In: Tudorancea C, Taylor WD (eds) Ethiopian Rift valley lakes. Backhuys Publishers, Leiden, pp 163–166

Matagi SV (2004) A biodiversity assessment of the flamingo lakes of eastern Africa. Biodiversity 5(1):13–26

Mavuti K (1975) Some aspects of the biology of the genera *Sigara* and *Micronectes* from Lake Nakuru. M.Sc. Thesis, University of Nairobi

McClanahan TR, Young TP, McClanahan T (1996) East African ecosystems and their conservation. Oxford University Press, Oxford

Melack JM (1976) Saline and freshwater lakes of the Kenyan Rift Valley. In: McClanahan TR, Young TP (eds) East African ecosystems and their conservation. Oxford University Press, Oxford, pp 171–190

Melack JM, Kilham P (1974) Photosynthetic rates of phytoplankton in East African Alkaline, saline lakes. Limnol Oceanogr 19:743–755

Mutanga JG, Mwatha W, Nasirwa O, Gichuru N (2000) Status and trends in the biodiversity of Eastern Rift Valley lakes in Kenya. In: Bennun LA, Ndede H, Gichuki NN (eds) Conservation and sustainable use of biodiversity in Eastern Rift Valley lakes, Kenya, GEF/UNDP Consultancy Report. National Museums of Kenya, Nairobi, pp 6–34

Mwatha WE (1991) Microbial ecology of Kenyan Soda Lakes. Dissertation, University of Leicester, UK

Oduor SO, Schagerl M (2007) Phytoplankton photosynthetic characteristics in three Kenyan Rift Valley saline-alkaline lakes. J Plankton Res 29:1041–1050

Ong'Ondo GO, Yasindi AW, Oduor SO, Jost S, Schagerl M, Sonntag B, Boenig J (2013) Ecology and community structure of ciliated protists in two alkaline, saline rift valley lakes in Kenya, with emphasis on Frontonia. J Plankton Res 35:759–771

Oyoo-Okoth EM, Muchiri M, Ngugi CC, Njenga EW, Ngure V, Orina PS, Chemoiiwa EC, Wanjohi BK (2011) Zooplankton partitioning in a tropical alkaline-saline endorheic Lake Nakuru, Kenya. Spatial and temporal trends in relation to the environment. Lake Reserv Res Manage 16:35–47

Paterson CG, Walker KF (1974) Seasonal dynamics and productivity of *Tanytarsus barbitarsis* Freeman (Diptera: Chironomidae) in the benthos of a shallow, saline lake. Aust J Mar Freshw Res 25:151–165

Rodriguez MA, Tundisi T (2002) Rotifer production in a shallow artificial lake (Lobo-Broa reservoir, Brazil). Braz J Biol 62:509–516

Santos-Wisniewski MJ, Rocha O (2007) Spatial distribution and secondary production of Copepoda in a tropical reservoir: Barra Bonita, SP, Brazil Distribuição espacial e produção secundária de Copepoda em um reservatório tropical em Barra Bonita, SP, Brasil. Braz J Biol 67:223–233

Sarunday WN (1999) Conservation and sustainable use of biodiversity in Eastern Rift Valley lakes in Tanzania. National Environment Management Council, Dar es Salaam, Tanzania

Talling JF (1992) Environmental regulation in African shallow lakes and wetlands. Rev Hydrobiol Trop 25:87–144

Talling JF, Wood RB, Prosser MV, Baxter RM (1973) The upper limit of primary productivity by phytoplankton: evidence from Ethiopian soda lakes. Freshw Biol 3:53–76

Tudorancea C, Taylor WD (2002) Ethiopian Rift Valley Lakes. Backhuys Publishers, Leiden

Uku JN, Mavuti KM (1994) Comparative limnology, species diversity and biomass relationship of zooplankton and phytoplankton in five freshwater lakes in Kenya. Hydrobiologia 272:251–258

Vareschi E, Jacobs J (1984) The ecology of lake Nakuru (Kenya) V. Production and consumption of consumer organisms. Oecologia 61:83–98

Vareschi E, Jacobs J (1985) The ecology of Lake Nakuru (Kenya) VI. Synopsis of production and energy flow. Oecologia 65:412–424

Vareschi E, Vareschi A (1984) The ecology of Lake Nakuru (Kenya) IV. Biomass and distribution of consumer organisms. Oecologia 61:70–82

Verschuren D, Cocquyt C, Tibby J, Roberts CN, Leavitt PR (1999) Long-term dynamics of algal and invertebrate communities in a small, fluctuating, tropical soda lake. Limnol Oceanogr 44:1216–1231

Winberg GG, Patalas K, Wright JC, Hillbricht-Ilkowska-A, Cooper WE, Mann KH (1971) Methods for calculating productivity. In: Edmondson WT, Winberg GG (eds) A manual on methods for the assessment of secondary productivity in fresh waters, IBP handbook no. 17. Blackwell Scientific Publications, Edinburgh, pp 296–317

Wondie A, Mengistou S (2006) Duration of development, biomass and rate of production of the dominant copepods (calanoida and cyclopoida) in Lake Tana, Ethiopia. SINET Ethiop J Sci 29(2):107–122

Wondie A, Mengistou S (2014) Seasonal variability of secondary production of Cladocerans and Rotifers and their trophic role in a large, turbid, tropical highland lake (Lake Tana, Ethiopia). Afr J Aquat Sci 39 (4):403–416

Wood RB, Talling JF (1988) Chemical and algal relationships in a salinity series of Ethiopian inland waters. Hydrobiologia 158:29–67

Yasindi AW, Taylor WD (2006) The trophic position of planktonic ciliate populations in the food web of some East African lakes. Afr J Aquat Sci 3(1):53–62

Fish Populations in East African Saline Lakes

9

Geraldine D. Kavembe, Axel Meyer, and Chris M. Wood

Abstract

The biodiversity of saline lakes is a topic of increasing interest among biologists, ecologists and conservation managers in East Africa. In spite of their extreme conditions and remote locations, East African saline lakes (EASL) support fish populations of considerable ecological, economic and biological importance. Among these are several endemic fish species that are highly specialized to survive in their individual lakes. Although there is growing concern that increasing human activities and projected adverse climatic conditions in the region may decimate these unique species, information on the status of individual fish populations remains scarce. In recognition of the important ecosystem services they provide, the EASL have been designated as World Heritage Sites (WHS) and protected by the Ramsar Convention. To complement these conservation efforts regarding EASL and to ensure full realization of the potential of their fisheries, there is need for up-to-date information on their fish population status. We present therefore an overview of the status of fish populations in EASL with special emphasis on the Magadi tilapia *Alcolapia grahami*, a teleost fish thriving in extreme hypersaline alkaline water conditions that would kill other fish in a matter of minutes. We show how several decades of research on this small cichlid fish inhabiting Lake Magadi, Kenya, reveal astonishing "snapshots" on how fish can survive under challenging environmental conditions, which, in this rare instance, are actually close to their physiological optima. As climatic models predict a decline in freshwater sources and an increment in adverse

G.D. Kavembe (✉)
Department of Biology, South Eastern Kenya University,
P.O. Box 170-90200 Kitui, Kenya
e-mail: dkavembe@seku.ac.ke

A. Meyer
Department of Biology, University of Konstanz,
Universitätsstrasse 10, 78457 Konstanz, Germany
e-mail: axel.meyer@uni-konstanz.de

C.M. Wood
Department of Zoology, University of British Columbia,
Vancouver, BC, CanadaV6T 1Z4
e-mail: woodcm@zoology.ubc.ca

M. Schagerl (ed.), *Soda Lakes of East Africa*, DOI 10.1007/978-3-319-28622-8_9

water conditions, studies on fish inhabiting saline lakes could aid scientists in modelling how species may evolve to adapt to extreme conditions in their changing habitats. We highlight conservation challenges facing the long-term existence of EASL fish populations. Finally, an integrated multidisciplinary approach is recommended to ensure the preservation and sustainable management of EASL fish populations and fisheries.

9.1 Introduction

Worldwide, the fauna of saline lakes are often neglected in research endeavours, management initiatives and conservation efforts. Fish communities in East African saline lakes (EASL) (Fig. 9.1) are no exception in this respect. Nevertheless, the EASL harbour several unique and endemic fish communities (Fig. 9.2 and Table 9.1) of great economic, aesthetic, ecological and biological value. Unlike their counterparts in freshwater lakes, fish communities in saline lakes remain largely unaltered by species introductions, overstocking and overexploitation (Melack 1996). These populations could thus provide a rare opportunity

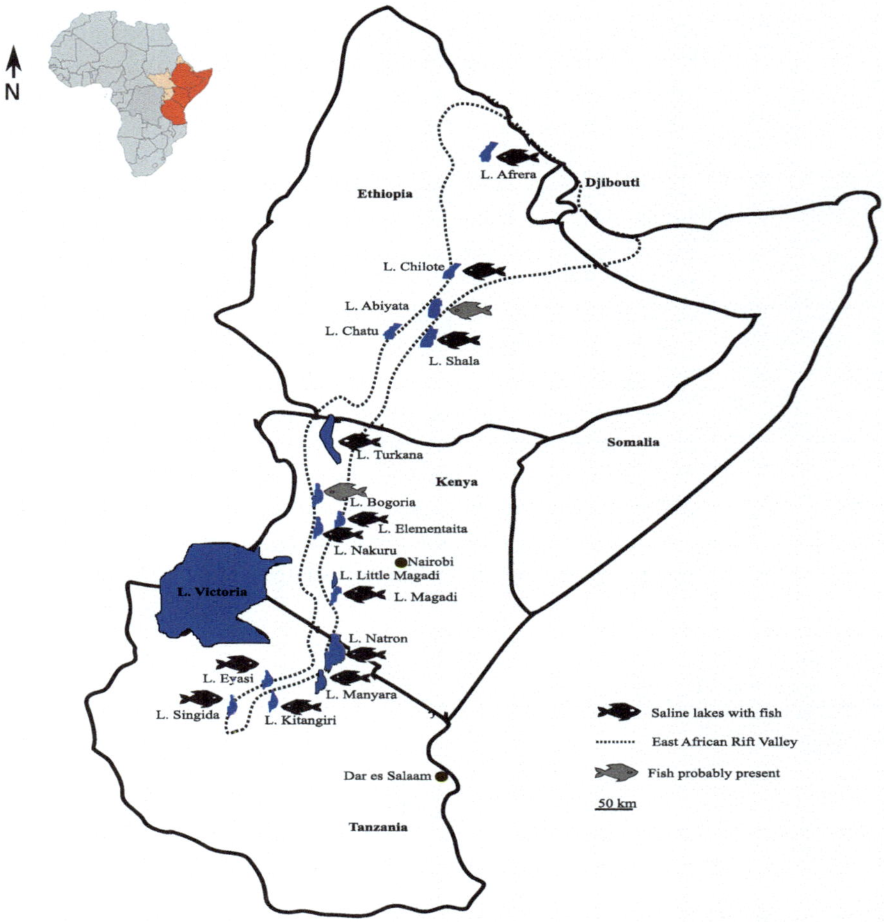

Fig. 9.1 Distribution of fish in East African saline lakes (EASL)

Fig. 9.2 Representative species of fish from East African saline lakes: (**a**) *Alcolapia grahami* from the southern end of Lake Magadi; (**b**) *Alcolapia alcalicus* from Shompole swamps, north of Lake Natron; (**c**) *Alcolapia* aff. *alcalica*, southwestern lagoon of Lake Natron; (**d**) *Alcolapia ndalalani* from Olomotony close to Ndalalani, southern Lake Natron; (**e**) *Alcolapia latilabris* from Olomotony, spring-fed creek, southern affluent of the Southern Lagoon, north of Ol Doinyo Lengai volcano, Lake Natron; (**f**) *Oreochromis amphimelas* from Mto wa Mbu, northern affluents of Lake Manyara; (**g**) *Oreochromis niloticus vulcani*, from springs at Loyangalani, eastern shore of Lake Turkana; (**h**) *Haplochromis rudolfianus* from south of Loyangalani, Lake Turkana; (**i**) *Hemichromis exsul* from south of Loyangalani, eastern shore of Lake Turkana; (**j**) *Clarias gariepinus* from Mto wa Mbu, northern affluents of Lake Manyara. Photographs kindly provided by Dr. Lothar Seegers

Table 9.1 Fish species in saline lakes of East Africa

Family	Species name	Occurrence	Status	Saline lake	Country
Cyprinidae	*Afronemacheilus kaffa*	Native	NE	Lake Turkana	Ethiopia
Cichlidae	*Alcolapia alcalicus* (Hilgendorf 1905)	Native	EN	Lake Natron	Tanzania
Cichlidae	*Alcolapia grahami* (Boulenger 1912)	Native	VU	Lake Magadi	Kenya
Cichlidae	*Alcolapia ndalalani* (Seegers and Tichy 1999)	Native	NE	Lake Natron	Tanzania
Cichlidae	*Alcolapia latilabris* (Seegers and Tichy 1999)	Native	NE	Lake Natron	Tanzania
Alestidae	*Alestes baremoze*	Native	LC	L. Turkana	Kenya, Ethiopia
Alestidae	*Alestes dentex*	Native	LC	L. Turkana	Kenya, Ethiopia
Poeciliidae	*Aplocheilichthys jeanneli*	Native	LC	L. Turkana	Kenya, Ethiopia
Poeciliidae	*Aplocheilichthys rudolfianus* (syn. *Haplochilichthys rudolfianus*/ *Micropanchax rudolfianus*)	Native	LC	L. Turkana	Kenya, Ethiopia
Claroteidae	*Auchenoglanis occidentalis*	Native	LC	L. Turkana	Kenya, Ethiopia
Bagridae	*Bagrus bajad*	Native	LC	L. Turkana	Kenya, Ethiopia
Bagridae	*Bagrus docmak*	Native	LC	L. Turkana	Kenya, Ethiopia
Cyprinidae	*Barbus bynni*	Native	LC	L. Turkana	Kenya, Ethiopia
Cyprinidae	*Barbus neumayeri*	Native	LC	L. Turkana	Kenya, Ethiopia
Cyprinidae	*Barbus turkanae*	Native	LC	L. Turkana	Kenya, Ethiopia
Cyprinidae	*Lebias stiassnyae* (Getahun and Lazara 2001)	Native	EN	L. Alfdera	Ethiopia
Alestidae	*Brycinus ferox*	Native	LC	L. Turkana	Kenya, Ethiopia
Alestidae	*Brycinus minutus*	Native	LC	L. Turkana	Kenya, Ethiopia
Alestidae	*Brycinus nurse*	Native	LC	L. Turkana	Kenya, Ethiopia
Cyprinidae	*Chelaethiops bibie*	Native	LC	L. Turkana	Kenya, Ethiopia
Claroteidae	*Chrysichthys auratus*	Native	LC	L. Turkana	Kenya, Ethiopia
Claroteidae	*Chrysichthys turkana*	Endemic	NE	Lake Turkana	Kenya, Ethiopia
Citharinidae	*Citharinus citharus*	Native	NE	Lake Turkana	Kenya, Ethiopia
Clariidae	*Clarias gariepinus*	Native	NE	Lake Turkana, Lake Manyara, Lake Bogoria drainage[a]	Kenya, Ethiopia
Distichodontidae	*Distichodus nefasch*	Native	NE	Lake Turkana	Kenya, Ethiopia

(*continued*)

Table 9.1 (continued)

Family	Species name	Occurrence	Status	Saline lake	Country
Gymnarchidae	*Gymnarchus niloticus*	Native	LC	Lake Turkana	Kenya, Ethiopia
Cichlidae	*Danakilia franchettii* (Trewavas 1983)	Native	EN	Lake Afdera	Ethiopia
Cichlidae	*Haplochromis macconneli*	Native	LC	Lake Turkana	Kenya, Ethiopia
Cichlidae	*Haplochromis rudolfianus*	Native	LC	Lake Turkana	Kenya, Ethiopia
Cichlidae	*Haplochromis turkanae*	Native	LC	Lake Turkana	Kenya, Ethiopia
Cichlidae	*Hemichromis exsul*	Native	LC	Lake Turkana	Kenya, Ethiopia
Cichlidae	*Hemichromis letourneuxi*	Native	LC	Lake Turkana	Kenya, Ethiopia
Clariidae	*Heterobranchus longifilis*	Native	LC	Lake Turkana	Kenya, Ethiopia
Arapaimidae	*Heterotis niloticus*	Native	LC	Lake Turkana	Kenya, Ethiopia
Alestidae	*Hydrocynus forskahlii*	Native	LC	Lake Turkana	Kenya, Ethiopia
Alestidae	*Hydrocynus vittatus*	Native	LC	Lake Turkana	Kenya, Ethiopia
Mormyridae	*Hyperopisus bebe*	Native	LC	Lake Turkana	Kenya, Ethiopia
Cyprinidae	*Labeo cylindricus*	Native	LC	Lake Bogoria drainage[a], Lake Turkana	Kenya, Ethiopia
Cyprinidae	*Labeo horie*	Native	NE	Lake Turkana	Kenya, Ethiopia
Cyprinidae	*Labeo niloticus*	Native	LC	Lake Turkana, Lake Chilotes (Hora Kilole)	Kenya, Ethiopia
Cyprinidae	*Labeobarbus intermedius*	Native	LC	Lake Turkana, Lake Bogoria drainage[a]	Kenya, Ethiopia
Poeciliidae: Procatopodinae	*Lacustricola jeanneli*	Native	NE	Lake Turkana	Kenya, Ethiopia
Latidae	*Lates longispinis*	Native	DD	Lake Turkana	Kenya, Ethiopia
Latidae	*Lates niloticus*	Native	LC	Lake Turkana	Kenya, Ethiopia
Malapteruridae	*Malapterurus electricus*	Native	LC	Lake Turkana	Kenya, Ethiopia
Lestidae	*Micralestes elongatus*	Native	NE	Lake Turkana	Kenya, Ethiopia
Mochokidae	*Mochokus niloticus*	Native	LC	Lake Turkana	Kenya, Ethiopia
Mormyridae	*Mormyrus kannume*	Native	LC	Lake Turkana	Kenya, Ethiopia
Cyprinidae	*Neobola bottegoi*	Native	DD	Lake Turkana	Kenya, Ethiopia
Cyprinidae	*Neobola stellae*	Native	LC	Lake Turkana	Kenya, Ethiopia

(*continued*)

Table 9.1 (continued)

Family	Species name	Occurrence	Status	Saline lake	Country
Cichlidae	*Oreochromis niloticus*	Native	NE	Lake Turkana, Lake Chilotes (Hora Kilole)	Kenya, Ethiopia
Cichlidae	*Oreochromis esculenta*	Native	CE	Lake Singida, Lake Kitangiri	Tanzania
Cichlidae	*Oreochromis niloticus vulcani*	Native	NE	Lake Turkana	Kenya, Ethiopia
Cichlidae	*Oreochromis spilurus*	Native	NE	Lake Turkana	Kenya, Ethiopia
Polypteridae	*Polypterus bichir*	Native	NE	Lake Turkana	Kenya, Ethiopia
Polypteridae	*Polypterus senegalus*	Native	NE	Lake Turkana	Kenya, Ethiopia
Protopteridae	*Protopterus aethiopicus aethiopicus*	Native	NE	Lake Turkana	Kenya, Ethiopia
Cyprinidae	*Raiamas senegalensis*	Native	LC	Lake Turkana	Kenya, Ethiopia
Cichlidae	*Sarotherodon galilaeus*	Native	NE	Lake Turkana	Kenya, Ethiopia
Schilbeidae	*Schilbe uranoscopus*	Native	LC	Lake Turkana	Kenya, Ethiopia
Mochokidae	*Synodontis frontosus*	Native	LC	Lake Turkana	Kenya, Ethiopia
Mochokidae	*Synodontis schall*	Native	LC	Lake Turkana	Kenya, Ethiopia
Tetraodontidae	*Tetraodon lineatus*	Native	LC	Lake Turkana	Kenya, Ethiopia
Cichlidae	*Tilapia zillii*	Native	NE	Lake Turkana	Kenya, Ethiopia
Cyprinidae	*Barbus intermedius*	Native	NE	Lake Abijata, Lake Shala	Kenya, Ethiopia
Cyprinidae	*Barbus paludinosus*	Native	LC	Lake Turkana, Lake Abijata, Lake Shala	Ethiopia
Clariidae	*Clarias gariepinus*	Native	NE	Lake Turkana, Lake Abijata, Lake Shala	Ethiopia
Cichlidae	*Aplocheilichthys antinorii*	Native	LC	Lake Turkana, Lake Abijata, Lake Shala	Ethiopia
Cichlidae	*Tilapia rendalli*	Native	LC	Lake Turkana, Lake Abijata[b], Lake Shala[b], Lake Kitangiri[b]	Ethiopia Kenya Tanzania
Cichlidae	*Oreochromis amphimelas* (Hilgendorf 1905)	Native	EN	Lake Manyara, Lake Kitangiri, Lake Eyasi and Lake Singida	Tanzania
Clariidae	*Clarias gariepinus*	Native	NE	Lake Manyara[b], Lake Chilotes (Hora Kilole)[b]	Tanzania, Ethiopia

EN endangered, *CN* critically endangered, *LC* least concern, *VU* vulnerable, *NE* not evaluated, *DD* data deficient
[a]Occasional occurrence of the species
[b]Introduced species. Data modified from Seegers et al. (2003)

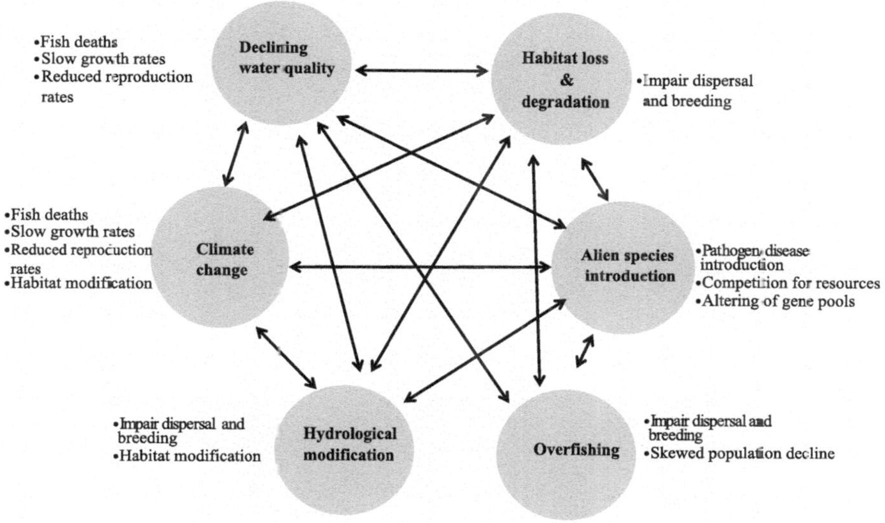

Fig. 9.3 Major threats and their impacts on saline lake fish communities in East Africa

to conserve fish communities in their original abundance and diversity. In recognition of the significance of saline lakes for in situ conservation of both threatened and endangered species, four Kenyan Rift Valley saline lakes (Lakes Turkana, Nakuru, Elementaita and Bogoria) are listed as World Heritage Sites (WHS) (http://whc.unesco.org/en/list/1060). Additionally, these lakes are protected by the Ramsar Convention (signed in Ramsar, Iran, in 1971) on Wetlands of International Importance (http://www.ramsar.org/pdf/sitelist.pdf), as well as by various government initiatives in close collaboration with local communities (Harper et al. 2003). In spite of these initiatives, some economically driven developments in and around the lakes such as mining, geothermal exploration and farming continue to exert considerable pressure on saline lake fish populations. Negative impacts associated with such activities include: habitat loss and fragmentation, pollution and introduction of pathogens and parasites (Figs. 9.3 and 9.4) (Jellison et al. 2004; Williams 2002; Zinabu 2002).

There is growing concern that, if not abated, current trends of human-mediated degradation of saline lakes could lead to massive species losses. Since saline lakes have been less studied, the magnitude of such losses may even be greater as they may affect species whose existence has not been documented (Jellison 2005). Additionally, the ecological and potential financial impacts of such losses may be irreversible (Constanza 1997; Harper et al. 2003). To strike a balance between sustainable economic development, scientific research and conservation of EASL fish populations, there is a need for scientists, conservation managers, governments, local communities, industry and other stakeholders to work together closely. New paradigms that promote collaborative efforts could indeed go a long way in averting many threats facing saline lake fisheries. To date, no such framework has been established for EASL. In fact, very little information exists on saline lake fish distribution, abundance, life history, stocking levels, threats and conservation status. Moreover, except for fish populations in a few lakes such as Lakes Magadi and Turkana, much of the current knowledge about fish communities in EASL lies with local residents, fishermen or in grey literature such as museum records and institutional reports.

A comprehensive review of this subject is important; here, we highlight the most important aspects of saline lake fish populations. This

Fig. 9.4 A local Maasai woman washes clothes in Fish Spring Lagoon of Lake Magadi, Kenya. The lagoon holds the largest fish population (*Alcolapia grahami*) relative to the other lagoons in Lake Magadi. Pressure on the available water resources in the lagoon and Lake Magadi as a whole is mounting as a result of competition for domestic, agricultural (livestock) and industrial use

chapter therefore (1) provides a brief overview of saline lakes of East Africa as far as their importance to fish and fisheries is concerned, (2) draws attention to the potentials of saline lake fishes, (3) reviews published information on saline lake fish populations in East Africa with special emphasis on the Magadi tilapia and (4) outlines challenges facing EASL fisheries and suggestions for better conservation and sustainable use.

9.2 East African Saline Lakes

The EASL consist of a series of alkaline-saline lakes located along the eastern arm of the African Great Rift Valley, which stretches from north-eastern Ethiopia through Kenya to Tanzania (Fig. 9.1) (Hughes and Hughes 1992; Melack 1996). The Ethiopian Rift Valley has Lakes Abijata, Shala, Alfdera and Chilotes (Grant 2004). The Kenyan Rift Valley contains five major saline lakes of varying salinity levels. In the order of increasing salinity they include Lakes Turkana, Bogoria, Nakuru, Elementaita and Magadi. Lake Turkana is the world's largest desert lake and the fourth largest lake in Africa after the Great Lakes (Lake Victoria, Tanganyika and Malawi) (Ferguson and Harbott 1982). In the Tanzanian Rift Valley, the main saline lakes are Lakes Eyasi, Manyara, Natron, Singida and Kitangiri (Hughes and Hughes 1992). Other small saline lakes are also present in the region (see Appendix). A comprehensive list of these and other saline lakes of the world is provided in Grant (2004). EASL are endorheic, and high temperatures associated with the Rift Valley lead to high evaporation rates of water, resulting in high concentrations of Na^+, HCO_3^- and CO_3^{2-} salts (Cole 1994). The highly alkaline nature of surrounding soils further adds to the saline levels of the EASL (Cole 1994). Unlike their freshwater counterparts, most saline lake basins along the Rift Valley do not constitute a single continuous mass of water but rather a series of isolated pools (Seegers et al. 1999). EASL are mainly recharged by direct precipitation and underground seepage facilitated by hot springs located along the margins or at the floor of the lake basins. In exceptional cases, the lakes may

Fig. 9.5 Schematic diagram of the simple food web in Lake Magadi. Each *arrow* represents a transfer of food or energy from one organism to another. *Arrows point* to the consumer. *Thick arrows* depict the food resources that are consumed by Magadi fish. Similar webs are present in the other EASL

acquire water through surface runoff and riverine delivery by affluent streams and rivers (Goerner et al. 2009).

As most of the EASL lack obvious outflows, they are considered as "amplifier lakes" for their dramatic response to climatic variability (Olaka et al. 2010; Street-Perrott and Roberts 1983). Recent studies have revealed extensive shifts in climatic patterns and accompanying changes in hydrology of most of these lakes following the onset of Holocene period ~13,000 years BP (Issar 2003; Legesse et al. 2002). These shifts have led to marked variation in sizes, salinity, temperature regimes, UV irradiation penetration, dissolved oxygen concentration, ionic composition and biotic assemblages of the lakes (Melack 1996; Oduor and Schagerl 2007; Verschuren 1996; Verschuren et al. 2000). Saline lakes are well known for their high primary productivity and large microbial and invertebrate diversity (Grant 2004; Harper et al. 2003). Nonetheless, their extreme conditions coupled with their rather simple food webs (Fig. 9.5) limit the abundance, composition and distribution of vertebrate species (Herbst 2001; Vareschi 1979; Williams et al. 1990). EASL are of particular interest for fish biologists because they represent all the described salinity classes (Cooper and Wissel 2012). As such, these lakes provide excellent opportunities to study many aspects of fish biology along the salinity continuum.

9.3 Fish Communities in East African Saline Lakes

Generally, compared to invertebrates, fish faunas in saline lakes are limited (Brauner et al. 2013). This has been attributed to lack of direct water links to other basins, which in turn limit the dispersal ability of their fish populations relative to invertebrates: the latter can disperse over land, mainly facilitated by dispersal vectors such as wind and migratory birds (Brauner et al. 2013; Frisch et al. 2007; Hammer 1986). Therefore, in the absence of human intervention, dispersal of fish relies entirely on rare chances such as during floods. This, however, may occur only in a few lakes (e.g. Lakes Natron and Turkana) that have temporary connections to rivers or streams. Species richness and abundance of saline fishes is therefore correlated to dispersal opportunities (Brauner et al. 2013), food availability and the physical complexity of individual lakes (Hammer 1986).

More often than not, less alkaline lakes (salinity 3–5 ‰ e.g. Lake Turkana, Kenya) have more species whereas hypersaline lakes (salinity ~21 ‰, e.g. Lakes Magadi and Little Magadi, Kenya) have fewer species (Hammer 1986; Seegers et al. 2003). The negative correlation of salinity to fish diversity has been reported in several other studies (Cooper and Wissel 2012; Sosa-López et al. 2007). Owing to the intimate physiological relationship fish have with their environment (Cossins and Crawford 2005), saline lake fishes are globally significant as biological models to study adaptation and response to a wide variety of natural and anthropogenic environmental conditions.

As EASL are geographically isolated and often lack inlets and outlets, the origins of the initial fish populations that seeded the lakes are not obvious. This often makes it difficult to state with certainty whether populations in individual lakes are natural or have even been stocked by humans in ancient times (Brauner et al. 2013; Hammer 1986). Nevertheless, due to relatively long periods of isolation, most saline lake fish faunas are evolutionarily quite divergent from their closest relatives in freshwater and have evolved unique adaptations to subsist in their individual lakes. Consequently, saline lake fish exhibit varied morphological, behavioural, physiological and ecological adaptations. Additionally, phenotypic diversifications have also occurred among populations of the same species within individual lakes as a result of occupying diverse environments (Seegers and Tichy 1999; Tichy and Seegers 1999). This may in turn lead to fish populations showing substantial intraspecific or interspecific differentiation even in the presence of gene flow (Zaccara et al. 2014). This has been reported for Lake Natron, Tanzania, where the lake's three morphologically distinct species, *Alcolapia alcalicus*, *A. ndalalani* and *A. latilabris*, exist in sympatry (Seegers and Tichy 1999). The most prevalent species *A. alcalicus* displays considerable intraspecific morphological variation (Seegers and Tichy 1999; Tichy and Seegers 1999). Nevertheless, genetic analyses (mtDNA and microsatellites) have revealed considerable gene flow among these species and morphotypes (Seegers et al. 1999; Zaccara et al. 2014).

Owing to their characteristic restricted ranges, small population sizes and stressful habitats, fish species in EASL are thought to be at a great risk of extinction. Model studies in other saline lakes of the world have predicted a decline in fish populations as a result of unprecedented environmental threats to their ecosystems (Jellison et al. 2004). This has been attributed to a decline in annual precipitation leading to deterioration of water quality and contraction of habitats, mainly driven by anthropogenic activities (Jellison et al. 2004).

9.3.1 Fish Species of Ethiopia's Saline Lakes

Although little is known about the fish communities in Ethiopian saline lakes, there have been suggestions that the lakes may hold a few endemic species (Table 9.1). The cichlid *Danakilia franchettii* (Trewavas 1983) and cyprinodontid *Lebias stiassnyae* (Getahun and Lazara 2001) are endemic to Lake Alfdera (Getahun and Stiassny 1998; Golubtsov et al. 2002). Lake Chilotes (also referred to as Hora Kilole) has an indigenous *Oreochromis niloticus* population. However, following recent diversions of freshwater from River Mori into Lake Chilotes, several riverine species (mainly *Barbus*) have colonized the lake within the last decade (Brook 2003).

9.3.2 Fish Species of Kenya's Saline Lakes

Compared to all other EASL, Lake Turkana has the highest number of species (Table 9.1). Slightly over 50 species have been described, of which 30 % are endemic to the lake (Table 9.1) (Hopson 1982; Kolding 1989; Seegers et al. 2003). The non-endemic species are derived from the Nile drainage (Kolding 1989, 1995) as a result of past connections of the lake basin to the River Nile (Johnson and Malala

2009). The high diversity of fishes in Lake Turkana is attributed to its mild saline conditions, a well-mixed water column and well-oxygenated waters among other factors (Ferguson and Harbott 1982; Kallqvist et al. 1988). Recent studies have, however, suggested a decline in the lake's fish populations due to fishing pressures (Muška et al. 2012) and fluctuations in water levels (Kolding 1995). In contrast to Lake Turkana, the other saline lakes of Kenya have very limited fish faunas. The hypersaline Lake Magadi contains only a single species, the endemic tilapia *Alcolapia grahami*. The Magadi tilapia was introduced to Lake Nakuru between 1952 and 1962 to combat mosquito larvae (Vareschi 1979) and probably also in Lake Elementaita (Okeyo 2006). Massive deaths of the introduced fish, however, were reported in 1991 for unknown reasons (Githaiga 1997). In a recent expedition (March, 2010) to Lake Elementaita, a thriving population of tilapia that resembled *Alcolapia grahami* but with slightly bigger body sizes were found (GD Kavembe, personal observation). The current population status of *Alcolapia grahami* species in Lake Nakuru remains unknown. Although Lake Bogoria does not have endemic species of its own (Harper et al. 2003), fish from affluent streams have been found to "stray" into the lake following heavy rains that make the water less saline.

9.3.3 Fish Species of Tanzania's Saline Lakes

Lake Manyara is the most important of the Tanzanian saline lakes in terms of fisheries. The lake is dominated by the endemic Lake Manyara tilapia *Oreochromis amphimelas*. The species is also found in other soda lakes in Tanzania such as Lakes Kitangiri, Eyasi and Singida (Froese and Pauly 2014). A sharp decline of fish catches in Lake Manyara has been recorded, from a high of 1800 t in the 1970s to a low of 0.5 t in 1990 (Mugisha et al. 1993). The decline was attributed to a drought that wiped out most of the populations of *O. amphimelas* (Mugisha et al. 1993). Nevertheless, more recent surveys suggest that Lake Manyara still supports a significant population of *O. amphimelas* (Bwathondi et al. 2000). Information on the status of this species in other Tanzanian saline lakes is scanty (Bwathondi et al. 2000). Catfish (*Clarias gariepinus*) have been reported in Lake Manyara (Yanda and Madulu 2005) and Lake Kitangiri (Bwathondi 2002). *Oreochromis esculentus* and *Tilapia rendalli* were introduced from Lake Victoria to Lakes Kitangiri and Singida (Thieme et al. 2005) (Table 9.1). Bwathondi (2002) reported sightings of *Protopterus aethiopicus* and *Oreochromis niloticus* as well as some unspecified haplochromine species in Lakes Kitangiri and Singida, respectively. Lake Natron, which displays extreme conditions (similar to but less extreme than those of Lake Magadi), is well known for its three endemic soda lake tilapia of the genus *Alcolapia*: *A. alcalicus*, *A. latilabris* and *A. ndalalani*. During the wet season, several riverine fish species have been reported to stray into Lake Natron from rivers (Ewaso Nyiro, Peninj, Moinik and Engare Sero) and seasonal streams that feed the lake (Seegers and Tichy 1999).

9.4 Potential Uses and Value of Saline Lake Fishes

9.4.1 Aquaculture Value

Compared to freshwater species, the fisheries sector of saline lakes is poorly developed and only scant information is available on the actual exploitation of the fish of these lakes. The global shortage of freshwater brought about by climate change and competition for the available freshwater resources by the ever-growing human populations has increased the need to develop aquaculture in brackish, saline and wastewaters (Sala et al. 2000; El-Zaeem et al. 2012, 2013). The need for such initiatives is even greater in East Africa, where unprecedented rates of population growth, often in combination with natural and human-mediated stressors, continue to put strain on available freshwater resources. Fish faunas of saline lakes could undoubtedly offer one of the solutions to dwindling freshwater fisheries. Apart from direct benefits from saline lake fish catches, saline lake fish stocks could also be used in breeding programmes for genetic improvement of common freshwater species. For instance, the freshwater tilapia *Oreochromis niloticus* is one of the world's leading aquaculture species. Like most teleost species, however, it excretes ammonia as its major nitrogenous waste, and ammonia buildup can cause sublethal toxicity manifested as impaired growth when water supplies are limited (El-Shafai et al. 2004). Possibly, it could be crossed with Magadi tilapia, which produces only urea, a much less toxic nitrogenous waste product (see below). For such attempts to be successful, there is a need to establish well-planned programmes including extensive research on the species biology, compatibility and prior assessment of the negative effects such programmes may have on the long-term survival of either species. Similar efforts using other stress-tolerant tilapia to improve common tilapia species have yielded impressive results (Kamal and Mair 2005).

9.4.2 Scientific and Educational Value

Over the past decades, many studies have examined fish species living in extreme environments. For instance, several studies are available on adaptation to hypersaline, exceptionally alkaline conditions in Magadi tilapia (*A. grahami*) (reviewed by Pörtner et al. 2010) and to hydrogen-sulphide-contaminated caves in the Shortfin Molly fish *Poecilia mexicana* (Tobler and Plath 2011). Global climate change scenarios predict a steady increase of salinity in freshwater ecosystems in East Africa. Saline lakes and their fish populations offer a natural laboratory for multidisciplinary studies that may provide much-needed management solutions. The simplicity of saline lake ecosystems and the less species-rich nature of their communities are among the aspects that make these systems good models for biological studies.

9.4.3 Cultural, Aesthetic and Economic Values

Most EASL are located in remote dry areas and, as such, they are the main source of livelihood for local communities. These communities are predominantly pastoralist, but dwindling land resources and declining annual precipitation have resulted in decreased livestock productivity. For instance, over 200,000 indigenous pastoralists and agro-pastoralists are estimated to depend entirely on Lake Turkana (Avery 2010). If exploited, fish populations could provide a complementary resource to improve food security of such communities. Almost inseparable from other values associated with the fish populations is the contribution of fish faunas to the economy of their respective countries. This could either be in the form of direct contributions from fish sales, proceeds from tourism or from indirect contributions such as increased local expenditures associated with research activities. To maximize returns from the few saline fish species and to ensure conservation of their biodiversity, there have been suggestions to develop recreational fisheries as opposed to commercial fisheries. For instance, in Issyk-Kul Lake in Kazakhstan, several studies have proposed that recreational fishing would impart less fishing pressure but yield major financial gains compared to industrial fishing (Alamanov and Mikkola 2011). The economically successful sport fishery for endemic Cutthroat Trout in saline, alkaline Lake Lahontan in Nevada, USA, managed by an aboriginal community, is a good example of this strategy (Coleman and Johnson 1988).

9.4.4 Ecological Value

Although it is the most difficult to quantify, the ecological value of saline lake fish is highly significant. The fish constitute an irreplaceable component of the saline lake food chain. For instance, the fish support significant populations of fish-eating birds (Vareschi 1979).

Additionally the fish prey on some saline lake organisms such as mosquito and dipteran larvae, as well as on other fish, invertebrates and microorganisms such as Cyanobacteria (Coe 1966; Vareschi 1979). Most importantly, most of these fish are endemic and adapted to specific lakes; hence, each single species represents an invaluable genetic resource that should be conserved.

9.5 Current Research on Fish Populations of Saline Lakes in East African Rift Valley

Although research activity is increasing, fish in EASL have attracted relatively limited attention from scientists, and their fisheries sectors are poorly developed, relative to freshwater fisheries (Kolding 1995; Muška et al. 2012). This may be attributed to several factors. First, there is generally a low awareness among relevant stakeholders on the importance of saline fish populations, particularly their potential in aquaculture production. Secondly, the remote geographical location of saline lakes and their associated adverse conditions may limit access to the lakes' fish populations by scientists. Lastly, research on saline lake fish may require sophisticated methods and equipment, particularly in cases where on-site or laboratory manipulations of populations are required. Most scientists may be wary of incurring huge expenses against high risks of failed projects, especially when it is often expensive, difficult and time-consuming to acquire the necessary permits from national and local regulatory authorities.

In spite of the relatively sparse previous research conducted on fish of EASL, there is growing interest among biologists, ecologists, fisheries and conservation managers, resulting in enhanced understanding of saline lake fish populations. For instance, recent taxonomic revisions of the Magadi–Natron Soda Tilapias by Lothar Seegers and Herbert Tichy (Seegers and Tichy 1999; Tichy and Seegers 1999) using modern genetic tools have greatly improved on the classical descriptions of the species by earlier ichthyologists such as Franz Hilgendorf (1905), George Boulenger (1912) and Ethelwynn Trewavas (1983). Several studies on other aspects of saline lake fish in EASL have also been undertaken in the recent years (Muška et al. 2012; Pörtner et al. 2010; Zaccara et al. 2014). As an example, we restrict our discussions to studies that have been undertaken on the single teleost species that thrives in Lake Magadi, Kenya, the Magadi tilapia, *Alcolapia grahami* (Boulenger 1912).

9.6 The Lake Magadi Tilapia as a Case Study

Lake Magadi is a hypersaline, highly alkaline, endorheic basin at the southern tip of the Kenyan Rift Valley section (Figs. 9.1 and 9.6). Historically, the lake has been joined and separated several times from Lake Natron, another saline lake in northern Tanzania (Vincens and Casanova 1987; Roberts 1993; Tichy and Seegers 1999). In fact the two lakes are believed to have been parts of large Paleolake Orolonga (Burtzer et al. 1972). This paleolake is thought to have held a large population of freshwater cichlids probably similar to *Alcolapia grahami* (Boulenger 1912; Tichy and Seegers 1999). Complete separation of Lake Magadi from Lake Natron probably occurred not less than 8000 years ago (Burtzer et al. 1972; Tichy and Seegers 1999). Lake Magadi further split ~7000 years ago to form a satellite Lake Little Magadi to its north (Figs. 9.1 and 9.6). Today, the Lake Magadi basin is a large shallow lake covered mainly (~80 %) by "trona" (a solid crust of precipitated sodium carbonate and bicarbonate (Fig. 9.6) separating shallow lagoons of varying physicochemical properties (Coe 1965; Wilson et al. 2004). Compared to Lake Natron, which harbours three species of tilapia *Alcolapia alcalicus*, *A. latilabris* and *A. ndalalani* (Table 9.1) (Seegers et al. 1999), the Lake Magadi basin has only one described tilapia species, *Alcolapia grahami*. In the older literature, this was variously called *Oreochromis alcalicus*

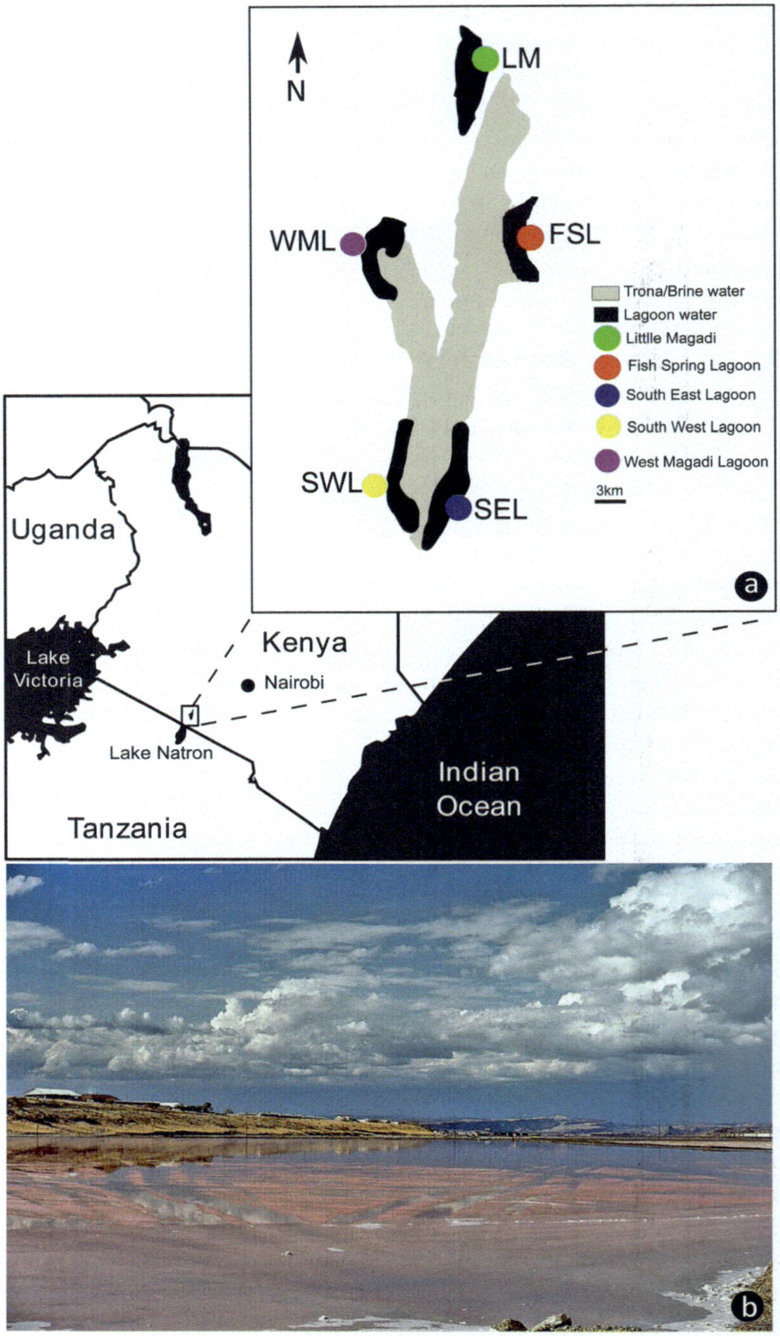

Fig. 9.6 (**a**) Map of Kenya showing the position of Lake Magadi; inset is an enlarged map of the lake showing individual lagoons. (**b**) Brine and trona pans at the Fish Spring Lagoon of Lake Magadi. The bright red colour of the brine is a result of halophilic archaebacteria

grahami, *Sarotherodon alcalicus grahami* or *Tilapia grahami*.

Magadi tilapia is the only teleost and indeed the only vertebrate known to inhabit the hypersaline lagoons of Lake Magadi and its satellite Lake Little Magadi (Coe 1966; Wilson et al. 2004). Some populations of the tilapia were transplanted to Lake Nakuru (Vareschi

1979) and probably to Lake Elementaita (Okeyo 2006). Throughout its distribution range, the Magadi tilapia is subjected to incredibly extreme conditions: pH ~10, alkalinity 300 mmol L^{-1}, temperature up to 42 °C, high UV radiation, daytime hyperoxia and night-time hypoxia, ionic concentrations equivalent to those of ~60 % sea water and intense predation by birds (Coe 1966, 1967; Johannsson et al. 2014; Narahara et al. 1996. Nevertheless, this species thrives in a number of isolated hot springs and lagoons around the edge of the lake. Indeed these conditions, which would quickly kill most other teleost fish, appear to be optimal for this species, as Magadi tilapia often deteriorate and die when transferred to other water qualities which "standard" teleosts inhabit (Laurent et al. 1995; Reite et al. 1974; Wood et al. 1989, 1994; Wright et al. 1990, our personal observations).

Dispersal of fish between individual lagoons is hampered by the presence of trona, uninhabitable water and in the case of Lake Magadi versus Little Lake Magadi, a land barrier separating the two lakes (Fig. 9.6) (Coe 1966, 1967; Wilson et al. 2004). There have been suggestions that occasional mixing of fish between these lagoons could be possible especially following floods (Coe 1966; White 1953), but several studies support the hypothesis of little or no gene flow (Kavembe et al. 2014; Wilson et al. 2004). Since the publication of the first monograph on Magadi tilapia (Coe 1966), the fish has been the subject of a range of studies on its physiology, taxonomy, morphology, behaviour, ecology and genetics, a selection of which are discussed below.

9.7 Physiological Studies

Because of its extreme environment, the Magadi tilapia is exposed to huge and diverse physiological challenges. Consequently, the fish has acquired several adaptive mechanisms to deal with physiological stresses associated with ionic balance, acid–base balance, nitrogenous waste management, gas exchange, high temperature and reactive oxygen species (ROS) (Johannsson et al. 2014; Pörtner et al. 2010; Wood et al. 2013).

9.7.1 Osmoregulatory and Ionoregulatory Adaptations

Magadi water has exceptionally high pH (10) (Johansen et al. 1975) and osmolality of ~580 mOsm (Wood et al. 1989). The dominant anion is HCO_3^-/CO_3^- (290 meq L^{-1}) rather than Cl^- (112 mmol L^{-1}). Similarly, the dominant cation Na^+ (356 mmol L^{-1}) is exceptionally high in concentration, while Ca^{2+} and Mg^{2+} are negligible <1 mmol L^{-1} (Wood et al. 1994, 2013). To minimize the high cost of acid–base regulation, which is about 50 % of its resting metabolism (Wood et al. 2002b), the Magadi tilapia maintains an exceptionally high blood pH (8–9) and tissue pH (7.6) (Wood et al. 1994). Plasma ions and osmolality are regulated at levels typical of other teleosts (Maloiy et al. 1978; Eddy et al. 1981; Eddy and Maloiy 1984; Wright et al. 1990; Wood et al. 2002a). Although the fish is ureotelic (see below), regulation of plasma urea plays only a quantitatively small role in its osmoregulatory strategy, unlike ureotelic elasmobranchs (Wood et al. 2002a). Instead, drinking of alkaline water at exceptionally high rates (8 mL g^{-1} h^{-1}) enables this fish to balance its osmolality by replacing water lost across the gills to the hypertonic environment (Bergman et al. 2003; Maloiy et al. 1978; Skadhauge et al. 1980; Wood et al. 2002a). To avoid neutralizing the stomach gastric acids by the alkaline water, Magadi tilapia's gut is uniquely modified to include a pyloric bypass (Bergman et al. 2003). This bypass also ensures that the fish can imbibe water regardless of a full stomach (Bergman et al. 2003). Most of the solutes in the imbibed water (Na^+, HCO_3^- and lesser amounts of Cl^-) appear to be absorbed in the gut and then excreted through the gills (Bergman et al. 2003). Intestinal fluid absorption occurs mainly in the anterior portion of the digestive tract, driven largely by $NaHCO_3$ absorption, with a contribution from NaCl absorption in

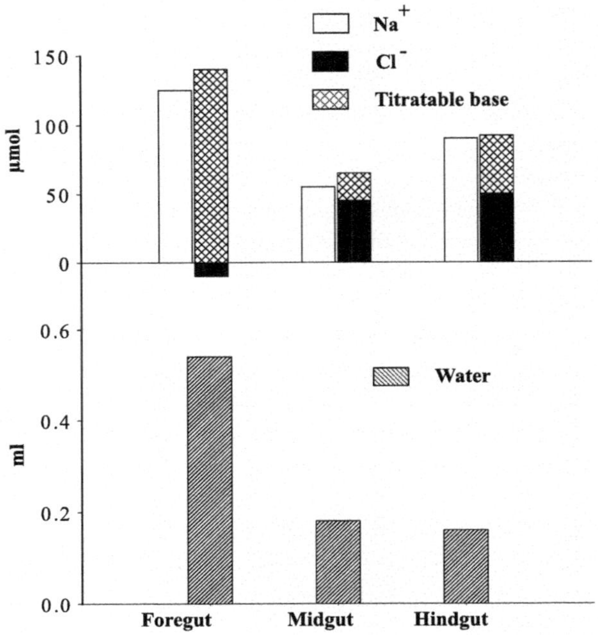

Fig. 9.7 Intestinal absorption (per 1 mL ingested) of hypersaline water and its major ionic constituents by Lake Magadi tilapia. Most of the consumed water (8 mL kg $^{-1}$ h^{-1}) is absorbed at the anterior portion of the digestive tract (Bergman et al. 2003)

more distal portions of the tract (Fig. 9.7). Building on the original model of Laurent et al. (1995), Wood et al. (2013) provided evidence that the active excretion of HCO_3^- could explain the electrogenic component of the transepithelial potential (TEP) across the gills observed in Magadi tilapia. Active excretion of HCO_3^- would in turn facilitate passive efflux of Na^+ across the gills by creating a positive TEP in the blood relative to the external water (Eddy et al. 1981; Maloiy et al. 1978; Wood et al. 2013). As Cl^- is actually in higher concentrations in the blood plasma than in Magadi lake water, it too probably effluxes passively across the gills (Wood et al. 2012).

9.7.2 Nitrogen Excretion

Nitrogenous waste excretion is a crucial aspect for survival of fish. The Magadi tilapia is a unique teleost in this aspect—it is the only 100 % ureotelic fish—it normally excretes all its nitrogenous waste as urea (Randall et al. 1989). Indeed, the ability of the Magadi tilapia to excrete urea has been equated to that of land mammals. Virtually all other adult teleosts excrete their nitrogenous wastes primarily as ammonia (Wright and Wood 2009). In fact, most embryonic stages of teleosts excrete urea as the primary nitrogenous waste, but switch to excrete ammonia at a relatively early stage (Wright and Fyhn 2001). The presence of ureotelism in adult Magadi tilapia thus appears to reflect the retention of an embryonic characteristic. Consequently, the Magadi tilapia has been shown to express all enzymes of the ornithine-urea cycle (OUC) (Randall et al. 1989; Walsh et al. 1993). A unique feature of urea production in Magadi tilapia is the expression of urea cycle enzymes primarily in the white muscle and to a lesser extent in the liver, the primary organ where the OUC enzymes are expressed in those few other fish that excrete substantial amounts of urea (Lindley et al. 1999). Continuous excretion of urea is achieved through a facilitated diffusion mechanism involving urea transporters in the gills (Walsh et al. 2001; Wood et al. 2013). This gives this species the highest gill urea permeability of any fish species (Fig. 9.8), even greater than that of the marine toadfish, which is a facultatively ureotelic species with similar

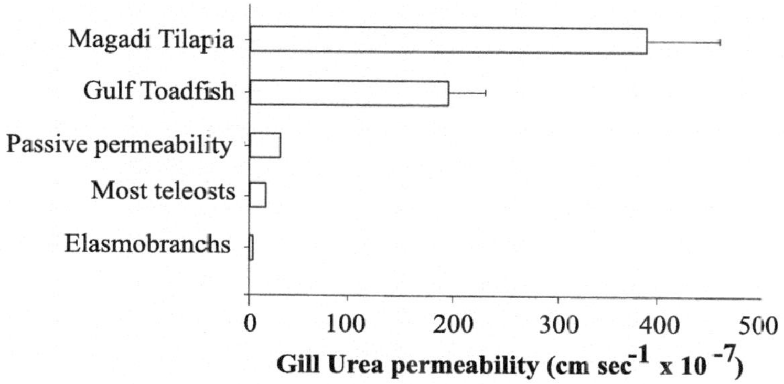

Fig. 9.8 A comparison of the gill permeability to urea in the obligatory ureotelic Magadi tilapia versus the facultatively ureotelic gulf toadfish, passive permeability through typical cell membranes, the gill permeability measured in standard ammonotelic teleosts and the gill permeability measured in most elasmobranchs (which actively retain urea). The very high branchial urea permeability of the Magadi tilapia is due to the high level of expression of UT-type facilitated urea transporters (Walsh et al. 2001)

gill urea transporters (Walsh et al. 2000). Although urea production is a metabolically demanding process, this remarkable adaptation ensures that the tilapia can excrete its nitrogenous waste across the gills into its highly alkaline and buffered aquatic habitat (Randall et al. 1989; Wood et al. 1989, 1994, 2002a, b). As originally pointed out by Randall et al. (1989), ammonia excretion should be theoretically impossible against an external pH of 10, at least by passive mechanisms. Nevertheless, it is remarkable that the Magadi tilapia expresses Rh proteins and an NH_4^+-activated Na^+-ATPase in its gills (Wood et al. 2013). Such mechanisms are thought to be associated with active ammonia excretion in other teleosts (Wright and Wood 2009), raising the prospect that the Magadi tilapia may actively excrete ammonia under ammonia-loading conditions. This, together with the ability of the OUC to detoxify ammonia by accelerating its conversion to urea (Wood et al. 1989, 2013), confers an exceptional ammonia tolerance on this species (Walsh et al. 1993). These conditions occur in areas of the lake where bacterial action converts the uric acid from flamingo guano deposits into ammonia (Wilson et al. 2004).

9.7.3 Respiratory Adaptations

Magadi tilapia has exceptionally high rates of oxygen consumption (Franklin et al. 1995; Narahara et al. 1996), reflecting its very active lifestyle (Coe 1966, 1967), the high cost of living in this hostile environment and the high environmental temperatures. The species mitochondria appear to be adapted to function optimally at these high temperatures (Johnston et al. 1994). Unsurprisingly, the fish has a relatively high capacity for sustained exercise (Walsh et al. 1993; Wood et al. 2002b). As mentioned above, Magadi tilapia is exposed to extremes of oxygen concentration, namely, daytime hyperoxia and night-time hypoxia (Johannsson et al. 2014; Narahara et al. 1996). These hyperoxic and hypoxic states result, respectively, from photosynthetic activity of Cyanobacteria during the day and respiration at night (Narahara et al. 1996). To deal with low oxygen concentrations, in Magadi tilapia, haemoglobin (the oxygen carrier protein) is adapted to extract oxygen even at near anoxic conditions (Lykkeboe et al. 1975). This is reflected in a blood with a very high oxygen affinity (Narahara et al. 1996).

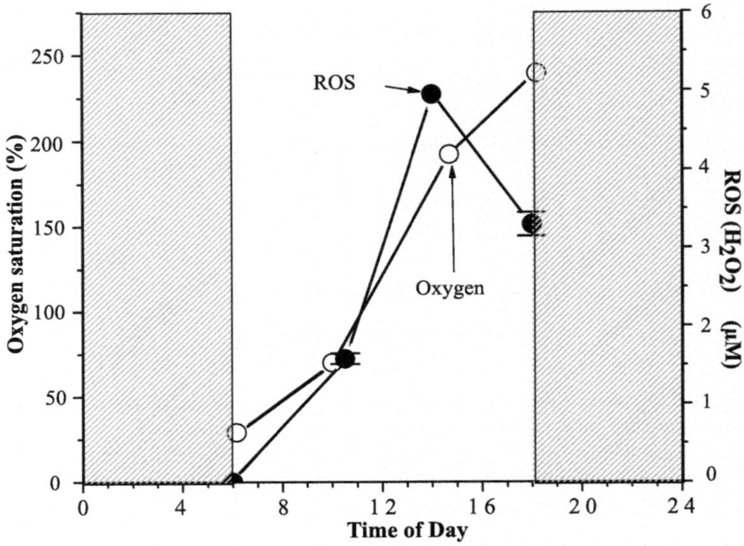

Fig. 9.9 Reactive oxygen species (ROS) levels of Lake Magadi water measured during the day. The lake has some of the highest ROS levels ever recorded in natural waters (Johannsson et al. 2014)

The species haemoglobin has a high thermostability (Franklin et al. 1994). Additionally, the gills have a very high surface area and a relatively thin water-to-blood diffusion barrier, making their branchial diffusing capacity for oxygen exceptionally high (Maina et al. 1996). The Magadi tilapia has also developed a highly vascularized physostomous swim bladder that allows the fish to perform facultative air breathing (Maina et al. 1995). A recent study observed that air breathing occurs not only under hypoxia but also under normoxic and hyperoxic conditions (Johannsson et al. 2014). Since air breathing was manifested more in fish exposed to sunlight, the authors suggested that the extremely high reactive oxygen species (ROS) in Magadi water during the peak of the day (Fig. 9.9) could irritate the gills, making air breathing a potential adaptive strategy in Magadi tilapia to avoid ROS damage (Johannsson et al. 2014). Air breathing, however, is costly as it may predispose the fish to greater risk of predation by birds (Johannsson et al. 2014; Narahara et al. 1996). Nevertheless, it seems to be a key adaptation uniquely exploited by Magadi tilapia to survive in its adverse environment. More studies are, however, required to test this hypothesis.

9.8 Genetic and Molecular Studies

Over the last two decades, several studies have investigated the population genetic aspects of Magadi tilapia. Seegers et al. (1999) used cytochrome b and control region (D-loop) markers to investigate the phylogenetic relationships of Magadi tilapia, its three sister species (Natron tilapia: *A. alcalicus*, *A. ndalalani* and *A. latilabris*) and those of the closely related genera *Oreochromis* (*O. niloticus* and *O. amphimelas*). Using both markers, Seegers et al. (1999) found a common shared haplotype in Lake Magadi and Lake Natron tilapia species. They proposed that the common haplotype could have been present in the ancestral fish population that occupied the Paleolake Orolonga that split to give rise to the present day Lake Natron and Lake Magadi basins ~13,000 years ago (Burtzer et al. 1972; Vincens and Casanova 1987; Roberts et al. 1993). Wilson et al. (2004) obtained similar results using the first hypervariable region of the mitochondrial D-Loop. An earlier study that used variable number tandem repeats (VNTRs), however, found significant differentiations between the two populations of Magadi tilapia (Wilson

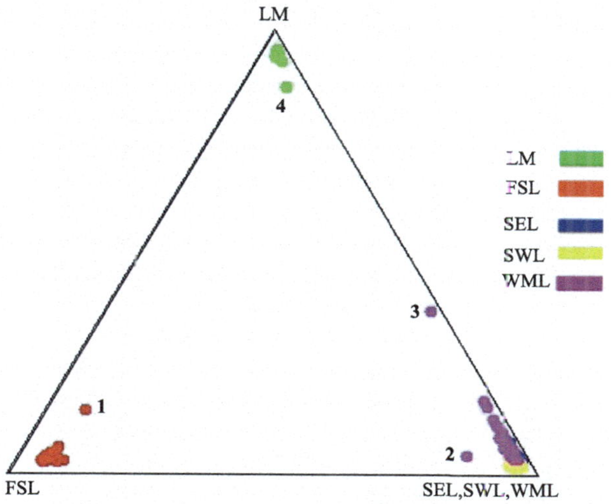

Fig. 9.10 *Structure triangle* plot showing the distribution of 259 Magadi tilapia samples collected in four isolated lagoons of Lake Magadi: Fish Spring (FSL), South East (SEL), South West (SWL) and West Magadi (WML) Lagoons and a single location in Little Magadi (LM). Each individual was genotyped at 10 microsatellite loci. The ancestry (similarity) of the individuals was estimated under a model in which individuals can have mixed ancestry. For this data set, the model with three populations (Little Magadi, Fish Spring Lagoon and the rest of Magadi) was highly supported. *Dots* represent each individual and the *colours* represent the localities. LM is detached completely from the rest of Magadi tilapia populations. FSL is also detached from the rest of Magadi tilapia but with a small distance compared with LM (Kavembe et al. 2014). Individuals with moderate posterior support for having migrant parents are labelled "1"–"4"

et al. 2000). In fact, these two populations were as distinct from each other as they were from samples of a Lake Natron tilapia population that were analysed in the same study (Wilson et al. 2000).

Most recently, Kavembe et al. (2014) investigated the genetic structure of Magadi tilapia throughout its distribution range using a data set of complete control region sequences and 10 microsatellite loci (Fig. 9.10). Except for the population from Little Magadi, all the other populations shared haplotypes. Little Magadi had one main haplotype and a few other haplotypes derived from this main haplotype (Kavembe et al. 2014). Despite a rather extreme environment, high levels of genetic variation were inferred in Magadi tilapia by all markers, in accord with the conclusions of earlier studies (Seegers et al. 1999; Wilson et al. 2004). This was interpreted to suggest an association of Magadi tilapia to a large ancestral population that may have inhabited the freshwater Paleolake Orolonga (Seegers et al. 1999). Based on the more variable microsatellite markers, the authors inferred three genetic clusters of Magadi tilapia: (1) Little Magadi cluster; (2) Fish Springs Lagoon cluster and (3) Rest of Lake Magadi cluster (Fig. 9.10) (Kavembe et al. 2014). Importantly, however, the genetic distance-based F static, another measure of population differentiation, found significant differences among all five populations studied. These included differences between three populations that had been grouped into a single cluster by a Bayesian model-based analysis (Fig. 9.10). Kavembe et al. (2014) proposed that the presence of a land barrier and impassable trona could limit the dispersal ability of Magadi tilapia, leading to formation of the distinct gene pools observed.

A question of great evolutionary importance has been the origin of the proto-*A. alcalicus* ancestral population that seeded the Magadi–Natron species flock (Seegers and Tichy 1999; Tichy and Seegers 1999). Owing to the close geographical proximity and similar water conditions in the Magadi–Natron basin and Lake Manyara, their species would be expected to have a common ancestry. In fact,

Table 9.2 Some differences between Lakes Magadi and Natron tilapia (genus *Alcolapia*) versus Lake Manyara tilapia (*Oreochromis amphimelas*)

Character	*Alcolapia* species	*Oreochromis amphimelas*
Vertebrae	27–30 (mode 28)	30–32 (mode 31)
Scales lateral line	27–30 (mode 27 or 28)	30–34 (mode 31–33)
Dorsal spines	Slender	Stout
Postorbital part of the head	Often <50 %	Often >50 %
Depth of preorbital bone	Deep	Shallow
Texture of preorbital bone	Naked	Scaly

Adopted from Trewavas (1983)

Trewavas (1983) reported the species shared some characters such as low gill raker numbers, absence of microbranchiospines and remarkably small, thin scales on the chest, belly and nape. But based on several striking differences between *O. amphimelas* and the *Alcolapia* genus (Table 9.2), she concluded that the two were indeed separate groups. She proposed that the shared characters may merely reflect the occurrence of both species groups in similar water conditions (Trewavas 1983). Other shared characteristics among saline lake fish thought to be associated with thermal and alkaline lake conditions include: small sizes at maturity and small maximum sizes, low number of vertebrae relative to putative parental species and low dorsal fin rays (Trewavas 1983).

A similar argument for independent evolution of similar phenotypes in *O. amphimelas* and members of the genus *Alcolapia* was put forward by Nagl et al. (2001) and by Kavembe et al. (2014). In these two studies, phylogenetic reconstructions of tilapiines based on mitochondrial DNA sequences revealed a rather distant relationship of *Alcolapia* species and *Oreochromis amphimelas* (Kavembe et al. 2014; Nagl et al. 2001). Several *Oreochromis* species lineages appeared between the branches leading to *Alcolapia* species and *Oreochromis amphimelas*. Instead, Kavembe et al. (2014), based on a phylogeny inferred from a data set of the highly conserved mitochondrial ND2 sequences, found *Oreochromis* (*Nyasalapia*) *variabilis* as the closest relative to *Alcolapia*. *O. variabilis* is a species endemic to Lake Victoria. Albrecht et al. (1968) reported similar breeding pit shapes between *Alcolapia grahami* and *Oreochromis variabilis* (Trewavas 1983). Lake Victoria is located more than 300 km away from the Magadi–Natron basin. However, at the closest points, the lake basins are separated by active volcanic rocks that could be younger than the fish populations themselves (Trewavas 1983). In a detailed revision of the taxonomy of the *Oreochromis* and *Sarotherodon* species, Trewavas (1983) noted the similarity in the snout shape of *Alcolapia* and *Oreochromis variabilis*. Nonetheless, the presence of tasselled male genital papillae only in the males of *Oreochromis variabilis* convinced her to retain them as separate genera.

Trewavas (1983) suggested that high levels of human-mediated dispersal of *Oreochromis variabilis* could have facilitated colonization of distant water bodies. In an experiment to test the dispersal ability of the species, females were captured and tagged at known brooding grounds (Fryer and Iles 1972). After shedding their young, the fish traversed long distances. In this particular experiment, the furthest individual was recaptured 28 days after tagging about 96 km (~3.4 km per day) away from the point of release (Fryer and Iles 1972). Nevertheless, given the absence of obvious water links between Lake Victoria and the Lake Magadi–Natron basin, the proposed close relationship of their species raises important questions as to the colonization history of the Magadi–Natron basin.

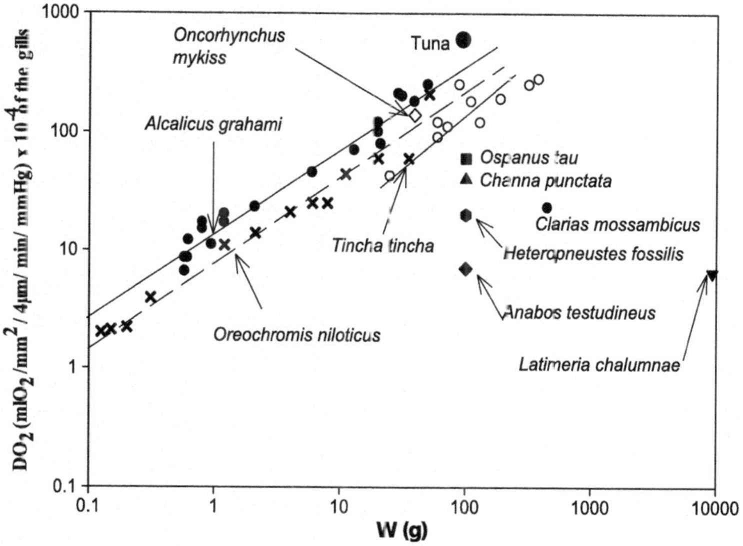

Fig. 9.11 The relationship between the morphometric oxygen-diffusing capacity (DO2) of the gills and body mass in a range of fish species. Detailed data for *Alcolapia grahami* (*solid line*) and *Oreochromis niloticus* (*broken line*) are shown. In this comparison, *A. grahami* displayed superior DO2 (due to relatively large gill size and thin water-to-blood diffusion distance, followed closely by *O. niloticus*), while all the other fish except the tuna had substantially lower DO2 (Maina et al. 1996)

9.9 Taxonomic and Morphological Studies

The systematics and classification of *Alcolapia*, the genus to which Magadi tilapia belongs, has been contested by many systematists and taxonomists (Seegers et al. 1999). Since the first record of Magadi tilapia as *Tilapia grahami* (Boulenger 1912) in the early twentieth century, the species taxonomy has been revised several times. Other names that have been used include: *Tilapia grahami* (Boulenger 1912), *Sarotherodon alcalicus grahami* (Boulenger 1912) and *Oreochromis alcalicus grahami* (Boulenger 1912). *Alcolapia grahami* (Boulenger 1912) is the currently accepted name for Magadi tilapia (Seegers and Tichy 1999). Although all fish populations in the Magadi basin are currently recognized as *Alcolapia grahami*, the consistent occurrence of a slanted mouth in the Little Magadi population relative to a terminal mouth in the Lake Magadi lagoon population calls for a close examination of the species for other lake-specific differences (Kavembe et al. 2014; Wilson et al. 2004).

Most morphological studies on Magadi tilapia have been on organs associated with its physiology such as the gills, gas bladder and the gut. The gills have been most studied owing to their importance in excretion, ionoregulation and respiration in this species (Johannsson et al. 2014; Laurent et al. 1995; Maina 1990, 1991; Maina et al. 1996). Unique features of its gills include: a relatively large number of mitochondria-rich cells with a profuse intracytoplasmic microtubular network (Maina 1990, 1991), the presence of accessory chloride cells (Laurent et al. 1995) and a very high surface area and thin water-to-blood diffusion distance, yielding a very high oxygen-diffusing capacity (Fig. 9.11; Maina et al. 1996). Interestingly, however, the presence of an interlamellar cell mass (ILCM), which would tend to restrict oxygen-diffusing capacity but protect against the harsh external environment was recently reported in fish in some lagoons (Johannsson et al. 2014). Other studies on Magadi tilapia have focused on the gut because of its dual role in food processing and osmoregulation (Bergman et al. 2003); the swim bladder (Maina et al. 1996; Maina 2000a) because of its role as an accessory air-breathing organ

(Narahara et al. 1996); the epithelium of the buccal cavity, both as a possible respiratory structure (Maina et al. 1995) and a secretory structure for nutrition of offspring (Maina 2000b); and the spermatozoa (Papah et al. 2013). The spermatozoon was shown to possess several adaptive morphological features thought to enable its function in the lake's extreme water (Papah et al. 2013).

9.10 Behavioural Studies

Several aspects of behaviour in Magadi tilapia have been described and may form a basis for more in-depth studies. These include air breathing and surface skimming, jumping out of water to feed, fighting in males (Narahara et al. 1996), feeding mainly during daylight hours (Bergman et al. 2003), evasive actions from avian predators and diurnal migrations within the lagoon systems (Coe 1966, 1967). Additionally, Coe (1966) provided extensive information on their reproductive behaviour (see next section). Recently, Johannsson et al. (2014) reported a form of cooperative social behaviour in which all individuals that performed air breathing formed intact groups that they referred to as pods. The authors noted aggressive attacks on those fish that attempted to air breathe outside the pods. The negative social interaction was useful in maintaining pod integrity. This behaviour may also be adaptive by minimizing chances of predation on air-breathing individuals by fish-eating birds (Johannsson et al. 2014).

9.11 Studies on Reproductive Aspects

Magadi tilapia is a mouth-brooder (Coe 1966). Breeding males and females display distinct colours during the breeding period. The most conspicuous feature is the fleshy brilliant white coloration of the lower lip in males (Fig. 9.12). Generally, however, both males and females display bright colours during courtship (Coe 1966). The males build pits at the floor of the lagoons and exhibit courtship behaviours similar to those of other tilapias. Territorial behaviours among males have been documented (Coe 1966). Bigger males have been reported to invade and replace young males from their breeding pits (Fryer and Iles 1972). Females may mate with several males, and a brood released by a single female may therefore consist of babies sired by different males (Trewavas 1983). In Fish Springs Lagoon, females have been observed to deposit their

Fig. 9.12 Ripe Magadi tilapia males. Note the *white* lips characteristic of breeding males of these species. Cyanobacteria, which are the main diet of Magadi tilapia, impart a *green* hue on the pools inhabited by fish. Magadi tilapia photos courtesy of Gudrun De Boeck

young in small pools around the springs where the water is cooler and is less concentrated with salts (GD Kavembe and CM Wood, personal observations). This may be a strategy to allow the young fish to develop in less harsh conditions. Alternately or additionally, it may serve to protect the offspring from being eaten by conspecifics, because snatching and foraging of eggs and young from mouth-brooding females by other fish of both sexes have been observed (Coe 1966). This cannibalistic behaviour has also been documented in Lake Natron cichlids (Seegers and Tichy 1999) and in other mouth-brooding cichlids (McKaye and Berge 1996). Since the waterways leading to these "nursery pools" are very shallow and exposed, this behaviour involves air exposure of the brooding females as they wriggle up to the pools. That behaviour was observed only in the evening and may be a female strategy to avoid the scorching UV radiations, dehydration and predator attacks, all of which would be higher during the day. Other observers have also reported brooding females voluntarily entering and remaining for long periods in water as hot as 44 °C, temperatures that are avoided by males and non-brooding females (cf. Albrecht et al. 1968; Fryer and Iles 1972).

Spermatogenesis in Magadi tilapia was recently examined by Papah et al. (2013). Despite the extreme conditions to which the fish are exposed and the occurrence of external fertilization (involving similar exposure of the sperm), the males exhibited normal spermatogenesis comparable to that of other mouth-brooding tilapia. However, one particular specialization was the sidepiece, with no cytoplasmic sheath, which ends blindly distally in a lobe-like pattern around the flagellum. The apparent absence of the cytoplasm sheath and indistinct side fins/ridges in the spermatozoa of the Magadi tilapia compared to other cichlids may help accelerate sperm swimming performance, which could be of adaptive significance in the harsh environment. Furthermore, Papah et al. (2013) speculated that the presence of a thin membrane heavily laden with a dark pigment on the parietal peritoneum of the male could serve to protect the testis against high levels of UV radiation.

9.12 Studies on Feeding

Only few studies have investigated food and feeding habits of fish populations in EASL. It is generally agreed that EASL support simple food webs (Fig. 9.5), with complexity increasing with decreasing salinity levels (Cooper and Wissel 2012; Harper et al. 2003). Owing to extreme water conditions in these lakes, their algal communities are restricted to a few species. The dominant benthic algal species are Cyanobacteria (notably *Arthrospira platensis*), and the pelagic algal species are dominated by free-floating *Dunaliella* spp. (Cooper and Wissel 2012). The diversities of zooplankton, large-bodied crustaceans, birds and vertebrates are also limited (Melack 1996). Stomach content analysis of Magadi tilapia revealed that it derives most of its diet (90 %) from Cyanobacteria and a small but significant proportion (10 %) from dipteran larvae and copepods (Coe 1966; Johannsson et al. 2014). The fish also obtain food from the mud, especially in the southern lagoons with their huge flamingo guano deposits. Coe (1966) provides a vivid description of Magadi tilapia feeding on Cyanobacteria. The fish fed within a certain temperature gradient bounded by a browse line. This browse line marked the maximum temperatures (about 41 °C) the fish could withstand while browsing (Coe 1966, 1967). Feeding was observed even in brooding females (Albrecht 1968). The species was also observed to display cannibalistic behaviour. Young individuals attempted to follow breeding females to collect and eat the deposited eggs (Trewavas 1983). Non-breeding conspecifics were also observed to snatch and consume eggs and young from breeding females (Coe 1966). More comprehensive studies on the food and feeding habits of saline lake fishes will provide information necessary for successful fisheries management of the lakes.

9.13 Studies on Diseases and Parasites

Magadi tilapia, like many other fish species, is susceptible to a range of diseases and parasites. To date, however, no detailed studies have been conducted on this important aspect of the fishes' wellbeing. A single parasitic worm species, *Contracaecum multipapillatum* (Drasche 1882), has been found to infest Magadi tilapia (Fig. 9.13) (John Maina, *unpublished results*). The nematode resides in the pericardial cavity (Fig. 9.13) but larval stages may be found in tissues or abdomen (Floria et al. 2009). Fish-eating birds that serve as definitive hosts for the parasite have been proposed as the main vectors for transmission from one water body to another. Surprisingly, in an expedition undertaken between July and August 2010, *C. multipapillatum* worms were observed only in fishes in Fish Spring Lagoon and were notably absent in fishes from other lagoons (John Maina, Geraldine D. Kavembe and Chris M. Wood, personal observations). This was the case despite an apparent movement of birds between the lagoons. This calls for further studies to determine the pattern, trends and possible vectors involved in parasite transmission and the prevalence of other diseases and parasites in Magadi tilapia. Given the conditions in which this tilapia subsists, the association between the fish and the nematode provides a model to study host-parasite coevolution in stressful environments. The expectation would be that this nematode would acquire local adaptations to their host. Since the nematode has also been reported in closely related tilapia species in freshwater lakes (Floria et al. 2009), comparisons of fish and worms from both systems would also be worthwhile.

9.14 Conservation Challenges for Saline Lake Fish

9.14.1 Climatic Change and Declining Water Quality

Very probably, fish in EASL are highly impacted by the ongoing global climate changes in their shallow saline basins, where they are exposed to increasing temperatures, salinity and high levels of UVB radiation. Lake fragmentation and salinization of habitable sections may reduce refuge sites for the fish. This may subsequently lead to

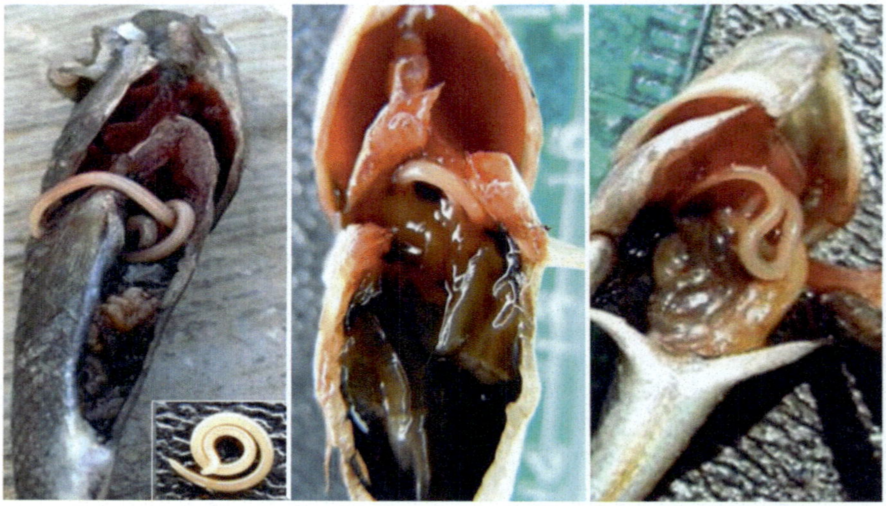

Fig. 9.13 Magadi tilapia (*Alcolapia grahami*) from Fish Spring Lagoon infested by the nematode *Contracaecum multipapillatum*. The parasitic worm resides in the pericardial cavity. Photo courtesy of John Maina

the disappearance of the populations. In recent years, soda lakes have been exploited for soda ash extraction, water extraction for agriculture and geothermal power generation and have become sites of active development. These human-mediated activities are associated with pollution and a decline of water quality, posing great challenges to the long-term existence of saline lake fisheries.

9.14.2 Migration and Movement

Saline lakes and their fish stretch across jurisdictional boundaries; often their occurrence and movement cut across political boundaries. This raises trans-boundary issues of mutual concern among the member states involved. Although the greatest portion of Lake Turkana lies in Kenya, most of its inflow (90 %) comes from the Omo River in Ethiopia (Avery 2010; Muška et al. 2012). Lake Natron, Tanzania, receives its major freshwater inflow from the Ewaso Nyiro River in Kenya. Proposals by the Ethiopian and Kenyan governments to dam the waters of the Rivers Omo (Ojwang et al. 2010) and Ewaso Nyiro, respectively, have raised concerns among conservationists over the effects both projects may have on biodiversity of those lakes. Clearly, there is need for countries to develop policies that ensure equitable sharing and conservation of lake resources. However, history has shown that cooperative management of trans-boundary resources is hindered by lack of common laws among member states. Another challenge to such mutual agreements is that too many institutional players are involved. This complicates formulating conservation strategies. Delineation of the movements of fish in EASL is also important in determining the dispersal ability and distribution ranges of individual species. Such information is in turn useful for conservation purposes and for assessing the effects of various species interactions.

9.14.3 Competing Interests

The need to strike a balance between economic development and conservation of wetlands remains a thorny issue among key stakeholders. For a long time, EASL have sustained the subsistence livelihood of the surrounding communities. In turn, the communities have protected the lakes to ensure a continued supply of their resources. In recent years, however, there has been an influx of local and foreign-owned companies into the saline lakes. For instance, there have been concerns that construction of the proposed Gibe 3 hydroelectric dam on the Omo River by the Ethiopian government may alter the river's seasonal flooding cycle (Ojwang et al. 2010). This may irreversibly change Lake Turkana, whose major water input comes from River Omo. Dwindling water resources in Lake Turkana have been shown to contribute to conflicts among farmers, pastoralists and fishermen who depend on the lake's water (Yongo et al. 2010). Lakes Abijata and Magadi are heavily mined for soda ash (Hughes 2008). Recently there was a proposal to establish a similar mining plant in Lake Natron. Many of the projects are executed following government approvals or by the governments themselves without full consideration of the potential impacts on fish populations or general aquatic ecosystem health. Drawing lessons from other studies, however, such projects may irreversibly destroy fish communities through pollution, loss of gene pools, habitat fragmentation and destruction (Avery 2010; Roberts 1993). Perceived benefits from such projects, e.g. employment creation and improved livelihoods, may further contribute to the socioeconomic gap between development and conservation.

9.14.4 Adverse Effects of Introducing Other Fish Species into Saline Lakes

Although saline lakes have been least affected by human-mediated species introductions, several alien species have been introduced in some saline lakes. The introduction of Magadi tilapia to Lake Nakuru during the period 1952–1962 to control mosquitoes is probably the best-known case (Vareschi 1979). The species has also been introduced to Lake Elementaita (Okeyo 2006). Other species introductions have also been reported in Lake Turkana. A good example of the detrimental effects of fish introductions to saline lakes is the case of Lake Chilotes in Ethiopia (Brook 1994). Lake Chilotes was a typical saline lake until 1990, when a government initiative through the Ministry of Agriculture diverted River Mori into the lake in an attempt to increase water for irrigation (Brook 2003). Consequently, the salinity of the lake dropped, favouring colonization by three *Barbus* species from River Mori. The introduced species have since led to a marked decrease of Lake Chilotes' native *O. niloticus* species (Brook 2003). In general such introductions risk destabilizing existing ecosystems.

9.15 Conclusion

The future of saline lakes and their fish populations in East Africa remain uncertain. In some cases, prompt interventions may be required to save the remaining local populations from extinction. This would require extensive surveys to collect data on species composition, distribution, life history traits and other aspects of individual fish populations. We therefore hope that the present information will serve as a guide in formulating conservation plans. As already shown in some alkaline lakes (e.g. Lake Turkana) (Kolding 1989; Hopson 1982), EASL can support more fish species than they normally contain. Given the challenges and threats associated with introducing alien fish species in EASL, we recommend management and conservation strategies aimed at preserving native fish populations. In lakes that have lost their fish populations or have no known fish populations of their own, introduction of fish from localities naturally connected to the lake in question may be an option. However, caution should be taken to ensure that only species with a high probability of survival and recruitment are introduced to EASL to maximize colonization success. Other important aspects to consider before such introductions include: food availability, disease and vectors, as well as the effects the fish may have on other components of the target lake's ecosystem.

Management and conservation of EASL and their fish populations require the commitment of multiple stakeholders including local communities, fisheries managers, scientists, regional and national governments and the international community at large. In particular, the East African Community should formulate and enforce strict policies and laws to govern utilization, management and conservation of EASL resources. Strategies aimed at forging a good working relationship among all potential users would be a major step towards a realistic and sustainable conservation framework.

Acknowledgement This work was supported by various grants. GDK is supported by a collaborative scholarship grant from the Deutscher Akademischer Austausch Dienst (DAAD) and Kenyan Ministry of Education, Science and Technology (MOEST). AM is supported by the University of Konstanz, grants of the Deutsche Forschungsgemeinschaft and the European Research Council. CMW is supported by the Canada Research Chair Program and grants from the Natural Sciences and Engineering Research Council of Canada (NSERC). We thank Papah Babak, Ora Johannsson, Daniel Msafiri Shayo and Ayub Oduor for useful discussions during the preparation of this manuscript. We appreciate the logistical support by Tata Chemicals Magadi (formerly Magadi Soda Company) during several research expeditions in which some of the observations and data reported in the chapter were collected.

References

Alamanov A, Mikkola H (2011) Biodiversity friendly fisheries management possible on Issyk-Kul Lake in the Kyrgyz Republic? Ambio 40:479–495

Albrecht H (1968) Freilandbeobachtungen an Tilapien (Pisces: Cichlidae) in Ostafrika. Z Tierpsychol 25:377–394

Albrecht H, Apfelbach R, Wickler W (1968) Über die Eigenständigkeit der Art *Tilapia grahami* Boulenger, ihren Grubenbau und die Zucht in reinem Süßwasser (Pisces, Cichlidae). Senck Biol 49:107–111

Avery S (2010) Hydrological impacts of Ethiopia's Omo basin on Kenya's Lake Turkana water levels and fisheries. Final report prepared for the African Development Bank, Tunis. http://www.afdb.org/. Accessed 23 Feb 2014

Bergman AN, Laurent P, Otiang'a-Owiti G, Bergman HL, Walsh PJ, Wilson P, Wood CM (2003) Physiological adaptations of the gut in the Lake Magadi tilapia, *Alcolapia grahami*, and alkaline and saline adapted fish. Comp Biochem Physiol A 136:701–715

Boulenger GA (1912) On a collection of fishes made by Mr. A. Blayney Percival in British East Africa to the East of Lake Baringo. Proc Zool Soc Lond 1912:672–676

Brauner CJ, Gonzalez RJ, Wilson JM (2013) Extreme environments: hypersaline, alkaline and ion poor waters. In: McCormick SD, Farrell A, Brauner CJ (eds) Fish physiology, vol 32. Academic, Boca Raton, pp 435–476

Brook L (1994) Changes in the limnological behavior of a tropical African explosion crater lake: L. Hora-Kilole, Ethiopia. Limnologica 24:57–70

Brook L (2003) Ecological changes of two Ethiopian lakes as caused by contrasting human interventions. Limnologica 33:44–53

Burtzer KW, Isaac GL, Richardson JL, Washbourn KC (1972) Radiocarbon dating of East African Lake levels. Science 175:1069–1076

Bwathondi POJ (2002) Fish diversity of the wild and aquaculture water bodies in Singida Region. Tanzania Fisheries Research Institute (TAFIRI), Dar es Salaam, Tanzania. http://aquaticcommons.org/4516/1/Bwathondi.pdf. Accessed 4 Mar 2014

Bwathondi POJ, Ngatunga BP, Mwambungu JA (2000) Biodiversity of the aquatic fauna of Tarangire river and Lake Manyara catchment basin, Tanzania. Unpublished consultancy report to WWF

Coe MJ (1965) *Tilapia grahami* Boulenger. A study in environmental extremes. J Appl Ecol 2:414

Coe MJ (1966) The biology of *Tilapia grahami* Boulenger in Lake Magadi, Kenya. Acta Trop 23:146–177

Coe MJ (1967) Local migration of *Tilapia grahami* Boulenger in Lake Magadi, Kenya in response to diurnal temperature changes in shallow water. Afr J Ecol 5:171–174

Cole GA (1994) Text book of limnology, 4th edn. Waveland Press, Prospect Heights

Coleman HE, Johnson VK (1988) Summary of trout management at Pyramid Lake, Nevada with emphasis on Lahontan cutthroat trout, 1954–1987. Am Fish Soc Symp 4:107–115

Constanza R (1997) The value of the world's ecosystem services and natural capital. Nature 387:253–260

Cooper RN, Wissel B (2012) Interactive effects of chemical and biological controls on food-web composition in saline prairie lakes. Aquat Biosyst 8:29. doi:10.1186/2046-9063-8-29

Cossins AR, Crawford DL (2005) Opinion – fish as models for environmental genomics. Nat Rev Genet 6:324–333

Eddy FB, Maloiy GMO (1984) Ionic content of body fluids and sodium efflux in *Oreochromis alcalicus grahami*, a fish living at temperatures above 30 °C and in conditions of extreme alkalinity. Comp Biochem Physiol A 78:359–361

Eddy FB, Bamford OS, Maloiy GMO (1981) Na^+ and Cl^- effluxes and ionic regulation in *Tilapia grahami*, a fish living in conditions of extreme alkalinity. J Exp Biol 91:349–353

El-Shafai SA, El-Gohary FA, Nasr FA, van der Steen NP, Gijzen HJ (2004) Chronic ammonia toxicity to duckweed-fed tilapia (*Oreochromis niloticus*). Aquaculture 232:117–127

El-Zaeem SY, Ahmed MM, Salama ME, Darwesh DMF (2012) Production of salinity tolerant tilapia through interspecific hybridization between Nile tilapia (*Oreochromis niloticus*) and red tilapia (*Oreochromis* sp.). Afr J Agric Res 7:2955–2961

El-Zaeem SY, Ahmed MM, Salama ME, El-Maremie HA (2013) Production of salinity tolerant Nile tilapia, *Oreochromis niloticus* through traditional and modern breeding methods: II. Application of genetically modified breeding by introducing foreign DNA into fish gonads. Afr J Biotechnol 10:684–695

Ferguson AJD, Harbott BJ (1982) In: Hopson AJ (ed) Geographical, physical and chemical aspects of lake Turkana. Overseas Development Administration, London, vol 1. pp 1–107

Floria D, Gustenelli A, Caffara M, Turci F, Quaglio F, Konecny R, Nilowitz T, Wathuta EM, Magana A, Otachi EO, Matolla GK, Warugu HW, Liti D, Mbaluka R, Thiga B, Munguti J, Akoll P, Mwanja W, Asaminew K, Tadesse Z, Fioravanti ML (2009) Veterinary and public health aspects in tilapia (*Oreochromis niloticus*) aquaculture in Kenya, Uganda and Ethiopia. Ittiopatologia 6:51–93

Franklin CE, Crockford T, Johnston IA, Kamunde C (1994) The thermostability of hemoglobins from the hot-spring fish *Oreochromis alcalicus grahami*: comparisons with Antarctic and temperate species. J Thermal Biol 19:277–280

Franklin CE, Crockford T, Johnston IA, Kamunde C (1995) Scaling of oxygen consumption in Lake

Magadi tilapia, *Oreochromis alcalicus grahami*: a fish living at 37 °C. J Fish Biol 46:829–834

Frisch D, Green AJ, Figuerola J (2007) High dispersal capacity of a broad spectrum of aquatic invertebrates via waterbirds. J Aquat Sci 69:568–574

Froese R, Pauly D (2014) FishBase. World Wide Web electronic publication. www.fishbase.org. Accessed Feb 2014

Fryer G, Iles TD (1972) The cichlid fishes of the great lakes of Africa. Their biology and evolution. Oliver and Boyd, Edinburgh

Getahun A, Lazara KJ (2001) *Lebias stiassnyae*: a new species of killifish from Lake Afdera, Ethiopia (Teleostei: Cyprinodontidae). Copeia 2001:150–153

Getahun A, Stiassny MLJ (1998) The freshwater biodiversity crisis: the case of the Ethiopian fish fauna. SINET Ethiop J Sci 21:207–230

Githaiga JM (1997) Utilization patterns and inter-lake movements of the lesser flamingo and their conservation in the saline lakes of Kenya. In: Howard G (ed) Conservation of the lesser flamingo in Eastern Africa and beyond Nairobi. IUCN, Kenya, pp 11–21

Goerner A, Jolie E, Gloaguen R (2009) Non-climatic growth of the saline Lake Beseka, Main Ethiopian Rift. J Arid Environ 73:287–295

Golubtsov AS, Dgebuadze YY, Mina MV (2002) Fishes of the Ethiopian rift valley. In: Tudorancea C, Taylor WD (eds) Ethiopian rift valley lakes, Biology of inland water series. Backhuys, Leiden, pp 167–258

Grant WD (2004) Introductory chapter: half a lifetime in soda lakes. In: Ventosa A (ed) Halophilic microorganisms. Springer, Heidelberg, pp 17–22

Hammer TU (1986) Saline lake ecosystems of the world. Dr. W. Junk, Dordrecht

Harper DM, Childress RB, Harper MM, Boar RR, Hickley P, Mills SC, Otieno N, Drane T, Vareschi T, Nasirwa O, Mwatha WE, Darlington JPEC, Escute-Gasulla X (2003) Aquatic biodiversity and saline lakes: Lake Bogoria National Reserve, Kenya. Hydrobiologia 500:259–276

Herbst DB (2001) Gradients of salinity stress, environmental stability and water chemistry as a template for defining habitat types and physiological strategies in inland salt waters. Hydrobiologia 466:209–219

Hilgendorf FM (1905) Fische von Deutsch und Englisch Ost-Afrika, gesammelt von Oskar Neumann, 1893–1895. Zool Jahrb Abt Syst Geogr Biol Tiere 22:405–418

Hopson AJ (1982) Lake Turkana. A report on the findings of the Lake Turkana project 1972–1975, vol 1–6. Overseas Development Administration, London, p 1614

Hughes L (2008) Mining the Maasai Reserve: the story of Magadi. J East Afr Stud 2:134–164

Hughes RH, Hughes JS (1992) A directory of African wetlands. IUCN/UNEP/WCMC, Gland/Nairobi/Cambridge

Issar AS (2003) Climate changes during the Holocene and their impact on hydrological systems. Cambridge University Press, Cambridge

Jellison R (2005) IX International conference on salt lake research: research opportunities and management challenges. Saline Syst 1:12

Jellison R, Zadereev YS, DasSarma PA, Melack JM, Rosen MR, Degermendzhy AG, DasSarma S, Zambrana G (2004) Conservation and management challenges of saline lakes: a review of five experience briefs. Lake Basin Management Initiative. Thematic Paper 1–28

Johannsson OE, Bergman HL, Wood CM, Laurent P, Kavembe DG, Bianchini A, Maina JN, Chevalier C, Bianchini LF, Papah MB, Ojoo RO (2014) Air breathing in Magadi tilapia *Alcolapia grahami*, under normoxic and hyperoxic conditions, and the association with sunlight and reactive oxygen species. J Fish Biol 84:844–863

Johansen K, Maloiy GM, Lykkeboe G (1975) A fish in extreme alkalinity. Respir Physiol 24:159–162

Johnson TC, Malala JO (2009) Lake Turkana and its link to the Nile. In: Dumond HJ (ed) The Nile: origin, environments, limnology and human use. Springer, New York, p 818

Johnston IA, Guderley H, Franklin CE, Crockford T, Kamunde C (1994) Are mitochondria subject to evolutionary temperature adaptation? J Exp Biol 195:293–306

Kallqvist T, Lien L, Liti D (1988) Lake Turkana – limnological study 1985–1988. Norwegian Institute for Water Research (NIVA), Kenya Marine and Fisheries Institute

Kamal AMM, Mair GC (2005) Salinity tolerance in superior genotypes of tilapia, *Oreochromis niloticus*, *Oreochromis mossambicus* and their hybrids. Aquaculture 247:189–201

Kavembe GD, Machado-Schiaffino G, Meyer A (2014) Pronounced genetic differentiation of small, isolated and fragmented tilapia populations inhabiting the Magadi soda lake in Kenya. Hydrobiologia 739:55–71

Kolding J (1989) The fish resources of Lake Turkana and their environment. Thesis for the Cand. Scient. degree in fisheries biology and final report of KEN 043 Trial Fishery 1986–1987. University of Bergen, Norway, Report to NORAD, Oslo

Kolding J (1995) Changes in species composition and fish populations in Lake Turkana. In: Pitcher TJ, Hart BH (eds) The impact of species changes in African lakes, vol 18, Fish and fisheries series. Chapman and Hall, New York, pp 335–363

Laurent P, Maina JN, Bergman HL, Narahara AN, Walsh PJ, Wood CM (1995) Gill structure of a fish from an alkaline lake: effect of short-term exposure to neutral conditions. Can J Zool 73:1170–1181

Legesse D, Gasse F, Radakovitch O, Vallet-Coulomb C, Bonnefille R, Verschuren D, Gibert E, Barker P (2002) Environmental changes in a tropical lake (Lake Abiyata, Ethiopia) during recent centuries. Palaeogeogr Palaeoclim Palaeoecol 187:233–258

Lindley TE, Scheiderer CL, Walsh PJ, Wood CM, Bergman HL, Bergman AN, Laurent P, Wilson P, Anderson PM (1999) Muscle as a primary site of urea cycle enzyme activity in an alkaline lake-adapted tilapia, *Oreochromis alcalicus grahami*. J Biol Chem 274:29858–29861

Lykkeboe G, Johanson K, Maloiy GMO (1975) Functional properties of hemoglobins in the teleost *Tilapia grahami*. J Comp Physiol 104:1–11

Maina JN (1990) A study of the morphology of the gills of an extreme alkalinity and hyperosmotic adapted teleost *Oreochromis alcalicus grahami* (Boulenger) with particular emphasis on the ultrastructure of the chloride cells and their modifications with water dilution. Anat Embryol 181:83–98

Maina JN (1991) A morphometric analysis of chloride cells in the gills of the teleosts *Oreochromis alcalicus* and *Oreochromis niloticus* and a description of presumptive urea-excreting cells in *O. alcalicus*. J Anat 175:131–145

Maina JN (2000a) Functional morphology of the gas-gland cells of the air-bladder of *Oreochromis alcalicus grahami* (Teleostei: Cichlidae): an ultrastructural study on a fish adapted to a severe, highly alkaline environment. Tissue Cell 32:117–132

Maina JN (2000b) The highly specialized secretory epithelium in the buccal cavity of the alkalinity adapted Lake Magadi cichlid, *Oreochromis alcalicus grahami* (Teleostei: Cichlidae): a scanning and transmission electron microscope study. J Zool 251:427–438

Maina JN, Wood CM, Narahara A, Bergman HL, Laurent P, Walsh PJ (1995) Morphology of the swim bladder of a cichlid teleost: *Oreochromis alcalicus grahami* (Trewavas 1983), a fish adapted to a hyperosmotic, alkaline and hypoxic environment: a brief outline of the structure and function of the swim bladder. In: Munshi JS, Dutta HM (eds) Horizons of new research in fish morphology in the 20th century. Oxford and IBH, New Delhi, pp 179–192

Maina JN, Kisia SM, Wood CM, Narahara A, Bergman HL, Laurent P, Walsh PJ (1996) A comparative allometry study of the morphology of the gills of an alkaline adapted cichlid fish, *Oreochromis alcalicus grahami* of Lake Magadi, Kenya. Int J Salt Lake Res 5:131–156

Maloiy GMO, Lykkeboe G, Johansen K, Bamford OS (1978) Osmoregulation in *Tilapia grahami*: a fish in extreme alkalinity. In: Schmidt-Nielsen K, Bolis L, Maddrell SHP (eds) Comparative physiology: water, ions and fluid mechanics. Cambridge University Press, Cambridge, pp 229–238

McKaye KR, van den Berghe EP (1996) Specialized egg feeding behavior by African and Central American cichlids. Ichthyol Explor Freshw 7:143–148

Melack JM (1996) Saline and freshwater lakes of the Kenyan rift valley. In: McClanahan TR, Young TP (eds) East African ecosystems and their conservation. Oxford University Press, New York, pp 171–190

Mugisha AG, Nhwani LB, Daffa MJ, Howell K (1993) Evaluation of selected wetlands for identification of one wetland to be dedicated to the Ramsar Convention by the Tanzania Government. In: Technical Report submitted to National Environment and Management Council (NEMC) Tanzania

Muška M, Vašek M, Modrý D, Jirků M, Ojwang WO, Malala JO, Kubečka J (2012) The last snapshot of natural pelagic fish assemblage in Lake Turkana, Kenya: a hydroacoustic study. J Great Lakes Res 38:98–106

Nagl S, Tichy H, Mayer WE, Takezaki N, Takahata N, Klein J (2001) The origin and age of haplochromine fishes in Lake Victoria, East Africa. Proc R Soc B 267:1049–1061

Narahara A, Bergman HL, Laurent P, Maina JN, Walsh PJ, Wood CM (1996) Respiratory physiology of the Lake Magadi tilapia (*Oreochromis alcalicus grahami*), a fish adapted to a hot, alkaline, and frequently hypoxic environment. Physiol Zool 69:1114–1136

Oduor SO, Schagerl M (2007) Temporal trends of ion contents and nutrients in the three Kenya rift valley saline-alkaline lakes and their influence on phytoplankton biomass. Hydrobiologia 548:59–68

Ojwang WO, Ojuok JE, Omondi R, Malala J, Abila R, Ikmat P (2010) Impacts of river impoundments: the case of hydro power projects on Omo River of Lake Turkana. SAMAKI News, Department of Fisheries, Kenya, vol 6, pp 26–28

Okeyo DO (2006) On the distribution of fishes of the Kenya's Great Rift Valley drainage system. Discov Innov 18:141–159

Olaka LA, Odada EO, Trauth MH, Olago DO (2010) The sensitivity of East African rift lakes to climate fluctuations. J Paleolimnol 44:629–644

Papah MB, Kisia SM, Ojoo RO, Makanya AN, Wood CM, Kavembe GD, Maina JN, Johannsson OE, Bergman HL, Laurent P, Chevalier C, Bianchini A, Bianchini LF, Onyango DW (2013) Morphological evaluation of spermatogenesis in Lake Magadi tilapia (*Alcolapia grahami*): a fish living on the edge. Tissue Cell 45:371–382

Pörtner HO, Schulte PM, Wood CM, Schiemer F (2010) Niche dimensions in fishes: an integrative view. Physiol Biochem Zool 83:808–826

Ramsar (2014) Ramstar site list. http://www.ramsar.org/pdf/sitelist.pdf. Accessed 15 Mar 2014

Randall DJ, Wood CM, Perry SF, Bergman HL, Maloiy GMO, Mommsen TP, Wright PA (1989) Urea excretion as a strategy for survival in a fish living in a very alkaline environment. Nature 337:165–166

Reite OB, Maloiy GMO, Aasehaug B (1974) pH, salinity and temperature tolerance of Lake Magadi *tilapia*. Nature 274:315–316

Roberts TR (1993) Just another dammed river? Negative impacts of Pak Mun Dam on fishes of the Mekong basin. Nat Hist Bull Siam Soc 41:105–133

Roberts N, Taieb M, Barker P, Damnati B, Icole M, Williamson D (1993) Timing of the younger Dryas event in East Africa from lake-level changes. Nature 366:146–148

Sala OE, Chapin FS, Armesto JJ, Berlow E, Bloomfield J, Dirzo R, Huber-Sanwald E, Huenneke LF, Jackson RB, Kinzig A, Leemans R, Lodge DM, Mooney HA, Oesterheld M, Poff NL, Sykes MT, Walker BH, Walker M, Wall DH (2000) Global biodiversity scenarios for the year 2100. Science 287:1770–1774

Seegers L, Tichy H (1999) The *Oreochromis alcalicus* flock (Teleostei: Cichlidae) from Lakes Natron and Magadi, Tanzania and Kenya, with descriptions of two new species. Ichthyol Explor Freshw 10:97–146

Seegers L, Sonnenberg R, Yamamoto R (1999) Molecular analysis of the *Alcolapia* flock from Lakes Natron and Magadi, Tanzania and Kenya (Teleostei: Cichlidae), and implications for their systematics and evolution. Ichthyol Explor Freshw 10:175–199

Seegers L, De Vos L, Okeyo DO (2003) Annotated checklist of the freshwater fishes of Kenya (excluding the lacustrine haplochromines from Lake Victoria). J East Afr Nat Hist 92:11–47

Skadhauge E, Lechene CP, Maloiy GMO (1980) *Tilapia grahami*: role of intestine in osmoregulation under conditions of extreme alkalinity. In: Lahlou B (ed) Epithelial transport in the lower vertebrates. Cambridge University Press, Cambridge, pp 133–142

Sosa-López A, Mouillot D, Ramos-Miranda J, Flores-Hernandez D, Do Chi T (2007) Fish species richness decreases with salinity in tropical coastal lagoons. J Biogeogr 34:52–61

Street-Perrott FA, Roberts N (1983) Fluctuations in closed-basin lakes as an indicator of past atmospheric circulation patterns. In: Street-Perrott A, Beran M, Ratcliffe R (eds) Variations in the global water budget. Reidel, Dordrecht, pp 331–345

Thieme ML, Abell R, Burgess N, Lehner B, Dinerstein E, Olson D, Teugels G, Kamdem-Toham A, Stiassny MLJS, Skelton P (2005) Freshwater ecoregions of Africa and Madagascar: a conservation assessment. Island Press, Washington, DC

Tichy H, Seegers L (1999) The *Oreochromis alcalicus* flock (Teleostei: Cichlidae) from lakes Natron and Magadi, Tanzania and Kenya, a model for the evolution of "new" species flocks in historical times? Ichthyol Explor Freshw 10:147–174

Tobler M, Plath M (2011) Living in extreme habitats. In: Evans J, Pilastro A, Schlupp I (eds) Ecology and evolution of poeciliid fishes. University of Chicago Press, Chicago, pp 120–127

Trewavas E (1983) Tilapiine fishes of the genera *Sarotherodon*, *Oreochromis* and *Danakilia*. British Museum of Natural History, London

Vareschi E (1979) The ecology of Lake Nakuru (Kenya). II. Biomass and spatial distribution of fish (*Tilapia grahami* Boulenger = *Sarotherodon alcalicum grahami* Boulenger). Oecologia 37:321–325

Verschuren D (1996) Comparative palaeolimnology in a system of four shallow, climate-sensitive tropical lake basins. In: Johnson TC, Odada E (eds) The limnology, climatology and palaeoclimatology of the east African lakes. Gordon and Breach, Newark, pp 559–572

Verschuren D, Tibby J, Sabbe K, Roberts N (2000) Effects of depth, salinity, and substrate on the invertebrate community of a fluctuating tropical lake. Ecology 81:164–182

Vincens A, Casanova J (1987) Modern background of Natron-Magadi Basin (Tanzania-Kenya): physiography, climate, hydrology and vegetation. Sci Geol Bull 40:9–22

Walsh PJ, Bergman HL, Narahara A, Wood CM, Wright PA, Randall DJ, Maina JN, Laurent P (1993) Effects of ammonia on survival, swimming, and activities of enzymes of nitrogen metabolism in the Lake Magadi tilapia *Oreochromis alcalicus grahami*. J Exp Biol 180:323–327

Walsh PJ, Heitz M, Campbell CE, Cooper GJ, Medina M, Wang YS, Goss GG, Vincek V, Wood CM, Smith CP (2000) Molecular identification of a urea transporter in gills of the ureotelic gulf toadfish (*Opsanus beta*). J Exp Biol 203:2357–2364

Walsh PJ, Grosell M, Goss GG, Bergman HL, Bergman AN, Wilson P, Laurent P, Alper SL, Smith CP, Kamunde C, Wood C (2001) Physiological and molecular characterization of urea transport by the gills of the Lake Magadi tilapia (*Alcolapia grahami*). J Exp Biol 204:509–520

White TH (1953) Some speculations on the sudden occurrence of floods in the history of Lake Magadi. J East Afr Nat Hist 22:69–71

Williams WD (2002) Environmental threats to salt lakes and the likely status of inland saline ecosystems in 2025. Environ Conserv 29:154–167

Williams WD, Boulton AJ, Taaffe RG (1990) Salinity as a determinant of salt lake fauna: a question of scale. Hydrobiologia 381:191–201

Wilson PJ, Wood CM, Maina JN, White BN (2000) Genetic structure of Lake Magadi tilapia populations. J Fish Biol 56:590–603

Wilson PJ, Wood CM, Walsh PJ, Bergman AN, Bergman HL, Laurent P, White BN (2004) Discordance between genetic structure and morphological, ecological, and physiological adaptation in Lake Magadi tilapia. Physiol Biochem Zool 77:537–555

Wood CM, Perry SF, Wright PA, Bergman HL, Randall DJ (1989) Ammonia and urea dynamics in the Lake Magadi tilapia, a ureotelic teleost fish adapted to an extremely alkaline environment. Respir Physiol 77:1–20

Wood CM, Bergman HL, Laurent P, Maina JN, Narahara A, Walsh PJ (1994) Urea production, acid-base regulation, and their interactions in the Lake Magadi tilapia, a unique teleost adapted to a highly alkaline environment. J Exp Biol 189:13–36

Wood CM, Wilson PW, Bergman HL, Bergman AN, Laurent P, Otiang'a-Owiti G, Walsh PJ (2002a) Iono-regulatory strategies and the role of urea in the Magadi tilapia (*Alcolapia grahami*). Can J Zool 80:503–515

Wood CM, Wilson PW, Bergman HL, Bergman AN, Laurent P, Otiang'a-Owiti G, Walsh PJ (2002b) Obligatory urea production and the cost of living in the Magadi tilapia revealed by acclimation to reduced salinity and alkalinity. Physiol Biochem Zool 75:111–122

Wood CM, Bergman HL, Bianchini A, Laurent P, Maina J, Johannsson OE, Bianchini LF, Chevalier C, Kavembe GD, Papah MB, Ojoo RO (2012) Transepithelial potential in the Magadi tilapia, a fish living in extreme alkalinity. J Comp Physiol 182:247–258

Wood CM, Nawata CM, Wilson JM, Laurent P, Chevalier C, Bergman HL, Bianchini A, Maina JN, Johannson OE, Bianchini LF, Kavembe GD, Papah MB, Ojoo RO (2013) Rh proteins and NH_4^+-activated Na^+ATPase in the Magadi tilapia (*Alcolapia grahami*), a 100 % ureotelic teleost fish. J Exp Biol 216:2998–3007

World Heritage Sites (2014). http://whc.unesco.org/en/list/1060. Accessed 15 Mar 2014

Wright PA, Fyhn JH (2001) Ontogeny of nitrogen metabolism and excretion. In: Anderson PA, Wright PA (eds) Fish physiology, vol 20. Academic, Orlando, pp 149–200

Wright PA, Wood CM (2009) A new paradigm for ammonia excretion in aquatic animals: role of Rhesus (Rh) glycoproteins. J Exp Biol 212:2303–2312

Wright PA, Perry SF, Randall DJ, Wood CM, Bergman H (1990) The effects of reducing water pH and total CO_2 on a teleost fish adapted to an extremely alkaline environment. J Exp Biol 151:361–369

Yanda PZ, Madulu NF (2005) Water resource management and biodiversity conservation in the Eastern rift valley lakes, Northern Tanzania. Phys Chem Earth Part A/B/C 30:717–725

Yongo EO, Abila RO, Lwenya C (2010) Emerging resource use conflicts between Kenyan fishermen, pastoralists and tribesmen of Lake Turkana. Aquat Ecosyst Health Manage 13:28–34

Zaccara S, Crosa G, Vanetti I, Binelli G, Harper DM, Mavuti KM, Balarin JD, Britton JR (2014) Genetic and morphological analyses indicate high population mixing in the endangered cichlid *Alcolapia* flock of East Africa. Conserv Genet 15:429–440

Zinabu G (2002) The Ethiopia rift valley lakes: major threats and strategies for conservation. In: Tudorancea C, Taylor WD (eds) Ethiopia rift valley lakes. Backhages, Leiden, pp 259–270

Lesser Flamingo as a Central Element of the East African Avifauna 10

Lothar Krienitz, Barbara Mähnert, and Michael Schagerl

Abstract

The East African Rift Valley lakes host an avifauna that is rich in biodiversity and individual numbers. Some of the water bodies are known as 'flamingo lakes'. They are renowned places even for non-biologists and attract hundreds of thousands of tourists from all over the world. Two flamingo species inhabit these lakes, the Greater and the Lesser Flamingo. The latter is classified as 'near threatened' by the IUCN. Flamingos, considered by birdlovers as a reincarnation of the mythic firebird *Phoenix*, play a key role in the circle of life in the harsh alkaline aquatic environments developed on the volcanic ashes of the African Rift Valley. We focus on the Lesser Flamingo, a flagship species of the East African avifauna, and provide information about the life cycle and phylogeny, feeding behaviour and inter-lake movements. Moreover, threats to the flamingo populations, including possible reasons for infrequent mass mortalities, are discussed.

10.1 Introduction

Visiting an East African saline lake (EASL) during a phase of high presence of Lesser Flamingos (*Phoeniconaias minor* Geoffroy Saint-Hilaire) and the associated avifauna definitely touches all senses (Brown 1959; Mari and Collar 2000). Hundreds of thousands of birds with their pinkish-coloured bodies attract our eyes (Fig. 10.1) and the monotonous chattering background noise sounds like music. This, together with the pronounced smell of decomposing Cyanobacteria and bird droppings, the high

L. Krienitz (✉)
Department of Experimental Limnology, Leibniz-Institute of Freshwater Ecology and Inland Fisheries, 16775 Stechlin, Germany
e-mail: krie@igb-berlin.de

B. Mähnert • M. Schagerl
Department of Limnology and Bio-Oceanography, University of Vienna, Althanstrasse 14, 1090 Vienna, Austria
e-mail: barbara1990@gmx.at; michael.schagerl@univie.ac.at

M. Schagerl (ed.), *Soda Lakes of East Africa*, DOI 10.1007/978-3-319-28622-8_10

Fig. 10.1 Flamingos of East African soda lakes. (**a**) Species comparison of Lesser Flamingos (*Phoeniconaias minor*) in the foreground and Greater Flamingos (*Phoenicopterus roseus*) in the background; (**b**) huge flocks visiting Lake Bogoria (Kenya); (**c**) feeding position of a Lesser Flamingo; (**d**) flamingos on the shore of Lake Chitu (Ethiopia)

Fig. 10.2 Raptors. (**a**, **b**) Steppe Eagle (*Aquila nipalensis orientalis*); (**c**) Rüppell's Griffon Vulture (*Gyps rueppellii*); (**d**) Long-crested Eagle (*Lophaetus occipitalis*); (**e**) Fish Eagle (*Haliaeetus vocifer*)

radiation burning the skin and caustic soda dust, makes such a visit an unforgettable experience.

A rich and spectacular avifauna inhabits the EASL (Figs. 10.1, 10.2, 10.3, 10.4, 10.5 and 10.6); waterfowl censuses yielded approximately

Fig. 10.3 Avifauna of East African soda lakes. (**a**, **b**) Marabou Stork (*Leptoptilos crumeniferus*); (**c**) African Spoonbill (*Platalea alba*); (**d**) Yellow-billed Stork (*Mycteria ibis*); (**e**) Little Egret (*Egretta garzetta*); (**f**) Cattle Egret (*Bubulcus ibis*)

200 different species. Depending on lake size and condition, bird occurrence fluctuates considerably. Pomeroy et al. (2003) studied waterfowl of small saline alkaline crater lakes in Uganda and found 72 species in total. The small crater lake Chitu (Ethiopia) is the most specific one, hosting exclusively one bird species, the Lesser Flamingo. For the nearby Lake Abijata

Fig. 10.4 Avifauna of East African soda lakes. (**a**) Hadada Ibis (*Bostrychia hagedash*); (**b**) Sacred Ibis (*Threskiornis aethiopicus*); (**c**) Black-winged Stilt (*Himantopus himantopus*); (**d**) Hamerkop (*Scopus umbretta*)

(Ethiopia), merely 22 species have been reported (Ewnetu et al. 2013). The Bogoria National Reserve (Kenya) hosts 223 bird species living at the lake shores and immediate surroundings (Harper et al. 2003). The avifauna can be assigned to their respective ecological niches. Nonetheless, our understanding of the ecological requirements of tropical water birds and the links

Fig. 10.5 Avifauna of East African soda lakes. (**a**) Common Ostrich (*Struthio camelus*) two males and one female (*centre*); (**b**) Egyptian Goose (*Alopochen aegyptiacus*); (**c**) bigger Egyptian Geese and smaller Cape Teals (*Anas capensis*); (**d**) the large colony of Great White Pelicans at Nakuru (*Pelecanus onocrotalus*)

between the different taxa remains rudimentary (Owino et al. 2001).

The Lesser Flamingo is the centrepiece of birdlife at EASL and a flagship species of the East African avifauna. The species is the focus of this compilation. A recent estimation of its abundance in the primary ranges of East Africa (Ethiopia, Kenya, Tanzania, Uganda) is between 832,260 and 2,126,834 individuals (Childress et al. 2008). Three smaller ranges of Lesser

Fig. 10.6 Fluctuating water levels of the amplifier Lake Nakuru result in different avifauna. (**a**, **b**) Large flocks of flamingos on the lake in 2004; whitish layers are trona, and pinkish areas along the shoreline are flamingos. (**c**, **d**, **e**) High water levels from 2013 onwards caused a shift in the bird community; (**c**) Red-billed Teals (*Anas erythrorhyncha*); (**e**) Grey-headed Gulls (*Larus cirrocephalus*). (**b**) and (**d**) show exactly the same area (please refer to the three trees in the centre of the photos)

Flamingos are known, which are South Africa (Botswana, Namibia, South Africa) with around 15,955–87,835 individuals, West Africa (Guinea, Mauritania) with 11,285–17,800 individuals and India, the only primary range state of the Asian continent, with 17,045–411,355 Lesser Flamingos (Childress et al. 2008).

Table 10.1 Comparison of the two flamingo species at soda lakes of East Africa

	Phoeniconaias minor Geoffroy Saint-Hilaire 1798	*Phoenicopterus roseus* Pallas 1811
	Lesser Flamingo	Greater Flamingo
Habitus		
Distribution	Africa, India	Africa, Middle East, South-East Europe
Number	2–3 million	600,000
Size (height)	<100 cm	110–150 cm
Weight	<2.5 kg	<3.5 kg
Colour	Feathers white or pink, legs pink, bill burgundy in different colour gradings	Feathers white or pale pink, legs pink, bill base pink, apex black
Bill, filter lamellae	Cross section deeply keeled, large areas of laminae, filter lamellae with fringed platelets	Cross section shallowly keeled, small areas of laminae, filter lamellae without fringed platelets
Eyes	Red	Pale yellow
Food	Surface feeder with a special diet: Cyanobacteria and algae	Bottom feeder with a more general diet: small molluscs, crabs, insect larvae, copepods, daphnids, mud, large algae

Observing the huge flocks at the lake shore reveals that a second flamingo species does occur at the EASL: the Greater Flamingo (*Phoenicopterus roseus* Pallas) (Fig. 10.1). About 90 % of the flamingo population at EASL is represented by Lesser Flamingos and 10 % by the Greater Flamingos; diacritic features are summarised in Table 10.1. The two species do not compete with one another because they have two different food selection tools and feeding strategies, featured as specialist versus generalist. Whereas the Lesser Flamingo is a primary consumer and preferably ingests Cyanobacteria and algae from the water surface, the Greater Flamingo is a secondary consumer and feeds mainly on small invertebrates from the lake bottom (Brown 1975; Jenkin 1957; Ridley et al. 1955).

Predators and scavengers are closely linked to flamingo life and death. Although predators are low in number, they are very efficient in hunting young and weakened flamingos. Predators are represented by Marabou Storks (*Leptoptilos crumeniferus* Lesson) and birds of prey such as Fish Eagles (*Haliaeetus vocifer* Daudin), Tawny Eagles (*Aquila rapax* Temminck) and Steppe Eagles (*Aquila nipalensis orientalis* Hodgson) (Figs. 10.2 and 10.3). Foraging areas of the predators are lake shores; their nests are usually built in large trees in the vicinity of the lakes. A number of vulture species such as the Egyptian Vulture (*Neophron percnopterus* L.) fly in from further away; they prey on eggs and chicks in breeding colonies, but also feed on carcasses (Brown and Root 1971). An interesting link between predators and flamingos are flocks of Grey-headed Gulls (*Larus cirrocephalus* Vieillot: Fig. 10.6e). Brown (1959) observed them circling along the lake shores with their characteristic shrieks. The gulls do not actively kill flamingos, but are in resource competition with eagles because they scavenge carcasses. Eagles are therefore forced to search for additional food and start increased hunting on young flamingos.

After introduction of salt-tolerant fish into several EASL, for example, in the early 1960s, the Magadi Tilapia *Alcolapia grahami* Boulenger into Lake Nakuru (Kenya), piscivorous birds (fish-eaters such as pelicans, cormorants, Hamerkop, terns, gulls, storks and herons; Figs. 10.4, 10.5 and 10.6) increased in numbers (Vareschi and Jacobs 1984). Amongst these secondary consumers, the pelicans (*Pelecanus onocrotalus* L.; Fig. 10.5d) have the strongest influence on the fish population, consuming about 90 % (Vareschi and Jacobs 1985). The abundance of fish-eating birds depends on fish density, which undergoes considerable fluctuations.

A large number of waterbirds depend directly on detritivorous chironomids (nonbiting midges), which are distributed in oxic zones of the water column. During a sampling survey in Lake Bogoria (Kenya) in the early 2000s, only a single taxon of chironomids was observed (cf. *Paratendipes* Kieffer). It was able to establish a wet biomass of about 3 t for the whole lake. 210 kg day^{-1} of adult chironomids emerged, which together with pupal exuvia and dead adults formed the food base for 50 different bird taxa such as teals, grebes, swifts and swallows (Harper et al. 2003). Also Ephydridae (shore and brine flies) show mass emergences at Lake Bogoria (see Chaps. 8 and 12). At other saline lakes, different species of chironomids may also act as a food source. For example, at Lake Nakuru, *Leptochironomus deribae* Freeman builds the food base for many waterbirds (Vareschi and Jacobs 1985).

At swampy lake banks, numerous waders pick invertebrates from the mud, some of them arriving from the Palearctics. Most species belong to the families Charadriidae (plovers, lapwings), Scolopacidae (sandpipers, snipes) and Recurvirostridae (avocets, stilts) (Fig. 10.4a–c). Their food is variable and comprises mainly arthropods including chironomids and other insects, worms, molluscs and plant parts such as seeds and decomposing shoots. Larger waterbirds wading around swampy shorelines belong to the Threskiornithidae (ibises, spoonbills; Figs. 10.3c and 10.4a, b) and Ciconiidae (storks; Fig. 10.3d). They prefer feeding on small fish, reptiles and mammals, but also feed on smaller animals such as insects. The shallow lagoons are home to Anatidae (ducks), living in pairs or flocks and grasping for grass and other herbs, submerged plants and tiny animals (Fig. 10.5b, c).

The number of the waterbird species recorded at individual EASL differs greatly. Waterfowl censuses between 1992 and 1999 revealed mean ratios of flamingos to non-flamingo counts: 108,081–10,093 (91:9 %) (Elmentaita), 232,275–20,275 (92:8 %) (Nakuru) and 521,925–5060 (99:1 %) (Bogoria). As expected from previous observations, Lake Bogoria hosted the highest number of flamingos but the lowest number of non-flamingos (Owino et al. 2001). This probably reflects the harsh lake conditions controlling the food sources. For Bogoria, growth of the preferred food source of Lesser Flamingos, *Arthrospira fusiformis* (Voronikhin) Komárek and J. W. G. Lund, is boosted in this extreme milieu (Fig. 10.7b). Piscivorous birds are excluded from this lake because fish are lacking.

These few examples show the complexity of the avifauna of EASL, which host in addition to the near-threatened Lesser Flamingo a number of protected and rare bird species such as Cape Teals (*Anas capensis* Gmelin) (Fig. 10.5c) and Black-necked Grebes (*Podiceps nigricollis* Boddaert). Last but not least, these lakes are of eminent importance as feeding and resting stations of Palearctic migratory birds. All these factors demand the highest priority for conservation of these unique and sensitive aquatic habitats.

10.2 Field Methods of Ornithologists

Identification of bird species, bird censuses and monitoring bird movements create the basis for further ornithological and ecological studies. The most popular methods are direct observations and counts by bird-watchers along the lake shorelines or from boats using binoculars. Small numbers of birds are counted individually, larger flocks are first divided into subsections and then

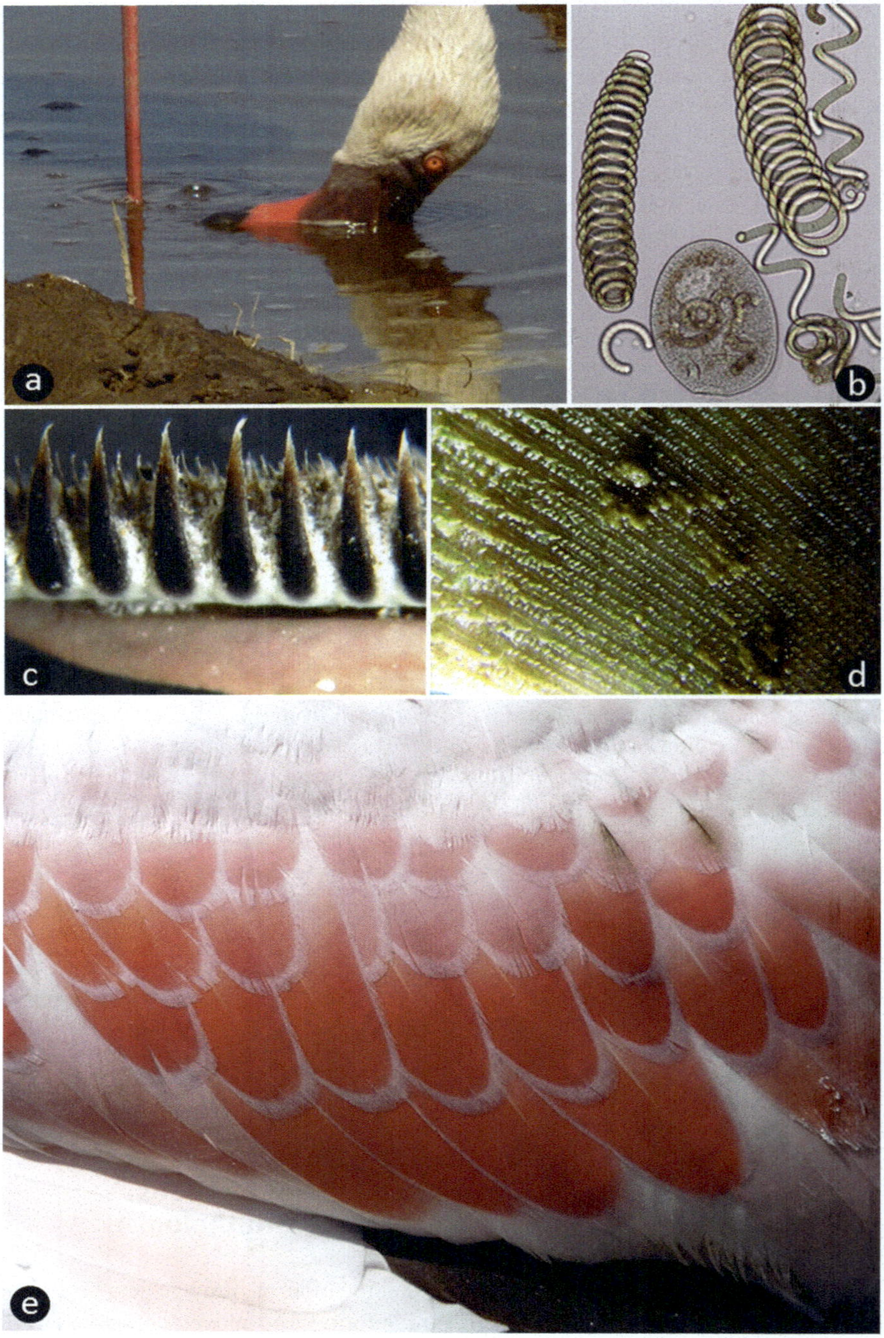

Fig. 10.7 Feeding technique of the Lesser Flamingo. (**a**) Characteristic feeding position for surface harvest; (**b**) the preferred main diet is the coiled cyanoprokaryote *Arthrospira fusiformis*; the ciliate *Frontonia* (*bottom*) also feeds on *A. fusiformis*; (**c**) excluders; (**d**) filter platelets inside the bill; (**e**) characteristic reddish plumage because of carotenoid inclusions

equivalent sections are estimated. This methodology was described in detail by Kihwele et al. (2014) and is applied in annual to biannual intervals by the African Waterfowl Census, which was initiated in Kenya in 1990 (Bennun and Nasirwa 2000). Also other countries follow this approach for flamingo counts (Ethiopia: Ewnetu et al. 2013; Tanzania: Baker and Baker 2002, 2006; Kihwele et al. 2014; Mlingwa and Baker 2006). It is coordinated worldwide by Wetlands International (Maina and Kyungu 2009).

Accurate countings of flamingos and their nests based on aerial photographs were introduced by Grzimek and Grzimek (1960) for Lake Natron. This methodology was later on also applied for Lake Nakuru (Vareschi 1978) and other EASL (Tuite 1979, 2000). More recently, 13 major lakes were studied in North Tanzania in 2002 (Raini and Ngowe 2009). At least in former times, analyses of the photographs were extremely time consuming (tagging single flamingos with pinholes, pers. comm. Chris Tuite). Additionally, this method is usually done in sequences lake by lake with some time gaps in between, during which huge flocks of flamingos might have moved. For reliable flamingo numbers and movement patterns, flamingo counts need to be synchronised at all flamingo sites, which is very difficult and expensive to realise. Fortunately, sophisticated methods of so-called reality mining are available nowadays, but they are not yet applied in the flamingo areas. Examples are the use of unmanned aerial aircraft (Sardà-Palomera et al. 2012; Chabot and Bird 2013) or satellite tracking; the latter was already successfully tested for few flamingos (~20) in South Africa (McCulloch et al. 2003) and East Africa (Childress et al. 2006; Salewski et al. 2009). Miniaturisation of animal-tracking devices and affordable running costs will enable tagging larger bird numbers in the very near future, which is the method of choice to study movement of flocks. Such automated techniques will allow in-depth analyses of social structures and other parameters of these enigmatic species in space and time (Krause et al. 2013).

Another promising approach is satellite-based remote sensing. To our knowledge, the resolution is still too low for counting single specimens, but in the near future this method will also be suitable for simultaneous counts. Remote sensing was already applied to evaluate flamingo food resources in five saline alkaline lakes in Kenya and Tanzania, in which in situ reflectance of the water and Landsat satellite imagery were combined to establish an optical classification scheme for Cyanobacteria and algae quantification based on chlorophyll-a (Tebbs et al. 2015). A basic requirement for such quantifications is calibration, which requires simultaneous chlorophyll-a measurements at the lakes. Additional analyses of the plankton community should also be considered because algal groups differ in their spectral properties.

Methods of capturing and handling of live Lesser Flamingos were described by Childress and Jarrett (2005). Standardised sampling of flamingo carcasses to analyse the causes of mass mortalities (see also below) are presented in detail by Cooper et al. (2014), who suggested suitable equipment and techniques, sample handling and standardised protocols.

10.3 The Phenotype of the Lesser Flamingo

Flamingos are considered by birdlovers as a reincarnation of the firebird *Phoenix*, which is also reflected in the scientific name of the family Phoenicopteridae ('Phoenix winged' or 'blood-red feathered'). Also the word 'flamingo', derived from the Spanish flamengo/flamenco, refers to the beautiful reddish plumage (Latin flamma = flame). Flamingos have peculiar features, which are difficult to explain in words: the special proportion of the tapered body and thin legs, the long and fragile neck, the huge crooked bill and the strikingly coloured plumage. The Lesser Flamingo is the smallest species of the family Phoenicopteridae.

The colour of feathers varies with taxon, age and also function. Juvenile Lesser Flamingos

Fig. 10.8 Emergence and decay in a flamingo life. (**a**) Nest building at Lake Oloidien (Kenya); (**b**) young Greater Flamingos; (**c**) Lesser Flamingos of different ages; (**d**) mass mortality at Lake Bogoria (Kenya) in August 2008, the underlying reasons are still under discussion

have a greyish plumage (Fig. 10.8c), which changes into gorgeous pink during adolescence (Fig. 10.1). Body feathers have a fine graduation of rose, wing feathers show stronger components of pink (Fig. 10.7e) and remiges (covered in non-flying situations) are black-brownish. The pink coloration originates from the carotenoids taken up via their major food, *Arthrospira fusiformis*. Mainly canthaxanthin (Thommen and Wackernagel 1963), astaxanthin, intermediate phoenicoxanthin (Fox et al. 1967) and phoenicopterone (Fox 1975), usually bound with proteins (Thommen 1975), are responsible for the pigmentation. Algal pigments are converted in the liver and skin.

10.4 Life Cycle of the Lesser Flamingo

Even a century ago, no one was able to explain the origin of flamingos. Folk tales of the Maasai suggested that young flamingos are children of the lake, arising mysteriously from the centre of Lake Natron. This enigma was solved by the famous researcher Leslie Brown (Brown 1955, 1959). The following description of the life cycle is mainly based on the excellent study of Lesser Flamingo breeding at lakes Natron and Magadi, which was done between 1954 and 1969 (Brown and Root 1971). More recent information and fascinating footage are provided in the documentary film 'The Crimson Wing' (Disneynature 2008).

Successful breeding of Lesser Flamingos has been reported for only a few sites worldwide: Lake Natron (Tanzania; the only site within the East African Rift Valley), Makgadikgadi Pan (Botswana), two salt pans of Etosha Pan (Namibia), an artificial breeding island in Kamfers Dam near Kimberley (South Africa) (Anderson and Anderson 2010), salt pans on the territory of the Rann of Kutch in Gujarat (India) (Childress et al. 2008) and small sites in Mauritania/Senegal (Fig. 10.9). Breeding areas require very special conditions because breeding success is highly dependent on solitude. The arid and shallow breeding grounds must be inaccessible for predators and humans and need to be flooded seasonally. In the further process of drying, islands of salty mud emerge from the shallow water; these serve as a basis for flamingo nests. In a figurative sense, flamingos—like the firebird *Phoenix*—arise from the ashes, which originated from the numerous volcanos in the African Rift Valley; the ash itself is used for building nests. From the breeding areas, the parents need to have access to feeding grounds and freshwater within a flying distance of <180 km (Childress et al. 2008; Simmons 1996; Tebbs et al. 2013). The breeding success of Lesser Flamingos fluctuates highly, but breeding failures lasting more than 10 years clearly have no effect on the general population dynamics (Parasharya et al. 2015). Estimates at lakes Natron and Magadi from 1954 to 1969 revealed highest breeding numbers in 1957, when 570,000 pairs accessed two breeding colonies at Natron, and 1962 when 935,000 pairs nested at Magadi. In 1954 and 1962, the total number of eggs including both sites was 3,190,000 and around 1,360,900 chicks fledged, resulting in a breeding success of 42.6 % (Brown and Root 1971).

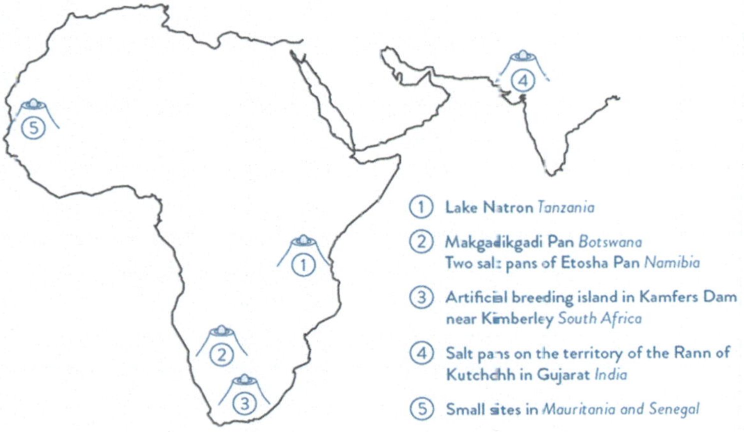

Fig. 10.9 Breeding areas of Lesser Flamingos. Main area is Lake Natron; Kamfers Dam is an artificial breeding ground

Breeding habitats of Lesser Flamingos are in an allegoric sense like the entrance to hell. Because of the hostile conditions, amongst which are heat and unstable surfaces, the pioneer of flamingo research in East Africa, Leslie Brown, almost paid his first walk to the breeding site of Lake Natron with his legs and life (Brown 1959). Hidden, in the centre of this remote soda lake, Lesser Flamingos breed without disturbance, but success depends on the actual precipitation and hydrological regime. During a phase of complete flooding of the mudflats of Lake Natron in 1962, the flamingo population escaped to the neighbouring Lake Magadi. There, the adult pairs built about one million nests and supported more than 350,000 chicks to develop fully fledged (Brown and Root 1971). Several thousand nests were recorded at Lake Abijata (Ethiopia) in 2005/2006, and 2500–3500 chicks of different age were counted (Bozic and Ewnetu 2008). On several other EASL, Lesser Flamingos also build nests. These include Bogoria, Manyara, Nakuru, Elmentaita, Oloidien and Turkana, but due to high disturbances, the breeding success remains zero (Brown 1955; own observations).

The big spectacle of courtship is frequently reported from lakes Bogoria and Nakuru, which are easier to access than Lake Natron (see also below). Displaying procedures last for hours and are repeated at the same site for days, weeks or even months. Nuptial displays finally result in copulation. After copulation, the flamingos start to accommodate at the breeding site. The breeding season at Lake Natron may start at any time of the year, but breeding is most successful during the short rainy season from October to December, when moderate precipitation drizzles the ground, producing viscid mud for nest building. If the bottom is completely flooded, breeding cannot take place. Ideal breeding conditions are directly after receding water levels, when a central island emerges and the rest of the lake area is covered by a patchwork of puddles and mudflats. As hydrological patterns vary erratically, each breeding event has an uncertain outcome and interplays of occasions.

Lesser Flamingos build nests resembling mini-craters arranged in large colonies of up to hundreds of thousands, with around 4–6 nests m^{-2} (Fig. 10.8a). Two main breeding sites are known in Lake Natron: the Gelai area with consistent mudflats and the Shompole colony, where the mud is covered by crystalline soda crusts. At Gelai, the nests are 'grouped' in tightly packed clusters with open areas in between. At Shompole, the nests are 'reticulated' in long strings along the weak lines of the soda crust polygons (see Chap. 1). The nest material differs in the ratio of mud to soda depending on the subsoil. At Magadi, the proportion of soda is higher, yielding a cement-like consistence of the mud and smaller nests. The nests weigh about 14–31 kg. In 1962, the breeding colony at Lake Magadi excavated about 20,000 t of soda mud for nest building, which was a considerably higher amount than the Magadi soda factory processed in the same year (for detailed measurements and calculations see Brown and Root 1971). Nests are not only for harbouring eggs and chicks; they also protect the offspring from excessive temperatures during daytime. Midday temperatures at the shoreline reach 40 °C in the scarce shade; at the sun-exposed white reflecting soda flats, 50–60 °C was measured. The temperature might even increase to 75 °C on dark mud. At the top of nests, temperatures were around 25 °C, which is explained by a chimney effect.

Usually, one chalky-white egg (size 72–86 × 48–51 mm) is laid per female and nest. The incubation time is ~28 days. The phase of roaming outside the nests is dangerous for the chicks because cement-like cuffs up to the size of a tennis ball may develop around the legs, considerably hampering the hatchlings. About 50,000 chicks born at Magadi in 1962 died because of this severe problem, but 27,000 were saved by volunteers who removed the anklets (Brown and Root 1971). The chicks fledge after ~70 days and are ready to discover other sites in the Rift Valley and further afar for 'everyday business'. The lifespan of Lesser Flamingos is 20–40 years.

10.5 Feeding

Lesser Flamingos are the only warm-blooded animals that feed exclusively on microphytes.

They are obligatory filter feeders and forage in either a standing or swimming position (Ridley et al. 1955). While wading in shallow water or on muddy areas, the neck is bent towards the bottom, and the head is tilted upside down, bringing the deeply keeled upper jaw of the bill in a downward direction (Figs. 10.1c and 10.7a). The birds sweep their heads from side to side just below the mud or water surface. The piston-shaped tongue vibrates up and down about 20 times per second, sucking in the suspension and retaining food particles inside the bill (Mari and Collar 2000). The bill flanks are equipped with a row of excluders (Fig. 10.7c), which preclude indigestible material (Burgis and Morris 1987). When the suspension is sucked inside the bill, hair-like fringed platelets of the filter lamellae inside the bill fold towards the pharynx; when the water is pumped outwards, the platelets erect and retain food particles (Fig. 10.7d). After enough food is trapped, the muscles in the distal part of the tongue press the material into the oesophagus. The filter lamellae of Lesser Flamingos are very fine and equipped with 25–40 fringed platelets cm^{-2} (Jenkin 1957), which fit perfectly to collect filaments of *Arthrospira*, benthic diatoms and other algae or particles of the same size (see Chap. 6; Kaggwa et al. 2013; Krienitz and Kotut 2010; Krienitz et al. 2016). Analyses of stomach contents of Lesser Flamingos revealed that the smallest particles ingested were pennate diatoms with a size between 15 and 70 µm, but the main food was coiled *Arthrospira* filaments between 40 and 800 µm (Jenkin 1957; Fig. 10.7b). In contrast to Lesser Flamingos, Greater Flamingos do not have fringed platelets on the filter lamellae, and their number of lamellae is considerably lower, namely, 12–15 cm^{-2} (Brown 1959). The entire number of lamellae in the bill of a Lesser Flamingo is up to 10,000 (Childress et al. 2008).

10.6 Behaviour

Flamingos are very social animals; they live together in large flocks (selfish herds) and are typical colonial breeders. Advantages of these bird agglomerations are seen in optimising food intake, avoiding pressure from predators and higher reproductive success; the latter has been proven for the Caribbean Flamingo (Stevens and Pickett 1994). Additional factors that have been discussed are stimulating synchronous nesting, facilitating pair formation and increased fertility/copulation success (Stevens 1991). These birds form strong pair bonds for one season, but may sometimes change their mates in very large colonies, presumably due to better mate variety (Studer-Thiersch 2000).

Both the Lesser Flamingo and the Greater Flamingo feed during the day and night (Ridley et al. 1955), but the Lesser Flamingo is more active during night-time. This feeding behaviour is probably related to wind occurring in the late afternoon, which makes feeding from the surface impossible due to strong wave action (Brown 1975). Usually, flamingos are very active at dusk and dawn and show somewhat reduced activity due to the high temperature around midday (Ridley et al. 1955). At dusk, Lesser Flamingos become more active, start circling around the lakes with complex manoeuvres (Fig. 10.10a) and thereby produce loud calls. They also need freshwater for drinking and preening (Fig. 10.10b). Flamingos therefore travel considerable distances to visit hot spring inflows, e.g. at Lake Bogoria (Krienitz et al. 2003; Ridley et al. 1955).

Flamingo courtship is famous and 11 ritual displays have been described in detail by Kahl (1975), some of which are listed below. The ritual displays are performed by both sexes, but usually initiated by the male. The males join together to flocks, bodies and necks erected, swinging their heads from side to side ('head flagging') with continuous calls to attract the females. They proceed with 'wing salutations' and 'inverted wing salutations', dancing and heads hooked in a crook-like posture ('broken neck display'). The feathers are partly erected (Fig. 10.10c), and crowded posing groups run in meanders along the shore, exhibiting powerful emotions ('marching'). 'False feeding' with heads swinging at the water surface also occurs during 'marching'. Courtship displays last for hours and are repeated at the same site for days, weeks or even months. Nuptial displays result in

Fig. 10.10 Lesser Flamingos in action. (**a**) Movement with characteristic flight silhouettes at Lake Oloidien; (**b**) hot springs at Lake Bogoria providing freshwater are overcrowded. Flamingos use them for drinking and for preening their plumage; (**c**) territorial displays

copulation, for which couples detach from the dense flocks towards the periphery. Sometimes posing also occurs on sites not used for breeding, and ritual displays can include 'alert posing' if the birds sense danger (Kahl 1975). This display starts in one individual, which stretches its neck from the usual 'S'-curved position straight upwards in a vertical position. This behaviour is contagious and within seconds is mimicked by many others. In case of disturbances by predators, flocks of flamingos immediately take flight and return after several minutes (Vareschi 1978).

Brood care is typical of colonial breeders. Both parents take turns for incubation, enabling the partner to search for food, which preferentially occurs at night to avoid predators. Parents feed only their own chick, which is probably recognised by its voice (Brown 1975). Breeding and hatching is synchronised within a colony, thus increasing the chance for survival in the shelter of the flock. Accordingly, several thousand chicks can hatch at the same time.

In the first week, the chicks are fed by the parents with a reddish crop-milk rich in lipids, proteins and glucose. In the second week, the chicks start to leave the nests. They establish groups called crèches roaming through the colony looking for 'real' flamingo diet. Congregating groups of the chicks increase from week to week and reach numbers of several thousand after 7–8 weeks. Then, the large flocks start to move away from the colony, accompanied by a few adults, which probably have lost their own chicks (Brown 1975). They head towards water bodies kilometres away from the breeding territory because the less saline waters and swamps fringing the shore offer better conditions for foraging. During this phase, the dark-greyish to brownish fluffy plumage is replaced stepwise by greyish feathers (Fig. 10.8c).

10.7 Movements

Flamingos move from one lake to another. With the exception of movements to breeding grounds, the reasons are not yet clarified for East Africa (Brown 1975), but may be related to fluctuating food availability (Kaggwa et al. 2013; Kihwele et al. 2014). At least for the American and the Greater Flamingo, longer flights in search of alternative food resources have been reported (Johnson and Cézilly 2007). The question arises whether the flamingos living in the EASL are migratory birds or not. From an ecological perspective, they are clearly moving and not migrating, because migration presumes periodical travels, which can either be diurnal or seasonal (Begon et al. 2005). The nomadic Lesser Flamingos perform unpredictable, erratic interlake movements in large flocks, mostly during the night to avoid eagle attacks (Brown 1975). Their flight speed was estimated to about 70 km h^{-1} and they can travel distances up to 450 km within one hop (Childress et al. 2004). These findings seem to make population interchange unlikely between EASL, West and South Africa (but see below). Flamingos at Ethiopian lakes most probably originate from the Natron breeding site, which was shown by ring recoveries and resightings at Magadi in 1962 (Brown 1975).

Supposedly there is no population exchange because of the geographical barriers of several thousand kilometres (Brown 1975) and the apparent lack of stopover places between the saline habitats needed for 'fueling'. A continuous, small gene flow was, however, reported between East and South Africa. In that study, 41 Lesser Flamingos from Lake Bogoria (Kenya) and Makgadikgadi Pan (Botswana) revealed a high degree of genetic similarity, reflecting the nomadic behaviour of that species (Zaccara et al. 2011). The estimated average number of individual flamingos exchanged was 3–4 per generation (10 years correspond with three generations). This genetic analysis for the first time confirms a commutation between African flamingo populations, as already supposed by Tuite (1981), Simmons (2000) and Childress et al. (2004). The assumption is that increased temporal rainfall is resulting in additional wetlands that serve as an intermediate link (Parasharya et al. 2015; Zaccara et al. 2008, 2011) and create a temporal corridor. The same methodological approach with 69 Lesser Flamingos from Bogoria and the Rann of Kutch (Gujarat) revealed genetic exchange between the African and Indian populations (Parasharya et al. 2015). Per generation, 2–3 migrants in both directions were detected. There is strong evidence that the population in East Africa represents the hub between the three other African and the Indian populations.

10.8 Phylogeny

The phylogeny of flamingos is poorly investigated and controversially discussed. Their ancestors probably belonged to the oldest lineages of birds. Oldest fossil discoveries of Phoenicopteriformes are from Eocene and date back to 45 Ma (genus *Juncitarsus* Olson and Feduccia). The closest relative of the 'modern' flamingo belongs to the 30-million-year-old genus *Elornis* Aymard from the Oligocene (Mayr 2005). Cladistic studies mainly based on bone morphology confirmed a sister-group relationship of Phoenicopteriformes and Podicipediformes (Mayr 2014). From the Upper Oligocene, the Palaelodidae, an extinct lineage of phoenicopteroid birds that combines morphological features of flamingo-like skulls and grebe-like foots, exhibited 11 anatomic peculiarities suggesting a link between Phoenicopteriformes and Podicipediformes. Especially the shape of the 'rami mandibulae' (parts of the lower bill bones) underlines the relationship of flamingos and grebes (Mayr 2005). Very recently, Torres et al. (2014) stressed the hypothesis that flamingos are amongst the oldest groups of birds. They showed by multilocus sequencing (three mitochondrial and ten nuclear markers) that modern flamingos did not evolve longer than 2–3 Ma. They probably developed in the New World and distributed later to the Old World.

Molecular phylogenetic analyses confirmed a joint lineage of flamingos and grebes (Hackett et al. 2008). Closely related to this separate phylogenetic line, the clades Galliformes (relatives of fowls and pheasants) and Phaethontidae (tropicbirds) evolved. The phylogenetic tree of birds revealed that different characteristics and skills, such as the ability to search for food in the mud of aquatic habitats, evolved polyphyletically in different lineages. Consequently, flamingos and grebes are not closely related with other 'waterbirds' such as ibises, spoonbills or teals.

The six living species of flamingos form two distinct clades: the *Phoenicopterus* clade including Greater, American and Chilean Flamingo and the *Phoenicoparrus* clade with Lesser, Andean and James Flamingo (Fig. 10.11). This topology is congruent with the morphology of the feeding apparatus. The *Phoenicopterus* group possesses shallow-keeled bills, whereas *Phoenicoparrus* species are characterised by deep-keeled bills. Torres et al. (2014) suggested placing the Lesser Flamingo into the genus *Phoenicoparrus* Bonaparte 1856. We, however, do not follow this suggestion because *Phoeniconaias* is the older name, which was already correctly described in 1798 by Geoffroy Saint-Hilaire.

A genetic study of parasitic lice of flamingos, grebes and ducks revealed fascinating insights (Johnson et al. 2006): generally, the louse fauna

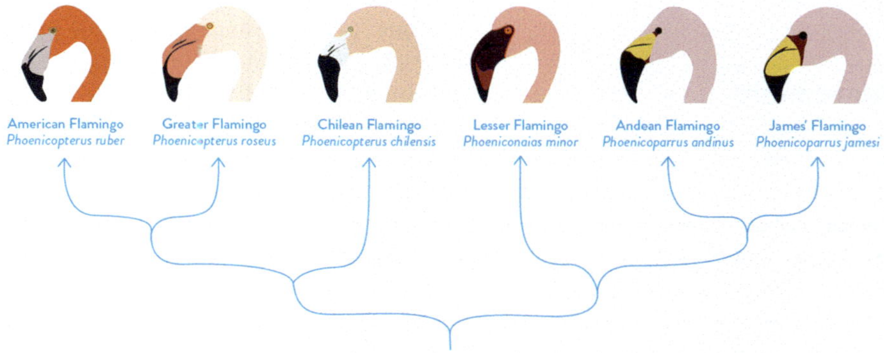

Fig. 10.11 Phylogenetic relationships of flamingo species (data from Torres et al. 2014). Key features are based on original photographs

of these three groups is closely related. Because of the flamingos and grebes share a common ancestor, their lice species are a result of co-speciation. In contrast, flamingos and ducks are phylogenetically more divergent, and their louse fauna is the result of host switching, preferably from (the taller) flamingos to ducks (Johnson et al. 2006).

10.9 Threats

The Lesser Flamingo is placed in the category 'near threatened' in the Red List of the International Union for Conservation of Nature (IUCN). This species is exposed to manifold threats. One of the most critical hazards is loss and/or degradation of habitats by (1) natural reasons such as flooding or drought due to climate and hydrological changes and/or (2) anthropogenic activities such as deforestation, farming, drainage, canalisation, infrastructural measures, pollution and industry in the vicinity of flamingo habitats (Childress et al. 2008; Figs. 10.6a, c). Unfortunately, many of these threats are combined at the primary breeding site, Lake Natron. This Ramsar habitat (classified by the Convention on Wetlands of International Importance) is not protected by law, and several proposals for industrial exploitation are under controversial discussion (Childress et al. 2006). A dam construction on the river Ewaso Ng'iro for hydroelectric power generation is planned; this Kenyan river is a major tributary into the lake. International protests also targeted a project to build a soda ash factory at Lake Natron (Baker 2011). A cost-benefit analysis demonstrated that mining of soda ash at Lake Natron is currently not economically feasible (BirdLife International 2012; Kadigi et al. 2014). A very bad example of habitat devastation is found at Lake Abijata (Ethiopia), which is caused by soda ash mining amongst other threats (Getaneh et al. 2015). As a result, the lake faces serious problems because of water abstraction, which resulted in a permanent decline of the lake water level. A connecting passage to Lake Shala for additional water input into Lake Abijata is currently being discussed to enable soda ash mining to continue, but these plans are strongly disapproved of by conservationists and other experts and will hopefully not be realised.

Further disturbances by humans include devastation of nests, disruption of breeding, salt panning, military exercises, hunting, pastoralists, wild bird trade and egg collection. Terrestrial predators (feral dogs, hyenas, jackals, mongooses, baboons) and avian predators (marabous, vultures, eagles) hunt for chicks and weakened adult birds. Competitors such as plankton-feeding fishes or algae-grazing zooplankton additionally reduce the standing crop of food for Lesser Flamingos (Childress et al. 2008), which ultimately leads to starvation and weakening of birds (Brown 1959; Kaggwa et al. 2013; Krienitz and Kotut 2010; Ton 2007).

Further threats are intoxication and bacterial infections, which both have the potential for inducing sudden deaths affecting tens of thousands of Lesser Flamingos. However, we are only at the beginning of understanding the mechanisms and consequences of such puzzling events. Episodic mass mortality of flamingos has been reported repeatedly since the 1970s in Kenya (summarised by Ndetei and Muhandiki 2005) and since the 2000s in Tanzania (Lugomela et al. 2006). The largest die-off in East Africa was recognised from Lake Bogoria in 1999–2000 (Harper et al. 2003). Calculations based on ~700 dead flamingos per day yielded ~200,000 carcasses during the 9-month-long mass mortality period. This corresponds to one fifth of the flamingo population visiting the lake. A big proportion of the loss was replenished by young birds arriving from Lake Natron to Bogoria in 2002–2003; more than 50 % of the population consisted of birds younger than 2 years (Harper et al. 2003). The reasons behind this mass mortality were never discovered, and multiple causes are assumed.

Intoxication of Lesser Flamingos by heavy metals and pesticides is reported by Greichus et al. (1978), Kairu (1996), Koeman et al. (1972), Motelin et al. (2000) and Ndetei and Muhandiki (2005). Concentrations of these contaminants increase over time and may act as one component that weakens the birds. Depending on industrialisation and agriculture in the catchment areas of the lakes, toxins are of anthropogenic origin or derive from natural sources; they are deposited in the sediment and in the biota. Anderson et al. (2005) compared the toxin concentrations with values from the Handbook of Chemical Risk Assessment (Eisler 2000) and concluded that the current levels are probably not distressing, but call for careful monitoring to enable Risk Assessment in future developments.

The main bacterial diseases of Lesser Flamingos are mycobacteriosis (avian tuberculosis), avian cholera, avian botulism and inflammations caused by opportunistic Bacteria (Anderson et al. 2005). Avian mycobacteriosis is caused by *Mycobacterium avium* Chester and was first reported in Lesser Flamingos at Lake Nakuru in 1970 (Koeman et al. 1972). Later, additional infections at lakes Nakuru and Bogoria were verified (Cooper et al. 1975; Kaliner and Cooper 1973; Kock et al. 1999; Sileo et al. 1979). Avian tuberculosis is of particular concern because of the gregarious lifestyle of Lesser Flamingos, which results in fast microbe transmissions within the population (Cooper et al. 2014). Tuberculosis is widely distributed in wild and domestic birds (Cooper 1990). Avian cholera is caused by *Pasteurella multocida* (Lehmann and Neumann) Rosenbusch and Merchant and most probably induced the mass mortality at Lake Bogoria in 2002 (Anderson et al. 2005). Avian botulism is widespread amongst waterfowl and caused by the toxin botulinum synthesised by *Clostridium botulinum* (van Ermengem) Bergey et al. living in maggots. The toxin is ingested via the food chain, so *Clostridium* itself is not contagious to birds. In 2013, flamingos at the artificial breeding grounds of Kamfers Dam died because of avian botulism. Opportunistic Bacteria inducing inflammations and infections in weakened flamingos are *Pseudomonas aeruginosa* Migula, *Corynebacterium* spp., *Proteus* spp. and *Escherichia coli* (Migula) Castellani and Chalmers (Fyumagwa et al. 2013; Nonga et al. 2011). It remains unclear whether these Bacteria are permanent residents in EASL or whether they are introduced directly before outbreaks of mass infections. Some members of the avifauna such as piscivorous birds, mainly pelicans, were suspected to act as germ vectors (Cooper et al. 1975). Even a link to birds of prey such as Fish Eagles was postulated (Kaliner and Cooper 1973). Marabou Storks guzzling on open litter deposits may also ingest food contaminated with *Pasteurella multocida* and transfer it to flamingo sites. All of these assumptions need to be investigated in further surveys; especially mechanisms of germ transmission need to be clarified.

Because of their very special diet, Lesser Flamingos may sometimes ingest toxic Cyanobacteria. The cyanotoxins microcystin and anatoxin-a were detected in Kenyan soda lakes in phytoplankton samples (Ballot et al. 2004, 2005; Ndetei and Muhandiki 2005), in cyanobacterial mats and fishes of hot springs (Krienitz et al. 2003, 2005b), and in sediments (Dadheech et al. 2009). Also lakes of North Tanzania (Fyumagwa et al. 2013; Lugomela et al. 2006; Nonga et al. 2011) and soda lakes of Ethiopia (Willén et al. 2011) had cyanotoxins. Even some strains of the main diet of Lesser Flamingos, *Arthrospira fusiformis*, are able to synthesise cyanotoxins such as microcystins and anatoxin-a (Ballot et al. 2004, 2005), and crude extracts of *Arthrospira*-dominated phytoplankton from Big Momela were toxic in mice tests (Lugomela et al. 2006). Cyanotoxins were also detected in flamingo tissues at lakes Bogoria and Nakuru in 2001–2003 (Krienitz et al. 2005a) and at Lake Manyara during a mass mortality in August 2008. That mass mortality was also documented by veterinarians who found enlarged, haemorrhagic visceral organs and livers with nodular lesions (Nonga et al. 2011), pointing to cyanotoxicosis. Moreover, bacterial infections were also detected (Fyumagwa et al. 2013). In contrast, Mussagy et al. (2006) did not found cyanotoxins in *Arthrospira* strains. Straubinger-Gansberger et al. (2014) screened phytoplankton samples for cyanotoxins originating from Nakuru and Bogoria in weekly intervals from July 2008 to November 2009 and also considered flamingo tissues. In that period, three flamingo mass mortalities were recognised at Bogoria, but no cyanotoxins were found, despite the application of different detection techniques.

Cyanotoxins undergo high fluctuations in a given habitat, and considerable changes of cyanotoxin concentrations are well documented worldwide; nonetheless, the reasons behind these erratic alterations are still poorly understood (Kurmayer and Christiansen 2009; Okello et al. 2011). The question arises whether flamingos are susceptible to cyanotoxins or whether they developed detoxification mechanisms. Toxic Cyanobacteria evolved 1500–2000 Ma (Rantala et al. 2004), but flamingos and their ancestors arose only 2–45 Ma (Mayr 2005; Torres et al. 2014). Hence, flamingos have always been exposed to cyanotoxins. Such a detoxification mechanism could be storage of toxins into the plumage, which was already shown by Metcalf et al. (2006); also the neurotoxic amino acids β-*N*-methylamino-L-alanine (BMAA) and 2,4-diaminobutyric acid (DAB) were detected in the same samples. Metcalf et al. (2013), however, assumed that the sequestration mechanisms were not effective enough in the weakened birds due to increased environmental stress.

Lesser Flamingos are also infected by intestinal worms independently of age, sex and body conditions. Especially, heavy infections by cestodes were shown (summarised by Anderson et al. 2005). Jones and Khalil (1980) described seven species of cestodes and two nematodes from Lesser Flamingos of Lake Nakuru. The contribution of these parasites to flamingo mass mortalities remains unclear.

10.10 Conservation

The conservation of Lesser Flamingos is part of the protection of the entire avifauna at the EASL and their whole ecosystem. This species, although still representing the most numerous flamingo species, is considered as 'near threatened' following the Red List of the IUCN and is furthermore listed in columns A and B of the Agreement on the Conservation of African-Eurasian Migratory Waterbirds (AEWA 2015) Action Plan, Appendix II of the Bonn Convention and Appendix II of the CITES convention due to wetland loss (feeding sites), steadily declining populations, low number of breeding habitats and their low reproductive rates (Childress et al. 2008; Parasharya et al. 2015). An international 'Single Species Action Plan' exists for the conservation of the Lesser Flamingo (Childress et al. 2008). It aims to reach the IUCN conservation status 'least concern' instead of 'near threatened' for all populations of the Lesser Flamingos by stabilising their size, abundance and inter-lake network (to conserve the metapopulation) by 2020.

Flamingo lakes in the East African Rift Valley are lakes Shala, Abijata and Chitu in Ethiopia (all belonging to the Abijata-Shala National Park) and Logipi, Bogoria (Lake Bogoria National Reserve), Elmentaita, Nakuru (Lake Nakuru National Park), Oloidien, Sonachi and Magadi in Kenya. The lakes Natron, Manyara (Lake Manyara National Park), Eyasi and the Momela are located in Tanzania. In South Africa, flamingos occur at the Makgadikgadi Pans (Botswana; Makgadikgadi Pans National Park), the Kamfers Dam (Kimberley, South Africa) and within the Etosha Pan (Namibia; Etosha National Park). In West Africa they are found at Vasieres de Khonibenki et Yongo Sale in Guinea as well as in the coastal lowlands Chott Boul in Aftout es Sâheli (Ramsar site) and Diawling National Park (Ramsar site) in Mauritania. Anecdotic occurrences were also reported for Lake Chad. In India, Lesser Flamingos occur at the two salt pans Zinzuwadia and Purabcheria in Gujarat and at Sambhar Salt Lake (Ramsar site) in Rajasthan. Lesser Flamingos move between lakes and continents, but the conservation status of these habitats differs distinctly. Lake Bogoria, Elmentaita, Natron (and its basin), Nakuru and Etosha Pan are Ramsar sites (Ramsar 2015), and the same applies for waterbodies within National Parks. Both, however, do not guarantee full protection (see Lake Abijata). Although Lake Natron provides the only long-term breeding habitat in the EASL, plans to build a hydropower and soda ash plant threaten this important distribution range. Lake Eyasi still remains unprotected and threatened by deforestation, agriculture and overgrazing (Baker and Baker 2002; Childress et al. 2004, 2008; Hill et al. 2013; Kumar 2008; Matagi 2004; Nasirwa 2000).

10.11 Perspectives

We definitely need a better understanding of population dynamics for an effective conservation of the Lesser Flamingo (Zaccara et al. 2008, 2011). Strict protection of its main distribution ranges is a key factor because this highly specialised species is strongly dependent on a limited number of saline lakes and their phytoplankton availability (Zaccara et al. 2011), with no alternative habitat. New advances in tracking techniques and genetics promise new insights into population dynamics and the movement ecology of these birds, but these methods are promising only if combined with applied conservation measures.

The suddenly occurring mass mortalities of flamingos are still an enigma that needs to be resolved. Clearly, multiple stressors and a combination of various factors weaken the birds and make them more susceptible to infections (Harper et al. 2003; Krienitz and Kotut 2010; Njuguna and Owuor 2006). The partly contradictory results strongly call for joint efforts of ornithologists, veterinarians, bacteriologists, phycologists, toxicologists, ecologists and conservationists. This is not only to thoroughly investigate the key factors of mass mortality, but also to determine the ecological consequences for other organisms. Moreover, flamingos are highly important for the socio-economy because they are a major tourist attraction. Without flamingos, tourist numbers would drop considerably.

One requirement of flamingo research is continuous monitoring of their habitats. Secchi-disc readings and installing a gauge are first steps, which can be done easily even by non-scientists. As the diet of Lesser Flamingos is restricted to Cyanobacteria and algae, samples should be taken whenever possible. A small bottle with plankton suspension and algal scums from the water surface, fixed with 2–3 drops of formaldehyde, is easy to collect and should be sent to specialists for further analyses. Detailed instructions for cyanotoxin sampling are already available (Metcalf et al. 2007). Also, proper and fast sampling of tissues of flamingo carcasses is extremely important and would provide information which would otherwise be irretrievably lost. All such studies need to consider the laws of the Convention on International Trade of Endangered Species of Wild Fauna and Flora (CITES, https://www.cites.org/) and the Nagoya Protocol on access to genetic resources and benefit sharing (Kamau et al. 2010).

A wealth of knowledge is held both by local scientists and researchers from abroad, but often they fail to share their knowledge and to make it jointly available. Coordinated monitoring and research, including local authorities and NGOs, would greatly facilitate new insights into these fascinating organisms and their ecological requirements to identify conservation demands.

References

AEWA (2015) Agreement on the conservation of African-Eurasian migratory waterbirds. http://www.unep-aewa.org/en/species. Accessed 6 Sept 2015

Anderson MD, Anderson TA (2010) Photospot: a breeding island for lesser flamingos *Phoeniconaias minor* at Kamfers Dam, Kimberley, South Africa. Bull ABC 17:225–228

Anderson MD, Beasley VR, Carmichael WW, Codd GA, Cooper JE, Cooper ME, Franson C, Kilewo MK, Kock R, Maddox CW, Manyibe T, Mlengeya TK, Motelin G, Nelson Y, Ruiz MO, Pessier AP, Sreenivasan AS, Sileo L (2005) Mass Die-offs of lesser flamingos (*Phoeniconaias minor*) in East and Southern Africa: current knowledge and priorities for research. Abstracts, transdisciplinary symposium at College of Veterinary Medicine, University of Illinois, 24–26 Sept 2004, p 28

Baker M (2011) Lake Natron, flamingos and the proposed soda ash factory. http://www.tnrf.org/en/natronbriefing. Accessed Nov 2015

Baker NE, Baker EM (2002) Important bird areas in Tanzania: a first inventory. WildLife Conservation Society in Tanzania. Dar es Salaam, Tanzania, pp 897–945

Baker NE, Baker EM (2006) Waterbirds in Tanzania: what we know and what we do not; where are the knowledge gaps? In: Boere GC, Galbraith CA, Stroud

DA (eds) Waterbirds around the world. The Stationery Office, Edinburgh, pp 245–249

Ballot A, Krienitz L, Kotut K, Wiegand C, Metcalf JS, Codd GA, Pflugmacher S (2004) Cyanobacteria and cyanobacterial toxins in three alkaline Rift Valley lakes of Kenya – Lakes Bogoria, Nakuru and Elmenteita. J Plankton Res 26:925–935

Ballot A, Krienitz L, Kotut K, Wiegand C, Pflugmacher S (2005) Cyanobacteria and cyanobacterial toxins in the alkaline crater lakes Sonachi and Simbi, Kenya. Harmful Algae 4:139–150

Begon M, Townsend CR, Harper JL (2005) Ecology: from individuals to ecosystems, 4th edn. Wiley-Blackwell, Oxford

Bennun L, Nasirwa O (2000) Trends in waterbird numbers in the southern rift valley of Kenya. Ostrich 71:220–226

BirdLife International (2012) Cost-benefit analysis demonstrates that mining of soda ash at Lake Natron is not economically viable. Presented as part of the BirdLife State of the world's birds website. http://www.birdlife.org/datazone/sowb/casestudy/507. Accessed 14 Mar 2015

Bozic L, Ewnetu M (2008) First confirmed breeding record of the lesser flamingo (*Phoeniconaias minor*) in Ethiopia. Flamingo, Bull IUCN-SSC Wetl Int 16:29–30

Brown LH (1955) Breeding of the lesser and greater flamingos in East Africa. J East Afr Nat Hist Soc 22:159–162

Brown L (1959) The mystery of the flamingos. Country Life, London

Brown LH (1975) East Africa. In: Kear J, Duplaix-Hall N (eds) Flamingos. T & A D Poyser, Berkhamstead, p 248

Brown LH, Root A (1971) The breeding behavior of the lesser flamingo *Phoeniconaias minor*. Ibis 113:147–172

Burgis MJ, Morris P (1987) The natural history of lakes. Cambridge University Press, Cambridge, p 218

Chabot D, Bird DM (2013) Small unmanned aircraft: precise and convenient new tools for surveying wetlands. J Unmanned Veh Syst 1:15–24

Childress B, Jarrett N (2005) Methods of capturing and handling wild lesser flamingos for research. Flamingo, Bull IUCN-SSC Wetl Int 13:25–29

Childress B, Harper D, Hughes B, Van den Bossche W, Berthold P, Querner U (2004) Satellite tracking lesser flamingo movements in the rift valley, East Africa: pilot study report. Ostrich 75(1&2):57–65

Childress B, Hughes B, Harper D, Van den Bossche W, Berthold P, Querner U (2006) Satellite tracking documents the East African flyway and key site network of the lesser flamingo *Phoenicopterus minor*. In: Boere GC, Galbraith CA, Stroud DA (eds) Waterbirds around the world. The Stationery Office, Edinburgh, pp 234–238

Childress B, Nagy S, Hughes B (2008) International single species action plan for the conservation of the lesser flamingo *Phoeniconaias minor*. CMS Technical Series 18, AEWA Technical Series 34, Bonn, p 59

CITES Convention on International Trade of Endangered Species of Wild Fauna and Flora. https://www.cites.org/. Accessed 26 Aug 2015

Cooper JE (1990) Birds and zoonoses. Ibis 132:181–191

Cooper JE, Karstad L, Boughton E (1975) Tuberculosis in lesser flamingos in Kenya. J Wildl Dis 11:32–36

Cooper JE, Deacon AE, Nyariki T (2014) Post-mortem examination and sampling of African flamingos (Phoenicopteridae) under field conditions. Ostrich 85:75–83

Dadheech PK, Krienitz L, Kotut K, Ballot A, Casper P (2009) Molecular detection of uncultured cyanobacteria and aminotransferase domains for cyanotoxin production in sediments. FEMS Microbiol Ecol 68:340–350

Disneynature (2008) The crimson wing. The mystery of the flamingos. http://nature.disney.com/the-crimson-wing. Accessed 3 Nov 2015

Eisler R (2000) Handbook of chemical risk assessment health hazards to humans, plants, and animals. Vol 1 Metals & Vol 3 Metalloids, radiation, cumulative index to chemicals and species. Lewis, Boca Raton

Ewnetu M, Mekonnen A, Yadeta T (2013) Water bird census in the wetlands of ethiopia by 2013. Ethiopian Wildlife Conservation Authority Report, p 23

Fox DL (1975) Carotenoids in pigmentation. In: Kear J, Duplaix-Hall N (eds) Flamingos. T & A D Poyser, Berkhamstead, p 248

Fox DL, Smith VE, Wolfson AA (1967) Carotenoid selectivity in blood and feathers of Lesser (African), Chilean and Greater (European) Flamingos. Comp Biochem Physiol 23:225–232

Fyumagwa RD, Bugwes Z, Mwita M, Kihwele ES, Nyaki A, Mdegela RH, Mpanduji DG (2013) Cyanobacterial toxins and bacterial infections are the possible cause of mass mortality of lesser flamingos in soda lakes in northern Tanzania. Res Opin Anim Vet Sci 3 (in press). https://www.researchgate.net/profile/Emilian_Kihwele/publications. Accessed 3 Nov 2015

Getaneh G, Temesgen G, Damena E (2015) Environmental degradation and its effect on terrestrial and aquatic diversity in the Abijata-Shala Lakes National Park, Ethiopia. Point J Agric Biotechnol Res 1:1–12

Greichus YA, Greichus A, Amman BB, Hopcraft J (1978) Insecticides, polychlorinated biphenyls and metals in African lake ecosystems. III. Lake Nakuru, Kenya. Bull Environ Contam Toxicol 19:454–461

Grzimek M, Grzimek B (1960) Flamingos censused in East Africa by aerial photography. J Wildl Manage 24:215–217

Hackett S, Kimball RT, Reddy S, Bowie RC, Braun EL, Braun MJ, Chojnowski JL, Cox WA, Han KL, Harshman J, Huddleston CJ, Marks BD, Miglia KJ,

Moore WS, Sheldon FH, Steadman DW, Witt CC, Yuri T (2008) A phylogenetic study of birds reveals their evolutionary history. Science 320:1763–1767

Harper DM, Childress RB, Harper MM, Boar RR, Hickley PH, Mills SC, Otieno N, Drane T, Vareschi E, Nasirwa O, Mwatha WE, Darlington JPEC, Escuté-Gasulla X (2003) Aquatic biodiversity and saline lakes: Lake Bogoria National Reserve, Kenya. Hydrobiologia 500:259–276

Hill LM, Bowerman WW, Roos JC, Bridges WC, Anderson MD (2013) Effects of water quality changes on phytoplankton and lesser flamingo *Phoeniconaias minor* populations at Kamfers Dam, a saline wetland near Kimberley, South Africa. Afr J Aquat Sci 38:287–294

Jenkin PM (1957) The filter feeding and food of flamingos (Phoenicopteri). Philos Trans R Soc Lond, Ser B 240:401–493

Johnson A, Cézilly F (2007) The greater flamingo. T & AD Poyser, Berkhamstead, UK

Johnson KP, Kennedy M, McCracken KG (2006) Reinterpreting the origins of flamingo lice: cospeciation or host-switching? Biol Lett 2:275–278

Jones A, Khalil LF (1980) The helminth parasites of the lesser flamingo, *Phoeniconaias minor* (Geoffroy), from Lake Nakuru, Kenya, including a new cestode, *Phoenicolepis nakurensis* n.g., n.sp. Syst Parasitol 2:61–76

Kadigi RMJ, Mwathe K, Dutton A, Kashaigili J, Kilima F (2014) Soda ash mining in Lake Natron: a reap or ruin for Tanzania? J Environ Conserv Res 2:37–49

Kaggwa MN, Gruber M, Oduor SO, Schagerl M (2013) A detailed time series assessment of the diet of lesser flamingos: further explanation for their itinerant behaviour. Hydrobiologia 710:83–93

Kahl (1975) 22 Ritual displays. In: Kear J, Duplaix-Hall N (eds) Flamingos. T & A D Poyser, Berkhamstead, p 248

Kairu JK (1996) Heavy metal residues in birds of Lake Nakuru, Kenya. Afr J Ecol 34:397–400

Kaliner G, Cooper JE (1973) Dual infection of an African fish eagle with acid-fast bacilli and an *Aspergillus* sp. J Wildl Dis 9:51–55

Kamau EC, Fedder B, Winter G (2010) The Nagoya protocol on access to genetic resources and benefit sharing: what is new and what are the implications for provider and user countries and the scientific community. Law Environ Dev J 2010:246–262

Kihwele ES, Lugomela C, Howell KM (2014) Temporal changes in the lesser flamingos population (*Phoenicopterus minor*) in relation to phytoplankton abundance in Lake Manyara, Tanzania. Open J Ecol 4:145–161

Kock ND, Kock RA, Wambua J, Kamau GJ, Mohan K (1999) *Mycobacterium avium*-related epizootic in free-ranging lesser flamingos in Kenya. J Wildl Dis 35:297–300

Koeman JH, Pennings JH, De Goeij JJ, Tijoe PS, Olindo PM, Hopcraft J (1972) A preliminary survey of the possible contamination of Lake Nakuru in Kenya with some metals and chlorinated hydrocarbon pesticides. J Appl Ecol 9:411–416

Krause J, Krause S, Arlinghaus R, Psorakis I, Roberts S, Rutz C (2013) Reality mining of animal social systems. Trends Ecol Evol 28:541–551

Krienitz L, Kotut K (2010) Fluctuating algal food populations and the occurrence of lesser flamingos (*Phoeniconaias minor*) in three Kenyan rift valley lakes. J Phycol 46:1088–1096

Krienitz L, Ballot A, Kotut K, Wiegand C, Pütz S, Metcalf JS, Codd GA, Pflugmacher S (2003) Contribution of hot spring cyanobacteria to the mysterious deaths of lesser flamingos at Lake Bogoria, Kenya. FEMS Microbiol Ecol 14:141–148

Krienitz L, Ballot A, Casper P, Codd GA, Kotut K, Metcalf JS, Morrison LF, Pflugmacher S, Wiegand C (2005a) Contribution of toxic cyanobacteria to massive deaths of lesser flamingos at saline-alkaline lakes of Kenya. Verhand Int Verein Limnol 29:783–786

Krienitz L, Ballot A, Casper P, Kotut K, Wiegand C, Pflugmacher S (2005b) Cyanobacteria in hot springs of East Africa and their potential toxicity. Algol Stud 117:297–306

Krienitz L, Krienitz D, Dadheech PK, Hübener T, Kotut K, Luo W, Teubner K, Versfeld WD (2016) Algal food for Lesser Flamingos: a stocktaking. Hydrobiologia 775:21–50

Kumar S (2008) Conservation of Sambhar Lake – an important waterfowl habitat and a ramsar site in India. In: Sengupta M, Dalwani R (eds) Proceedings of Taal 2007. The 12th world lake conference, pp 1509–1517

Kurmayer R, Christiansen G (2009) The genetic basis of toxin production in Cyanobacteria. Freshw Rev 2:31–51

Lugomela C, Pratap HB, Mgaya YD (2006) Cyanobacteria blooms – a possible cause of mass mortality of lesser flamingos in Lake Manyara and Lake Big Momela, Tanzania. Harmful Algae 5:534–541

Maina C, Kyungu S (2009) The African waterfowl census. Museum News, Quarterly newsletter of the National Museums of Kenya, March–June 2009, p 8

Mari C, Collar N (2000) Pink Africa. The Harvill Press, London

Matagi SV (2004) A biodiversity assessment of the flamingo Lakes of eastern Africa. Biodiversity 5:13–26

Mayr G (2005) The Paleogene fossil record of birds in Europe. Biol Rev 80:515–542

Mayr G (2014) The Eocene *Juncitarsus* – its phylogenetic position and significance for the evolution and higher-level affinities of flamingos and grebes. Comptes Rendus Palevol 13:9–18

McCulloch GP, Aebischer A, Irvine K (2003) Satellite tracking of flamingos in southern Africa: the importance of small wetlands for management and conservation. Oryx 37:480–483

Metcalf JS, Morrison LF, Krienitz L, Ballot A, Krause E, Kotut K, Pütz S, Wiegand C, Pflugmacher S, Codd GA (2006) Analysis of cyanotoxins anatoxin-a and microcystins in lesser flamingo feathers. Toxicol Environ Chem 88:159–167

Metcalf JS, Ballot A, Kotut K, Codd GA (2007) Guidance for collection of samples for assessment of lesser flamingo *Phoenicopterus minor* exposure to cyanotoxins. Flamingo, Bull IUCN-SSC Wetl Int 15:32–35

Metcalf JS, Banack SA, Kotut K, Krienitz L, Codd GA (2013) Amino acid neurotoxins in feathers of the lesser flamingo. Chemosphere 90:835–839

Mlingwa C, Baker N (2006) Lesser flamingo *Phoenicopterus minor* counts in Tanzanian soda lakes: implications for conservation. In: Boere GC, Galbraith CA, Stroud DA (eds) Waterbirds around the world. The Stationery Office, Edinburgh, pp 230–233

Motelin GK, Thampy R, Doros D (2000) An ecotoxicological study of the potential roles of metals, pesticides and algal toxins on the 1993/5 lesser flamingo mass die-offs in Lake Bogoria and Nakuru, Kenya; and the health status of the same species of birds in the rift valley lakes during the 1990s. In: Proceedings of the East African Environmental Forum, Nairobi, May 2000. East African Environmental Network, Nairobi, pp 11–12

Mussagy A, Annadotter H, Cronberg G (2006) An experimental study of toxin production in *Arthrospira fusiformis* (Cyanophyceae) isolated from African waters. Toxicon 48:1027–1034

Nasirwa O (2000) Conservation status of flamingos in Kenya. Conservation biology of flamingos. Waterbirds 23(Spec Publ 1):47–51

Ndetei R, Muhandiki VS (2005) Mortalities of lesser flamingos in Kenyan rift valley saline lakes and the implications for sustainable management of the lakes. Lakes Reserv Res Manage 10:51–58

Njuguna M, Owuor O (2006) Lake Nakuru: flamingo death camp. Daily Nation Nairobi, Horizons, Sept 7, 2006, pp 1–2

Nonga HE, Sandvik M, Miles CO, Lie E, Mdegela RH, Mwamengele GL, Semuguruka WD, Skaare JU (2011) Possible involvement of microcystins in the unexplained mass mortalities of lesser flamingo (*Phoeniconaias minor* Geoffroy) at Lake Manyara in Tanzania. Hydrobiologia 678:167–178

Okello W, Ostermaier V, Portmann C, Gademann K, Krmayer R (2011) Spatial isolation favours the divergence in microcystin net production by *Microcystis* in Ugandan freshwater lakes. Water Res 44:2803–2814

Owino AO, Oyugi JO, Nasirwa OO, Bennun LA (2001) Patterns of variation in waterbird numbers on four rift valley lakes in Kenya, 1991–1999. Hydrobiologia 458:45–53

Parasharya BM, Rank DN, Harper DM, Crosa G, Zaccara S, Patel N, Joshi CG (2015) Long-distance dispersal capability of lesser flamingo *Phoeniconaias minor* between India and Africa: genetic inferences for future conservation plans. Ostrich 86:1–9

Pomeroy D, Byaruhanga A, Wilson M (2003) Waterbirds of alkaline lakes in Western Uganda. J East Afr Nat Hist 92:63–79

Raini J, Ngowe N (2009) Aerial census of lesser flamingos (*Phoeniconaias minor*) on the rift valley lakes of northern Tanzania, January 2002. Flamingo, Bull IUCN-SSC Wetl Int 17:30–34

Ramsar (2015) The list of wetlands of international importance, Published 25.06.2015; http://www.ramsar.org/document/the-list-of-wetlands-of-international-importance-the-ramsar-list. Accessed 03 Nov 2015

Rantala A, Fewer DP, Hisbergues M, Rouhianinen L, Vaitomaa J, Börner T, Sivonen K (2004) Phylogenetic evidence for early evolution of microcystin synthesis. Proc Natl Acad Sci USA 101:568–573

Ridley MW, Moss BL, Percy LRC (1955) The food of flamingos in Kenya Colony. J East Afr Nat Hist Soc 22:147–158

Salewski V, Childress B, Wikelski M (2009) Investigating the potential connectivity of lesser flamingo populations by satellite-telemetry. Flamingo, Bull IUCN-SSC Wetl Int 17:26–30

Sardà-Palomera F, Bota G, Viñolo C, Pallares O, Sazatornil V, Brotons L, Gomàriz S, Sardà F (2012) Fine scale bird monitoring from light unmanned aircraft systems. Ibis 154:177–183

Sileo L, Grootenhuis JG, Tuite CH, Hopcraft JB (1979) Mycobacteriosis in the lesser flamingos of Lake Nakuru, Kenya. J Wildl Dis 15:387–389

Simmons RE (1996) Population declines, viable breeding areas, and management options for flamingos in southern Africa. Conserv Biol 10:504–514

Simmons RE (2000) Declines and movements of lesser flamingos in Africa. In: Baldassarre GA, Arengo F, Bildtstein KL (eds) Conservation biology of flamingos, Waterbirds, vol 23 (Spec Publ 1), pp 40–46

Stevens EF (1991) Flamingo breeding: the role of group displays. Zoo Biol 10:53–63

Stevens EF, Pickett C (1994) Managing the social environments of flamingos for reproductive success. Zoo Biol 13:501–507

Straubinger-Gansberger N, Lawton L, Gruber M, Kaggwa MN, Oduor SO, Schagerl M (2014) Cyanobacterial toxins in East African rift valley lakes – do they really cause flamingo mass mortality? Wildl Biol 20:185–189

Studer-Thiersch A (2000) What 19 years of observation on captive greater flamingos suggests about adaptations to breeding under irregular conditions. Waterbirds, vol 23 (Spec Publ 1), pp 150–159

Tebbs EJ, Remedios JJ, Avery ST, Harper DM (2013) Remote sensing the hydrological variability of Tanzania's Lake Natron, a vital lesser flamingo breeding site under threat. Ecohydrol Hydrobiol 13:148–158

Tebbs EJ, Remedios JJ, Avery ST, Rowland CS, Harper DM (2015) Regional assessment of lake ecological states using Landsat: a classification scheme for alkaline-saline flamingo lakes in the East African rift valley. Int J Appl Earth Observ Geoinform 40:100–108

Thommen H (1975) Carotenoid metabolism in ornamental birds. In: Kear J, Duplaix-Hall N (eds) Flamingos. T & A D Poyser, Berkhamstead, p 248

Thommen H, Wackernagel H (1963) Isolierung und Nachweis von Canthaxanthin im Kleinen Flamingo (*Phoeniconaias minor*). Biochim Biophys Acta 69:387–396

Ton T (2007) Pink starvation: flamingos in Africa malnourished. Epoch Times. http://en.epochtimes.com/news/7-1-23/50812.html. Accessed 27 Sept 2010

Torres CR, Ogawa LM, Gillingham MAF, Ferrari B, Van Tuinen M (2014) A multi-locus inference of the evolutionary diversification of extant flamingos (Phoenicopteridae). BMC Evol Biol 14:36

Tuite CH (1979) Population size, distribution and biomass density of the lesser flamingo in the Eastern rift valley, 1974–76. J App Ecol 16:765–775

Tuite CH (1981) Standing crop densities and distribution of *Spirulina* and benthic diatoms in East African alkaline saline lakes. Freshw Biol 11:345–360

Tuite CH (2000) The distribution and density of lesser flamingos in East Africa in relation to food availability and productivity. Waterbirds 23:52–63

Vareschi E (1978) The ecology of Lake Nakuru (Kenya). I. Abundance and feeding of the lesser flamingo. Oecologia 32:11–35

Vareschi E, Jacobs J (1984) The ecology of Lake Nakuru V. Production and consumption of consumer organisms. Oecologia 61:83–98

Vareschi E, Jacobs J (1985) The ecology of Lake Nakuru VI. Synopsis of production and energy flow. Oecologia 65:412–424

Willén E, Ahlgren G, Tilahun G, Spoof L, Neffling MR, Meriluoto J (2011) Cyanotoxin production in seven Ethiopian rift valley lakes. Inland Waters 1:81–91

Zaccara S, Crosa G, Childress B, McCulloch G, Harper DM (2008) Lesser flamingo *Phoenicopterus minor* populations in eastern and southern Africa are not genetically isolated. Ostrich 79:165–170

Zaccara S, Crosa G, Vanetti I, Binelli G, Childress B, McCulloch G, Harper DM (2011) Lesser flamingo *Phoeniconaias minor* as a nomadic species in African shallow alkaline lakes and pans: genetic structure and future perspectives. Ostrich 82:95–100

Vascular Plants in Eastern Africa Rift Valley Saline Wetlands

11

Julius Kipkemboi

Abstract

Plants form a major component of aquatic biota by virtue of their trophic position. Most inland saline ecosystems, particularly lakes in the East African Rift System, are known as spectacular avian habitats. The population density of Lesser Flamingos is especially high due to the frequently high algal biomass productivity, which forms the main food. These environments also provide habitats to a unique vegetation of vascular plants adapted to saline environments. Vascular plants in the littoral zone and the associated floodplains of EARS lakes are dominated by two families—Poaceae and Cyperaceae—but about ten other families are commonly observed in shoreline areas with mild salinity. The two halophytes *Cyperus laevigatus* and *Sporobolus spicatus* are common along the shores of most East African saline lakes. Although the contribution of these plants to allochthonous input into the open water may not be significant, they play a significant role in providing nutrition to terrestrial herbivores associated with these ecosystems. The open-water and littoral zones of highly saline lakes are devoid of aquatic macrophytes with a few exceptions where freshwater percolates into the system. In most East African saline wetlands, salinity to some extent limits higher plant diversity. This prevents these ecosystems from being choked by noxious vascular aquatic weeds, particularly floating macrophytes that prefer low salinity conditions. Nonetheless, in some saline wetlands in Australia and parts of Europe, higher plant diversity and biomass occur.

11.1 Introduction

Inland saline aquatic habitats comprise the open-water areas of shallow athalassic lakes, their adjacent floodplains and pockets of saline areas in palustrine systems. They also include lake

J. Kipkemboi (✉)
Department of Biological Sciences, Egerton University, P.O. Box 536-20115, Egerton, Kenya
e-mail: jkipkemboi@egerton.ac.ke

M. Schagerl (ed.), *Soda Lakes of East Africa*, DOI 10.1007/978-3-319-28622-8_11

margins of some freshwater lakes, especially in regions with excessive evapotranspiration occasioned by aridity. Lakes of the East African Rift System are some of the most spectacular wetlands and are well known for their avian densities, particularly the flamingos. Accordingly, most of them can easily meet the criteria for Ramsar sites. The avian population and diversity are supported by the microflora and macrofauna—mainly phytoplankton and macroinvertebrate communities in the lake. In the case of piscivorous fish, this support is often indirect via secondary production. Vascular plants are restricted by salinity and poorly documented in these ecosystems (Brock and Shiel 1983; Brock et al. 2005; Wetzel 1983). Vascular aquatic plants comprise life forms adapted to hydrophytic conditions characterized by two major stressors: low oxygen availability and salinity. Aquatic macrophytes adapted to saline environments are referred to as halophytes. Using the conventional aquatic macrophyte life form classification, these plants are categorized into emergent, submerged, rooted floating leaved and floating plants (Denny 1985). Vascular plant composition and consequently the contribution to aquatic primary productivity are largely determined by the prevailing environmental conditions such as the nutrient supply, turbidity or salinity. Unlike in marine ecosystems, inland saline ecosystems lack constancy in the environmental conditions due to unpredictable seasonality-driven changes in hydrology. Aquatic plants here have evolved physiological, structural and biochemical means of dealing with saline environments and hence exhibit different levels of adaptation to salinity (Haller 1974).

11.2 Stress Gradients in Saline Wetlands

The aquatic environment is characterized by factors that limit the dispersal and growth of terrestrial organisms. In wetlands, the main stressors are anoxia, wide variability in salinity and water level fluctuations. This chapter focuses on salinity as a forcing factor in the occurrence of vascular plants in inland saline wetlands. High salinities are naturally associated with coastal environments and are uncommon in inland waters. However, due to continued accumulation of salts washed out from the catchments, some lakes and adjacent wetlands tend to develop elevated salinity over time. This is particularly true in situations where water loss due to evapotranspiration is exceedingly high. Vascular plants are limited in saline inland waters due to the inability of most macrophytes to overcome salt stress. Salinity stress restricts the growth of typical inland wetland macrophytes (Brock 1981, 1982; Sim et al. 2006). Where salinity exceeds 1000 mg L^{-1}, this factor becomes the main regulator of species occurrence (Denny 1985). Some aquatic macrophytes, however, grow in salinities of up to over 100,000 mg L^{-1} (Brock and Lane 1983; Brock et al. 2005) This is one reason why most littoral zones of saline lakes are often devoid of aquatic plants suitable for microphytes adapted to saline conditions. In such cases, there is a reduced autochthonous biomass production contribution from littoral zones to the adjacent open-water ecosystems (McKee and Mendelsohn 1989).

11.3 Causes of Salinity in Inland Wetlands

Salinity in inland water originates from the ionic contribution from the rocks and soils of the drainage basin. Salinity as indicated by total ionic concentration is closely correlated with conductivity. The dominant ions in most East African soda lakes (EASL) are sodium and hydrogen carbonate, but in some systems HCO_3^- is replaced by chloride (Talling and Talling 1965 in Denny 1985). Conductivity strongly correlates with total alkalinity ($HCO_3^- + CO_3^{2-} + OH^-$). pH increases as alkalinity increases.

Elevated salinity is induced by the prevailing climatic conditions, particularly by a negative water balance. This phenomenon is common in shallow tropical lakes with no outlets. Here, salts tend to accumulate as a result of evaporation of rain or fresh groundwater and/or evapotranspiration by aquatic plants. Ions are then concentrated in the substrate, leading to sharp salinity gradients. The lithology in the catchment area coupled with climatic conditions largely influences the degree of salinization.

Salinity can also be caused by anthropogenic activities such as discharge of wastewater containing high levels of salts to wetlands (Kitzhner et al. 2011). Examples of such wastewater sources are aquaculture mines and floriculture. Irrigation in arid areas has also been known to create suitable conditions for salinization of soils (McEwan et al. 2009). With increasing demand for agricultural produce and unreliable rainfall, irrigated agriculture, which depends largely on groundwater, is taking preference over rain-fed production in many low rainfall areas of the East African Rift Valley.

11.4 Occurrence of Inland Saline Environments

Inland saline wetlands are mostly associated with shallow EASL, which experience major hydrological changes. These changes occur over varied time scales, potentially even centuries or longer. A persistent negative water balance is thought to have resulted in separation of previously large freshwater lakes into smaller ones. For instance, Lake Elementaita and Lake Nakuru were once joined, forming a single large freshwater lake (Nilsson 1932 cited in Harper et al. 2003), but are currently separate, shallow and saline ecosystems. The salinity can vary within an order of magnitude, and occasionally these lakes have dried. Inundation of the shoreline by saline and alkaline water creates saline soils, thus offering an environment conducive to salt-

Fig. 11.1 *Cyperus laevigatus* on the shore of Lake Nakuru, Kenya, (**a**) and growing in a pocket of sodic soil at the shore of Lake Victoria, Kenya (**b**) (Photos: J. Kipkemboi)

tolerant macrophytes (Fig. 11.1a). In some freshwater lakes, pockets of sodic saline soils are common in floodplain wetlands, especially in locations characterized by aridity. For instance, on the northeastern shores of Lake Victoria in Kenya off Winam Gulf, there are pockets of sodic areas where *Cyperus laevigatus* is the only macrophyte (Fig. 11.1b) (J. Kipkemboi, personal observation).

11.5 Vascular Plant Adaptations to Salinity

Salinity is one of the forcing factors that determine species composition in wetlands. Other determinants that affect plant growth are rainfall, soil characteristics, nutrient supply, flooding depth and flooding duration. All these variables interact individually or through interactions among factors and between plants themselves. In freshwater environments, the osmotic concentration of the cytoplasm in living cells is usually higher than that of the surrounding medium. This enables the cells to develop turgor, i.e. absorb water until the turgor pressure of the cytoplasm is balanced by the resistance of their cell membrane (Mitsch and Gosselink 2000). At the cell level, plants overcome salinity stress in the same way as Bacteria. In addition, however, vascular plants have evolved adaptations that take advantage of their structural complexity The principal mechanism of adaptation to salinity by aquatic macrophytes is exclusion, excretion and induced compatible solute biosynthesis (Bohnert et al. 1995). Some plants do not exclude salts at the roots but have secretory organs at the leaves. The leaves of many marsh grasses are covered by crystalline salt particles excreted by specialized, embedded salt glands (Khan and Weber 2006; Ramadan 2000 and references therein). Rout and Shaw (2001) proposed a possible role of antioxidative enzymes in salt tolerance of submerged macrophytes. At pH > 8.3, inorganic carbon is available as hydrogen carbonate and carbonate; free carbon dioxide is no longer present. Submerged plants can therefore be categorized as bicarbonate and non-bicarbonate users (Denny 1985).

Eastern African saline lakes exhibit a great diversity in salinity characteristics. The littoral wetlands of African athalassic lakes have elevated salinities sometimes in excess of 160 mS cm^{-1} (Talling and Talling 1965 in Denny 1985). Salinity exerts selection pressure and restricts plant community development. Information on morphological, anatomical and physiological adaptations is scanty. According to Flowers et al. (1986), the most tolerant families are Chenopodiaceae and Poaceae. Even within the same genus, single species exhibit different tolerance levels. For instance, *Sporobolus spicatus* can tolerate much higher salinities than *S. robustus* (Finlayson and Moser 1991). The slightly less alkaline floodplain surrounding the most EASL is dominated by freshwater emergent macrophytes and by *Sesbania sesban*, with scattered *Acacia xanthophloea* trees (White 1983).

11.6 Effects of Salinity on Vascular Plant Physiological Functioning and Diversity

Vascular plant species richness is low in inland saline wetlands compared to their freshwater and marine counterparts (Keddy 2010; Nielsen et al. 2003; Wetzel 1983). The effect of climate-induced salinity can be traced not only in the open-water environment but also in wetland soils. Inland saline wetlands may contain excess soluble salts (saline soils), excess exchangeable sodium (sodic soils) or both (saline-sodic soils). Salinity in inland wetlands is attributed to a mixture of cations of sodium, calcium, magnesium and potassium along with anions of chloride, sulfate, bicarbonate and carbonate. The amount of total salt, individual salt or combination of salts in saline wetlands is usually high enough to retard plant growth, injure plant tissue and/or decrease productivity (Ogle 2010). The intricate relationship between salinity, alkalinity and pH in determining the biogeographical distribution of euhydrophytes has been discussed in detail by

Denny (1985). What is often unclear is whether it is the chloride ion concentration, alkalinity or pH that is the overriding factor in determining vascular aquatic plant distribution. In his description, Denny (1985) demonstrated that even lakes with similar chloride ion concentration have different floristic compositions. He further indicated that alkalinity values >10 meq L^{-1} limit athalassic euhydrophyte tolerance.

Soil salinity can affect plant growth both physically (osmotic effect) and chemically (nutrient and/or toxicity effect). As salinity increases, it becomes more difficult for plants to take up water. Salt-intolerant plants appear drought stricken even at fairly low salt concentrations. Growth and productivity usually progressively decline as salinity levels increase. High concentrations of specific ions, which characterize salinity conditions in many cases, can cause disorders in mineral nutrition. For example, high sodium concentrations may cause deficiencies of other elements such as potassium and calcium, and high levels of sulfate and chloride diminish the rate of nitrate absorption. Specific ions such as sodium and chloride may have toxic effects on plants, reducing growth or causing damage to cells and cell membranes (USDA 2010). Very high pH values and accompanying alkalinity damage plant tissues (Denny 1985). If alkalinity is considered as the major component of salinity, we can conclude that it largely affects vascular plant distribution in saline wetlands.

How does salinity relate to vascular plant diversity? Salinity causes stress to physiological functions and is a limiting factor to the diversity of plants in such areas. Salinity is therefore an overriding factor in species occurrence in littoral zones of athalassic lakes and temporarily flooded depressions in arid areas. Highly saline lakes restrict the establishment of vascular emergent and euhydrophyte plants in the littoral zone. The floodplain may have salt-tolerant genera comprising *Sporobolus*, *Cyperus*, *Diplacne* and *Odyssea*, whilst the periphery with mild salinity hosts less tolerant taxa such as *Scirpus*, *Typha* and *Aeschynomene* (Denny 1985). Table 11.1 presents a list of vascular plants commonly

Table 11.1 Some common vascular plant life forms in East African Rift Valley lakes and other African saline ecosystems categorized by salinity tolerance (Denny 1985; Hughes and Hughes 1992)

	Salinity level tolerance		
Life form	High	Moderate	Low
Emergent macrophytes	*Sporobolus spicatus* Kunth	*Sporobolus marginatus* A. Rich.	*Phragmites mauritanus* Kunth
	Cyperus laevigatus L.	*Sporobolus pyramidalis* (L.) Pers.	*Ipomoea* sp.
	Diplachne fusca L.	*Cynodon dactylon* L.	*Leersia hexandra* Sw.
	Odyssea jaegeri (Pilg.)	*Hydrocotyle* sp.	*Oenanthe* sp.
	Scirpus maritimus Rottb.	*Juncus* sp.	*Paspalidium geminatum* (Forssk.) Stapf
	Scirpus littoralis L.	*Typha* spp.	*Hyparrhenia rufa* (Nees) Stapf
	Scirpus holoschoenus L.	*Commelina* sp.	
Submerged macrophytes		*Ceratophyllum demersum* L.	
Rooted floating leaved macrophytes		*Potamogeton pectinatus* L. *Najas marina*	*Nymphaea* sp.
Floating macrophytes		*Azolla nilotica* L.	*Pistia stratiotes* L.
		Lemna gibba L.	*Ottelia ulvifolia* (R.Br.) Rich
Trees and shrubs		*Aeschynomene elaphroxylon* L.	*Acacia xanthophloea* Benth.
		Pluchea sp.	*Sesbania sesban* (Jacq.) W. Wight

Highly tolerant plants include certain genera of Poaceae, Cyperaceae and Juncaceae. Other families that show appreciable tolerance are Araliaceae, Asteraceae, Ceratophyllaceae, Salviniaceae, Araceae, Fabaceae, Apiaceae and Nymphaeaceae

Fig. 11.2 Inundated macrophyte zone (**a**) and floating mats of *C. laevigatus* in Lake Nakuru, Kenya (**b**, **c**) (Photos: J. Kipkemboi)

found in saline wetlands in Africa. In highly saline ecosystems such as lakes Bogoria, Nakuru and Elementaita, euhydrophytes are completely absent. In lakes with mild salinity such as Lake Turkana and Lake Chad, studies have revealed a correlation between salinity and the presence of euhydrophytes (Denny 1985 and references therein).

Although appreciable plant diversity has been recorded in some saline lakes, much fewer typical vascular plants are usually adapted to such an environment. This is often circumstantial: opportunistic plant species exploit environments conducive for succession, particularly when salinity declines as a result of positive water balance during the wet seasons. For example, in Lake Bogoria in Kenya, less than 10 % of the total plant species encountered were true halophytes (Harper et al. 2003). Out of the 27 species of grasses recorded at the lake floodplain, the true halophytes comprised one grass (*Sporobolus spicatus*) and one sedge (*Cyperus laevigatus*). Some emergent macrophytes occur predominantly in coastal wetlands, but sometimes expand to inland saline systems. In the salt pans of southern Africa (the Makgadikgadi), *Scirpus maritimus* and *S. littoralis* occur along southern shores of Lake Chilwa. *S. holoschoenus* is common in northern Africa saline depressions. Information on the microbial flora in the littoral zone of saline environments is scanty, and this is an area worthy of further research.

11.7 Unusual Floating Mats in a Saline Lake?

Floating vegetation is uncommon in the open water of saline lakes. Nonetheless, in exceptional circumstances, floating mats can be observed particularly when the water volume in the lake has increased drastically within a short time, inundating the littoral areas dominated by emergent plants. This phenomenon has been observed in Lake Nakuru in 2011–2014 when the lake volume increased, causing detachment of some *C. laevigatus* mats to float over the lake surface. These mats are distributed by wind to various parts of the lake which previously did not have the macrophytes (Fig. 11.2a–c).

11.8 Ecological Importance of Interactions Between Vascular Plants and Saline Environments

Emergent macrophytes in saline wetlands contribute to primary production and nutrient dynamics and provide habitats for fauna. The diversity and contribution of vascular plants in inland saline wetlands to aquatic productivity is generally lower compared to marine salt marshes. This is partly explained by the exclusion of non-salt-tolerant plants by environmental conditions created by salinity. Salt uptake "spices" the plant biomass and encourages grazing by herbivores. Some salt-tolerant grasses such as *S. spicatus* and *C. dactylon* are highly palatable for herbivores. Palatability affects herbivore preference and consequently determines above-ground biomass of salt-tolerant macrophytes at any given time (Fig. 11.3). These plants are also resilient and can tolerate flooding regime variability and are robust to grazing pressure.

The effect of salinity stress on macrophytes restricts proliferation of noxious macrophytes, particularly floating vegetation, which comprises the most problematic weeds in freshwater wetlands. Thus, it is notable that whereas freshwater environments such as Lakes Victoria and Naivasha have been chocked by floating aquatic weeds, saline systems such as Lakes Turkana, Nakuru, Natron and Bogoria, among others, are devoid of vegetation or have sparse submerged vegetation. Salinity restrictions on vascular plants in saline lakes have both negative and positive ecological significance.

Fig. 11.3 Above-ground biomass in heavily grazed *S. spicatus* (**a**) and less preferred *C. laevigatus* along the shore of Lake Nakuru (**b**) (Photos: J. Kipkemboi)

References

Bohnert HJ, Nelson DE, Jensen RG (1995) Adaptations to environmental stresses. Plant Cell 7:1099–1111

Brock MA (1981) The ecology of halophytes in salt lakes in the south-east of South Australia. Hydrobiologia 81:23–32

Brock MA (1982) Biology of the salinity tolerant genus *Ruppia* L. in salt lakes in South Australia. I. Morphological variation within the between species and ecophysiology. Aquat Bot 13:219–248

Brock MA, Lane JAK (1983) The aquatic flora of saline wetlands in Western Australia in relation to salinity and permanence. Hydrobiologia 105:63–76

Brock MA, Shiel RJ (1983) The composition of aquatic communities in saline wetlands in Australia. Hydrobiologia 105:77–84

Brock MA, Nielsen DR, Crossle C (2005) Changes in biotic communities developing from freshwater wetland sediments under experimental salinity and water regimes. Freshw Biol 50(8):1376–1390

Denny P (ed) (1985) The ecology and management of African wetlands. Dr. W. Junk Publishers, Dordrecht

Finlayson CM, Moser M (1991) Wetlands. International Waterfowl and Wetlands Research Bureau, Facts on File, New York

Flowers TJ, Hajibagheri MA, Clipson NJW (1986) Halophytes. Q Rev Biol 61:313–337

Haller WT (1974) Effects of salinity on growth of several aquatic macrophytes. Ecology 55:891–894

Harper DM, Childress RB, Harper MM, Boar RR, Hickleys P, Mills SC, Otieno N, Drane T, Vareschi E, Nasirwa O, Mwatha WE, Darlington JPEC, Escute-Gasulla X (2003) Aquatic biodiversity and saline lakes: Lake Bogoria National Reserve, Kenya. Hydrobiologia 500:259–276

Hughes RH, Hughes JS (1992) A directory of African wetlands. IUCN/UNEP/WCMC, Gland/Cambridge, p 820

Keddy P (2010) Wetland ecology: principles and conservation, 2nd edn. Cambridge University Press, Cambridge

Khan MA, Weber DJ (eds) (2006) Ecophysiology of high salinity tolerant plants. Springer, Dordrecht
Kitzhner K, Zimmels Y, Gafni A (2011) Effects of evapotraspiration on salinity of wastewater treated by aquatic plants. Rev Environ Health 23(2):149–166
McEwan K, Jolly I, Holland K (2009) Groundwater-surface interactions in arid/semi-arid wetlands and the consequences of salinity for wetland ecology. CSIRO Land and Water Science Report 56/06, CSIRO, Australia
McKee KM, Mendelsohn IA (1989) Response of freshwater marsh community to increased salinity and increased water level. Aquat Bot 34(4):301–316
Mitsch WJ, Gosselink JG (2000) Wetlands, 3rd edn. Wiley, New York
Nielsen DL, Brock MA, Rees GN, Baldwin DS (2003) Effects of increasing salinity on freshwater ecosystems in Australia. Aust J Bot 51:655–665
Ogle D (2010) Plants for saline to sodic soils conditions, Technical Note, Plant Materials No 9A USDA-Natural Resources Conservation Service, Boise, ID
Ramadan T (2000) Dynamics of salt secretion by *Sporobolus spicatus* (Vahl) Kunth from sites of differing salinity. Ann Bot 87:259–266
Rout NP, Shaw BP (2001) Salt tolerance in aquatic macrophytes: possible involvement of the antioxidative enzymes. Biol Plant 44(1):95–99
Sim LL, Davis JA, Chambers JM, Strehlow K (2006) What evidence exists for alternative ecological regimes in salinising wetlands? Freshw Biol 51 (7):1229–1248
Talling JF, Talling IB (1965) The chemical composition of African lake waters. Int Rev Hydrobiol Hydrogr 50:421–463
Wetzel RG (1983) Limnology, 2nd edn. Saunders College, New York
White F (1983) The vegetation of Africa, a descriptive memoir to accompany the UNESCO/AETFAT/UNSO Vegetation Map of Africa (3 Plates, Northwestern Africa, Northeastern Africa, and Southern Africa, 1:5,000,000). UNESCO, Paris

The Ecology of African Soda Lakes: Driven by Variable and Extreme Conditions

12

Michael Schagerl and Alfred Burian

Abstract

East African soda lakes are extreme environments that force organisms to cope with high salinities, pH levels and turbidity, but are nonetheless amongst the most productive ecosystems worldwide. We provide an overview of the species inventory in these lakes, which, amongst other fascinating species, host the well-known cyanoprokaryote *Arthrospira fusiformis*, commercially sold under the name 'Spirulina', and large flocks of Lesser Flamingos sometimes reaching densities of over one million individuals per lake. We further discuss physiological adaptations of organisms to the harsh environment as well as the variability of population dynamics and the food web structure of soda lakes. An emphasis is put on the description of environmental drivers and autogenic internal biological dynamics triggering shifts in plankton community composition and the consequences for higher trophic levels. We also elaborate on energy budgets and the importance of aquatic–terrestrial coupling in soda lakes. These are some of the key aspects pointing to promising new avenues for future research on these fascinating ecosystems.

12.1 The Environment

Saline lakes have salinities ≥3.0 ‰, and fresh lake waters show a salinity <0.5 ‰ Hammer (1986). The range from 0.5 to 3.0 ‰, termed subsaline, is a transitional zone in which both freshwater organisms and halophiles can coexist. Saline lakes are categorized as hyposaline (3 < 20 ‰), mesosaline (20 < 50 ‰) and hypersaline systems (≥50 ‰). Most saline lakes are closed basins and this also applies to the saline alkaline lakes located in the East

M. Schagerl (✉)
Department of Limnology and Bio-Oceanography, University of Vienna, Althanstrasse 14, 1090 Vienna, Austria
e-mail: michael.schagerl@univie.ac.at

A. Burian (✉)
Department of Ecology, Environment and Plant Sciences, Stockholm University, 106 91 Stockholm, Sweden
e-mail: flinserl@hotmail.com

M. Schagerl (ed.), *Soda Lakes of East Africa*, DOI 10.1007/978-3-319-28622-8_12

African Rift Valley. Evaporation is the main driver of water loss from such water bodies, also referred to as endorheic or terminal lakes, thus causing an increase in salt concentration over time (Burgis and Morris 1987). The catchment area of East African saline lakes (EASL) is of volcanic origin, which is responsible for high SiO_2, HCO_3^- and Na^+ concentrations in inflows (Jones et al. 1977, Chap. 2). The high carbonate concentration of soda lakes implies a high buffer capacity and elevated pH values of commonly >9.5 (see Chap. 4). Na^+ represents the prevalent cation in the water bodies, as Ca^{2+} and Mg^{2+} precipitate under such highly alkaline conditions. Evaporites mainly consist of gaylussite ($CaCO_3 \times Na_2CO_3 \times 5H_2O$) and trona ($Na_3H(CO_3)_2 \times 2H_2O$) (Deocampo and Jones 2014), the latter being mined in lakes Abijata (Ethiopia) and Magadi (Kenya) and exported for glass production. The brines are visible as whitish layers on lake shores and sometimes cover the lake completely (Fig. 12.1).

Fig. 12.1 Lake Magadi (Kenya); (**a**) a thick coat of trona covers the lake; (**b**) Lesser Flamingo buried in trona, (**c**) Tata Chemicals Magadi (formerly Magadi Soda Company) and its salterns

Climate variability has a strong impact on saline endorheic lakes because lake levels are defined by the balance between inflows and evaporation. If evaporation over longer periods exceeds water inputs, endorheic lakes will ultimately dry out (Fig. 12.2a). High rainfall will result in rising water levels linked to a decrease of salinity levels, until a spillway is reached (Deocampo and Jones 2014; Hammer 1986). Saline lakes are therefore very sensitive to climate change and altered hydrological conditions.

The climate in the Rift Valley is arid to semi-arid and largely influenced by the Intertropical Convergence Zone leading to two rainy and two dry seasons per year. In the lake areas, annual rainfall is <800 mm (Matagi 2004), whereby interannual variations greatly affect water levels of certain lakes (Figs. 12.2 and 12.3). The sensitivity of water levels to climate variations depends on the bathymetry of lake basins, evaporation rates and the interannual variability of water inflows (Talling 2001; Wood and Talling 1988). Besides climatic variations, also changes of anthropogenic influences can trigger shifts in lake levels. This turns population growth and intensified resource utilization into major drivers of future lake level shifts in closed-basin catchments (Odada et al. 2006). Water bodies prone to climate change such as Lake Abbe in Ethiopia and Lake Nakuru in Kenya (Trauth et al. 2010) are so-called amplifier lakes (Street 1980). They can be regarded as climate sentinels that react, similar to glaciers, with a lag time to environmental changes. Especially soda lakes with a high surface to volume ratio respond strongly to climatic changes (Olaka et al. 2010). Ephemeral lakes, which dry out from time to time such as lakes Nakuru, Amboseli (Kenya) and Eyasi (Tanzania Fig. 12.2), are sometimes also called playas and vary largely in size and depth, with major impacts on biota. In contrast, lakes with less variable surface to volume ratios are termed reservoir lakes (e.g. Lake Turkana). The amplifier and reservoir states are not stable and may shift from one to the other within short geological time (Williams 2014).

Renaut et al. (2013) demonstrated that also hot springs and adjacent habitats can be influenced by small variations in lake levels and lake-marginal groundwater. These springs and inflows into the lakes are extreme biotopes and constitute transitions zones (ecotones) offering a distinct habitat with often higher temperature and lower ion content compared to the lakes (Dadheech et al. 2013; Owen et al. 2004). The different environmental conditions are reflected by species composition and can promote speciation of microorganisms (Dadheech et al. 2013).

Lake levels and chemistry change very rapidly in terms of geological timescale: the whole area between lakes Ziway and Shala (Ethiopia) was covered by a big lake about 9000 years ago (see Chap. 1). The lakes Nakuru and Elmentaita were also connected to a single lake with water levels about 180 m higher than present (Olaka et al. 2010), and the depth of Lake Turkana varied by about 100 m within the last 10,000 years (Johnson and Malala 2009; Junginger and Trauth 2013). There is increasing evidence that these ephemeral deep-water lakes were a major driving force for hominid evolution and speciation (Maslin et al. 2014).

12.2 Organisms and Their Physiological Adaptations

Soda lakes provide a very distinct environment when compared to other aquatic ecosystems. Organisms must cope with multiple environmental stressors such as hypertonic and alkaline conditions, high turbidity and high temperature. Without protection measures, cells would become shrivelled or plasmolysed because of water loss through osmosis. Additionally, alkaline conditions of the surrounding water lower the proton motive force and thereby the energy utilization efficiency in Prokaryotes. Organisms inhabiting extreme environments are commonly termed extremophiles, and, depending on their specific environmental conditions, classification systems have been developed for Prokaryotes (Table 12.1) and also for Eukaryotes such as diatoms (Kolbe 1927) and aquatic animals (Williams 1978). A general classification for Eukaryotes includes non-halophilic (freshwater

Fig. 12.2 Lake level variations in amplifier lakes of the East African Rift Valley. (**a**) Eyasi (Tanzania); (**b–d**) Nakuru (**b** courtesy of M. Gruber-Dorninger)

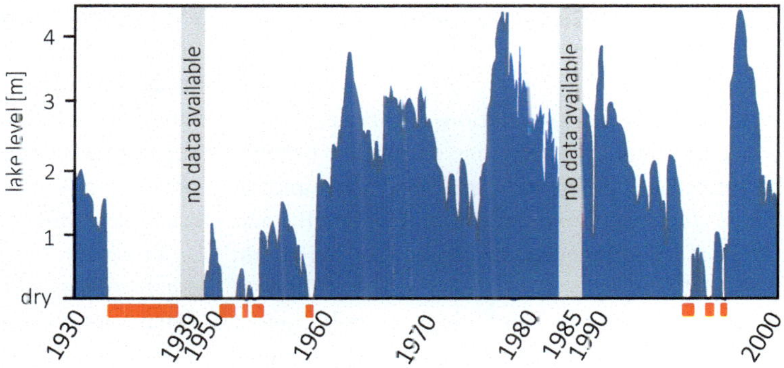

Fig. 12.3 Variability of water levels of Lake Nakuru (Kenya). Red bars indicate times of desiccation. Data from Odada et al. (2006) and Vareschi (1987)

Table 12.1 Classification of extremophiles, which may occur in soda lakes of East Africa. Data combined from Banciu and Sorokin (2013), Madigan et al. (2011), Mesbah and Wiegel (2011) and Oren (2006). Combinations of more than one criterion result in polyextremophiles. Conversion factor 1 M NaCl to salinity = 58.5 ‰

Main types	Range	Optimum
Psychrotolerant	3–35 °C	Tolerate low temp
Thermotolerant	<60 °C	Tolerate high temp
Thermophile	45–80 °C	55–75 °C
Extreme thermophile	35–85 °C	65–85 °C
Hyperthermophile	70–120 °C	80–90 °C
Alkalitolerant	pH 7.5–11.0	Tolerate high pH
Slight (=facultative) alkaliphile	pH 7.5–11.0	pH 8.0–10.0
True (=obligate) alkaliphile	pH 9.0–11.5	pH 9.5–10.5
Non-halophilic	<10 ‰	
Halotolerant	<90 ‰	Tolerate salt
Extreme halotolerant	Up to 150 ‰	Tolerate salt
Slight (=facultative) halophile	10–35 ‰	6–12 ‰
Moderate halophile	30–150 ‰	70–120 ‰
True (=obligate) extreme halophile	90–300 ‰	150–250 ‰
Natronophile	Growth at elevated Na^+ concentrations	

species <0.5 ‰), halotolerant (which prefer freshwater, but tolerate salt <20 ‰), halophilic ('salt loving' with a preference between 10 and 60 ‰) and halobiontic species ('salt living' with salinities >50 ‰).

The alkaline and saline environment shape the habitats of organisms living in soda lakes turning them into haloalkaliphiles. Three major strategies exist to handle extreme environments, namely, regulation, conformity and avoidance (Willmer et al. 2004). Regulators maintain constant internal conditions (homeostasis) by investing energy into regulatory mechanisms. Conformers lack regulatory mechanisms, and their internal conditions resemble those of the external environment; they need to adjust intracellular metabolic processes to be functional under extreme conditions. Avoiders obviate unfavourable conditions by, e.g. diapause or migration. Due to rapid and substantial changes of environmental conditions in the EASL, most organisms are regulators.

12.2.1 Adaptations to Increased Salinity

Without special adaptations, increased ion concentrations disturb the osmotic balance between the cytoplasm and the surrounding water, leading to water efflux from cells and influx of ions into cells (Erdmann and Hagemann 2004). Bacteria and Archaea often follow a mixed approach, combining morphological adaptations of cell walls and plasma membranes to counteract osmotic pressure (Banciu and Sorokin 2013) with regulatory mechanisms. Two strategies to adapt intracellular conditions to the hypersaline environment are known for these domains: the 'all-salt-in' strategy (Oren 2013) and the 'low-salt-organic-solute-in' strategy (Mesbah and Wiegel 2011). 'All-salt-in' is mainly found in halophilic Archaea and includes a mechanism that accumulates KCl in the cytoplasm for maintaining the osmotic balance (Oren 2006). This strategy, however, requires adaptation of intracellular enzymes and a highly acidic proteome (Oren 2008; Oren 2013). 'Low-salt-organic-solute-in' is found in Bacteria (Mesbah and Wiegel 2011; Oren 2011) and other microorganisms, where so-called compatible solutes (osmolytes) such as glycine betaine, glutamine, proline and hydroxyectoine are synthesized (Severin et al. 1992; Sorokin et al. 2014). These substances are hydrophilic organic compounds of low molecular weight and can be accumulated in high concentration without hampering cell metabolism (Erdmann and Hagemann 2004) or creating the need for adapted enzymes.

Special attention has been paid to adaptations in *Arthrospira fusiformis* (Voronikhin) Komárek and Lund because of its commercial importance (Hosseini Tafreshi and Shariati 2009; Vonshak and Tomaselli 2000). In *Arthrospira*, glucosyl–glycerol and trehalose are synthesized a osmoprotectants (Warr et al. 1985), in other Cyanobacteria also sucrose (Reed et al. 1984) and glycine betaine (Erdmann and Hagemann 2004). *Arthrospira fusiformis* tolerates high salinities (Fig. 12.4) and elevated temperatures of >35 °C (Vonshak and Tomaselli 2000). In laboratory experiments, Kebede (1997) studied growth of *A. fusiformis* strains over a wide range of salinities and different salt compositions and found an inverse relationship between high salinity and growth, although positive growth rates were still maintained at a salinity of 90 ‰.

Protozoa have three osmoregulatory mechanisms: (1) active water uptake and ion expulsion via contractile vacuoles (Buchmann and Becker 2009; Patterson 1980), (2) accumulation of compatible solutes (Reese et al. 2014) and (3) active ion transport across cell membranes (Hauer and Rogerson 2005). Little is known about osmoregulation in rotifers (Walker 1981), which can reach extremely high abundances in soda lakes (Burian et al. 2014, 2016): *Brachionus plicatilis* Müller was formerly considered an osmoconformer, adjusting the osmolarity of its body fluid to the surrounding medium (Epp and Winston 1977). This was refuted by Lowe et al. (2005), who found an increased activity of Na^+/K^+-ATPase at elevated salinity. The brine fly *Ephydra*, which has not been listed for EASL so far (see Chap. 8), although at times mass emergences in Lake Bogoria have been observed (unpublished data, Fig. 12.5b), develops a unique Malpighian tubule system for osmoregulation; it is partly modified to a lime gland, where HCO_3^- and CO_3^{-2} are precipitated and excreted (Bradley 2009; Herbst and Bradley 1989).

The well-studied cichlid *Alcolapia grahami* Boulenger, also referred to as the Magadi Tilapia, is adapted to extremely high salinity levels (see Chap. 9). Further, the pH in its blood is adjusted to extremely high levels compared to other teleosts (Wood et al. 1994). This minimizes energy costs for maintaining the acid–base balance, which accounts for about 50 % of total energy requirements (Wood et al. 2002). *Alcolapia* drinks exceptionally large amounts of water to maintain its ion balance, and a so-called pyloric bypass in the gut ensures water uptake irrespective of stomach fullness (Bergman et al. 2003). Flamingos developed bilateral supraorbital salt glands to excrete excess ions

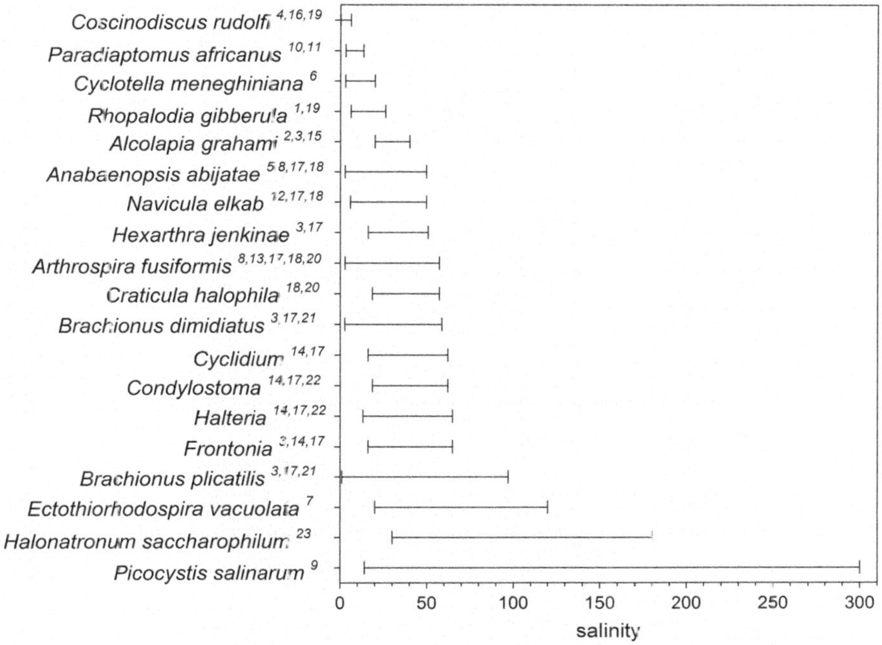

Fig. 12.4 Salinity tolerance of selected organisms from EASL. Data combined from various sources ([1]Ashley et al. 2002; [2]Beadle 1981; [3]Burian et al. 2014; [4]Chalié and Gasse 2002; [5]Dadheech et al. 2013; [6]Hammer et al. 1983; [7]Imhoff et al. 1981; [8]Kotut et al. 2006; [9]Krienitz et al. 2012; [10]LaBarbera and Kilham 1974; [11]Melack 1988; [12]Melack and Kilham 1974; [13]Ogato et al. 2014; [14]Ong'Ondo et al. 2013; [15]Reite et al. 1974; [16]Roubeix et al. 2014; [17]Schagerl unpublished data; [18]Schagerl and Oduor 2008; [19]Telford and Lamb 1999; [20]Tewodros and Afework 2014; [21]Yasindi 2001; [22]Yasindi and Taylor 2006; [23]Zhilina et al. 2001)

(Almansour 2007). Although they use freshwater for bathing, they do not require freshwater for drinking (Hammer 1986).

12.2.2 Adaptations to Increased pH

pH homeostasis is essential for living cells because many enzymatic reactions are limited to a narrow pH range. The cytoplasm of Eukaryotes is usually close to neutral, whereas organelles feature lower (e.g. chloroplasts) or higher pH (e.g. mitochondria) (Casey et al. 2010). For aquatic Prokaryotes, the pH of the surrounding medium is crucial because ATPases located in the cell membranes work along a proton gradient, which together with the transmembrane electrical potential is called the proton motive force (Gimmler and Degenhardt 2004). Acidophilic and neutralophilic Prokaryotes require a lower pH in the surrounding solution compared to the intracellular pH (Krulwich et al. 2011). Soda lakes, however, offer an alkaline environment. Here the pH gradient across the cell membranes is inverted, resulting in a lowered proton motive force (Hicks et al. 2010), which is partially compensated by an increased transmembrane potential (Krulwich et al. 2011). Cation/H+ antiporters with a high specificity to sodium play a central role in maintaining pH homeostasis (Padan et al. 2005).

The cyanoprokaryote *Arthrospira fusiformis* is an alkaliphilic species (Grant et al. 1990; Mesbah and Wiegel 2011), which thrives even at pH beyond 11.0 (Belkin and Boussiba 1991). Like most alkaliphilic organisms, *Arthrospira* is strongly dependent on elevated sodium concentrations to maintain pH homoeostasis (Pogoryelov et al. 2003; Schlesinger et al. 1996), though growth with different anion supplies is still possible (Kebede 1997).

Fig. 12.5 Mass occurrences of organisms in EASL. (**a**) Cyanobacteria form scums floating on the lake surface; (**b**) Ephydridae; (**c**) plankton sample during *Brachionus* bloom—left dish = *Branchionus* (brownish), right dish *Arthrospira* (greenish); (**d**) flocks of pelicans attracted by *Alcolapia grahami*; (**e**) flocks of flamingos. (**a**, **b**, **e**) around Lake Bogoria; (**c**, **d**) Lake Nakuru, both located in Kenya

Alcolapia grahami is obligatory ureotelic, which is interpreted as an adaptation to the alkaline conditions because high pH levels impede excretion of ammonia via diffusion (Pörtner et al. 2010). Further, the pyloric bypass enables ingested alkaline water to be transported around the stomach, thus avoiding neutralization of stomach acids (Bergman et al. 2003).

12.2.3 Adaptations to High Turbidity

Alkaline–saline lakes are characterized by very high turbidity compared to many other aquatic systems (see Chap. 3). This is a result of extremely high plankton densities (Oduor and Schagerl 2007a), increased sediment resuspension due to their shallowness, salt precipitation (whiting) and high loads of dissolved organic matter (Jirsa et al. 2013). In three Kenyan soda lakes, mean light attenuation coefficients of 13 have been calculated (Oduor and Schagerl 2007a), limiting the euphotic zone to the uppermost 0.35 m (Fig. 12.6).

The ratio of the mixing depth to the euphotic zone is referred to as the critical mixing depth and constitutes a key factor for primary production in turbid waters (Dokulil 1984; Grobbelaar 1989). According to Talling (1971), a critical mixing depth >5 will exclude many phytoplankton taxa such as most pelagic diatoms and other large nonmotile eukaryotic algae. High critical mixing depths are common in the EASL and, together with the alkaline–saline conditions, contribute to the exclusion of many primary producers from these habitats. As a further consequence, a significant contribution of phytobenthos to overall primary productivity is limited to periods of increased light penetration during breakdowns of dominant phytoplankton communities (Melack 1988; Tuite 1981; Vareschi and Jacobs 1985).

The dominating primary producers in EASL developed strategies to cope with the poor underwater light supply. Gas vesicles produced in Cyanobacteria such as *Anabaenopsis*, *Arthrospira* and *Cyanospira* (see Chap. 6) facilitate buoyancy and result in floating algae layers with a pudding-like consistency along shoreline areas (Fig. 12.5a). The photosynthetic characteristics resemble those of low-light acclimated algae, with a strong response of primary production to increased light intensity (steep initial slopes of photosynthetic–irradiance

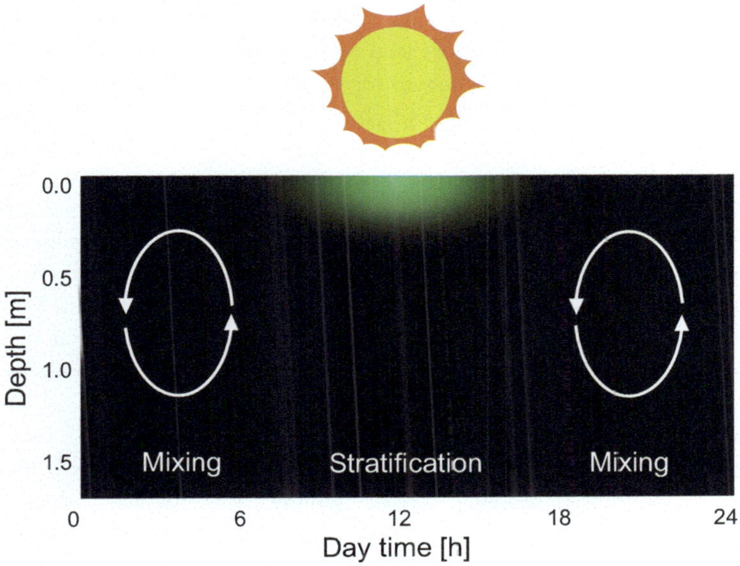

Fig. 12.6 Light availability in the water column of a turbid soda lake. Light attenuation coefficients >13 and critical mixing depths >5 are common. During daytime, stratification occurs due to high irradiation; cooling during nighttime causes partial or total mixing

Table 12.2 *Arthrospira fusiformis* photosynthesis characteristics. Onset of saturation = Ik in μmol photons m^{-2} s^{-1}; Pmax = maximum productivity in mg O_2 L^{-1} h^{-1}

Ik	Pmax	Data	References
185–325		Culture	Vonshak et al. (1996)
40–100	0.6–1.3	Field	Oduor and Schagerl (2007a)
150–200		Culture	Vonshak and Tomaselli (2000)
140–200	3.2	Culture	Kebede and Ahlgren (1996)
150–200		Culture	Bocci et al. (1980)

curves), an onset of light saturation at low-light intensities (I_k; Table 12.2) and high cellular pigment concentrations often surpassing concentrations of 1 % chlorophyll-*a* per unit biomass (Oduor and Schagerl 2007a).

12.3 Biomass and Diversity

Many organisms in EASL build up very high biomass levels (Fig. 12.7). The elevated nutrient concentrations in EASL, e.g. a total phosphorus content of usually several 100 μg L^{-1} (Kebede and Willén 1998; Oduor and Schagerl 2007b; Vareschi 1982), and the good carbon supply boost growth of photoautotrophs, which results in very high primary productivity. Whereas total alkalinities between 2 and 4 meq L^{-1} are common in well-buffered freshwater biotopes, alkalinities exceeding 1000 meq L^{-1} were recorded for Kenyan soda lakes (Jirsa et al. 2013). All these factors lead to exceptionally high biomass of photoautotrophs, at times exceeding 1 mg L^{-1} chlorophyll-*a* (Schagerl et al. 2015), a value comparable to open-pond cultivation systems. In contrast to shallow freshwater ecosystems, submerged macrophytes are absent (Fig. 12.7a; Harper et al. 2003). Accordingly, phytoplankton is the main driver of high production rates (Fig. 12.7b).

Other essential components of the aquatic environment, which so far have been neglected in regards to energy and nutrient fluxes, are Bacteria and viruses (Fig. 12.7). With amounts peaking at around 1000 mg L^{-1}, EASL are extremely rich in dissolved organic carbon (Jirsa et al. 2013), thus providing optimal conditions for the microbes. Bacteria abundances reach $>10^8$ mL^{-1} (Kilham 1981; Yasindi et al. 2002; Zinabu and Taylor 1997), and aquatic virus counts yielded densities of up to 1.4×10^{10} mL^{-1} (Gruber-Dorninger 2015). These values range amongst the highest abundances worldwide (Wilhelm and Matteson 2008) and imply a large quantitative importance of the microbial loop and the viral shunt (Fig. 12.7), because the microbial community is strongly involved in sulphur, nitrogen and carbon biogeochemical cycles (Sorokin et al. 2014).

Exceptionally high densities were also recorded for micro- and metazooplankton (ciliates 1.4×10^4 L^{-1}, Yasindi et al. 2002; rotifers 1.2×10^5 L^{-1}, Burian et al. 2013, 2016). According to Zinabu and Taylor (1997), the high densities of ciliates may be responsible for a low number of heterotrophic nanoflagellates (HNF) and, in a cascading effect, lead to the low chlorophyll-*a*: Bacteria ratios in EASL.

The high community biomass of photoautotrophs provides a richly set table for flamingos and other primary consumers. Flamingos show extremely variable densities at single lakes, and flock sizes can vary from a few individuals to hundreds of thousands of birds on a single lake (Mlingwa and Baker 2006; Owino et al. 2001; Vareschi 1978). Such flocks are often, but not always, coupled to the occurrence of their preferred food, the cyanoprokaryote *Arthrospira* (Fig. 12.8; Kaggwa et al. 2013; Krienitz and Kotut 2010).

Microbial diversity is exceptionally high at low to moderately saline alkaline lakes (see Chap. 5; Lanzén et al. 2013; Sorokin et al. 2014). Besides the special salt composition of the EASL that offers ideal living conditions for the haloalkaliphilic microbial community (Sorokin et al. 2014), the very high organic matter content also plays an important role (Jirsa et al. 2013). Primary productivity rates range

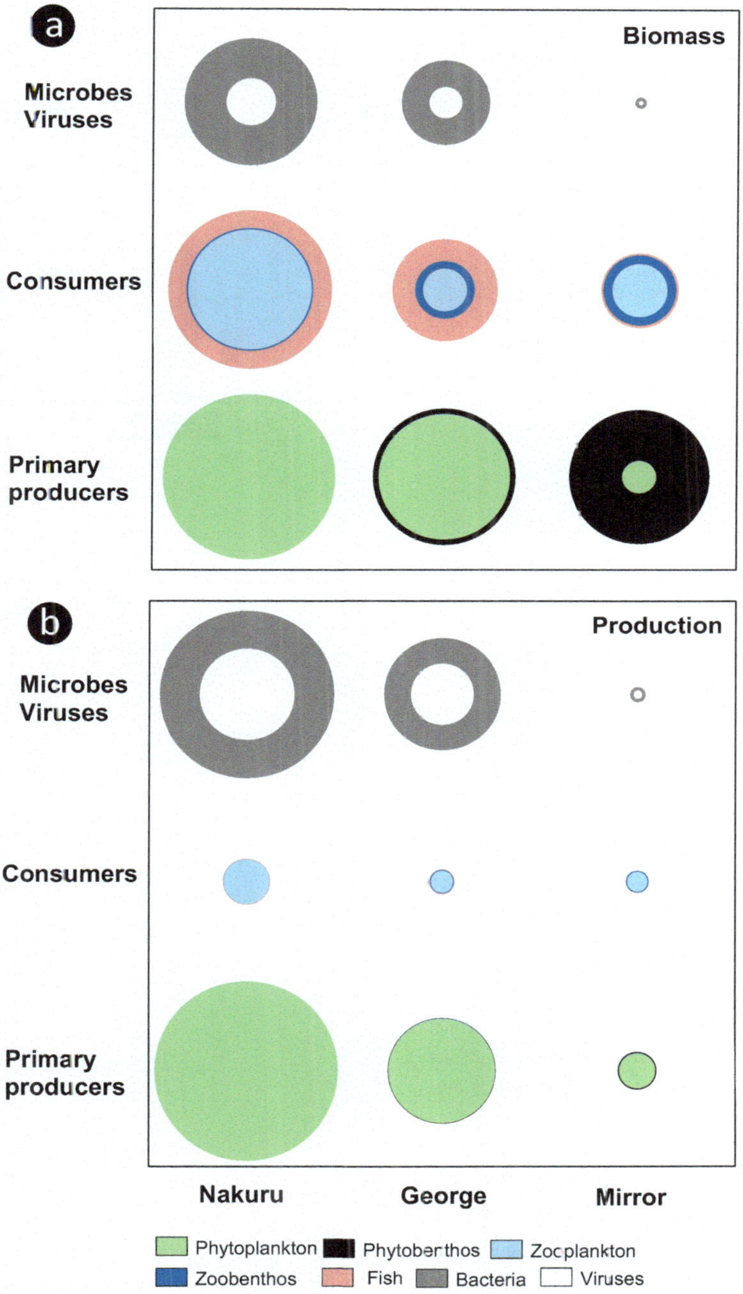

Fig. 12.7 Comparison of food web compartments in terms of biomass (**a**) and productivity (**b**). Lake Nakuru (Kenya) is saline alkaline, Lake George (Uganda) is a tropical, highly eutrophic lake, and Mirror Lake (United States) is a temperate mesotrophic lake. Data combined (Ganf 1974; Vareschi 1979; Vareschi and Jacobs 1984; Vareschi and Vareschi 1984; Hecky 1984; Wetzel 2001; unpublished results). Circle areas correspond to relative contributions of the food web compartments; bacterial productivity was estimated based on biomass. Total biomass is 21.3, 11.3 and 7.4 g C m^{-2} for lakes Nakuru, George and Mirror, overall productivity is 1900, 740 and 65 g C m^{-2} a^{-2} for the respective lakes

amongst the highest worldwide (Table 12.3), attracting many terrestrial organisms, e.g. around 500 bird species (Matagi 2004); Lake Bogoria alone hosts 223 different bird species (Harper et al. 2003). Although the lakes are threatened by climate change and water abstraction for various purposes (see Chap. 14), human activities have sometimes also increased diversity: *Alcolapia grahami* was introduced into Lake Nakuru during the 1950s and 1960s to fight mosquitoes (Matagi 2004). The fish population attracted piscivores such as pelicans, swifts and

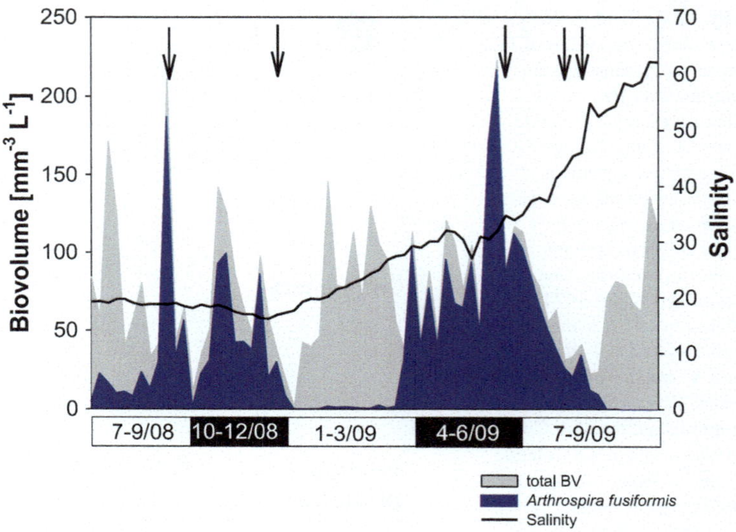

Fig. 12.8 Phytoplankton biomass sampled in weekly intervals from Lake Nakuru between July 2008 and September 2009. Total algal biovolume (Cyanobacteria included), *Arthrospira fusiformis* biovolume and salinity levels are displayed. Crashes of *Arthrospira* coinciding with cyanophage infection are indicated with vertical arrows. Data combined (Jirsa et al. 2013; Kaggwa et al. 2013; Peduzzi et al. 2014)

Table 12.3 Comparison of annual gross primary production rates of terrestrial and aquatic ecosystems

Ecosystem	Gross primary production ($g\ O_2\ m^{-2}\ year^{-1}$)	References
Arctic tundra[a]	445	Turner et al. (2006)
Conifer forest[a]	2840	Turner et al. (2006)
Tallgrass prairie[a]	1877	Turner et al. (2006)
Corn/soya bean fields[a]	1621	Turner et al. (2006)
Hardwood forest[a]	3547	Turner et al. (2006)
Tropical moist forest[a]	6317	Turner et al. (2006)
Temperate estuary (Chesapeake Bay)	1666	Kemp et al. (1997)
Floodprone river (Switzerland)	1825	Uehlinger (2006)
Wadden Sea (Germany)	469	Tillmann et al. (2000)
Data set of 25 temperate lakes[b]	<2000	Solomon et al. (2013)
Tropical lagoon (Gulf of Mexico)	4450	Ziegler and Benner (1998)
Tropical sea grass mead. (Indonesia)	4177	Erftemeijer et al. (1993)
Tropical freshwater lake (L. Tana)	887	Wondie et al. (2007)
L. Elementaita (African soda lake)	4282	Oduor and Schagerl (2007a)
L. Bogoria (African soda lake)	4875	Oduor and Schagerl (2007a)
L. Nakuru (African soda lake)	6854	Oduor and Schagerl (2007a)

[a]Estimates based on MODIS measurements
[b]Lakes ranged between 25 and 60° latitude

other fish-eating birds, which built stable populations around the lake.

Because of the harsh living conditions in the alkaline–saline lakes, the biodiversity of photoautotrophs and heterotrophic Eukaryotes is considered to be low compared to freshwater (Dudgeon et al. 2006) and marine systems (Fig. 12.9a; Costello et al. 2010). Whereas transition zones between freshwater and marine environments (brackish water) show a low diversity with minimum taxa number between salinities of 5–7 ‰ and increases with higher salinities (U-shaped species–salinity curve; Remane and Schlieper 1971), species numbers in saline alkaline systems are assumed to drop with increasing salinity (L-shaped curve; Fig. 12.9a, b; Cooper and Wissel 2012; Hammer 1986; Vareschi 1987). Salinity is, however, not

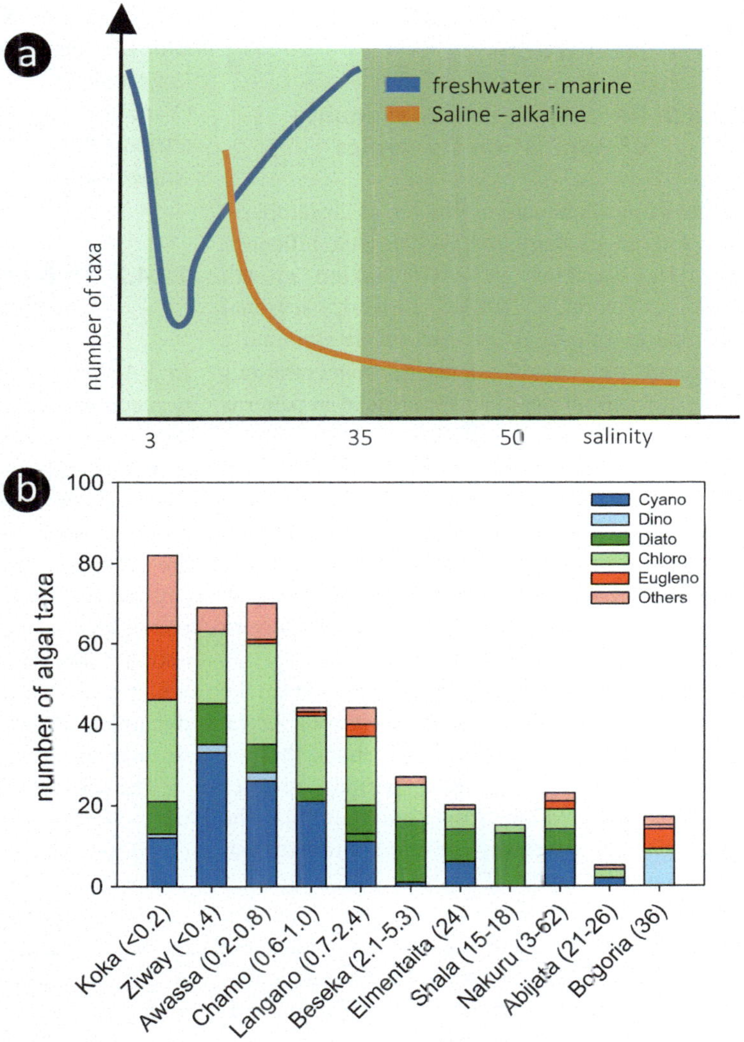

Fig. 12.9 Biodiversity and salinity. (**a**) Brackish waters show a minimum species number at a salinity between 5 and 7 (Remane and Schlieper 1971), saline inland waters show a sharp decline in species number with increasing salinity (Hammer 1986; Vareschi 1987). Salinity levels of 3 ‰ (limit freshwater and saline systems) and 35 (marine systems) are indicated. (**b**) Phytoplankton composition along a salinity gradient of Ethiopian and Kenyan lakes. Data from Kebede (2002) and unpublished results. Numbers in brackets indicate salinity ranges

necessarily the most important variable affecting biodiversity; biological interactions have been stated to have a stronger effect on species richness (Williams 1998). In a meta-analysis, Williams (1998) found salinities >50 ‰ to have only a small impact on community structure; only at salinity ranges <50 ‰, does this variable take on a key role in defining the communities. This probably reflects a lower salinity tolerance of organisms growing at low salinities (Fig. 12.4). However, the traditional opinion of low diversity at high salinity levels needs to be scrutinized because the use of molecular tools has provided a completely different result with an unexpectedly high diversity of Archaea, Bacteria (Duckworth et al. 1996; Lanzén et al. 2013; Sorokin et al. 2014) and Eukaryotes (Luo et al. 2013).

12.4 Communities and Their Interactions

12.4.1 Periodicity and Seasonality of Population Dynamics

In the tropics, annual amplitudes of air temperature and radiation are low, reducing the influence of these variables on phytoplankton growth cycles (Twombly 1983). Instead, seasonal fluctuations of primary production are commonly triggered by variable nutrient concentrations linked to rainfall, wind and stratification patterns (Ndebele-Murisa et al. 2010; Talling 1986). The elevated nutrient levels commonly found in EASL (Oduor and Schagerl 2007b; Sorokin et al. 2014; Sorokin and Kuenen 2005) weaken the influence of annual environmental variability on plankton populations and result in the apparent uncoupling of primary production rates from seasonal cycles (Talling 1986).

Exceptions are deep soda lakes and lake sections near river deltas. In deep soda lakes, such as Lake Turkana and Lake Shala, thermal stratification over lengthier periods reduces nutrient recirculation from sediments to the pelagic zone, increasing the importance of seasonal nutrient inputs (Ferguson 1982). In river deltas, in contrast, large amounts of suspended sediment particles can enter lake ecosystems and significantly influence turbidity. Consequently, changes in light attenuation can have strong seasonal impacts on primary production (Lind et al. 1992) and potentially affect higher trophic levels by alternating food quantity and quality (Arruda et al. 1983; Kirk and Gilbert 1990).

Limited or even lacking seasonal cycles are also found in key consumers. The bloom formation of rotifers, which belong to the quantitatively most important consumer groups in EASL (Vareschi and Jacobs 1984), is driven by sediment resuspension and resting egg hatching (Burian et al. 2016). These factors are more dependent on interannual than on seasonal variation (Fig. 12.3). Likewise, Lesser Flamingos are known to frequently move between EASL, following patterns of food availability rather than seasonal cycles (Childress et al. 2007; Chaps. 10 and 13). Nevertheless, species densities show large temporal fluctuations at all trophic levels (Vareschi and Jacobs 1985), which can partly be attributed to shifts in species clusters and changes in the food web structure of EASL.

12.4.2 Shifts in Community Composition

In EASL, phytoplankton community composition can change frequently, affecting higher trophic levels such as zooplankton and vertebrates and potentially also nutrient and energy flows (Peduzzi et al. 2014; Vareschi and Jacobs 1985). Processes driving community alternations are not yet fully understood but can be categorized into (1) changes in environmental conditions leading to shifts of dominant taxa and (2) autogenic biological processes, which are potentially triggered by chaotic biological dynamics (Beninca et al. 2009) or short external disturbance (Burian et al. 2016).

The water level of EASL is a key environmental variable shaping the physico-chemical conditions and influencing plankton community composition (Fig. 12.10). Changes in water levels impact the hydrochemistry, the stability of the sediment–water interface and the underwater light supply, thus defining the habitat suitability of EASL for many species (Verschuren et al. 2000). Interannual variations in lake levels have a great potential to influence the outcome of competition within species-rich plankton communities (Krienitz et al. 2013a) and the resulting changes in size–composition and food quality can cascade up the food chain to higher trophic levels (Peduzzi et al. 2014).

Amongst the environmental factors influenced by changing water levels, salinity has been shown to have a strong impact on the species composition and association of littoral and pelagic communities (Iltis 1974; Krienitz et al. 2013a; Verschuren et al. 1999). During low salinities, which occur at high water levels but also after the refilling of lakes when salt

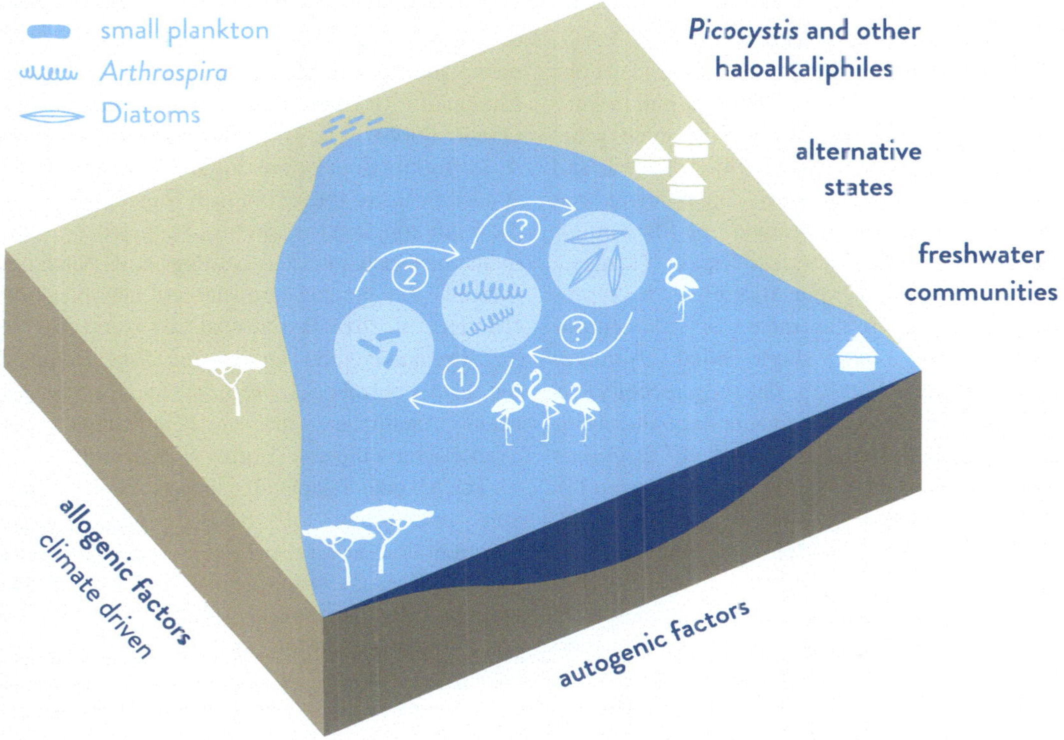

Fig. 12.10 Model of plankton states in East African soda lakes, which are defined either by allogenic or autogenic factors. Between the two extremes 'flooding' (dominated by various freshwater species) and 'near dryness' (with the prasinophyte *Picocystis* and other halophilic taxa), three states are possible at intermediate salinity. These three states are mainly driven by autogenic factors such as viral infections, grazing pressure and competition. For further explanation of (1) and (2) see text; (?) key factors still unknown

crystals are still stuck in the sediment, diverse but little-studied freshwater plankton communities predominate (Luo et al. 2013). Typical vertebrate consumers of soda lakes such as the Magadi Tilapia or the Lesser Flamingo, however, suffer under high costs of osmoregulation and/or food limitation (Krienitz and Kotut 2010; Wood et al. 2012). A salinity increase from about 3 to 10 ‰ causes a gradual shift from freshwater species to mesohaline plankton communities (Krienitz et al. 2013b), which are often low in diversity and dominated by the filamentous Cyanobacteria *Arthrospira* and *Anabaenopsis*. During blooms of these filamentous Cyanobacteria, planktonic consumers are dominated by relatively small mesozooplankton taxa, typically the rotifers *Brachionus plicatilis* and *Brachionus dimidiatus* Pallas and the large, Cyanobacteria-feeding ciliates *Frontonia* and *Condylostoma* (Burian et al. 2013; Ong'Ondo et al. 2013; Yasindi et al. 2002). A further increase of salinity >50 ‰ triggers the replacement of *Arthrospira* and *Anabaenopsis* species by halophilic algae such as the small Eukaryote *Picocystis salinarum* Lewin or picocyanobacteria (Schagerl et al. 2015). Little is known about the responses of key consumers to salinity increases >50 ‰ and concomitant shifts of primary producers, but it seems that at least Lesser Flamingos are negatively affected due to lack of appropriate food (Kaggwa et al. 2013).

Internal biological dynamics are the second potential trigger of changes in community composition in EASL (Schagerl et al. 2015; Schagerl and Oduor 2008). The transition from filamentous Cyanobacteria and nanophytoplankton states during mesohaline conditions strongly affects the food web structure and energy flows in EASL (Schagerl and Oduor 2008; Vareschi and Jacobs 1985). Whereas *A. fusiformis* blooms can be very stable and prevail for months and even years (Ballot et al. 2004; Iltis 1974; Schagerl and Oduor 2008), swift biomass breakdowns can cause a complete disappearance of filamentous Cyanobacteria within a few weeks (Kaggwa et al. 2013). Several explanations have been proposed for the replacement of filamentous Cyanobacteria by single-celled primary producers (Melack 1988), but only recently has it been demonstrated that cyanophage infections can cause swift breakdowns of *A. fusiformis* blooms (Peduzzi et al. 2014; Fig. 12.10 arrow 1)).

In the absence of filamentous Cyanobacteria, single-celled phytoplankton taxa quickly reach high biomass comparable to that of filamentous Cyanobacteria. Nanoplankton communities show much higher species richness (Schagerl et al. 2015) and support a large and diverse community of protozoan zooplankton (Burian et al. 2016). Lesser Flamingos, however, are unable to feed on nanoplankton (Krienitz and Kotut 2010; Vareschi 1978), which negatively affects their numbers (Kaggwa et al. 2013; Kihwele et al. 2014). Similarly, the Magadi Tilapia is also limited to alternative diets in the absence of filamentous Cyanobacteria, one of its main food sources (Vareschi 1979).

The stability of nanoalgae communities as an alternative state is not yet well understood, but might be connected to the food web structure (Fig. 12.11) and the occurrence of the Soda Tilapia. In soda lakes free of fish such as Lake Bogoria, nanoalgae communities are merely a transitional state that is outcompeted by filamentous Cyanobacteria within a few weeks (Kaggwa et al. 2013; Schagerl and Oduor 2008). In contrast, high densities of the Soda Tilapia might extend the persistence of nanoplankton communities. Although the high production and standing stock of *A. fusiformis* blooms make them insensitive to high grazing rates of fish and other consumers (Oduor and Schagerl 2007a; Vareschi and Jacobs 1985), Cyanobacteria densities below a threshold level might be susceptible to top-down control (Lampert and Sommer 1999). Hence, the presence of fish may explain longer-lasting dominance of nanoplankton in Lake Nakuru and Lake Elmentaita (Ballot et al. 2004; Schagerl and Oduor 2008). The re-establishment of dense *A. fusiformis* populations might be supported by sudden peaks of rotifer biomass (Fig. 12.10 arrow 2). Such zooplankton blooms, triggered by sediment resuspension and resting egg hatching, increase the grazing pressure on nanoplankton (Burian et al. 2014; Burian et al. 2013) and provide an alternative food source for fish. Reduced top-down control on *A. fusiformis* and less competition facilitates the re-establishment of cyanobacteria blooms (Burian et al. 2016).

The Magadi Tilapia also affects the behaviour and occurrence of benthic and pelagic invertebrates. Several pelagic insect larvae such as Culicidae, Ceratopogonidae and Ephydridae and the large omnivorous ciliate *Condylostoma* sp. are very common in fish-free EASL but strongly reduced in the presence of fish (Ong'Ondo et al. 2013; Vareschi 1979). Moreover, chironomid larvae, which are frequently found in the pelagic zone of lakes without fish (unpublished data), are restricted to the benthic zone in lakes with higher fish densities (Vareschi and Vareschi 1984). The Magadi Tilapia has a mixed effect on the avifauna of EASL. On one hand, fish support piscivorous species such as pelicans and swifts, increasing the bird diversity around lakes. On the other hand, the fish negatively affect invertebrate biomass, an important resource for 50–100 bird species of Lake Bogoria (Harper et al. 2003).

Besides filamentous Cyanobacteria and nanoalgae blooms, a state dominated by benthic diatoms has been reported during mesohaline conditions (Kebede and Willén 1997; Tuite 1981; Vareschi and Jacobs 1985; Fig. 12.10). The factors causing low pelagic biomass and facilitating benthic diatom blooms have not yet been studied, although low lake levels apparently coincide with their occurrence (Tebbs et al. 2015; Tuite 1981). Benthic algae blooms will, from a consumer perspective, limit fish to feeding on benthic chironomids and low densities of pelagic zooplankton, which may

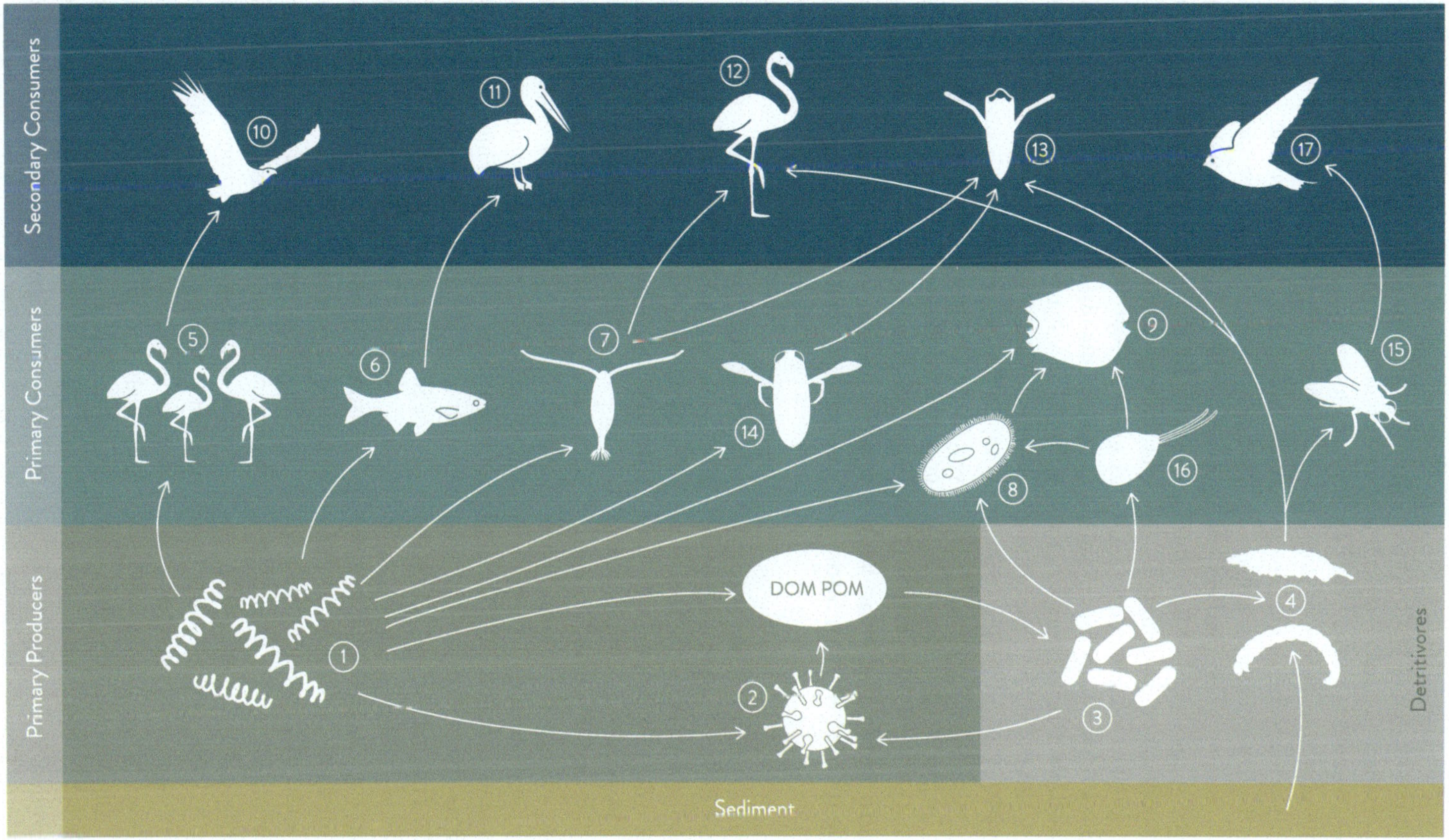

Fig. 12.11 Food web of a saline alkaline lake in East Africa. *Arthrospira* dominates the phytoplankton community (1). The huge DOM-pool has been largely neglected until now, as has been the microbial loop (3), the viral shunt (2) and heterotrophic and mixotrophic nanoflagellates (16). Detritivorous and omnivorous insect larvae such as chironomids and ephydrids (4) can emerge in mass hatchings (15). Primary consumers are Lesser Flamingos (5), *Alcolapia* (6), copepods (*Paradiaptomus*; 7), water boatmen (Corixidae; 14), ciliates (8) and rotifers (*Brachionus*; 9). Ciliates and rotifers feed also on HNF and Bacteria. Secondary consumers are raptors (10), pelicans (11), Greater Flamingos (12) and water bugs (*Anisops*; 13)

result in food limitation. Lesser Flamingos are able to feed on benthic algae (Melack 1988; Tuite 2000), but the production of benthic algae is 1–2 orders lower than that of *Arthrospira* blooms (Tuite 2000), leading to a lower carrying capacity of lakes dominated by benthic diatoms.

Whether alternatives to blooms of filamentous Cyanobacteria can be considered as stable or merely transitional remains to be clarified and might depend on the species pool of individual lakes. At least nanoplankton communities in mesohaline, fish-free lakes seem to occur only for short time periods of weeks to few months after disturbance of filamentous Cyanobacteria. This potentially turns the succeeding *A. fusiformis* blooms into a pelagic climax vegetation.

12.4.3 Production Rates and Food Web Structure of East African Soda Lakes

Primary production rates of EASL are extremely high (Fig. 12.7b and Table 12.3) and close to the theoretical maximum for primary production in aquatic ecosystems, when taking photosynthetic efficiencies and self-shading into account (Melack 1979; Talling and Wood 1973). Yearly production rates are amongst the highest values found in aquatic and terrestrial ecosystems worldwide (Oduor and Schagerl 2007a); only tropical rain forests bear comparison with EASL (Table 12.3). The *Arthrospira* state is the most productive, and highest rates were measured in sheltered crater lakes characterized by stable Cyanobacteria blooms (Melack 1981). A substantial decrease of depth-integrated net production rates from 102 to 16 kJ m^{-2} d^{-1} (equivalent to 7.9 to 1.2 g O_2 m^{-2} d^{-1}) was measured in Lake Nakuru after a switch from an *Arthrospira* to the nanoplankton state (Vareschi 1982, Vareschi and Jacobs 1985). A similar variability over time has been shown for Lake Sonachi (Melack 1981), where seasonal gross productivity varied between 0.8 and 0.2 g O_2 m^{-2} h^{-1}, and Lake Elmentaita, where primary production dropped by >70 % after a shift from a phytoplankton bloom to a microphytobenthic community (Melack 1981, 1988). Tuite (2000) stated that a lower but more constant community production seems to be a common characteristic of the benthic diatom state.

In contrast to production by photoautotrophs, production rates of heterotrophic Bacteria and Archaea and the importance of the microbial loop have been neglected so far. Based on the extremely high abundances of heterotrophic Bacteria, a large production can be assumed (Fig. 12.7). Unfortunately, no measurements have been performed and no information exists about vertical and temporal patterns. Similar to Bacteria and Archaea, virioplankton also shows high abundance, but very little is known about the importance of viruses for food webs in EASL (Gruber-Dorninger 2015). Another research gap are protozoan communities, which are mainly dominated by ciliates and characterized by low densities of HNF (Finlay et al. 1987; Zinabu and Taylor 1997). Also ciliate taxa seem to be coupled to certain phytoplankton states: species below an equivalent spherical diameter of 45 μm prevail during the nanoplankton state (Burian et al. 2016), and larger species dominate during *A. fusiformis* blooms (Ong'Ondo et al. 2013). Food of ciliates is diverse and includes detritus, Bacteria and Archaea, other protozoans, all types of phytoplankton and even rotifers (Burian et al. 2013; Finlay et al. 1987; Yasindi et al. 2002). The production of ciliates was roughly estimated to about 3.2 mg C L^{-1} in Lake Nakuru (Yasindi et al. 2002), demonstrating their importance for energy flows within the ecosystem.

A common characteristic of metazoan consumers is their large density fluctuation and therefore also variations in community consumption and production. Rotifers, for example, develop short-lived blooms of up to 6×10^5 individuals L^{-1} with exceedingly high production rates (Burian et al. 2016; Iltis and Riou-Duwat 1971), whereas non-bloom densities are often about 10^4 individuals L^{-1}. Other zooplankton taxa such as the calanoid *Paradiaptomus africana* Daday, the cladoceran *Moina* sp. and

Ephydra larvae show a stochastic occurrence. The factors leading to their emergence or their extinction from the pelagic zone are poorly understood.

Vertebrate consumption during *Arthrospira* blooms is often dominated by feeding activity of Lesser Flamingos. An average stock size of 9×10^5 individuals at Lake Nakuru has been calculated to consume 305 g dry mass m^{-2} d^{-1} and thereby nearly four times as much as the Magadi Tilapia population (Vareschi and Jacobs 1985). Consumption rates of the fish are, however, much more constant because flamingos may move to other lakes once densities of suitable food sources decrease (Childress et al. 2007, Chaps. 10 and 13). Energy flows from primary to secondary consumers are relatively small in EASL, resulting in a quickly narrowing food-web pyramid. Whereas piscivorous birds consume about 1 % of total fish stock per day, Lesser Flamingos are only minimally predated by Fish Eagles or hyenas, leading to an overall trophic efficiency (sensu Lindeman 1942, consumption of tropic level *n*/consumption of trophic level $n + 1$) in Lake Nakuru of about 3–6 % between trophic levels 2 and 3 (Vareschi and Jacobs 1985).

12.4.4 Land–Water Interconnectivity

Food webs in almost all EASL are largely dependent on autochthonous production; allochthonous matter input is mostly restricted to areas near river inflows (e.g. Lake Bogoria, Tebbs et al. 2015) and constitutes a food source of low quality compared to readily available pelagic organisms (Taipale et al. 2014). Small local whirlwinds have the potential to introduce material from barren lake shores, although their quantitative importance is difficult to determine and much of the introduced material will likely be derived from the lake itself. Nevertheless, land–water interconnectivity plays an important role in EASL. Especially the introduction and deposition of inorganic suspended solids can affect the turbidity and bathymetry of shallow lakes (Ferguson and Habott 1982; Harper et al. 2003). For example, increased sediment input due to land-use changes in the catchment area has caused Lake Turkana to shrink, shifting the delta of the Omo River from Ethiopia to Kenya and increasing political tensions about fishing rights between different tribes (Avery 2012). Also in Lake Nakuru, increased terrestrial sediment input has affected the bathymetry of the lake (Fred Omengo, pers. comm.). This may affect surface–volume ratios, the variability of salinity levels and thereby the ecological stability of the lake.

The high biomass production of aquatic organisms in EASL supports various terrestrial consumers, fuelling the food webs around lakes (Harper et al. 2003; Vareschi 1978). Major taxa involved in exporting nutrients and energy from lakes are (i) insects with aquatic larvae, (ii) flamingos and (iii) fish-eating birds. Larvae of insect taxa, mainly Chironomidae but also Ephydridae, Ceratopogonidae and Lepidoptera, are important food sources for the rich avifauna. They also support invertebrate predators such as spiders or dragonflies. Harper et al. (2003) reported that out of the 223 bird species of the National Reserve around Lake Bogoria, about 100 prey on either aquatic larvae, exuvia or adult insects emerging from the lake. Furthermore, high densities of Magadi Tilapia constitute a main food source for many fish-eating birds (>50 species in Lake Nakuru, Vareschi 1979). Pelicans, one of the quantitatively most important fish predators in the EASL, have also been shown to contribute to energy and nutrient flows between lakes. Populations that feed mainly at Lake Nakuru but breed close to Lake Elmentaita (Brown 1973) have been estimated to export about 9 t fresh mass per day and substantial amounts of nitrogen and phosphorus to Lake Elmentaita (Vareschi 1979). Flamingos, feeding on algae, Cyanobacteria and invertebrates, are only a minor food source for other terrestrial consumers (Vareschi and Jacobs 1985). Their main contribution to aquatic–terrestrial energy transfer is largely due to their feeding activity: about 30 % of the energy consumed by Lesser Flamingos is returned to the lakes as faeces, urinary wastes and dead carcasses (Vareschi 1978), 70 % is either metabolically consumed or invested in breeding.

Overall, the nature and extent of organic matter transport from aquatic to terrestrial ecosystems strongly depend on species composition and population sizes of EASL. Large variation of Lesser Flamingo densities, which play an important role in exporting nutrients and energy, is not only contingent on the community composition of primary producers at a given lake but also on the conditions at other lakes in their fly network. This interdependency and the potential transmission of environmental and anthropogenic disturbances between EASL, which can arguably be seen as a supra-ecosystem, bear important consequences for management and conservation.

12.5 Outlook

After Ekkehard Vareschi's pioneering work on the ecology of Lake Nakuru in the 1970s, ecological research on EASL was again intensified from around 2005 onwards. An exciting research topic that has been addressed in recent years is shifts in plankton communities. A better understanding of the triggers of such shifts, which can take place within a few days, calls for more detailed investigation of community changes with short sampling intervals over longer periods. A future challenge will be to interlink biological interactions between different trophic levels, examine the impact of frequently changing allogenic conditions and analyses both from evolutionary and ecological perspectives. In our opinion, this field of research would benefit considerably from a network of scientists from various research fields and different East African countries. The conditions of amplifier lakes change fast, and coordinated research between countries will certainly provide new insights into the functioning of these unique ecological systems. We hope that the examples mentioned below will further encourage scientists to delve into this fascinating field of research.

Bacteria and Archaea densities in EASL turned out to be amongst the highest found in the aquatic environment worldwide, but data on their importance for element cycling are still lacking. Viral abundance was shown to be extraordinarily high. Cyanophages are involved in *Arthrospira* breakdowns, but very little is known about the triggers of cyanophage attacks. Viral infections provide an interesting setting for the investigation of co-evolutionary dynamics and the role of viruses in sustaining intraspecific diversity. Although the few data available are promising, there are no further studies on the microbial loop/viral shunt.

Most investigations focused on the pelagic zone, but the benthic zone is also very important for EASL. Mechanisms of benthic–pelagic coupling including denitrification processes in microanaerobic zones and the ecological memory of the benthic zone in the form of resting stages are still widely unexplored.

Little knowledge is available on the variability of fish abundances and the reasons for density fluctuations, although there is growing evidence that fish are important drivers for changes in plankton communities and that fish influence the stability of food webs. The fish population in some lakes has attracted fish-eating birds, which together with other birds act as vectors transporting aquatic organisms, including pathogens such as cyanophages, from one lake to the next. The interconnectivity between lakes is still a black box, which is linked to another interesting field of research: flamingo movement between EASL as well as between East and South Africa and India. Although genetic exchange between the flamingo populations has already been demonstrated, accurate and confident data on flamingo numbers inhabiting the EASL are still lacking. Here, the state-of-the-art methods such as high-resolution remote sensing, telemetry and unmanned aerial vehicle systems are promising tools to solve this question. The phenomenon of flamingo mass mortality has given rise to much speculation, but an explanation based on scientific facts is still missing. Regular monitoring and the installation of an international scientific task force with short reaction time would help to solve this open question.

Another hot topic is biodiversity of the EASL. Contrary to what was generally assumed earlier, biodiversity in the lakes might be very high not only for Archaea but also for other domains. The application of modern genetic and biochemical tools may well yield a completely different picture of functional diversity and species richness.

All these knowledge gaps cover essential components of ecosystem functioning, but they are only remotely related to lake management. We therefore need to mention as a last point the urgent need for installing a monitoring programme, which ideally should be coordinated between countries. We were able to show that simple-to-obtain parameters (salinity, water level, turbidity, water temperature) have a big impact on the lake ecosystem and help to define the current state and future changes. Such data are still not at hand over longer time intervals, but they would greatly facilitate both a very specific and generally valid understanding of these unique systems.

References

Almansour MI (2007) Anatomy, histology and histochemistry of the salt glands of the greater flamingo *Phoenicopterus ruber roseus* (Aves, Phoenicopteridae). Saudi J Biol Sci 14:145–152

Arruda JA, Marzolf GR, Faulk RT (1983) The role of suspended sediments in the nutrition of zooplankton in turbid reservoirs. Ecology 64:1225–1235

Ashley GM, Goman M, Hover VC, Owen RB, Renaut RW, Muasya AM (2002) Artesian blister wetlands, a perennial water resource in the semi-arid rift valley of East Africa. Wetlands 22:686–695

Avery S (2012) Lake Turkana and the lower Omo: hydrological impacts of major dam and irrigation developments. African Studies Centre, University of Oxford, Oxford

Ballot A, Krienitz L, Kotut K, Wiegand C, Metcalf JS, Codd CA, Pflugmacher S (2004) Cyanobacteria and cyanobacterial toxins in three alkaline rift valley lakes of Kenya – Lakes Bogoria, Nakuru and Elmenteita. J Plankton Res 26:925–935

Banciu H, Sorokin D (2013) Adaptation in haloalkaliphiles and natronophilic bacteria. In: Seckbach J, Oren A, Stan-Lotter H (eds) Polyextremophiles, vol 27, Series cellular origin, life in extreme habitats and astrobiology. Springer, Dordrecht, pp 121–178

Beadle LC (1981) The inland waters of tropical Africa, an introduction to tropical limnology, 2nd edn. Longman, London, 475 p

Belkin S, Boussiba S (1991) Resistance of *Spirulina platensis* to ammonia at high pH values. Plant Cell Physiol 32:953–958

Beninca E, Johnk KD, Heerkloss R, Huisman J (2009) Coupled predator-prey oscillations in a chaotic food web. Ecol Lett 12:1367–1378

Bergman AN, Laurent P, Otiang'a-Owiti G, Bergman HL, Walsh PJ, Wilson P, Wood CM (2003) Physiological adaptations of the gut in the lake Magadi tilapia, *Alcolapia grahami*, an alkaline- and saline-adapted teleost fish. Comp Biochem Physiol A Mol Integr Physiol 136:701–715

Bocci F, Ferrari F, Materassi R, Mannelli D (1980) Prime ricerche sulla crescita di *Spirulina platensis* e *Spirulina maxima* in coltura fotolimitata. In: Abstracts of the XIX Congr Naz Soc Ital Microbiol (SIM), Catania, Italy, 5–8 July

Bradley TJ (2009) Animal osmoregulation, Oxford animal biology series. Oxford University Press, New York, p 320

Brown LH (1973) The mystery of the flamingo. East African Publishing House, Nairobi, p 130

Buchmann K, Becker B (2009) The system of contractile vacuoles in the green alga *Mesostigma viride* (Streptophyta). Protist 160:427–443

Burgis MJ, Morris P (1987) The natural history of lakes. Cambridge University Press, Cambridge, p 218

Burian A, Schagerl M, Yasindi A (2013) Microzooplankton feeding behaviour: grazing on the microbial and the classical food web of African soda lakes. Hydrobiologia 710:61–72

Burian A, Kainz MJ, Schagerl M, Yasindi A (2014) Species-specific separation of lake plankton reveals divergent food assimilation patterns in rotifers. Freshw Biol 59:1257–1265

Burian A, Schagerl M, Yasindi AW, Singer G, Kaggwa M, Winder M (2016) Benthic-pelagic coupling drives non-seasonal zooplankton blooms and restructures energy flows in shallow tropical lakes. Limnol Oceanogr 61:795-805

Casey JR, Grinstein S, Orlowski J (2010) Sensors and regulators of intracellular pH. Nat Rev Mol Cell Biol 11:50–61

Chalié F, Gasse F (2002) Late Glacial-Holocene diatom record of water chemistry and lake level change from the tropical East African Rift Lake Abiyata (Ethiopia). Palaeogeogr Palaeoclimatol Palaeoecol 187:259–283

Childress B, Hughes B, Harper D, van den Bossche W (2007) East African flyway and key site network of the lesser flamingo (*Phoenicopterus minor*) documented through satellite tracking. Ostrich 78:463–468

Cooper R, Wissel B (2012) Interactive effects of chemical and biological controls on food-web composition in saline prairie lakes. Aquat Biosyst 8:29

Costello MJ, Coll M, Danovaro R, Halpin P, Ojaveer H, Miloslavich P (2010) A census of marine biodiversity knowledge, resources, and future challenges. PLoS One 5:e12110

Dadheech PK, Glockner G, Casper P, Kotut K, Mazzoni CJ, Mbedi S, Krienitz L (2013) Cyanobacterial diversity in the hot spring, pelagic and benthic habitats of a tropical soda lake. FEMS Microbiol Ecol 85:389–401

Deocampo DM, Jones BF (2014) Geochemistry of saline lakes. In: Turekian HD, Holland KK (eds) Treatise on geochemistry, 2nd edn. Elsevier, Oxford, pp 437–469

Dokulil M (1984) Assessment of components controlling phytoplankton photosynthesis and bacterioplankton production in a shallow, alkaline, turbid lake (Neusiedlersee, Austria). Int Rev Ges Hydrobiol Hydrogr 69:679–727

Duckworth AW, Grant WD, Jones BE, van Steenbergen R (1996) Phylogenetic diversity of soda lake alkaliphiles. FEMS Microb Ecol 19:181–191

Dudgeon D, Arthington AH, Gessner MO, Kawabata Z, Knowler DJ, Leveque C, Naiman RJ, Prieur-Richard AH, Soto D, Stiassny ML, Sullivan CA (2006) Freshwater biodiversity: importance, threats, status and conservation challenges. Biol Rev Camb Philos Soc 81:163–182

Epp RW, Winston PW (1977) Osmotic regulation in the brackish-water rotifer *Brachionus plicatilis* (Müller). J Exp Biol 68:151–156

Erdmann N, Hagemann M (2004) Salt acclimation of algae and cyanobacteria: a comparison. In: Rai LC, Gaur JP (eds) Algal adaptation to environmental stresses. Springer, Berlin, pp 323–361

Erftemeijer PLA, Osinga R, Mars AE (1993) Primary production of seagrass beds in South Sulawesi (Indonesia) – a comparison of habitats, methods and species. Aquat Bot 46:67–90

Ferguson AJD (1982) Studies on algal dynamics and primary productivity in lake Turkana. In: Hopson AJ (ed) A report on the findings of the lake Turkana project, vol 1. UK Overseas Development Administration, London, pp 109–162

Ferguson AJ, Habott BJ (1982) Geographical, physical and chemical aspects of lake Turkana. In: Hopson AJ (ed) A report on the findings of the lake Turkana project, vol 1. UK Overseas Development Administration, London, pp 1–110

Finlay BJ, Curds CR, Bamforth SS, Bafort JM (1987) Ciliated protozoa and other microorganisms from 2 African soda lakes (lake Nakuru and Lake Simbi, Kenya). Arch Protistenkunde 133:81–91

Ganf GG (1974) Incident solar irradiance and underwater light penetration as factors controlling the chlorophyll a content of a shallow equatorial lake (Lake George, Uganda). J Ecol 62:593–609

Gimmler H, Degenhardt B (2004) Alkaliphilic and alkali tolerant algae. In: Rai LC, Gaur JP (eds) Algal adaptation to environmental stresses. Springer, Berlin, pp 291–322

Grant WD, Mwatha WE, Jones BE (1990) Alkaliphiles: ecology, diversity and applications. FEMS Microbiol Lett 75:255–269

Grobbelaar J (1989) The contribution of phytoplankton productivity in turbid freshwaters to their trophic status. Hydrobiologia 173:127–133

Gruber-Dorninger M (2015) East African soda lakes – biogeochemistry and interactions between organismic compartments. Dissertation, University of Vienna

Hammer UT (1986) Saline lake ecosystems of the world. Dr. W. Junk Publishers, Dortrecht, p 616

Hammer UT, Shamess J, Haynes RC (1983) The distribution and abundance of algae in saline lakes of Saskatchewan, Canada. Hydrobiologia 105:1–26

Harper DM, Childress RB, Harper MM, Boar RR, Hickley P, Mills SC, Otieno N, Drane T, Vareschi E, Nasirwa O, Mwatha WE, Darlington JPEC, Escuté-Gasulla X (2003) Aquatic biodiversity and saline lakes: lake Bogoria National Reserve, Kenya. Hydrobiologia 500:259–276

Hauer G, Rogerson A (2005) Heterotrophic protozoa from hypersaline environments. In: Gunde-Cimerman N, Oren A, Plemenitaš A (eds) Adaptation to life at high salt concentrations in archaea, bacteria, and eukarya. Springer, Dordrecht, pp 519–539

Hecky R (1984) African lakes and their trophic efficiencies: a temporal perspective. In: Meyers DG, Strickler JR (eds) Trophic interactions within aquatic ecosystems, AAAS selected symposium 85. Westview Press, Colorado, pp 405–448

Herbst DB, Bradley TJ (1989) A malpighian tubule lime gland in an insect inhabiting alkaline salt lakes. J Exp Biol 145:63–78

Hicks DB, Liu J, Fujisawa M, Krulwich TA (2010) F(1)F(0)-ATP synthases of alkaliphilic bacteria: lessons from their adaptations. Biochim Biophys Acta 1797:1362–1377

Hosseini Tafreshi A, Shariati M (2009) Dunaliella biotechnology: methods and applications. J Appl Microbiol 107:14–35

Iltis A (1974) Le phyoplancton des eaux natroniees du Kandem (Tchad): influence de la teneur en sels dissous sur le peuplement algal. Dissertation, University of Paris

Iltis A, Riou-Duwat S (1971) Variations saisonnières du peuplement en rotifères des eaux natronées du Kanem (Tchad). Cah O.R.S.T.O.M., ser Hydriobiol 2:101–112

Imhoff JF, Tindall BJ, Grant WD, Trüper HG (1981) *Ectothiorhodospira vacuolata* sp. nov., a new phototrophic bacterium from soda lakes. Arch Microbiol 130:238–242

Jirsa F, Gruber M, Stojanovic A, Omondi SO, Mader D, Körner W, Schagerl M (2013) Major and trace element geochemistry of lake Bogoria and Lake Nakuru, Kenya, during extreme draught. Chem Erde – Geochem 73:275–282

Johnson T, Malala J (2009) Lake Turkana and its link to the Nile. In: Dumont H (ed) The Nile, vol 89. Springer, Dordrecht, pp 287–304

Jones BF, Eugster HP, Rettig SL (1977) Hydrochemistry of the lake Magadi basin, Kenya. Geochim Cosmochim Acta 41:53–72

Junginger A, Trauth MH (2013) Hydrological constraints of paleo-Lake Suguta in the Northern Kenya rift during the African humid period (15–5 ka BP). Global Planet Change 111:174–188

Kaggwa MN, Gruber M, Oduor SO, Schagerl M (2013) A detailed time series assessment of the diet of lesser flamingos: further explanation for their itinerant behaviour. Hydrobiologia 710:83–93

Kebede E (1997) Response of *Spirulina platensis* (=*Arthrospira fusiformis*) from lake Chitu, Ethiopia, to salinity stress from sodium salts. J Appl Phycol 9:551–558

Kebede E (2002) Phytoplankton distribution in lakes of the Ethiopian rift valley. In: Tudorancea C, Taylor WD (eds) Ethiopian rift valley lakes. Backhuys, Leiden, pp 61–93

Kebede E, Ahlgren G (1996) Optimum growth conditions and light utilization efficiency of *Spirulina platensis* (=*Arthrospira fusiformis*) (Cyanophyta) from lake Chitu, Ethiopia. Hydrobiologia 332:99–109

Kebede E, Willén E (1997) Phytoplankton composition and biomass in a salinity-alkalinity series of lakes in the Ethiopian rift valley. Phycologia 36:48–49

Kebede E, Willén E (1998) Phytoplankton in a salinity-alkalinity series of lakes in the Ethiopian rift valley. Arch Hydrobiol/Algol Stud 89(Suppl 124):63–96

Kemp WM, Smith EM, MarvinDiPasquale M, Boynton WR (1997) Organic carbon balance and net ecosystem metabolism in Chesapeake Bay. Mar Ecol Prog Ser 150:229–248

Kihwele ES, Lugomela C, Howell KM (2014) Temporal changes in the lesser flamingos population (*Phoenicopterus minor*) in relation to phytoplankton abundance in lake Manyara, Tanzania. OJE 4:1–17

Kilham P (1981) Pelagic bacteria – extreme abundances in African Saline Lakes. Naturwissenschaften 68:380–381

Kirk KL, Gilbert JJ (1990) Suspended clay and the population-dynamics of planktonic rotifers and cladocerans. Ecology 71:1741–1755

Kolbe RW (1927) Zur Ökologie, Morphologie und Systematik der Brackwasser-Diatomeen: Die Kieselalgen des Sperenberger Salzgebiets. Pflanzenforschung 7:1–146

Kotut K, Ballot A, Krienitz L (2006) Toxic cyanobacteria and their toxins in standing waters of Kenya: implications for water resource use. J Water Health 4:233–245

Krienitz L, Kotut K (2010) Fluctuating algal food populations and the occurrence of lesser flamingos (*Phoeniconaias minor*) in three Kenyan rift valley lakes. J Phycol 46:1088–1096

Krienitz L, Bock C, Kotut K, Luo W (2012) *Picocystis salinarum* (Chlorophyta) in saline lakes and hot springs of East Africa. Phycologia 51:22–32

Krienitz L, Dadheech PK, Kotut K (2013a) Mass developments of the cyanobacteria *Anabaenopsis* and *Cyanospira* (Nostocales) in the soda lakes of Kenya: ecological and systematic implications. Hydrobiologia 703:79–93

Krienitz L, Dapheech PK, Kotut K (2013b) Mass developments of a small sized ecotype of *Arthrospira fusiformis* in lake Oloidien, Kenya, a new feeding ground for lesser flamingos in East Africa. Fottea 13:215–225

Krulwich TA, Sachs G, Padan E (2011) Molecular aspects of bacterial pH sensing and homeostasis. Nat Rev Microbiol 9:330–343

LaBarbera MC, Kilham P (1974) The chemical ecology of copepod distribution in the lakes of East and Central Africa. Limnol Oceanogr 19:459–465

Lampert W, Sommer U (1999) Limnoökologie. Thieme, Stuttgart, NY

Lanzén A, Simachew A, Gessesse A, Chmolowska D, Jonassen I, Øvreås L (2013) Surprising prokaryotic and eukaryotic diversity, community structure and biogeography of Ethiopian soda lakes. PLoS One 8: e72577

Lind OT, Doyle R, Vodopich DS Trotter BG, Limon JG, Davaloslind L (1992) Clay turbidity – regulation of phytoplankton production in a large, nutrient-rich tropical lake. Limnol Oceanogr 37:549–565

Lindeman RL (1942) The trophic-dynamic aspect of ecology. Ecology 23:399–418

Lowe CD, Kemp SJ, Bates AD, Montagnes DJS (2005) Evidence that the rotifer *Brachionus plicatilis* is not an osmoconformer. Mar Biol 146:923–929

Luo W, Kotut K, Krienitz L (2013) Hidden diversity of eukaryotic plankton in the soda lake Nakuru, Kenya, during a phase of low salinity revealed by a SSU rRNA gene clone library. Hydrobiologia 702:95–103

Madigan MT, Martinko JM, Stahl DA, Clark DP (2011) Brock biology of microorganisms, 13th edn. Pearson Education, San Francisco, p 1152

Maslin MA, Brierley CM, Milner AM, Shultz S, Trauth MH, Wilson KE (2014) East African climate pulses and early human evolution. Q Sci Rev 101:1–17

Matagi SV (2004) A biodiversity assessment of the flamingo lakes of eastern Africa. Biodiversity 5 (1):13–26

Melack JM (1979) Photosynthesis and growth of *Spirulina platensis* (Cyanophyta) in an equatorial lake (Lake Simbi, Kenya). Limnol Oceanogr 24:753–760

Melack JM (1981) Photosynthetic activity of phytoplankton in tropical African soda lakes. Hydrobiologia 81:71–85

Melack JM (1988) Primary producer dynamics associated with evaporative concentration in a shallow, equatorial soda lake (Lake Elmenteita, Kenya). Hydrobiologia 158:1–14

Melack JM, Kilham P (1974) Photosynthetic rates of phytoplankton in East African alkaline, saline lakes. Limnol Oceanogr 19:743–755

Mesbah N, Wiegel J (2011) Halophiles exposed concomitantly to multiple stressors: adaptive mechanisms of halophilic alkalithermophiles. In: Ventosa A, Oren A, Ma Y (eds) Halophiles and hypersaline environments. Springer, Berlin, pp 249–273

Mlingwa C, Baker N (2006) Lesser flamingo *Phoenicopterus minor* counts in Tanzanian soda lakes: implications for conservation. In: Boere GC, Galbraith CA, Stroud DA (eds) Waterbirds around the world. The Stationery Office, Edinburgh, pp 230–233

Ndebele-Murisa MR, Musil CF, Raitt L (2010) A review of phytoplankton dynamics in tropical African lakes. S Afr J Sci 106:13–18

Odada EO, Raini J, Nedetei R (2006) Lake Nakuru. Experience and lessens leared briefs. http://www.worldlakes.org/uploads/18_Lake_Nakuru_27February2006.pdf. Accessed 30 Sept 2015

Oduor SO, Schagerl M (2007a) Phytoplankton primary productivity characteristics in response to photosynthetically active radiation in three Kenyan rift valley saline-alkaline lakes. J Plankton Res 29:1041–1050

Oduor SO, Schagerl M (2007b) Temporal trends of ion contents and nutrients in three Kenyan rift valley saline-alkaline lakes and their influence on phytoplankton biomass. Hydrobiologia 584:59–68

Ogato T, Kifle D, Fetahi T, Sitotaw B (2014) Evaluation of growth and biomass production of *Arthrospira (Spirulina) fusiformis* in laboratory cultures using waters from the Ethiopian soda lakes Chitu and Shala. J Appl Phycol 26:2273–2282

Olaka LA, Odada EO, Trauth MH, Olago DO (2010) The sensitivity of East African rift lakes to climate fluctuations. J Paleolimnol 44:629–644

Ong'Ondo GO, Yasindi AW, Oduor SO, Jost S, Schagerl M, Sonntag B, Boenigk J (2013) Ecology and community structure of ciliated protists in two alkaline-saline rift valley lakes in Kenya with special emphasis on Frontonia. J Plankton Res 35:759–771

Oren A (2006) Life at high salt concentrations. In: Dworkin M, Falkow S, Rosenberg E, Schleifer K-H, Stackebrandt E (eds) The prokaryotes. Springer, New York, pp 263–282

Oren A (2008) Microbial life at high salt concentrations: phylogenetic and metabolic diversity. Saline Syst 4:2

Oren A (2011) Diversity of halophiles. In: Horikoshi K (ed) Extremophiles handbook. Springer, Tokyo, pp 309–325

Oren A (2013) Life at high salt concentrations, intracellular KCl concentrations, and acidic proteomes. Front Microbiol 4:315

Owen RB, Renaut RW, Hover VC, Ashley GM, Muasya AM (2004) Swamps, springs and diatoms: wetlands of the semi-arid Bogoria-Baringo rift, Kenya. Hydrobiologia 518:59–78

Owino AO, Oyugi JO, Nasirwa OO, Bennun LA (2001) Patterns of variation in waterbird numbers on four rift valley lakes in Kenya, 1991–1999. Hydrobiologia 458:45–53

Padan E, Bibi E, Ito M, Krulwich TA (2005) Alkaline pH homeostasis in bacteria: new insights. Biochim Biophys Acta 1717:67–88

Patterson DJ (1980) Contractile vacuoles and associated structures: their organization and function. Biol Rev 55:1–46

Peduzzi M, Gruber M, Gruber M, Schagerl M (2014) The virus's tooth: cyanophages affect an African flamingo population in a bottom-up cascade. ISME J 8:1346–1351

Pogoryelov D, Sudhir PR, Kovacs L, Gombos Z, Brown I, Garab G (2003) Sodium dependency of the photosynthetic electron transport in the alkaliphilic cyanobacterium *Arthrospira platensis*. J Bioenerg Biomembr 35:427–437

Pörtner HO, Schulte PM, Wood CM, Schiemer F (2010) Niche dimensions in fishes: an integrative view. Physiol Biochem Zool 83:808–826

Reed RH, Richardson DL, Warr SRC, Stewart WDP (1984) Carbohydrate accumulation and osmotic stress in cyanobacteria. Microbiology 130:1–4

Reese BK, Koester JA, Kirkpatrick J, Konotchick T, Allen LZ, Dziallas C (2014) How extreme is extreme? In: Kemp PF (ed) Eco-DAS X symposium proceedings, ASLO, pp 69–87

Reite OB, Maloiy GMO, Aasehaug B (1974) pH, salinity and temperature tolerance of lake Magadi Tilapia. Nature 247:315

Remane A, Schlieper C (1971) Biology of brackish water. Schweizerbart'sche Verlagsbuchhandlung (Nägele u. Obermiller), Stuttgart

Renaut RW, Owen RB, Jones B, Tiercelin J-J, Tarits C, Ego JK, Konhauser KO (2013) Impact of lake-level changes on the formation of thermogene travertine in continental rifts: evidence from lake Bogoria, Kenya rift valley. Sedimentology 60:428–468

Roubeix V, Chalié F, Gasse F (2014) The diatom *Thalassiosira faurii* (Gasse) Hasle in the Ziway-Shala lakes (Ethiopia) and implications for palaeoclimatic reconstructions: case study oft he Glacial-Holocene transition in East Africa. Palaeogeogr Palaeoclimatol Palaeoecol 402:104–112

Schagerl M, Oduor SO (2008) Phytoplankton community relationship to environmental variables in three Kenyan rift valley saline-alkaline lakes. Mar Freshw Res 59:125–136

Schagerl M, Burian A, Gruber-Dorninger M, Oduor SO, Kaggwa MN (2015) Algal communities of Kenyan soda lakes with a special focus on *Arthrospira fusiformis*. Fottea 15:245–257

Schlesinger P, Belkin S, Boussiba S (1996) Sodium deprivation under alkaline conditions causes rapid death of the filamentous cyanobacterium *Spirulina platensis*. J Phycol 32:608–613

Severin J, Wohlfarth A, Galinski EA (1992) The predominant role of recently discovered tetrahydropyrimidines for the osmoadaptation of halophilic eubacteria. Microbiology 138:1629–1638

Solomon CT, Bruesewitz DA, Richardson DC, Rose KC, Van de Bogert MC, Hanson PC, Kratz TK, Larget B, Adrian R, Babin BL, Chiu CY, Hamilton DP, Gaiser EE, Hendricks S, Istvanovics V, Laas A, O'Donnell DM, Pace ML, Ryder E, Staehr PA, Torgersen T, Vanni MJ, Weathers KC, Zhu GW (2013) Ecosystem respiration: drivers of daily variability and background respiration in lakes around the globe. Limnol Oceanogr 58:849–866

Sorokin DY, Kuenen JG (2005) Chemolithotrophic haloalkaliphiles from soda lakes. FEMS Microbiol Ecol 52:287–295

Sorokin DY, Berben T, Melton ED, Overmars L, Vavourakis CD, Muyzer G (2014) Microbial diversity and biogeochemical cycling in soda lakes. Extremophiles 18:791–809

Street FA (1980) The relative importance of climate and local hydrogeological factors in influencing lake level fluctuations. Palaeoecol Afr 12:137–158

Taipale SJ, Brett MT, Hahn MW, Martin-Creuzburg D, Yeung S, Hiltunen M, Strandberg U, Kankaala P (2014) Differing *Daphnia magna* assimilation efficiencies for terrestrial, bacterial, and algal carbon and fatty acids. Ecology 95:563–576

Talling JF (1971) The underwater light climate as a controlling factor in the production ecology of freshwater phytoplankton. Mitt Verh Int Verein theor angew Limnol 19:214–243

Talling JF (1986) The seasonality of phytoplankton in African lakes. Hydrobiologia 138:139–160

Talling JF (2001) Environmental controls on the functioning of shallow tropical lakes. Hydrobiologia 458:1–8

Talling JF, Wood RB (1973) The upper limit of photosynthetic productivity by phytoplankton: evidence from Ethiopian soda lakes. Freshw Biol 3:53–76

Tebbs EJ, Remedios JJ, Avery ST, Rowland CS, Harper DM (2015) Regional assessment of lake ecological states using landsat: a classification scheme for alkaline-saline, flamingo lakes in the East African rift valley. Int J Appl Earth Obs Geoinf 40:100–108

Telford RJ, Lamb HF (1999) Groundwater-mediated response to Holocene climatic change recorded by the diatom stratigraphy of an Ethiopian crater lake. Quatern Res 52:63–75

Tewodros K, Afework B (2014) Phytoplankton composition and physico-chemical parameters study in water bodies of Abijata-Shalla Lakes National Park (ASLNP), Ethiopia. Greener J Biol Sci 4:69–76

Tillmann U, Hesse KJ, Colijn F (2000) Planktonic primary production in the German Wadden Sea. J Plankton Res 22(7):1253–1276

Trauth MH, Maslin MA, Deino AL, Junginger A, Lesoloyia M, Odada EO, Olago DO, Olaka LA, Strecker MR, Tiedemann R (2010) Human evolution in a variable environment: the amplifier lakes of Eastern Africa. Q Sci Rev 29:2981–2988

Tuite CH (1981) Standing crop densities and distribution of *Spirulina* and benthic diatoms in East-African alkaline saline lakes. Freshw Biol 11:345–360

Tuite CH (2000) The distribution and density of lesser flamingos in East Africa in relation to food availability and productivity. Waterbirds 23(Spec Publ 1):52–63

Turner DP, Ritts WD, Cohen WB, Gower ST, Running SW, Zhao MS, Costa MH, Kirschbaum AA, Ham JM, Saleska SR, Ahl DE (2006) Evaluation of MODIS NPP and GPP products across multiple biomes. Remote Sens Environ 102:282–292

Twombly S (1983) Seasonal and short-term fluctuations in zooplankton abundance in tropical lake Malawi. Limnol Oceanogr 28:1214–1224

Uehlinger U (2006) Annual cycle and inter-annual variability of gross primary production and ecosystem respiration in a floodprone river during a 15-year period. Freshw Biol 51:938–950

Vareschi E (1978) Ecology of lake Nakuru (Kenya). 1. Abundance and feeding of lesser flamingo. Oecologia 32:11–35

Vareschi E (1979) Ecology of lake Nakuru (Kenya). 2. Biomass and spatial-distribution of fish (*Tilapia-Grahami* Boulenger = *Sarotherodon-Alcalicum-*Grahami Boulenger). Oecologia 37:321–335

Vareschi E (1982) The ecology of lake Nakuru (Kenya). 3. Abiotic factors and primary production. Oecologia 55:81–101

Vareschi E (1987) Saline lake ecosystems. In: Schulze E-D, Zwölfer H (eds) Potentials and limitations of ecosystem analysis, vol 61. Springer, Berlin, pp 347–364

Vareschi E, Jacobs J (1984) The ecology of lake Nakuru (Kenya). 5. Production and consumption of consumer organisms. Oecologia 61:83–98

Vareschi E, Jacobs J (1985) The ecology of lake Nakuru. 6. Synopsis of production and energy-flow. Oecologia 65:412–424

Vareschi E, Vareschi A (1984) The ecology of lake Nakuru (Kenya). 4. Biomass and distribution of consumer organisms. Oecologia 61:70–82

Verschuren D, Cocquyt C, Tibby J, Roberts CN, Leavitt PR (1999) Long-term dynamics of algal and invertebrate communities in a small, fluctuating tropical soda lake. Limnol Oceanogr 44:1216–1231

Verschuren D, Tibby J, Sabbe K, Roberts N (2000) Effects of depth, salinity, and substrate on the invertebrate community of a fluctuating tropical lake. Ecology 81:164–182

Vonshak A, Tomaselli L (2000) *Arthrospira* (*Spirulina*): systematics and eco-physiology. In: Whitton BA, Potts M (eds) The ecology of cyanobacteria. Kluwer, Dordrecht, pp 505–522

Vonshak A, Torzillo G, Accolla P, Tomaselli L (1996) Light and oxygen stress in *Spirulina platensis* (cyanobacteria) grown outdoors in tubular reactors. Phys Plant 97:175–179

Walker KF (1981) A synopsis of ecological information on the saline lake rotifer *Brachionus plicatilis* Müller 1786. In: Williams WD (ed) Salt lakes. Proceedings of the international symposium on athalassic (Inland) salt lakes, held at Adelaide, Oct 1979, Springer, pp 159–167

Warr SRC, Reed RH, Chudek JA, Foster R, Stewart WDP (1985) Osmotic adjustment in *Spirulina platensis*. Planta 163:424–429

Wetzel RG (2001) Limnology. Lakes and rivers ecosystems. Academic, San Diego, p 1006

Wilhelm SW, Matteson AR (2008) Freshwater and marine virioplankton: a brief overview of commonalities and differences. Freshw Biol 53:1076–1089

Williams WD (1978) Limnology of Victorian salt lakes, Australia. Verh Int Ver Limnol 20:1165–1174

Williams WD (1998) Salinity as a determinant of the structure of biological communities in salt lakes. Hydrobiologia 381:191–201

Williams M (2014) Climate change in deserts – past, present and future. Cambridge University Press, New York

Willmer P, Stone G, Johnston I (2004) Environmental physiology of animals, 2nd edn. Blackwell, Malden, p 768

Wondie A, Mengistu S, Vijverberg J, Dejen E (2007) Seasonal variation in primary production of a large high altitude tropical lake (Lake Tana, Ethiopia): effects of nutrient availability and water transparency. Aquat Ecol 41:195–207

Wood RB, Talling JF (1988) Chemical and algal relationships in a salinity series of Ethiopian inland waters. Hydrobiologia 158:29–67

Wood C, Bergman H, Laurent P, Maina J, Narahara A, Walsh P (1994) Urea production, acid-base regulation and their interactions in the lake Magadi tilapia, a unique teleost adapted to a highly alkaline environment. J Exp Biol 189:13–36

Wood CM, Wilson P, Bergman HL, Bergman AN, Laurent P, Otiang'a-Owiti G, Walsh PJ (2002) Obligatory urea production and the cost of living in the Magadi Tilapia revealed by acclimation to reduced salinity and alkalinity. Physiol Biochem Zool 75:111–122

Wood CM, Bergman HL, Bianchini A, Laurent P, Maina J, Johannsson OE, Bianchini LF, Chevalier C, Kavembe GD, Papah MB, Ojoo RO (2012) Transepithelial potential in the Magadi tilapia, a fish living in extreme alkalinity. J Comp Physiol B Biochem Syst Environ Physiol 182:247–258

Yasindi AW (2001) The ecology of planktonic ciliates in tropical lakes of East Africa. Dissertation, University of Waterloo, Waterloo

Yasindi AW, Taylor WD (2006) The trophic position of planktonic ciliate populations in the food webs of some East African lakes. Afr J Aquat Sci 31:53–62

Yasindi AW, Lynn DH, Taylor WD (2002) Ciliated protozoa in Lake Nakuru, a shallow alkaline-saline lake in Kenya: seasonal variation, potential production and role in the food web. Arch Hydrobiol 154:311–325

Zhilina TN, Garnova ES, Tourova TP, Kostrikina NA, Zavarzin GA (2001) *Halonatronum saccharophilum* gen. nov. sp. nov.: a new haloalkaliphilic bacterium of the order *Haloanaerobiales* from lake Magadi. Microbiology 70:64–72

Ziegler S, Benner R (1998) Ecosystem metabolism in a subtropical, seagrass-dominated lagoon. Mar Ecol Prog Ser 173:1–12

Zinabu GM, Taylor WD (1997) Bacteria-chlorophyll relationships in Ethiopian lakes of varying salinity: are soda lakes different? J Plankton Res 19:647–654

Part III

Utilization, Threats and Conservation

The soda lakes of East Africa are not only fascinating landscape elements, they are also in the focus of industry. Soda-ash mining was established about 100 years ago at Lake Magadi (Kenya) and 25 years ago at Lake Abijata (Ethiopia). A soda-ash factory at Lake Natron (Tanzania) is planned, but for the present not realized because of vigorous protests by conservationists and scientists. This lake is the only permanent breeding site of the Lesser Flamingo, a flagship species of East African soda lakes. Some lakes and their organisms also face serious threats because of pollution and deterioration of their catchment area. Because they are closed systems, they are very prone to changes in water quality, supply and climate change. The following chapters highlight potential uses but also list problems of these sensitive ecosystems. Coordinated management plans, continuous monitoring and clear regulations for researchers and for exploitation are urgently needed to protect these unique and rich genetic resources, which are already being tapped for biotechnological applications.

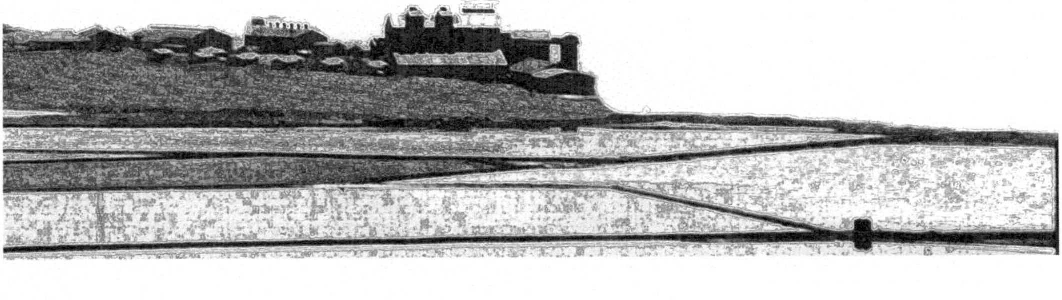

An Underexplored Resource for Biotechnology: Selected Microphytes of East African Soda Lakes and Adjacent Waters

13

Lothar Krienitz, Christina Bock, Pawan K. Dadheech, Kiplagat Kotut, Wei Luo, and Michael Schagerl

Abstract

Microphytes (Cyanobacteria and algae) are amongst the most promising sustainable resources for high-quality compounds, but high expenditures for culturing, harvesting and subsequent processing complicate their exploitation. An economic way to reduce costs of algal mass cultures is to install open-pond culture systems in tropical regions, which guarantee high solar irradiance throughout the year, but this approach is still waiting to be introduced in East Africa. One essential but often neglected requirement to carry out successful microphytean mass cultures is strain selection. We provide a selection of high-yielding microphytes with high potential for biotechnology use collected in the soda lakes and adjacent systems such as salt evaporation ponds or sewage oxidation ponds. Microphytes were detected as field clones or cultured strains and identified via microscopy and molecular methods; SSU rRNA and/or ITS sequences were entered in the NCBI database. Besides the

L. Krienitz (✉)
Department of Experimental Limnology, Leibniz-Institute of Freshwater Ecology and Inland Fisheries, 16775 Stechlin, Germany
e-mail: krie@igb-berlin.de

C. Bock
Department of Biodiversity, University of Duisburg-Essen, Universitätsstraße 5, 45141 Essen, Germany
e-mail: christina.bock@uni-due.de

P.K. Dadheech
Department of Microbiology, Central University of Rajasthan, NH 8, Bandar Sindri, 305817 Ajmer, Rajasthan, India
e-mail: pdadheech@curaj.ac.in

K. Kotut
Department of Biological Sciences, Embu University College, P.O. Box 6-60100, Embu, Kenya
e-mail: kkotut@gmail.com

W. Luo
SOA Key Laboratory for Polar Science, Polar Research Institute of China, Shanghai 200136, China
e-mail: luowei@pric.org.cn

M. Schagerl
Department of Limnology and Bio-Oceanography, University of Vienna, Althanstrasse 14, 1090 Vienna, Austria
e-mail: michael.schagerl@univie.ac.at

M. Schagerl (ed.), *Soda Lakes of East Africa*, DOI 10.1007/978-3-319-28622-8_13

well-known and already commercially exploited cyanobacterium *Arthrospira*, promising relatives of the famous "green balls" *Chlorella*, members of the green algal genera *Selenastrum*, *Picocystis* and *Mychonastes* and the eustigmatophytes *Nannochloropsis* and *Microchloropsis* are presented. All of them are suitable for ex situ conservation in strain collections, which is the basis for future mass cultivation and commercial use.

13.1 Introduction

Algal cultivation has received substantial attention in recent decades. Microalgae have come to the fore as a potential source of energy and high-quality compounds (Bosma et al. 2014; Chisti 2007; Draaisma et al. 2013; Georgianna and Mayfield 2012; Henrikson (2013); Mata et al. 2010; Stephens et al. 2010; Wijffels and Barbosa 2010) because they show several advantages. Contrary to land plants, microalgae do not need to invest their biomass into reinforcement structures: they can invest most of their energy into their offspring. Compared to land plants, this strategy results in a fivefold increased biomass yield (Posten 2012). Photosynthetic efficiencies of about 20 % have been attained in cultures of *Chlorella* spp., *Phaeodactylum tricornutum* Bohlin, *Porphyridium purpureum* (Bory) Drew et Ross and *Tetraselmis suecica* (Kylin) Butcher (Csögör et al. 2001; Packer 2009), which have been calculated as the theoretical maximum (Kirk 1994; Wilhelm and Jakobs 2012). Moreover, algae offer an enormous potential for high-quality products, some of which are already in use. They are sources of natural dyes, polysaccharides and vitamins (Cannel 1990; Lu et al. 2001; Solovchenko et al. 2008; Vílchez et al. 1997) along with other substances of pharmaceutical interest (Faulkner 2000). Also, polyunsaturated fatty acids (PUFA) from algae are in the focus of research groups and enterprises (Sukenik 1991; Yap and Chen 2001). Research into other applications such as hydrogen production (Doebbe et al. 2010) is ongoing. Finally, algae cultivation systems are not in competition with agricultural areas—they can also be installed on nonarable land.

For commercial purposes, microalgae have to be cultivated in either open-pond systems or closed photobioreactors (Fig. 13.1; Richmond and Hu 2013; Shen et al. 2009). Mass cultivation in open-pond systems is usually restricted to extremophiles [*Arthrospira* (syn. "Spirulina"), *Dunaliella*], which minimises the risks of contamination by other organisms (Wang et al. 2014a, b). Compared to photobioreactors, open-pond systems result in lower productivity (Posten 2009) because of poorer light and carbon supply. Another disadvantage is that they require more area than photobioreactors for the same biomass yield. On the other side, open ponds are less expensive and easier to maintain. In fact, they are currently the most economic algal mass culture systems (Borowitzka and Moheimani 2013). Open ponds are highly dependent on the local climate and are therefore not suitable for continuous operation in temperate regions, but ideally placed in the tropics. Those advantages for outdoor production, including a low-cost management of culture plants on nonarable land (Najafi et al. 2011), raise the economic potential of arid and salinity-affected regions (Chisti 2007) such as the East African Rift Valley. Besides synthesis of high-quality compounds, extremophiles are of high interest for use in biorefineries because they contain extremozymes (enzymes which still function at extreme conditions). Alkaline-saline environments are rich sources of such organisms and are already in the focus of microbiologists (Chakkiath et al. 2013; Horikoshi 2006). Possible applications are laundry detergents, detoxicants and decontaminants (Sorokin et al. 2014); two enzymes isolated from microbes originating from Kenyan soda lakes are commercially

Fig. 13.1 Algal mass cultivation. (**a**, **b**) greenhouses with tube photobioreactors for *Chlorella* cultivation, Klötze, Germany (courtesy Roquette Klötze GmbH & Co. KG, photos Jörg Ullmann); (**c**) *Arthropira* mass production, Kona, Hawaii (courtesy Cyanotech); (**d–f**) *Arthropira* microfarm "Manjul Spirulina, Sanwardhan Sansthan", Burthal village, Bassi Panchayat Samiti, 21 km from Jaipur, Rajasthan, India. This microfarm was constructed under the sponsorship of the Department of Biotechnology of the Government of India. The daily yield of biomass is 8–12 g m^{-2} (Srivastava and Gajraj 1996). (**d**) the fresh harvest of *Arthrospira* biomass is dried on tissues, which is later ground and used for therapeutic and food additive purpose for humans and animals; (**e**) open ponds of this microfarm; (**f**) fresh harvest. (**g**, **h**) *Arthropira* microfarms in other parts of the world (copyright Antenna Technologies France); (**g**) Togo, West Africa; (**h**) Cambodia

available since years (Grant and Heaphy 2010). So far, large-scale biotechnological exploitation of microalgae obtained from the native area has largely been underestimated and remains unexploited in East Africa.

One of the initial steps for establishing algal mass cultures is strain selection (Borowitzka 2013), which includes assessing the strain potential and studying the places of origin. The East African Rift Valley offers an ideal environment for algal biotechnology: high incoming irradiance, nonarable land over large areas and saline alkaline lakes occupied by extremophiles. We explored microphytes of East African soda lakes (EASL) with a high potential for biotechnological applications and also studied their phylogenetic relationships. In a next step, these taxa should be isolated, deposited and maintained for ex situ conservation in culture collections. Such algal culture collections already serve as a backup in many industrialised countries and comprise taxa with high potential for industrial exploitation of valuable compounds such as pigments, PUFAs, proteins and other bioactive components.

13.2 Soda Lakes: Habitats of Cyanobacteria and Algae Tailor Made for Mass Cultivation

The extreme living conditions in EASL with pH ranging above 10 and salinity levels of more than the sea make them promising systems for algal mass cultivation. Phytoplankton of EASL comprises only few taxa of Cyanobacteria and algae, some of them with an exceptionally high growth rate (Melack and Kilham 1974; Oduor and Schagerl 2007; Talling et al. 1973; Schagerl et al. 2015). These extremophiles are well adapted to increased salinity and alkalinity (see Chap. 12). Additionally, the lakes host many other species, some of them discovered only recently by applying modern molecular techniques, with an exceptionally high diversity of Eukaryotes (Luo et al. 2013), Archaea and Bacteria (Lanzén et al. 2013). Amongst the microphytes thriving under the harsh conditions are organisms already commercially exploited on other continents (Fig. 13.1). The cyanobacterium *Arthrospira* is sold under the name "Spirulina" and often predominates the community (Ballot et al. 2004; Vareschi 1978).

Nonetheless, the physical and chemical properties of EASL can fluctuate considerably and are reflected by shifts in the community composition with a complete failure of *Arthrospira* (see Chap. 12). Some lakes such as Lake Nakuru (Kenya) are called amplifier lakes because they are very sensitive to climate change and react strongly with water level changes (Trauth et al. 2010). These, in turn, are accompanied with shifts in water chemistry; occasionally such lakes might dry out, at times they offer freshwater conditions. Such different salinity phases trigger changes in the phytoplankton dominance structure (Harper et al. 2003; Schagerl and Oduor 2008; Schagerl et al. 2015; Verschuren et al. 1999). For example, under hypersaline conditions (salinity > 50 ‰), *Arthrospira* can be outcompeted by the picoplanktonic prasinophyte *Picocystis salinarum* Lewin (Krienitz and Kotut 2010). Under hyposaline conditions (salinity 3–20 ‰), coccoid green algae related to *Chlorella*, *Mychonastes* and *Selenastrum* can replace the blooms of *Arthrospira* (Luo et al. 2013).

Lake communities are strongly influenced by the catchment area, which serves as a rich species source and which is considered to be a genetic pool for adaptive changes in lake communities. Some lake shores harbour hot springs and geysers (e.g. Shala—Ethiopia, Bogoria and Magadi—Kenya). Rising lake levels submerge these biota, enabling direct exchange of organisms between hot springs and lake water (Dadheech et al. 2013). Overall, large fluctuations of hydrologic conditions in EASL and their close interaction with the environment have created dynamic systems that support a higher degree of microphytean biodiversity (Krienitz et al. 2016). Saline alkaline inland waters are hot spots for evolution, acting as links between freshwaters and marine systems. Species adaptation to the saline environment is the start of evolutionary processes forcing the "adjustment" of ecophysiological entities to genetic fixation of the new characteristics and survival strategies (Kirkwood et al. 2008).

13.3 Microphytes as a Potential Source of Biotechnology

13.3.1 *Arthrospira fusiformis*: The Blue-Green Harvest

Wild populations of the cyanobacterium *Arthrospira* were independently discovered as nutrition for human beings in South America (Mexico) and western Africa (Ciferri 1983). Already the Aztecs in ancient Mexico used *Arthrospira maxima* Setchell et Gardner from Lake Texcoco to produce flatbread (Sánchez et al. 2003). In small water bodies along Lake Chad, West African tribes collected quasi-monoculture suspensions of *Arthrospira fusiformis* (Voronikhin) Komárek and J. W. G. Lund, sun-drying them on the shores to hard "cakes" before consuming them (Léonard 1966). Nowadays, *Arthrospira* is cultivated in large raceway ponds (e.g. Cyanotech, Hawaii, Fig. 13.1c), but also in simple and small systems in third world countries (Fig. 13.1d–h; Hug and von der Weid 2012).

The coiled filaments of *Arthrospira fusiformis* exhibit major morphological variability triggered by the prevailing environmental conditions (Kaggwa et al. 2013). Phylogenetic studies covering a wide ecological range of EASL have shown that all these "ecomorphotypes" belong to the same species (Fig. 13.2; Dadheech et al. 2010; Krienitz et al. 2013). Compared to other algae and Cyanobacteria taxa, *Arthrospira* is one of the most promising species for biotechnological use. The first large-scale production facility for *Arthrospira* was established at Lake Texcoco using evaporation ponds of a soda production facility (Ciferri 1983; Vonshak and Tomaselli 2000). Subsequently, open-pond mass cultivation of *Arthrospira* was established in many countries, mainly in America and Asia. The potential and mass culture of *Arthrospira* has been the subject of several publications (summarised in Sánchez et al. 2003; Shimamatsu 2004; Vonshak and Richmond 1988), the main focus being health food and food supplements (Belay et al. 1993; Fox 1996; Holman and Malau-Aduli 2013; Masojídek and Prášil 2010). In addition to its nutritional value, *Arthrospira fusiformis* contains a significant amount of proteins equivalent to 60–70 % per unit dry mass (Belay 2002). Some remarkable health benefits such as anti-inflammatory properties because of linolenic acid were reviewed by Howe et al. (2006). Chu et al. (2010) found that phycocyanin is an important antioxidant for destroying reactive oxygen species (ROS). Other positive effects of *Arthrospira* are linked to improvements in animal fertility, growth and aesthetic and nutritional product quality (Belay et al. 1996; Holman et al. 2014). *Arthrospira* products have antiviral and anticancer activities, reduce cholesterol, blood sugar and hypertension, and can promote malnutrition recovery and help protect the immune system.

For introducing *Arthrospira* mass production in East Africa, microfarms in India can be used as models because the conditions are comparable (Fig. 13.1d–f). This suggestion is also justified by the close phylogenetic relationship of the strains isolated from Africa and India (Dadheech et al. 2010) and the shared "homeland" of Lesser Flamingos, which are also encountered in western India apart from the East African core region (Childress et al. 2008). A team headed by Pushpa Srivastava from the Algal Biotechnology Laboratory of the University of Rajasthan has reported on the rural *Arthrospira* mass production in Rajasthan (Khan and Srivastava 2006; Srivastava and Gajraj 1996; Srivastava and Sharma 2002, 2004). In 2007, specialists from India, Kenya, Ethiopia, Mexico and Germany organised a joint workshop sponsored by UNESCO on the potential of *Arthrospira* at the Government College Ajmer in Rajasthan (UNESCO 2007). The outcome of this workshop was the endorsement for implementation of rural *Arthrospira* mass production plants in East Africa (see also Fig. 13.1d–f).

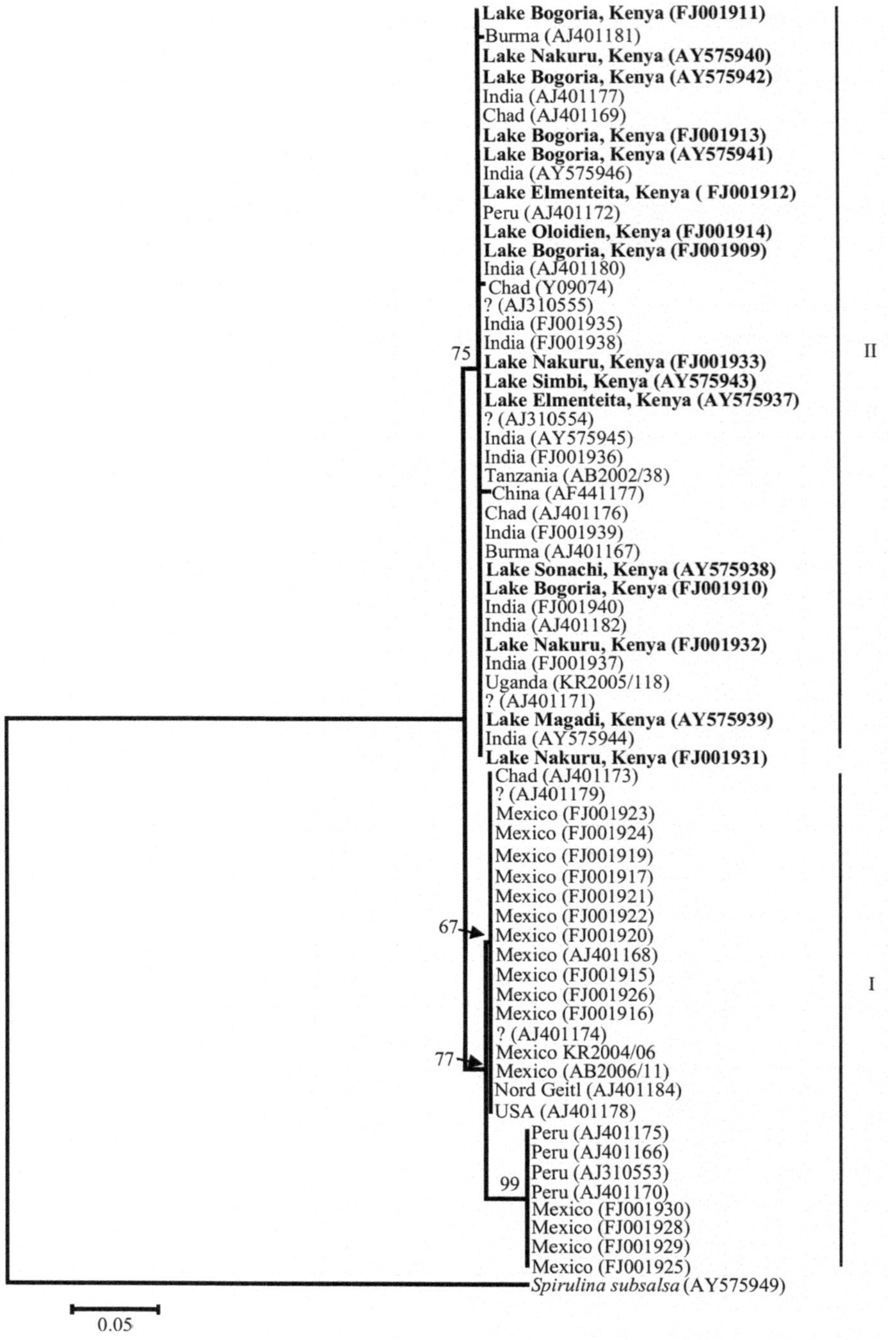

Fig. 13.2 Neighbour-joining tree of *Arthrospira* strains based on sequences of the *cpc*BA-IGS region of the phycocyanin operon (after Dadheech et al. 2010) as inferred with the MEGA v4.0 software. The nodes of the tree were tested by bootstrap analysis of 1000 replicates using NJ method. The bootstrap values >50 % are shown. An alignment of 347 nucleotides comprising positions of beta (139 nt) and alpha (97 nt) subunits including intergenic spacer (111 nt) of phycocyanin operon was used for phylogenetic study. The tree was constructed using Tamura-Nei model (Tamura et al. 2007) for nucleotide sequences available in the MEGA phylogenetic software. *Spirulina subsalsa* Oersted ex Gomont (AY575949) was used as out-group. Sixteen strains of *Arthrospira* from soda lakes of Kenya are in bold font. GenBank (NCBI) accession numbers are given in brackets. In clace I, different strains from America are included which belong to the species *Arthrospira maxima*. In clade II, different strains of *Arthrospira fusiformis* from Africa and Asia are unified

So far, *Arthrospira* has been neglected for biofuel production because of its low lipid content. However, given that *Arthrospira* belongs to a group of microorganisms that are easiest to cultivate and widely resistant against contamination, this use might become of interest in the future. Lipid synthesis depends on culture conditions, and certain modifications result in higher oil content (Baunillo et al. 2012). For example, phosphorus limitation increases carbohydrate content to a level suitable for biofuel generation (Markou et al. 2012). A comparison of microalgal lipid contents with other biofuel feedstocks such as macrophytes revealed considerably higher oil yield of microalgae (Mata et al. 2010). Nonetheless, modelling of the overall cultivation process showed that conventional microalgal cultures have more negative environmental impacts than crops (Clarens et al. 2010). An overview of countries where *Arthrospira* is produced, marketed and consumed is available from Spirulinasource (2015).

13.3.2 *Chlorella vulgaris* and Relatives: The Green Survivalists

The spherical cells of the green alga *Chlorella* have a diameter of 2–10 μm and are commonly referred to as the "green balls" (Krienitz et al. 2015). *Chlorella* is often used for plant physiology experiments, in screening tests and mass cultivation systems (Fig. 13.1a, b; Falkowski and Raven 1997). After the first description of the type species *Chlorella vulgaris* Beijerinck (Beijerinck 1890), more than one hundred globular-shaped "*Chlorella*" taxa have been described (Komárek and Fott 1983), but re-evaluations based on modern analytical tools using biochemical and molecular data resulted in a significantly lower number of true species between 4 (Huss et al. 1999) and 14 (Bock et al. 2011). Phylogenetic analyses revealed that the Chlorellaceae have evolved in two different sister clades, which are the *Chlorella* and the *Parachlorella* clade (Krienitz et al. 2004).

The family Chlorellaceae has more members than previously assigned. Besides the ancient spherical phenotype, several morphologically distinct taxa such as needle-shaped types or spined coenobia previously assigned to other families belong to this group (Krienitz et al. 2015; Luo et al. 2010). Consequently, the image of the *Chlorella* relationship has changed from simple green spheres towards more complicated and variable morphologies. Of great interest for biotechnological application are the taxa with mucilaginous envelopes (*Dictyosphaerium* and relatives) or extended microfibrils on the surface (*Planktochlorella*; Škaloud et al. 2014). Formation of these structural components depends on both environmental conditions and biotic interactions such as grazing pressure and resource competition (Luo et al. 2006; Reynolds 2007; Van Donk 2005). Mucilaginous envelopes also act as a substrate for biofilms that produce substances with nutritional or stimulatory effects (Cole 1982) and that serve as a nutrient depository (Decho 1990). Mucilage and microfibrils affect cell buoyancy (Boney 1981) and largely influence handling of cultures (Škaloud et al. 2014), e.g. through changes in viscosity.

How is the high diversity of Chlorellaceae reflected in EASL and adjacent water bodies? We conducted two sets of phylogenetic analyses, one derived from 18S rRNA gene sequences of field clones and strains (Fig. 13.3) and a more detailed set based on concatenated 18S rRNA genes (SSU) and ITS sequences of strains (Fig. 13.4). We analysed five field clones from Lake Nakuru (Fig. 13.3 upper part of the phylogenetic tree), which are closely related to *Chlorella* sp. and *Chlorella sorokiniana* Shihira et Krauss. Two more *Chlorella* species were found, one in a small dam in Nakuru National Park (*C. volutis* Bock et al.) and the other in the Nakuru sewage oxidation ponds (Fig. 13.5b) of the Nakuru town sewage treatment plant (*C. singularis* Bock et al.). The Nakuru sewage oxidation ponds are frequently colonised by dense populations of *Chlorella* relatives (Kotut et al. 2010) and are one of the main sources of coccoid green algae for Lake Nakuru. The oxidation ponds were found to host the highest biodiversity of Chlorellaceae in Kenyan waters

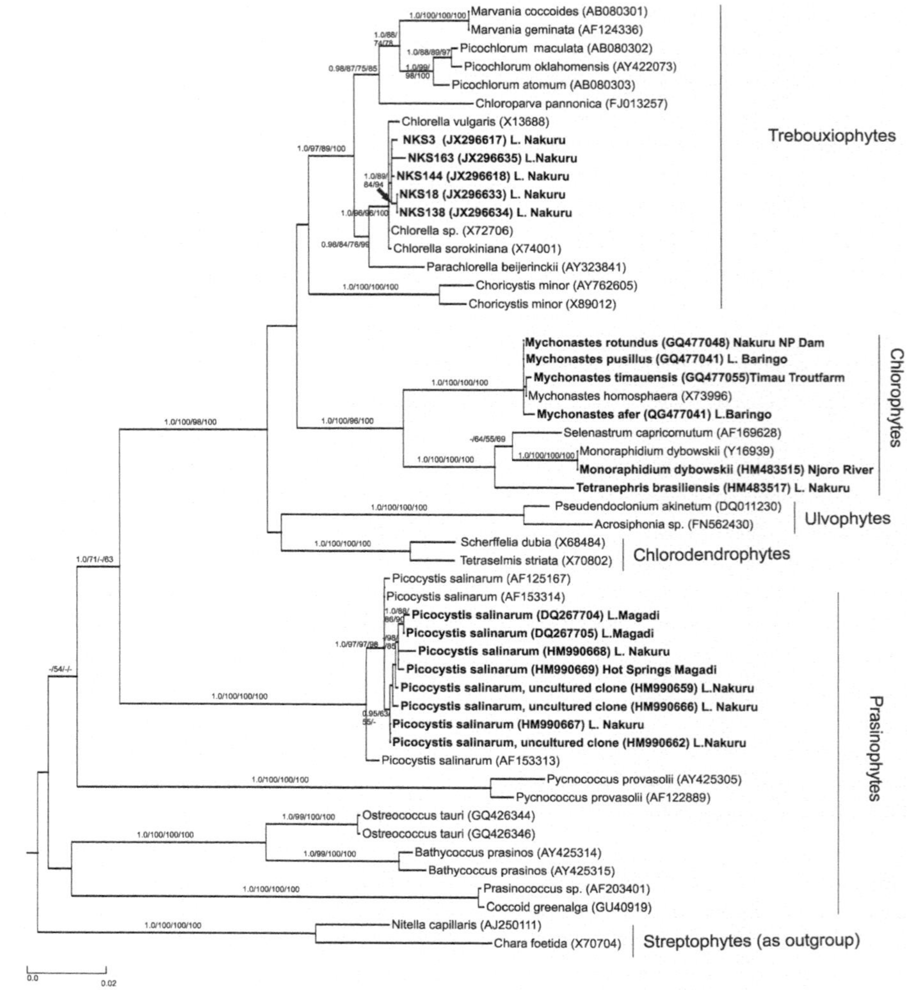

Fig. 13.3 Molecular phylogeny of selected green algae based on SSU rRNA gene sequence comparisons. The phylogenetic tree shown was inferred using the maximum likelihood (ML) method [with substitution model: TN (Optimum, Empirical): G (Optimum): 5], based on 1594 aligned positions of 51 taxa using Treefinder (Jobb 2008). Bayesian values ([0.95)] (MB) were calculated by MrBayes 3.1 using GTR settings (Posada and Buckley 2004; Ronquist and Huelsenbeck 2003). The stationary distribution was assumed after two million generations when the average standard deviations of split frequencies between two runs were lower than 0.01. To test the tree confidence, bootstrap values (>50 %) ML (1000 replicates, settings same as above), neighbour-joining (NJ) (1000 replicates) and maximum parsimony (MP) (1000 replicates) calculated using Paup 4.0 (Swofford 2002) were determined. Support values are shown at the branches in the order: MB, ML, MP, NJ. Scale bar indicates substitutions per site. The sequences were obtained from GenBank [National Center for Biotechnology Information (NCBI)]. For each taxon, the NCBI accession number is given in brackets

studied so far (Krienitz et al. 2012b). After the final oxidation pond, a spillway discharges into Lake Nakuru; its water acts as inoculum for the lake.

Four new Chlorellaceae genera were described in Kenya's inland waters: *Marasphaerium* (solitary, with mucilage), *Compactochlorella* (in large, compact colonies,

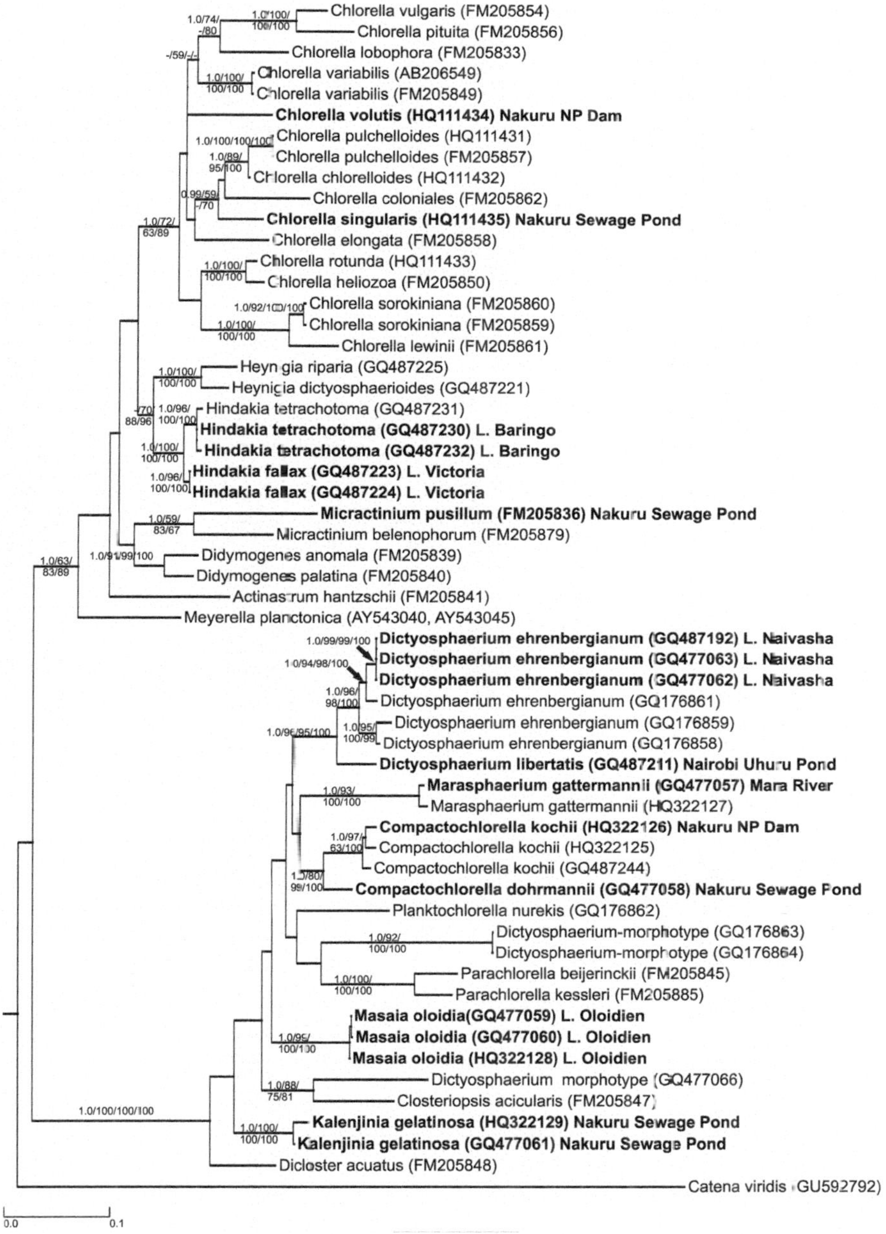

Fig. 13.4 Molecular phylogeny of the relationship of the green alga *Chlorella* based on a partitioned dataset of SSU, ITS1, 5.8S and ITS2 gene sequences (after Krienitz et al. 2012b; Krienitz and Bock 2012). The phylogenetic tree was inferred using the maximum likelihood (ML) method (with substitution model: 18S: J2:G:5; ITS1: J1:G:5, 5.8S: HKY; ITS2: GTR:G:5) based on 2693 aligned positions of 57 taxa using Treefinder (Jobb 2008). Bayesian values ([0.95] (MB) were calculated by MrBayes 3.1 using GTR settings (Posada and Buckley 2004; Ronquist and Huelsenbeck 2003). The stationary distribution was assumed after two million generations when the average standard deviations of split frequencies between two runs were lower than 0.01. To test the tree confidence, bootstrap values (>50 %) ML (1000 replicates; settings same as above above), NJ (1000 replicates) and MP (1000 replicates) calculated using Paup 4.0 (Swofford 2002) were determined. Support values are shown at the branches in the order: MB, ML, MP, NJ. Scale bar indicates substitutions per site. The sequences were obtained from GenBank [National Center for Biotechnology Information (NCBI)]. For each taxon, the NCBI accession number is given in brackets

Fig. 13.5 (**a**) Lake Magadi, salt evaporation ponds of the soda ash factory Tata Chemicals Magadi (TCM); (**b**) Sewage oxidation ponds of Nakuru town in the Lake Nakuru National Park

with wide mucilaginous envelope), *Masaia* (small colonies with mucilage) and *Kalenjinia* (small colonies enveloped with mucilage). All these genera belong to the *Parachlorella* clade (Krienitz et al. 2012b). In the Lake Nakuru area, we detected several species of the following chlorellacean genera: *Chlorella*, *Compacto-chlorella*, *Dictyosphaerium*, *Kalenjinia* and *Micractinium* (Fig. 13.4). *Masaia oloidea* Krienitz et al. was found in Lake Oloidien. The remaining Chlorellaceae were discovered in Kenyan freshwaters such as the lakes Victoria, Naivasha and Baringo (*Hindakia*, *Dictyosphaerium*) and in the Mara River (*Marasphaerium*) (Fig. 13.4).

The biotechnological potential of all these strains is yet to be investigated, but studies on relatives of *Chlorella* yielded promising results. *Chlorella* is the most widely cultivated eukaryotic alga for biotechnological applications (Masojídek and Prášil 2010; Safi et al. 2014). Relatives of *Chlorella* have been identified as ideal sources of fatty acids and the most effective organisms for biofuel production (Hempel et al. 2012; Zhou et al. 2013). *Chlorella vulgaris* (the former, invalid name was *Chlorella pyrenoidosa* Chick) established a high number of robust strains with top biomass yields (Kessler and Huss 1992) and is commercially already available (Roquette Klötze GmbH & Co; Fig. 13.1a, b). It is able to synthesise long-chain PUFAs such as docosahexaenoic acid and eicosapentaenoic acid (Elumalai et al. 2011).

Green algae taxonomy is currently undergoing substantial changes, which also affects applied phycology. Strains sold commercially are

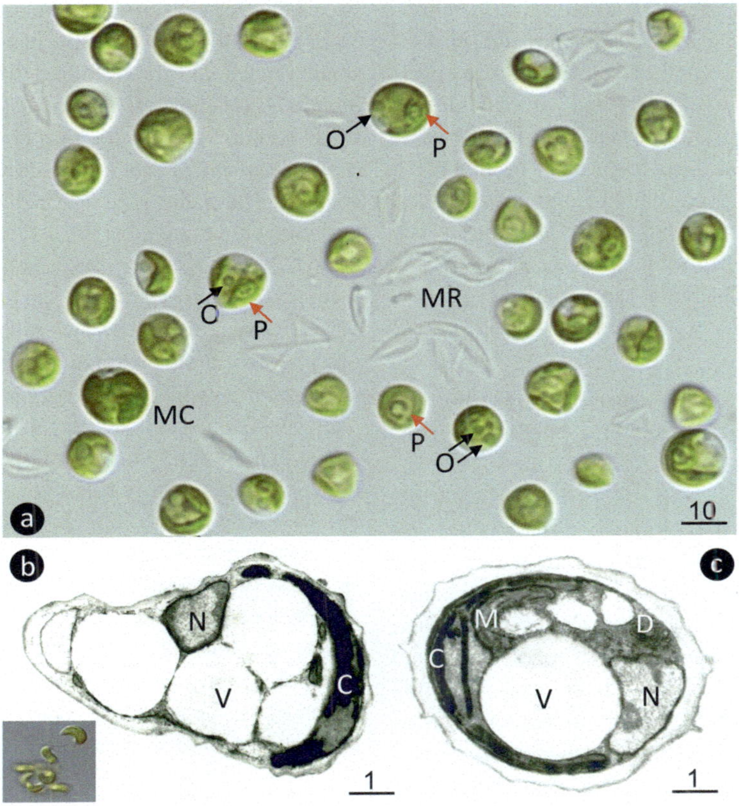

Fig. 13.6 Green algae from EASL with biotechnological potential. (**a**) *Parachlorella beijerinckii* Krienitz et al. culture under the light microscope; (**b**) *Tetranephris brasiliensis* Leite et Bicudo (the insert represents a light microscopical photo) under the transmission electron microscope showing large lipid vacuoles in the cell; (**c**) *Mychonastes afer* Krienitz et al. under the transmission electron microscope showing a large lipid vacuole in the centre. *C* chloroplast, *D* dictyosome, *M* mitochondrion, *MC* mother cell, *MR* empty mother cell wall remnant, *N* nucleus, *O* oil droplet, *P* pyrenoid, *V* vacuole. Scale bars in μm

assigned to *Chlorella*, but in the meantime it turned out that some of the cultures belong to other genera or even families. For example, the lipid and biofuel production potential of mucilaginous members of *Parachlorella* (Fig. 13.6a) has recently been reported (Dyo et al. 2013; Xiuling et al. 2013). Other members of *Parachlorella* reduce the risk of common diseases such as hypertension, arteriosclerosis and hyperlipidaemia (Noguchi et al. 2013). The mucilage has the ability to retain water and also has gelatinising, viscosifying and stabilising features. Moreover, the thick, mucilaginous envelope of *Parachlorella* absorbs and accumulates radioactive nuclides, which can be applied to decontaminate water and soil (Shimura et al. 2012). In Argentina, a *Parachlorella* strain was detected in acidic, geothermal environments, suggesting the importance of such mucous matrixes for survival under extreme conditions (Juárez et al. 2011).

Another newly described taxon of biotechnological importance is *Planktochlorella nurekis* Škaloud et al., originally isolated from a reservoir in Tajikistan and applied successfully in water purification and animal farming (Škaloud et al. 2014). The striking feature of this alga is the outer cell wall layer, which contains extended microfibrils facilitating cell buoyancy. We isolated a strain from this species in East Africa from the Kazinga Channel in Uganda.

13.3.3 *Selenastrum capricornutum* and Relatives: The High-Energy "Microbeans"

Selenastrum is a common chlorophyte with capricorn-shaped cells and placed in the family Selenastraceae. The genus occurs in the pelagic zone of nearly all types of inland waters and produces algal blooms in lakes, ponds, pools and rivers as solitary cells or in colonies (Krienitz et al. 2011b). The archetype of crescent-, croissant-, half-moon-, sickle-, bean- or capricorn-shaped Selenastraceae is represented by "*Selenastrum capricornutum* Printz"; this species is one of the classic organisms used in many experiments and bioassays, similar to *Arabidopsis* for vascular plants. Nevertheless, the taxonomic designation was shown to be incorrect and was renamed to *Raphidocelis subcapitata* (Korshikov) Nygaard et al. (Nygaard et al. 1986). The cells of Selenastraceae contain many lipid droplets and are suitable for biotechnological applications because of their high-value byproducts (Gutierrez-Wing et al. 2012; Song et al. 2013). The strong potential of these "high-energy microbeans" for biofuel production was recently confirmed by Hong and Xu (2013), Song et al. (2013), and Yee (2016) based on the remarkable lipid content of the cells. The algae can also be used to detoxify airport runoffs through pollutant bioaccumulation (Corsi et al. 2009).

We collected two strains of Selenastraceae from EASL, *Monoraphidium dybowskii* (Woloszynska) Hindák from the mouth of Njoro River into Lake Nakuru and *Tetranephris brasiliensis* Leite et Bicudo directly from Lake Nakuru (Fig. 13.6b). Vareschi (1978) frequently observed crescent-shaped coccoid green algae ("*Monoraphidium minutum* (Nägeli) Komárková-Legnerová") in Lake Nakuru, especially during shifts from an *Arthrospira fusiformis* community towards other species. The salinity of Lake Nakuru is highly variable with a range between 17 and 64 ‰ (Jirsa et al. 2013; Krienitz and Kotut 2010); very recently, salinities <1 ‰ (January 2015) and 5.5 ‰ (November 2015) were measured (pers. obs.). Crescent-shaped taxa are also present with a comparatively higher diversity in freshwater Rift Valley lakes such as Naivasha (salinity 0.1–0.3 ‰; Ballot et al. 2009), which confirms the broad ecological niche of this morphological group. Ferroni et al. (2007) conducted an experiment with a crescent-shaped strain ("*Kirchneriella* Schmidle sp". AW15) isolated from the freshwater Lake Awasa located in the Ethiopian part of the Rift Valley. In batch cultures, the alga was grown under different concentrations of NaCl between 0.4 and 19.4 ‰ and survived without any damage to its photosynthetic apparatus.

13.3.4 *Picocystis salinarum*: A High Performer in Extreme Saline Waters

Picocystis salinarum Lewin (Lewin et al. 2000) is a unique chlorophyte that forms a separate phylogenetic lineage within the prasinophytes (Guillou et al. 2004; Hepperle et al. 2001). The locus classicus of *Picocystis* is a saline pond (salinity ~ 100 ‰) at the San Francisco Salt Works, California, USA. The species was later found in Mono Lake (~85 ‰), USA (Hollibaugh et al. 2001) and in an alkaline lake in Inner Mongolia, P. R. China (Hollibaugh et al. 2001). Another strain of *Picocystis* (DGN-Z1) was isolated from a picoplankton bloom in Dagenoer Soda Lake (188 ‰) in Inner Mongolia (Fanjing et al. 2009). Recently, this alga was found to be very abundant in EASL (Krienitz et al. 2012a; Schagerl et al. 2015). It is probably more widespread than expected.

Picocystis is characterised by a very specific set of pigments with an unusual carotenoid pattern (Lewin et al. 2000; Roesler et al. 2002). The tiny cells (about 2 µm in diameter) are difficult to recognise and were probably often overlooked or incorrectly identified (Chap. 6). Because of its key position as a primary producer with a high production rate in food webs of saline habitats, special attention should be paid to its distribution. In lakes Bogoria and Nakuru, *Picocystis* outcompeted *Arthrospira* in 2006 and 2010, which also affects top consumers such as Lesser Flamingos because the cells are too small for ingestion (Krienitz and Kotut 2010).

Picocystis has a high potential for biotechnological use. In *Picocystis* cultures, lipid droplets have been observed during periods of nitrogen limitation, suggesting a promising source of biofuel (Wang et al. 2014a, b). Krienitz et al. (2012a) reported that this prasinophyte survives under the extreme saline conditions of Lake Magadi and the associated evaporation ponds of the Tata Chemicals Magadi soda factory (Fig. 13.5a). Hence, outdoor *Picocystis* mass production in some of the evaporation ponds can be established with little effort. The high salinity, temperatures and incoming irradiance facilitate dominance of *Picocystis* and lower the risk of contamination by competitors or predators.

13.3.5 *Mychonastes* div. spec.: The Ubiquitous, Lipid-Rich All-Rounders

Mychonastes was described from brackish waters as a monospecific genus with the type species *M. ruminatus* Simpson et Van Valkenburg (Simpson and Van Valkenburg 1978). Later phylogenetic studies revealed more than 15 species, which belong to the most common chlorophycean algae of the picoplankton and small-sized nanoplankton of fresh and brackish waters (Krienitz et al. 2011a). The presence of *Mychonastes* in inland waters is such a common occurrence that it is rare to collect a sample devoid of this alga.

In contrast to *Picocystis*, which occurs under hypersaline conditions, the picoplankton *Mychonastes afer* Krienitz et al. inhabits Kenyan inland waters during periods of low salinity; the genus is also prevalent in freshwater. The genus *Mychonastes* comprises autosporine spherical, ovoid or ellipsoidal algae that are solitary or arranged in small mucilaginous colonies resembling those of *Dictyosphaerium* (previously named as *Pseudodictyosphaerium*) (Fig. 13.6c).

From Kenya's subsaline waters, we isolated four *Mychonastes* species: *M. afer*, *M. pusillus* Krienitz et al., *M. rotundus* Krienitz et al. and *M. timauensis* Krienitz et al. These strains could be of interest for future biotechnological applications because recent studies focusing on Chinese isolates closely related to *M. afer* showed a high content of lipids; the total lipid content of the cultured material exceeded 50 % of the dry mass, with an interesting fatty acid composition (Guo et al. 2013; Yuan et al. 2011).

13.3.6 *Nannochloropsis* and *Microchloropsis*: The Fish Waste Alternatives for PUFA Supplements and Promising Biofuel Resources

The unicellular, spherical or ovoid members of the genus *Nannochloropsis* (Eustigmatophyceae) are widely distributed in marine habitats. Six species have been described from the sea: *N. australis* Fawley et al., *N. gaditana* Lubián, *N. granulata* Karlson et Potter, *N. oceanica* Suda et Miyashita, *N. oculata* (Droop) Hibberd and *N. salina* Hibberd (Andersen et al. 1998; Fawley et al. 2015; Karlson et al. 1996; Suda et al. 2002). From freshwaters, only *Nannochloropsis limnetica* Krienitz et al. is known (Fawley and Fawley 2007; Krienitz et al. 2000). Because of its simple morphology and tininess (2–5 µm), *Nannochloropsis* is often confused with other picoplankton. The morphological species concept of this taxon was questioned already by Andersen et al. (1998), who suggested replacing it by a molecular phylogenetic concept. Each species exhibit a wide salinity range (Martinez-

Roldán et al. 2014; Sukenik et al. 2009; L Krienitz unpubl. results); therefore the species cannot be grouped according to their salt tolerance levels. Recently, two former *Nannochloropsis* species, *N. salina* and *N. gaditana*, were transferred into a new genus *Microchloropsis* Fawley et al. based on molecular phylogenetic analyses (Fawley et al. 2015).

Peculiarities of *Nannochloropsis* are fast growth in open-pond mass cultures and an extraordinarily high production of fatty acids, especially of the PUFA eicosapentaenoic acid. This offers many options for aquaculture, bioprospecting and as an alternative energy source (Khozin-Goldberg et al. 2013; Volkman et al. 1993). The high potential for biotechnological applications and in particular for biofuel generation prompted numerous genomic studies, summarised by Jinkerson et al. (2013), suggesting to treat *Nannochloropsis* as an oleaginous model alga. Wang et al. (2014a, b) provided a whole genome-based phylogeny and investigated the evolution of oleaginous traits as well as the driving evolutionary mechanisms in five species of *Nannochloropsis*.

The species distribution of *Nannochloropsis* and *Microchloropsis* in saline inland waters is not well known. In EASL, *Microchloropsis salina* (Hibberd) Fawley et al. was observed in high abundance in Lake Sonachi under the name *Monallantus salina* Bourrelly (Melack 1982) and—like *Picocystis*—might have been overlooked in other studies because of its small size. This taxon was already successfully cultivated in small-scaled open-pond systems and showed high concentrations of eicosapentaenoic acid (Bellou and Agglis 2012; Quinn et al. 2012; Sheets et al. 2014). In January 2015, under almost freshwater conditions, *N. limnetica* was found in Lake Nakuru by microscopy and proved by molecular analysis (L Krienitz and W Luo, unpubl. results). Using 454 pyrosequencing, an operational taxonomic unit (OTU) was identified which fitted to 99 % with parts of the plastide genome of strain CCMP505 from brackish ponds at Morehead City, USA (Fig. 13.7). This strain was originally designated as *Nannochloropsis* sp. (Andersen et al. 1998), but later classified to *N. limnetica* (Krienitz et al. 2000). Strains which are able to grow under freshwater conditions are becoming increasingly important for aquaculture and biotechnology because of elevated fatty acid content, sometimes higher than in marine/saline strains under particular cultivation conditions (Krienitz and Wirth 2006).

A challenge for marine and freshwater aquaculture is the high and increasing demand for fish meal and oil for the farmed fishes (Steffens 1997), often produced using fish waste (FAO 1986). *Nannochloropsis* and *Microchloropsis* have the potential to provide a valuable alternative to the fish-based fish food. Furthermore, the biomass from these algae is an interesting alternative to fish oil produced from fatty fish for medical applications to reduce myocardial illnesses (Singer and Wirth 2004).

Pioneering surveys on *Nannochloropsis* mass cultivation in large outdoor tanks filled with enriched seawater in Japan (Okauchi 1991) or in large-scale outdoor shallow ponds of about 60 m^2 and 3000 m^2, respectively, were conducted by Sukenik et al. (1993) and Sukenik (1999) in Israel. Since then, many publications on exploiting *Nannochloropsis* outdoor mass cultivation and their optimisation for high yield in fatty acids have become available (Chini Zittelli et al. 1999; Quinn et al. 2012; San Pedro et al. 2015; Sheets et al. 2014; Sukenik et al. 2009). In the last decade, fundamental research on algal biomass production has increased considerably, and *Nannochloropsis* entered the group of most promising algae for biotechnology and biofuel production (Benvenuti et al. 2015). Moody et al. (2014) published a global evaluation model of biofuel potential of microalgae based on the example of *Nannochloropsis*. The model showed that, on one side, many assessments dramatically overestimated the yield, but, on the other side, considering climate conditions and economic factors in the different regions, countries of East Africa belong to the most promising regions for outdoor mass production of microalgae in general and *Nannochloropsis* applications in particular.

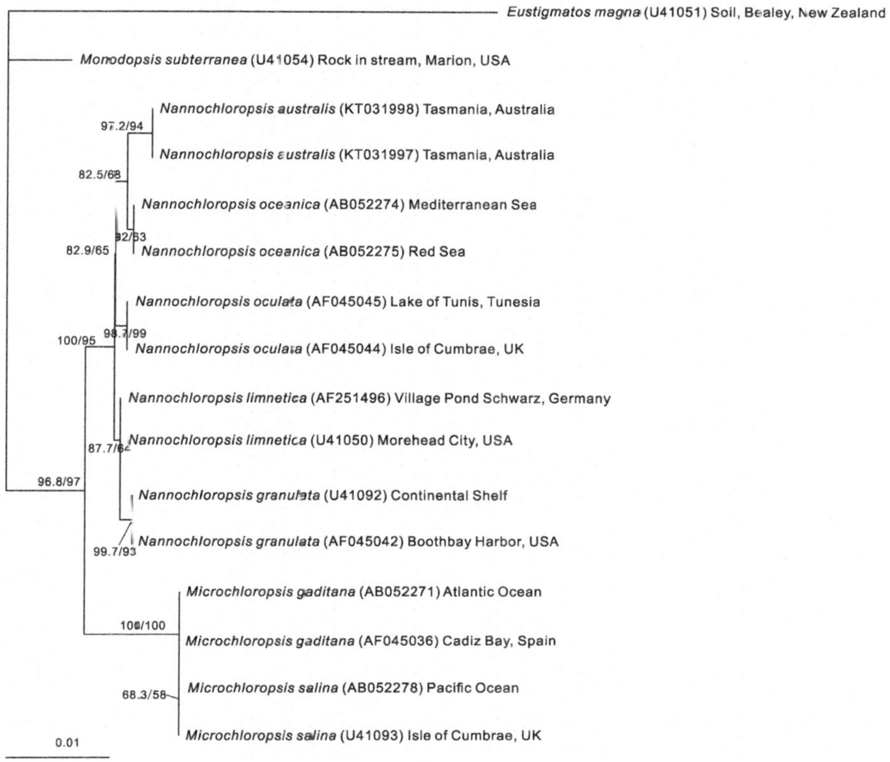

Fig. 13.7 Molecular phylogeny of the eustigmatophytes *Nannochloropsis* and *Microchloropsis* based on SSU rRNA gene sequence comparisons using a total of 1721 aligned positions. The tree analysis was calculated by the neighbour-joining method (NJ) with Kimura's two-parameter corrections (Kimura 1980). To test the tree confidence, bootstrap values (>50 %) of NJ (1000 replicates) and maximum parsimony (MP) (1000 replicates) were calculated using Phylip 3.695 (Felsenstein 1993). Bootstrap values >50 for NJ and MP were indicated at the branches. Scale bar indicates substitutions per site. The sequences were obtained from GenBank [National Center for Biotechnology Information (NCBI)]. For each taxon, the NCBI accession number is given in brackets

13.4 Concluding Remarks

Microphytes of EASL and their catchment areas feature a very high potential for biotechnological use. Although Archaea and Bacteria have been the main focus for years, Cyanobacteria and algae are still waiting to be explored in this region. We listed a few promising taxa in this article, which can be seen as an initial step to get the ball rolling. A recent study conducted by Luo et al. (2013) revealed an unexpectedly high diversity of algae, some of them promising for mass cultivation. To explore the rich genetic pool, we suggest establishing a culture collection of cyanobacterial and algal strains and screening for their suitability for biotechnological applications. In a further process, small-scale open-pond systems can be installed, followed by a scale-up for industrial production. The scale-up must be accompanied by modelling considering the whole footprint of microphyte cultures (Clarens et al. 2010). Algal farms should be near wastewater treatments, because the receiving water might be used for nutrient supply. All this work requires well-trained scientific and technical staff, which can be introduced by international experts. We suggest to enter into contracts with non-profit academic institutions including student scholarships to transfer essential knowledge to the East African countries. Partnerships at eye level, e.g. in research and development and with enterprises of algal biotechnology properly considering the rules of benefit sharing (Kamau 2013), are essential for successful long-term cooperation. Clean laboratory facilities with permanent water and power supply are also prerequisites, but sometimes ignored.

At present, the rich genetic pool of EASL can be tapped successfully only with international collaborations. One central challenge will be the transfer of biological material to institutions abroad, which are capable of applying modern techniques for screening and sequencing. The current situation is unsatisfactory and we urgently need new regulations that are simple and enable rapid action. This unresolved issue hampers knowledge exchange, especially in basic science. Clearly, third world countries must protect themselves against biopiracy (Sheridan 2004), but a workable differentiation between basic science and knowledge transfer, benefit sharing and commercial exploitation is strongly needed. The policymakers should be aware that scientists will focus their research on comparable environments elsewhere if no clear solution is offered in the near future. Other countries on the continent, such as South Africa, already provide manageable, easily accessible templates (Sanbi 2015).

References

Andersen RA, Brett RW, Potter D, Sexton JP (1998) Phylogeny of the Eustigmatophyceae based upon the 18S rRNA gene, with emphasis on *Nannochloropsis*. Protist 149:61–74

Ballot A, Krienitz L, Kotut K, Wiegand C, Metcalf JS, Codd GA, Pflugmacher S (2004) Cyanobacteria and cyanobacterial toxins in three alkaline Rift Valley lakes of Kenya – Lakes Bogoria, Nakuru and Elmenteita. J Plankt Res 26:925–935

Ballot A, Kotut K, Novelo E, Krienitz L (2009) Changes of phytoplankton communities in Lakes Naivasha and Oloidien, examples of degradation and salinization of lakes in the Kenyan Rift Valley. Hydrobiologia 632:359–363

Baunillo KE, Tan RS, Barros HR, Luque R (2012) Investigations on microalgal oil production from *Arthrospira platensis*: towards more sustainable biodiesel production. RSC Advances 2:11267–11272

Beijerinck MW (1890) Culturversuche mit Zoochlorellen, Lichenengonidien und anderen niederen Algen I-III. Bot Ztg 48:726–740

Belay A (2002) The potential application of *Spirulina* (*Arthrospira*) as a nutritional and therapeutic supplement in health management. J Am Nutraceutical Ass 5:27–48

Belay A, Ota Y, Miyakawa K, Shimamatsu H (1993) Current knowledge on potential health benefits of *Spirulina*. J Appl Phycol 5:235–241

Belay A, Kato T, Ota Y (1996) *Spirulina* (*Arthrospira*): potential application as an animal feed supplement. J Appl Phycol 8:303–311

Bellou S, Agglis G (2012) Biochemical activities in *Chlorella* sp. and *Nannochloropsis salina* during lipid and sugar synthesis in a lab-scale open pond simulating reactor. J Biotechnol 164:318–329

Benvenuti G, Bosma R, Klok AJ, Ji F, Lamers PP, Barbosa MJ, Wijffels RH (2015) Microalgal triacylglycerides production in outdoor batch-operated tubular PBRs. Biotechnol Biofuels 8:100. doi:10.1186/s13068-015-0283-2

Bock C, Krienitz L, Pröschold T (2011) Taxonomic reassessment of the genus *Chlorella* (Trebouxiophyceae) using molecular signatures (barcodes), including description of seven new species. Fottea 11:293–312

Boney AA (1981) Mucilage: the ubiquitous algal attribute. Brit Phycol J 16:115–132

Borowitzka MA (2013) Species and strain selection. In: Borowitzka MA, Moheimani NR (eds) Algae for biofuels and energy. Springer, Dordrecht, pp 77–89

Borowitzka MA, Moheimani NR (2013) Open pond culture systems. In: Borowitzka MA, Moheimani NR (eds) Algae for biofuels and energy. Springer, Dordrecht, pp 133–152

Bosma R, de Vree JH, Slegers PM, Janssen M, Wijffels RH, Barbosa MJ (2014) Design and construction of the microalgal pilot facility AlgaePARC. Algal Res 6:160–169

Cannel RJP (1990) Algal biotechnology. Appl Biochem Biotechnol 26:85–105

Chakkiath PA, Kumaresan D, Hunger S, Drake HL, Murrell JC, Shouche YS (2013) Microbiology of Lonar Lake and other soda lakes. ISME J 7:1252

Childress B, Nagy S, Hughes B (Compilers) (2008) International single species action plan for conservation of the lesser flamingo (*Phoenicopterus minor*). CMS Technical Series No. 18, AEWA Technical Series No. 34, Bonn, 59 pp

Chini Zittelli G, Lavista F, Bastianini A, Rodolfi L, Vincenzini M, Tredici MR (1999) Production of eicosapentaenoic acid by *Nannochloropsis* sp. cultures in outdoor tubular photobioreactors. J Biotechnol 70:299–312

Chisti Y (2007) Biodiesel from microalgae. Biotechnol Adv 25:294–306

Chu WL, Lim YW, Radhakrishnan AK, Lim PE (2010) Protective effect of aqueous extract from *Spirulina platensis* against cell death induced by free radicals. BMC Complement Altern Med 10:53

Ciferri O (1983) *Spirulina*, the edible microorganism. Microbiol Rev 47:551–578

Clarens AF, Resurreccion EP, White MA, Colosi LM (2010) Environmental life cycle comparison of algae to other bioenergy feedstocks. Environ Sci Technol 44:1813–1819

Cole JJ (1982) Interactions between bacteria and algae in aquatic ecosystems. Ann Rev Ecol Evol Syst 13:291–314

Corsi SR, Geis SW, Bowman G, Failey GG, Rutter TD (2009) Aquatic toxicity of airfield-pavement deicer materials and implications for airport runoff. Environ Sci Technol 43:40–46

Csögör Z, Herrenbauer M, Schmidt K, Posten C (2001) Light distribution in a novel photobioreactor – modeling for optimization. J Appl Phycol 13:325–333

Dadheech PK, Ballot A, Casper C, Kotut K, Novelo E, Lemma B, Pröschold T, Krienitz L (2010) Phylogenetic relationship and divergence among planktonic strains of *Arthrospira* (Oscillatoriales, Cyanobacteria) of African, Asian and American origin deduced by 16S-23S ITS and phycocyanin operon sequences. Phycologia 49:361–372

Dadheech PK, Glöckner G, Casper P, Kotut K, Mazzoni CJ, Mbedi S, Krienitz L (2013) Cyanobacterial diversity in the hot spring, pelagic and benthic habitats of a tropical soda lake. FEMS Microbiol Ecol 85:389–401

Decho AW (1990) Microbial exopolymer secretions in ocean environments – their role(s) in food webs and marine processes. Ocean Mar Biol 28:73–153

Doebbe A, Keck M, La Russa M, Mussgnug JH, Hankamer B, Tekçe E, Niehaus K, Kruse O (2010) The interplay of proton, electron, and metabolite supply for photosynthetic H2 production in *Chlamydomonas reinhardtii*. J Biol Chem 285:30247–30260

Draaisma R, Wijffels RH, Slegers PM, Brentner LB, Roy A, Barbosa MJ (2013) Food commodities from microalgae. Curr Opin Biotechnol 24:169–177

Dyo YM, Vonlanthen SE, Purton S, Zayadan BK (2013) Evaluating new isolates of microalgae from Kazakhstan for biodiesel production. Russ J Plant Physiol 60:549–554

Elumalai S, Prakasam V, Selvarajan R (2011) Optimization of abiotic conditions suitable for the production of biodiesel from *Chlorella vulgaris*. Indian J Sci Technol 4:91–97

Falkowski PG, Raven JA (1997) Aquatic photosynthesis. Blackwell Scientific, Oxford

Fanjing K, Qinxian J, Jia E, Mianping Z (2009) Characterization of a eukaryotic picoplankton alga, strain DGN-Z1, isolated from a soda lake in Inner Mongolia, China. Nat Res Environ Issues 15:185–189

FAO (1986) Fisheries Tech Paper 142. Food and Agriculture Organization of the United Nations, Rome. http://www.fao.org/docrep/003/x6899e/x6899e00.htm. Accessed 6 Dec 2015

Faulkner DJ (2000) Marine natural products. Nat Prod Rep 17:7–55

Fawley KP, Fawley MW (2007) Observations on the diversity and ecology of freshwater *Nannochloropsis* (Eustigmatophyceae), with descriptions of new taxa. Protist 158:325–336

Fawley MW, Jameson I, Fawley KP (2015) The phylogeny of the genus *Nannochloropsis* (Monodopsidaceae, Eustigmatophyceae), with descriptions of *N. australis* sp. nov. and *Microchloropsis* gen. nov. Phycologia 54:545–552

Felsenstein J (1993) Phylogeny Inference Package (PHYLIP), Version 3.5. University of Washington, Seattle, WA

Ferroni L, Baldisserotto C, Pantaleoni L, Billi P, Fasulo MP, Pancaldi S (2007) High salinity alters chloroplast morpho-physiology in a freshwater *Kirchneriella* species (Selenastraceae) from Ethiopian Lake Awasa. Am J Bot 94:1972–1983

Fox RD (1996) *Spirulina*: production and potential. EDISUD, Aix-en-Provence

Georgianna DR, Mayfield SP (2012) Exploiting diversity and synthetic biology for the production of algal biofuels. Nature 488:329–335

Grant WD, Heaphy S (2010) Metagenomics and recovery of enzyme genes from alkaline saline environments. Environ Tech 31:1135–1143

Guillou L, Eikrem W, Chrétiennot-Dinet M-J, Le Gall F, Massana R, Romari K, Pedrós-Alió C, Vaulot D (2004) Diversity of picoplanktonic prasinophytes assessed by direct nuclear SSU rDNA sequencing of environmental samples and novel isolates retrieved from oceanic and coastal marine ecosystems. Protist 155:193–214

Guo H, Daroch M, Liu L, Qiu G, Geng S, Wang G (2013) Biochemical features and bioethanol production of microalgae from coastal waters of Pearl River Delta. Bioresour Technol 127:422–428

Gutierrez-Wing MT, Benson BC, Rusch KA (2012) Impact of light quality and quantity on growth rate kinetics of *Selenastrum capricornutum*. Eng Life Sci 12:79–88

Harper DM, Childress RB, Harper MM, Boar RR, Hickley PH, Mills SC, Otieno N, Drane T, Vareschi E, Nasirwa O, Mwatha WE, Darlington JPEC, Escuté-Gasulla X (2003) Aquatic biodiversity and saline lakes: Lake Bogoria National Reserve, Kenya. Hydrobiologia 500:259–276

Hempel N, Petrick I, Behrendt F (2012) Biomass productivity and productivity of fatty acids and amino acids of microalgae strains as key characteristics of suitability for biodiesel production. J Appl Phycol 24:1407–1418

Henrikson R (2013) Algae microfarms for home, school, community and urban gardens, rooftop, mobile, and vertical farms and living buildings. Ronore Enterprises, Richmond, p 147

Hepperle D, Krienitz L, Hollibaugh T (2001) Molecular phylogeny of picoplanktonic chlorophytes and the discovery of a new evolutionary lineage. Phycologia 40:36

Hollibaugh JT, Wong PS, Bano N, Pak SK, Prager EM, Orrego C (2001) Stratification of microbial assemblages in Mono Lake, California, and response to a mixing event. Hydrobiologia 466:45–60

Holman BWB, Malau-Aduli AEO (2013) *Spirulina* as a livestock supplement and animal feed. J Anim Physiol Anim Nutr 97:615–623

Holman BWB, Kashani A, Malau-Aduli AEO (2014) Effects of Spirulina (*Arthrospira platensis*) supplementation level and basal diet on live weight, body conformation and growth traits in genetically divergent Australian dual purpose lambs during simulated drought and typical pasture grazing. Small Ruminant Res 120:6–14

Hong Y, Xu K (2013) Co-existing growth relationships of a lipid-producing alga with three microalgae. Allelopathy J 32:301–314

Horikoshi K (2006) Alkaliphiles: genetic properties and applications of enzymes. Springer, New York, p 256

Howe P, Meyer B, Record S, Baghurst K (2006) Dietary intake of long-chain omega-3 polyunsaturated fatty acids: contribution of meat sources. Nutrition 22:47–53

Hug C, von der Weid D (2012) Algae as an approach to combat malnutrition in developing countries. In: Posten C, Walter C (eds) Microalgal biotechnology: integration and economy. De Gruyter, Berlin, pp 101–110

Huss VAR, Frank C, Hartmann EC, Hirmer M, Kloboucek A, Seidel BM, Wenzeler P, Kessler E (1999) Biochemical taxonomy and molecular phylogeny of the genus *Chlorella* sensu lato (Chlorophyta). J Phycol 35:587–598

Jinkerson RE, Radakovits R, Posewitz MC (2013) Genomic insights from the oleaginous model alga *Nannochloropsis gaditana*. Bioengineered 4:37–43

Jirsa F, Gruber M, Stojanovic A, Odour SO, Mader D, Körner W, Schagerl M (2013) Major and trace element geochemistry of Lake Bogoria and Lake Nakuru, Kenya, during extreme drought. Chem Erde-Geochem 73:275–282

Jobb G (2008) Treefinder, version of October 2008. Munich, Germany. www.treefinder.de. Accessed 6 Dec 2015

Juárez A, Vélez CG, Iñiguez AR, Martínez DE, Rodríguez MC, Vigna MS, de Molina MDCR (2011) A *Parachlorella kessleri* (Trebouxiophyceae, Chlorophyta) strain from an extremely acidic geothermal pond in Argentina. Phycologia 50:413–421

Kaggwa MN, Burian A, Oduor SO, Schagerl M (2013) Ecomorphological variability of *Arthrospira fusiformis* (Cyanoprokaryota) in African soda lakes. Microbiol Open 2:881–891

Kamau CE (2013) Exploring bases for building common pools in Eastern Africa. In: Kamau CE, Winter G (eds) Common pools of genetic resources: equity and innovation in international biodiversity law. Earthscan-Routledge, London, pp 373–398

Karlson B, Potter D, Kuylenstierna M, Andresen RA (1996) Ultrastructure, pigment composition, and 18S rRNA gene sequence for *Nannochloropsis granulata* sp. nov. (Monodopsidaceae, Eustigmatophyceae), a marine ultraplankter isolated from the Skagerrak, northeast Atlantic Ocean. Phycologia 35:235–260

Kessler E, Huss VAR (1992) Comparative physiology and biochemistry and taxonomic assignment of the *Chlorella* (Chlorophyceae) strains of the Culture Collection of the University of Texas at Austin. J Phycol 28:550–553

Khan MA, Srivastava P (2006) Ramgarh lake isolates. Indian J Appl Pure Biol 21:339–346

Khozin-Goldberg I, Solovchenko A, Pal D, Cohen Z, Boussiba S (2013) Omega-3 and Omega-6 LC-PUFA from photosynthetic microalgae: studies on *Parietochloris incisa* and *Nannochloropsis* sp. In: Catala A (ed) Polyunsaturated fatty acids: sources, antioxidant properties and health benefits. Nova Science, Hauppauge, pp 1–22

Kimura M (1980) A simple method for estimating evolutionary rates of base substitutions through comparative studies of nucleotide sequences. J Mol Evol 16:111–120

Kirk JTO (1994) Light and photosynthesis in aquatic ecosystems, 2nd edn. Cambridge University Press, Cambridge, p 509

Kirkwood AE, Buchheim JA, Buchheim MA, Henley WJ (2008) Cyanobacterial diversity and halotolerance in a variable hypersaline environment. Microb Ecol 55:453–465

Komárek J, Fott B (1983) Chlorophyceae (Grünalgen) Ordnung: chlorococcales. In: Huber-Pestalozzi G (ed) Das Phytoplankton des Süßwassers 7. Teil, 1. Hälfte, E. Schweizerbart'sche Verlagsbuchhandlung (Nägele u. Obermiller), Stuttgart, p 1044

Kotut K, Ballot A, Wiegand C, Krienitz L (2010) Toxic cyanobacteria at Nakuru sewage oxidation ponds – a potential threat to wildlife. Limnologica 40:47–53

Krienitz L, Bock C (2012) Present state of the systematics of planktonic green algae of inland waters. Hydrobiologia 698:295–326

Krienitz L, Kotut K (2010) Fluctuating algal food populations and the occurrence of lesser flamingos (*Phoeniconaias minor*) in three Kenyan Rift Valley lakes. J Phycol 46:1088–1096

Krienitz L, Wirth M (2006) The high content of polyunsaturated fatty acids in *Nannochloropsis limnetica* (Eustigmatophyceae) and its implication for food web interactions, freshwater aquaculture and biotechnology. Limnologica 36:204–210

Krienitz L, Hepperle D, Stich H-B, Weiler W (2000) *Nannochloropsis limnetica* (Eustigmatophyceae), a new species of picoplankton from freshwater. Phycologia 39:219–227

Krienitz L, Hegewald EH, Hepperle D, Huss VAR, Rohr T, Wolf M (2004) Phylogenetic relationship of *Chlorella* and *Parachlorella* gen. nov. (Chlorophyta, Trebouxiophyceae). Phycologia 43:529–542

Krienitz L, Bock C, Dadheech PK, Pröschold T (2011a) Taxonomic reassessment of the genus *Mychonastes* (Chlorophyceae, Chlorophyta) including the description of eight new species. Phycologia 50:89–106

Krienitz L, Bock C, Nozaki H, Wolf M (2011b) SSU rRNA gene phylogeny of morphospecies affiliated to the bioassay alga "*Selenastrum capricornutum*" recovered the polyphyletic origin of crescent-shaped Chlorophyta. J Phycol 47:880–893

Krienitz L, Bock C, Kotut K, Luo W (2012a) *Picocystis salinarum* (Chlorophyta) in saline lakes and hot springs of East Africa. Phycologia 51:22–32

Krienitz L, Bock C, Kotut K, Pröschold T (2012b) Genotypic diversity of *Dictyosphaerium*-morphospecies (Chlorellaceae, Trebouxiophyceae) in African inland waters, including the description of four new genera. Fottea 12:231–253

Krienitz L, Dadheech PK, Kotut K (2013) Mass developments of a small sized ecotype of *Arthrospira fusiformis* in Lake Oloidien, Kenya, a new feeding ground for Lesser Flamingos in East Africa. Fottea 13:215–225

Krienitz L, Huss VAR, Bock C (2015) *Chlorella*: 125 years of the green survivalist. Trends Plant Sci 20:67–69

Krienitz L, Krienitz D, Dadheech PK, Hübener T, Kotut K, Luo W, Teubner K, Versfeld WD (2016) Algal food for Lesser Flamingos: a stocktaking. Hydrobiologia 775:21–50

Lanzén A, Simachew A, Gessesse A, Chmolowska D, Jonassen I, Øvreås L (2013) Surprising prokaryotic and eukaryotic diversity, community structure and biogeography of ethiopian soda lakes. PLoS One 8: e72577

Léonard J (1966) The 1964–65 Belgian trans-Saharan expedition. Nature 209:126–128

Lewin RA, Krienitz L, Goericke R, Takeda H, Hepperle D (2000) *Picocystis salinarum* gen. et sp. nov. (Chlorophyta) – a new picoplanktonic green alga. Phycologia 39:560–565

Lu C, Rao K, Hall D, Vonshak A (2001) Production of eicosapentaenoic acid (EPA) in Monodus subterraneus grown in a helical tubular photobioreactor as affected by cell density. J Appl Phycol 13(6):517–522

Luo W, Pflugmacher S, Pröschold T, Walz N, Krienitz L (2006) Genotype versus phenotype variability in *Chlorella* and *Micractinium* (Chlorophyta, Trebouxiophyceae). Protist 157:315–333

Luo W, Pröschold T, Bock C, Krienitz L (2010) Generic concept in *Chlorella*-related coccoid green algae (Chlorophyta, Trebouxiophyceae). Plant Biol 12:545–553

Luo W, Kotut K, Krienitz L (2013) Hidden diversity of eukaryotic plankton in the soda lake Nakuru, Kenya, during a phase of low salinity revealed by a SSU rRNA gene clone library. Hydrobiologia 702:95–103

Markou G, Chatzipavlidis I, Georgakakis D (2012) Carbohydrates production and bio-flocculation characteristics in cultures of *Arthrospira* (*Spirulina*) *platensis*: improvements through phosphorus limitation process. Bioenergy Res 5:915–925

Martinez-Roldán AJ, Perales-Vela HV, Cañizares-Villanueva RO, Torzillo G (2014) Physiological response of *Nannochloropsis* sp to saline stress in laboratory batch cultures. J Appl Phycol 26:115–121

Masojídek J, Prášil O (2010) The development of microalgal biotechnology in the Czech Republic. J Ind Microbiol Biotechnol 37:1307–1317

Mata TM, Martins AA, Caetano NS (2010) Microalgae for biodiesel production and other applications: a review. Renew Sustain Energy Rev 14:217–232

Melack JM (1982) Photosynthetic activity and respiration in an equatorial African soda lake. Freshwater Biol 12:381–400

Melack JM, Kilham P (1974) Photosynthetic rates of phytoplankton in East African alkaline, saline lakes. Limnol Oceanogr 19:743–755

Moody JW, McGinty CM, Quinn JC (2014) Global evaluation of biofuel potential from microalgae. Proc Natl Acad Sci USA 111:8691–8696

Najafi G, Ghobadian B, Yusaf TF (2011) Algae as a sustainable energy source for biofuel production in Iran: a case study. Renew Sustain Energy Rev 15:3870–3876

Noguchi N, Konishi F, Kumamoto S, Maruyama I, Ando Y, Yanagita T (2013) Beneficial effects of *Chlorella* on glucose and lipid metabolism in obese rodents on a high-fat diet. Obes Res Clin Pract 7:e95–e105

Nygaard G, Komárek J, Kristiansen J, Skulberg OM (1986) Taxonomic designations of the bioassay alga NIVA-CHL 1 ("*Selenastrum capricornutum*"). Opera Bot 90:1–46

Oduor SO, Schagerl M (2007) Phytoplankton photosynthetic characteristics in three Kenyan Rift Valley saline-alkaline lakes. J Plankton Res 29:1041–1050

Okauchi M (1991) The status of phytoplankton as food organisms in Japan. In: Fulks W, Main KL (eds) Rotifers and microalgae culture system. Proceedings of the US-Asia workshop, Honolulu, p 247–256

Packer M (2009) Algae capture of carbon dioxide; biomass generation as a tool for greenhouse gas mitigation with reference to New Zealand energy strategy and policy. Energy Policy 37:3428–3437

Posada D, Buckley TR (2004) Model selection and model averaging in phylogenetics: advantages of akaike information criterion and Bayesian approaches over likelihood ratio tests. Syst Biol 53:793–808

Posten C (2009) Design principles of photo-bioreactors for cultivation of microalgae. Eng Life Sci 9:165–177

Posten C (2012) Introduction – discovering microalgae as source for sustainable biomass. In: Posten C, Walter C (eds) Microalgal biotechnology: potential and production. De Gruyter, Berlin, pp 1–10

Quinn JC, Yates T, Douglas N, Weyer K, Butler J, Bradley TH, Lammers PJ (2012) *Nannochloropsis* production metrics in a scalable outdoor photobioreactor for commercial applications. Bioresour Technol 117:164–171

Reynolds CS (2007) Variability in the provision and function of mucilage in phytoplankton: facultative responses to the environment. Hydrobiologia 578:37–45

Richmond A, Hu Q (2013) Handbook of microalgal culture, 2nd edn. Wiley Blackwell, Chichester, UK, 719p

Roesler CS, Culbertson CW, Etheridge SM, Goericke R, Kiene RP, Miller LG, Oremland RS (2002) Distribution, production, and ecophysiology of *Picocystis* strain ML in Mono Lake, California. Limnol Oceanogr 47:440–452

Ronquist F, Huelsenbeck JP (2003) MrBayes 3: Bayesian phylogenetic inference under mixed models. Bioinformatics 19:1572–1574

Safi C, Zebib B, Merah O, Pontalier P-Y, Vaca-Garcia C (2014) Morphology, composition, production, processing and applications of *Chlorella vulgaris*: a review. Renew Sustain Energy Rev 35:265–278

San Pedro A, González-López CV, Acién FG, Molina-Grima E (2015) Outdoor pilot production of *Nannochloropsis gaditana*: influence of culture parameters and lipid production rates in raceway ponds. Algal Res 8:205–213

Sanbi (2015) Biodiversity for life. http://www.sanbi.org/information/documents. Accessed 6 Dec 2015

Sánchez M, Bernal-Castillo J, Rozo C, Rodríguez I (2003) *Spirulina* (*Arthrospira*): an edible microorganism: a review. Univ Sci 8:7–24

Schagerl M, Oduor SO (2008) Phytoplankton community relationship to environmental variables in three Kenyan Rift Valley saline-alkaline lakes. Mar Freshw Res 59:125–136

Schagerl M, Burian A, Gruber-Dorninger M, Oduor SO, Kaggwa MN (2015) Algal communities of Kenyan soda lakes with a special focus on *Arthrospira fusiformis*. Fottea 15:245–257

Sheets JP, Ge X, Park SY, Li Y (2014) Effect of outdoor conditions on *Nannochloropsis salina* cultivation in artificial seawater using nutrients from aerobic digestion effluent. Bioresour Technol 152:154–161

Shen Y, Yuan W, Pei ZJ, Wu Q, Mao E (2009) Microalgae mass production methods. Transact ASABE 52:1275–1287

Sheridan C (2004) Kenyan dispute illuminates bioprospecting difficulties. Nat Biotechnol 22:1337

Shimamatsu H (2004) Mass production of *Spirulina*, an edible microalga. Hydrobiologia 512:39–44

Shimura H, Itoh K, Sugiyama A, Ichijo S, Ichijo M, Furuya F, Nakamura Y, Kitahara K, Kobayashi K, Yukawa Y, Kobayashi T (2012) Absorption of radionuclides from the Fukushima nuclear accident by a novel algal strain. PLoS One 7:e44200

Simpson PD, Van Valkenburg SD (1978) The ultrastructure of *Mychonastes ruminatus* gen. et sp. nov., a new member of the Chlorophyceae isolated from brackish water. Brit Phycol J 13:117–130

Singer P, Wirth M (2004) Can n-3 PUFA reduce cardiac arrhythmias? Results of a clinical trial. Prostaglan, Leukotr Essent Fatty Acids 71:153–159

Škaloud P, Němková Y, Pytela J, Bogdanov NI, Bock C, Pickinpaugh SH (2014) *Planktochlorella nurekis* gen. et sp. nov. (Trebouxiophyceae, Chlorophyta), a novel coccoid green alga carrying significant biotechnological potential. Fottea 14:53–62

Solovchenko AE, Khozin-Goldberg I, Didi-Cohen S, Cohen Z, Merzlyak MN (2008) Effects of light intensity and nitrogen starvation on growth, total fatty acids and arachidonic acid in the green microalga *Parietochloris incisa*. J App Phycol 20:245–251

Song M, Pei H, Hu W, Ma G (2013) Evaluation of 10 microalgal strains for biodiesel production. Bioresour Technol 141:245–251

Sorokin DY, Berben T, Melton ED, Overmars L, Vavourakis CD, Muyzer G (2014) Microbial diversity and biogeochemical cycling in soda lakes. Extremophiles 18:791–809

Spirulinasource (2015) Resource Center for Spirulina, Algae and Green Superfoods. http://www.spirulinasource.com. Accessed 6 Dec 2015

Srivastava P, Gajraj RS (1996) *Spirulina* biotechnology of Rajasthan-Jaipur. Recent Adv Biotech 1996:47–56

Srivastava P, Sharma A (2002) *Spirulina* and human health in new millennium. In: Anand N (ed) Algological research in India. Bishen Singh Mahendra Pal Singh, Dehra Dun, pp 337–343

Srivastava P, Sharma A (2004) Therapeutic application of *Spirulina fusiformis* in hypercholesterolemia. In: Triveda PC (ed) Herbal drugs and biotechnology. Pointer Publication, Jaipur, pp 67–78

Steffens W (1997) Effects of variation in essential fatty acids in fish feeds on nutritive value of freshwater fish for humans. Aquaculture 151:97–119

Stephens E, Ross IL, Mussgnug JH, Wagner LD, Borowitzka MA, Posten C, Kruse O, Hankamer B (2010) Future prospects of microalgal biofuel production systems. Trends Plant Sci 15:554–564

Suda S, Atsumi M, Miyashita H (2002) Taxonomic characterization of a marine *Nannochloropsis* species, *N. oceanica* sp. nov. (Eustigmatophyceae). Phycologia 41:273–279

Sukenik A (1991) Ecophysiological considerations in the optimization of eicosapentaenoic acid production by *Nannochloropsis* sp. (Eustigmatophyceae). Bioresour Technol 35:263–269

Sukenik A, Zmora O, Carmeli Y (1993) Biochemical quality of marine unicellular algae with special emphasis on lipid composition II *Nannochloropsis* sp. Aquaculture 117:313–326

Sukenik A (1999) Eicosapentaenoic acid by the marine eustigmatophyte *Nannochloropsis*. In: Cohen Z (ed) Chemicals from microalgae. Taylor & Francis, London, pp 41–56

Sukenik A, Beardall J, Kromkamp JC, Kopeck J, Masojídek J, van Bergeijk S, Gabai S, Yamshon A (2009) Photosynthetic performance of outdoor *Nannochloropsis* mass cultures under a wide range of environmental conditions. Aquat Microbiol Ecol 56:297–308

Swofford DL (2002) Phylogenetic analysis using parsimony (* and other methods), version 4.0b 10. Sinauer Associates, Sunderland, MA

Talling JF, Wood RB, Prosser V, Baxter RM (1973) The upper limit of photosynthetic productivity by phytoplankton: evidence from Ethiopian soda lakes. Freshw Biol 3:53–76

Tamura K, Dudley J, Nei M, Kumar S (2007) MEGA4: molecular evolutionary genetics analysis (MEGA) software v4.0. Mol Biol Evol 24:1596–1599

Trauth MH, Ma M, Deino AL, Junginger A, Lesoloyia M, Odada EO, Olago D, Olaka LA, Strecker MR, Tiedemann R (2010) Human evolution in a variable environment: the amplifier lakes of Eastern Africa. Q Sci Rev 29:2981–2988

UNESCO (2007) Blaualgen für die Regionalentwicklung? Potenzial und Risiko des Cyanobakteriums Arthrospira. http://www.unesco.de/2180.html. Accessed 6 Dec 2015

Van Donk E (2005) Planktonic interactions: developments and perspectives. Verh Internat Verein Limnol 29:61–72

Vareschi E (1978) The ecology of Lake Nakuru (Kenya) I. Abundance and feeding of the Lesser Flamingo Oecologia 32:11–35

Verschuren D, Cocquyt C, Tibby J, Roberts CN, Leavitt PR (1999) Long-term dynamics of algal and invertebrate communities in a small, fluctuating tropical soda lake. Limnol Oceanogr 44:1216–1231

Vílchez C, Garbayo I, Lobato ML, Vega JM (1997) Microalgae-mediated chemicals production and wastes removal. Enzyme Microb Technol 20:562–572

Volkman JK, Brown MR, Dunstan GA, Jeffrey SW (1993) The biochemical composition of marine microalgae from the class Eustigmatophyceae. J Phycol 29:69–78

Vonshak A, Richmond A (1988) Mass production of the blue-green alga *Spirulina*: an overview. Biomass 15:233–247

Vonshak A, Tomaselli L (2000) *Arthrospira* (*Spirulina*): systematics and eco-physiology. In: Whitton BA, Potts M (eds) The ecology of cyanobacteria: their diversity in time and space. Kluwer Academic, Dordrecht, pp 505–522

Wang D, Ning K, Li J, Hu J, Han D, Wang H, Zeng X, Jing X, Zhou Q, Su X, Chang X, Wang A, Wang W, Jia J, Wei L, Xin Y, Qiao Y, Huang R, Chen J, Han B, Yoon K, Hill RT, Zohatr Y, Chen F, Hu Q, Xu J (2014a) *Nannochloropsis* genomes reveal evolution of microalgal oleaginous traits. PLoS Genet 10:e1004094.

Wang S, Lambert W, Giang S, Goericke R, Palenik B (2014b) Microalgal assemblages in a poikilohaline pond. J Phycol 43:275–283

Wijffels RH, Barbosa MJ (2010) An outlook on microalgal biofuels. Science 329:796–799

Wilhelm C, Jakobs T (2012) Balancing the conversion efficiency from photon to biomass. In: Posten C, Walter C (eds) Microalgal biotechnology: potential and production. De Gruyter, Berlin, pp 39–54

Xiuling L, Přibyl P, Bišová K, Kawano S, Cepák V, Zachleder V, Čížková M, Brányiková I, Vítová M (2013) The microalga *Parachlorella kessleri* – a novel highly efficient lipid producer. Biotechnol Bioenerg 110:97–107

Yap CY, Chen F (2001) Polyunsaturated fatty acids: biological significance, biosynthesis, and production by microalgae and microalgae-like organisms. In: Chen F, Jiang Y (eds) Algae and their biotechnological potential. Springer Science, Dortrecht, pp 1–32

Yee W (2016) Microalgae from the Selenastraceae as emerging candidates for biodiesel production: a minireview. World J Microbiol Biotechnol 32:64. doi:10.1007/s11274-016-2023-6

Yuan C, Liu J, Fan Y, Ren X, Hu G, Li F (2011) *Mychonastes* afer HSO-3-1 as a potential new source of biodiesel. Biotechnol Biofuels 4:47

Zhou XP, Ge HM, Xia L, Zhang D, Hu CX (2013) Evaluation of oil-producing algae as potential biodiesel feedstock. Bioresour Technol 134:24–29

Conservation and Management of East Africa's Soda Lakes

14

David M. Harper, Emma Tebbs, Olivia Bell, and Victoria J. Robinson

Abstract

The conservation value of East African saline alkaline lakes is identified based on the spatial and temporal gradients of regional variability, landscapes, biodiversity and population changes. The present extent of their management is evaluated along a spatial gradient from international to local, through the perspective of international agreements, management plans and policies. Gaps in present-day conservation management are identified and the theoretical possibilities for improved management and more secure conservation outlined. The final section suggests ways in which improved management policies can be achieved in the next decade by simple knowledge collection at each lake and sharing using social media.

14.1 Introduction: The Value of Soda Lakes

The previous chapters of this book should have left you in no doubt that East African saline alkaline lakes (EASL) are globally unique. This chapter seeks to put that value into the perspectives of space and time, with the overarching concept of their usefulness to people. It is now common in environmental writings and evaluations to talk about ecosystems in terms of their 'services' to humankind and their 'natural capital', in order to place an economic value upon them. Such classifications are very valuable when one is forced to defend biodiversity-rich areas against short-term exploitative proposals, such as biofuel plantations in East Africa, but are more limited in placing monetary equivalence

D.M. Harper (✉)
Department of Zoology, National Museums of Kenya, Nairobi, Kenya

Department of Biology, University of Leicester, Leicester LE1 7RH, UK
e-mail: dmh@le.ac.uk

E. Tebbs
Department of Geography, King's College London, Strand Campus, London WC2R 2LS, UK
e-mail: emma.tebbs@gmail.com

O. Bell • V.J. Robinson
Department of Biology, University of Leicester, Leicester LE1 7RH, UK
e-mail: olivia-b@hotmail.co.uk; vjrobinson28@gmail.com

M. Schagerl (ed.), *Soda Lakes of East Africa*, DOI 10.1007/978-3-319-28622-8_14

upon the sheer majesty of nature. Many of the EASL are uniquely majestic, both in their landscape setting of the Rift Valley and their enormous concentrations of that unique species, the Lesser Flamingo (*Phoeniconaias minor* Geoffroy Saint-Hilaire). Hence, economic valuation makes little sense, even though it is necessary in our present society. Several lakes play a vital role in the culture of indigenous peoples and it is equally difficult to place a realistic monetary value upon such roles.

14.1.1 Regional Value

EASL are highly variable in their chemistry, as first shown by that great limnologist Jack Talling with his wife 50 years ago (Talling and Talling 1965) and revisited in greater detail in the earlier chapters of this book. Whilst we know something about the chemistry and biology of all of them, fewer have any kind of protected status and only a small number have documented conservation management. This chapter will concentrate, by necessity, on what is known, but emphasise the importance of some lakes where little is known.

The regional value of EASL ecosystems can best be examined through their value and use by the Lesser Flamingo. The birds are confined to these lakes due to their extreme specialisation for filter-feeding planktonic Cyanobacteria (see Chaps. 10 and 13), which reach enormous biomass because the lakes support some of the highest rates of aquatic primary production in the world (see Chaps. 6 and 12; Talling et al. 1973).

P. minor can also feed on benthic algal mats (predominantly diatoms) in very shallow water, which populations in their other three locations outside East Africa do more extensively (Gujarat, India; southern Africa and western Africa). They feed in this way inside East Africa sometimes out of necessity, sometimes, as at Lake Natron lagoons, because the biomass can be incredibly high and meets their daily requirements. Their population is around one and a half million birds (and perhaps it has always been at about this level, because earlier estimates were guesses by ornithologists amazed at the spectacle, but recent ones are actual counts). In theory, this population could be sustained on a single lake with its optimal biomass of *Arthrospira fusiformis*; in the past 20 years, over a million birds have been counted together at least five times—in 1994 about 1.4 million birds were counted at the three accessible Kenyan lakes, Bogoria, Nakuru and Elmenteita; in 1999–2000 around the same number were counted, almost all at Bogoria; in 2003 about 1.2 million were counted, almost all at Nakuru; in 2006 almost 1.5 million, all at Nakuru and in 2010 about 1.3 almost all at Bogoria (unpublished data, African-Eurasian Water Bird Census; National Museums of Kenya). The fluctuation in numbers between these three lakes has been very erratic in the past 4 years (2011–2015) and will be referred to later in this chapter.

The unpredictable nature of *P. minor*'s population locations is a result of erratic fluctuations in the amount of food available in each lake, which itself is the result of an as-yet poorly understood combination of hydrological changes in the catchments and biological changes in each lake. These closed basin lake systems have no surface outlet and are intrinsically linked to climate, with both water chemistry and lake level resulting from the delicate balance between inputs from direct precipitation, surface runoff, springs and groundwater flow—with losses through evaporation and groundwater recharge (Mason et al. 1994). East Africa lies in the Intertropical Convergence Zone (ITCZ). Its rainy seasons are a consequence of Indian Ocean monsoons coming south-west and Atlantic monsoons coming north-east; these meet because they are held by the mountain masses on either side of the Rift Valley. The relative strength of each monsoon over the year results in two rainy and dry seasons a year in this central part of East Africa and determines their intensity. Figure 14.1 shows the extent of such natural changes in the surface area of several lakes, together with photographs taken from the same spot at Lake Bogoria, showing the differences. Such climatic unpredictability has substantial effects on lake

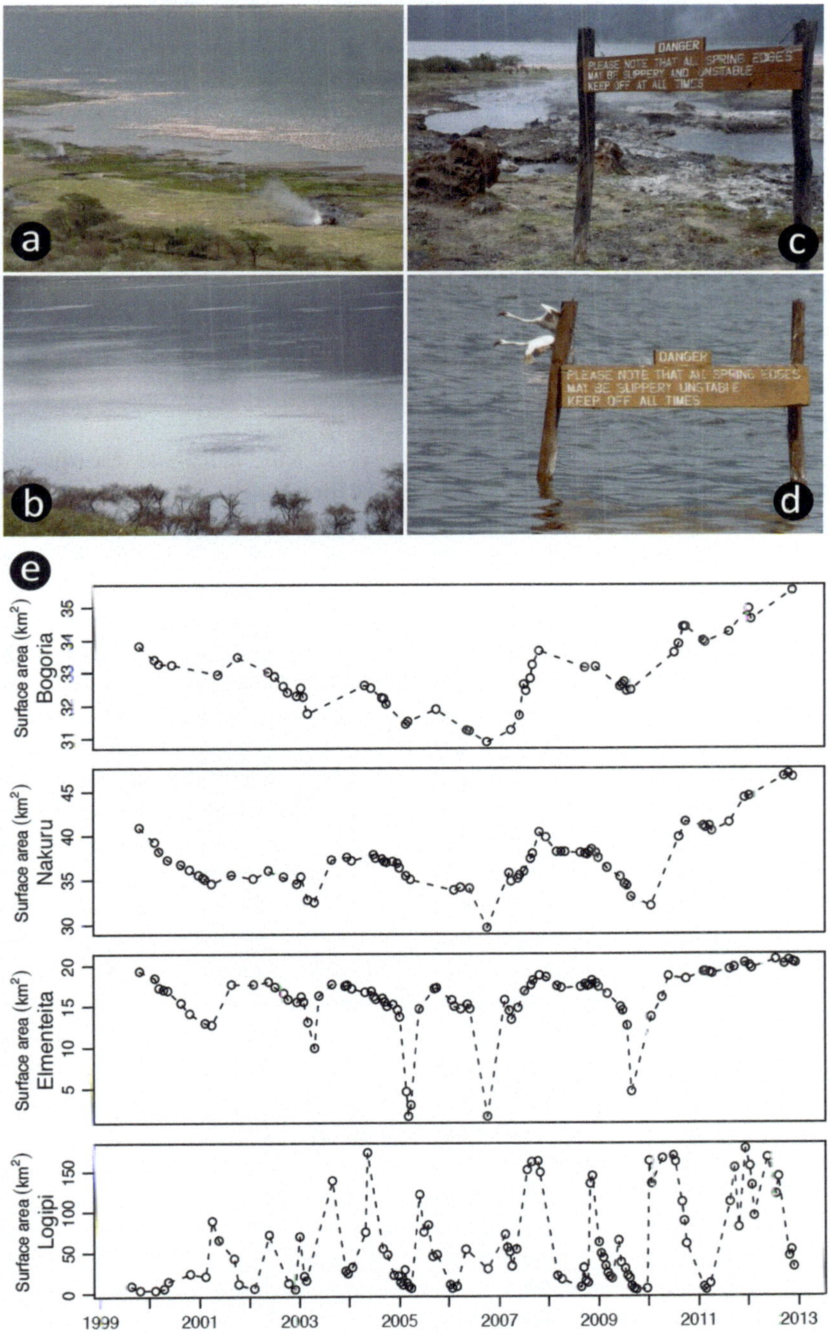

Fig. 14.1 Pairs of photographs that show lake level differences in Bogoria along with a graph of lake area change over the past decade for four Rift Valley lakes. (**a**) Lake Bogoria taken from the viewpoint above the main hot springs group, 11 years apart, the first in 2003; (**b**) the second in 2014; (**c**) taken from beside the same hot spring, at adjacent low lake level (2009) and (**d**) high level (2012; courtesy of M. Schagerl); (**e**) graph of lake area changes over a decade, derived from satellite image interpretation. The most recent rise in lake level is very clear in the three central lakes (Elmenteita, Nakuru, Bogoria), but fluctuating in the more northern, very shallow, lake Logipi

parameters, including the biomass of their phytoplankton (Smol and Stoermer 2010).

No two lakes have the same limnological characteristics due to their different histories and morphometries, and no one lake is stable enough to remain consistently highly productive. *P. minor* have to feed almost every day, so their long-term survival depends upon the regional network of lakes, enabling them to always find one with adequate food by moving away from lakes whose food supply unexpectedly drops.

14.1.2 Landscape Value

Almost all the saline alkaline lakes lie within the eastern part of the Rift Valley that comes down through Kenya and Tanzania. A very few are volcanic crater lakes outside the Rift, but all of the lakes owe their formation to those tectonic activities which gave rise to the Rift and, as outlined above, which caused its unpredictable climate. The Rift was formed in the Cenozoic era (approximately 40 Ma). It has experienced a long-term climate of alternating periods of high humidity and high aridity, which dates back at least 2.7–0.9 Ma (Trauth et al. 2005) after the onset of the glacial cycles in the Northern Hemisphere in the Plio-Pleistocene (de Menocal 1995). Diatomite evidence suggests the presence of very large lakes, exceeding 250 m in depth, between 1.9 and 0.9 Ma, one of which was still in existence 9200 years ago (Trauth et al. 2010). This lake shrank over the following years, being considerably smaller 3000 years ago (Richardson and Richardson 1972) and the present lakes are all that remain. The most well known of these paleo-lakes is the one which joined Naivasha to Nakuru and on whose shorelines *Homo habilis* made stone tools at Kariandusi, just outside the present town of Gilgil (Bergner et al. 2009).

All the lakes have undergone extreme and erratic fluctuations in both water level and environmental conditions over the last few hundred years (Verschuren et al. 2000), more so than the present fluctuations shown in Fig. 14.1. The present changes are dramatic, especially in the shallow ones such as Nakuru, which are dry about once a decade. All the lakes differ from one another considerably in terms of hydrology and ecology, ranging in depth from about 15 cm in basins of Lake Natron to over 16 m at Lake Bogoria's deepest point. The conductivity of the deeper lakes is characteristically high and semi-stable at around 70,000 μS cm^{-1} (Harper et al. 2003), whilst the shallow lakes experience great chemical variability, ranging from 11,000 to 160,000 μS cm^{-1} as their levels fluctuate (Verschuren et al. 2004).

The lakes all have very spectacular and varied landscapes, derived from their past. The most striking are probably those that lie along the base of Rift walls and escarpments, such as Lake Bogoria in Kenya and Lake Manyara in Tanzania, or that have a backdrop of volcanoes, such as Natron in Tanzania. More open are the shallow basins in the middle of the Rift floor, surrounded by savannah grassland, such as Lake Nakuru (Kenya) or Magadi (Tanzania). Where protected status has been declared, the landscape can include elephant (Manyara), rhinoceros (Nakuru), lion (Magadi), hippopotamus (Oloidien) or buffalo—all against the backdrop of a soda lake with its pink edge of flamingos.

Very large, shallow lakes at low altitude show great fluctuations in horizontal extent (e.g. Lake Logipi, Fig. 14.1), with extensive salt flats when the lakes are low. These landscapes are very open and harsh, exemplified by the survival of human communities (e.g. Maasai at Lake Natron) on the one hand or the encrusting of dead animals that have succumbed to the alkaline waters as 'mummified statues' on the other hand (National Geographic 2013).

Some lakes have local landscape features such as springs or geysers, which are also a consequence of their turbulent original and history. Cool-water springs, often at the base of an escarpment or volcano, have water fresh enough to serve as water holes for animals from the surrounding areas and sources of drinking water for *P. minor* (Bogoria, Natron, Nakuru). Warm-water springs may be more saline than mammals can tolerate, but still fresh enough for *P. minor* to drink and bathe in (Bogoria, Manyara, Natron). A few have hot springs and geysers similar in salinity to the lake waters (some of which flamingos can still drink from), forming a feature of geological value in addition to the lake's biological value (Bogoria).

14.1.3 Biodiversity Value: Species Richness

The biodiversity value has been referred to in the previous section, but here the goal is to emphasise the value of the regional network of lakes, using Lesser Flamingos as an example. This 'network'—largely a chain of lakes running from north to south—also has enormous value as a series of 'stepping stones' for species that are motile as well as providing habitat for separate populations of species that are sedentary. The value to motile species is best seen in migrating birds down the Palearctic-African flyway, none more so than members of the Charadriiformes (waders or shorebirds)—predominant species in the families Scolopacidae (sandpipers and allies), Recurvirostridae (avocets and allies) and Charadriidae (plovers and lapwings)—from October to March (winter in the Northern Hemisphere). The extremely high primary productivity of soda lakes results in high secondary productivity not only of filter-feeders (Lesser Flamingos) but also of detritivores (a few species of Diptera, numerically dominated by Chironomidae midges). Their larvae and pupae in shallow water (in tens of thousands per square metre; Harper et al. 2003) and their pupal cases collecting on the surface tension, blown to lake edges after adult emergence, provide a plentiful source of high-energy food. They enable birds to increase their weight either after or before the long migration between Europe and Africa.

The lakes' high productivity provides food not only for migrants but also for native aquatic insectivorous and small piscivorous birds, which congregate in high numbers when chironomid densities are high. Lake Bogoria has a conductivity usually over 70,000 μS cm^{-2} (it has recently halved as a result of excessive rainfall between 2012 and 2015) and is devoid of fish. Nonetheless, it is an important Kenyan site (Bennun and Nasirwa 2000) for the Black-necked Grebe *Podiceps nigricollis* Brehm and Cape Teal *Anas capensis* Gmelin. During one African Waterbird Census, for example, Lake Bogoria held all but two of the Black-necked Grebes and 96 % of the Cape Teal counted on all of those Kenyan Rift Valley lakes censused (Nasirwa and Bennun 1994). Black-necked Grebe dives for chironomid larvae and pupae in the plankton, whilst Cape Teal skims the water surface for pupal exuviae and stranded adults; their numbers reached a maximum of 2125 (teal) and 3694 (grebe), which was coincident with the high phytoplankton biomass in the mid-1990s (Tebbs et al. 2013b).

It is not only the biodiversity of water birds that is supported by the high productivity of soda lakes; there is an extensive transfer of energy from secondary producers in the lake to the terrestrial environment. In Lake Bogoria again, for example, the density of adult midges (family Chironomidae) emerging from larvae at less than 5 m depth (lower depths were devoid of oxygen and the mud lifeless) was about 3000, or 210 kg day^{-1} between 2001 and 2003. This is calculated from an estimated area of 1.4×10^4 m^{-2}, with larval density up to 40,000 m^{-2}, assuming that adult emergence was continuous, with an estimated life cycle of 2 weeks at adult size, and based on lake temperature (Harper et al. 2003). This biomass supported about half a million Apodidae (swifts) and Hirundinidae (swallows and martins), families with five and seven species, respectively. Thus, the lake is an important feeding station for the four Palaearctic migrants, particularly *Hirundo rustica* Linnaeus (swallow), *Riparia riparia* Linnaeus [Sand Martin (UK) or Bank Swallow (USA)] and *Apus apus* Linnaeus (swift), in both directions, being at the southern edge of Kenya's extensive arid semi-desert region.

P. minor, which can account for well over half of a soda lake's primary production (Vareschi and Jacobs 1985), also support the terrestrial environment in several other ways. The edges of soda lakes are fringed by a continuous strand line of flamingo feathers, droppings and accumulated vegetation debris, overlying volcanic sandy mud. The strand line is inhabited by terrestrial Coleoptera, particularly Tenebrionidae such as *Gonocephalum* sp., *Sepidium* spp., *Vietomorpha* spp. and *Rhytinota praelonga* Koch, which scavenge by night and shelter by day under stones. The lake margin is dominated by predatory *Zophosis* spp. and Tiger Beetles such as *Lophyra boreodilatata* W. Horn, during the day. The lake edge also supports many Cerambycidae (Longhorn Beetles) such as *Tithoes confinis* Laporte de Castelnau and *Macrotoma palmata* Fabricius, which are primary agents in breaking down organic plant debris (Harper et al. 2003).

The debris can be enhanced from time to time by *P. minor* carcasses, particularly in some lakes when a 'mass mortality' has been recorded (see below). This considerably increases the density, diversity and biomass of terrestrial scavengers. These range from several species of Dermaptera (earwigs) to large mammals such as *Phacochoerus africanus* Gmelin (warthog) and *Papio anubis* Lesson (baboon) and large birds such as *Aquila nipalensis* Hodgson (Steppe Eagle) and *Leptoptilos crumenifer* Lesson (Marabou Stork), all feeding on *P. minor* carcasses (DM Harper, pers. obs.). All help to break the carcasses down into the smaller debris processed by the invertebrates described above.

14.1.4 Biodiversity Value: Unique Species

Soda lakes are all highly extreme environments, as earlier chapters have made clear, and those few organisms that have evolved to tolerate them physically and chemically can reach enormous population concentrations in the absence of competitors. Thus, in the lakes—as opposed to the land around them—few species are highly abundant. This gives the modern phrase 'biodiversity value' a very different meaning to the previous section, which highlighted 'biodiversity' as equivalent to 'species richness'. The unique species that are best known scientifically are microbes (see Chap. 5), where there are many species new to science and endemic. They also include the Lesser Flamingo (confined to soda lakes within East Africa), which feeds largely upon the dominant photosynthetic microbe, *Arthrospira fusiformis. A. fusiformis* and the invertebrates which feed upon it, such as the benthic chironomid *Leptochironomus deribae* Freeman (Vareschi and Jacobs 1985), are not endemic as such, but specialised to the chemical range displayed by soda lakes. This enables them to become overwhelmingly dominant due to the lack of competition from conspecifics that they may encounter in less extreme aquatic environments (Thiemann's law of Biocenoses; Remmert 1980).

14.1.5 Ecosystem Services

Soda lakes have a number of economic values (which are the most widely understood measurements of ecosystem services) based upon their original and present environmental state, some more obvious than others. Tourism and recreational value (cultural services) are the most widespread. Initially, this evaluation was based upon the spectacle of hundreds of thousands of pink flamingos. This was considered one of the top ten natural spectacles in the world 50 years ago and resulted in the purchase of Lake Nakuru from its private farming owners, for the Kenyan government as a National Park, by WWF in 1963. Ten years later, WWF also purchased an extension of land around the park, bringing terrestrial mammals alongside flamingos and other shorebirds into the tourists' checklist. Nakuru is the main Kenyan National Park which includes a soda lake (and the flamingos are a major contribution to it; Amboseli NP also has a shallow soda lake, frequently dry, unimportant to Lesser Flamingos). Ethiopia has two and Tanzania four parks where both species of flamingos can be seen, sometimes in high numbers. Ethiopia has the Abijatta-Shala Lakes National Park, whereby the former lake is shallow (maximum 14 m) and the latter lake is a deep extinct volcanic crater (over 200 m). Unfortunately, the land surrounding the lakes is severely degraded by tree removal for fuelwood and overgrazing by livestock, and Lake Abijatta itself has shrunk due to industrial abstraction of water by a soda ash company. Nonetheless, up to a quarter of a million Lesser Flamingos were recorded here 20–25 years ago. Ethiopia's other income-generating soda lake is Lake Beseka within Awash National Park, but this is also under threat from enlargement—from 3 to 42 km^2—due to inflow from irrigation water via surface and groundwater from surrounding cotton fields. None of the Ethiopian lakes have ever had such large flocks of Lesser Flamingos as recorded in some of the lakes in Kenya or Tanzania.

Tanzania contains a large lake—Lake Manyara—inside a National Park of the same name, well established on its northern tourist circuit. Here, flamingo numbers rivalling those of Lake Nakuru were recorded in the 1990s (Birdlife International 2015). Tanzania also has smaller lakes—two in one conservation area and several lakes in two other National Parks—where impressive flocks have been seen. These are Lake Magadi (not to be confused with Lake Magadi in Kenya), which is a shallow pan inside Ngorongoro Crater, and Lake Embaki, a crater lake outside Ngorongoro crater to the north (both inside Ngorongoro Conservation Area). Additional lakes include Lagaja (also known as Ndutu) inside Serengeti National Park and the Momella Lakes, a series of crater lakes of varying salinity within the volcanic landscape of Arusha National Park (which includes Mt Meru, the 5th highest peak in Africa).

In addition to the main Kenyan saline alkaline lake protected as a National Park (Nakuru), another one Lake Bogoria is a National Reserve (managed by the County of Baringo) and yet another one Lake Elmenteita is largely within Soysambu Conservancy, a private conservation charity. The range of Kenya's soda lakes demonstrates what lakes in the other countries may also show in smaller quantities—a full range of ecosystem services, beyond merely bird or wildlife tourism. Bogoria is visited as much for its hot springs, around the western shore of the lake, as for its flamingos; these springs are reputed to be the most visually impressive and extensive in Africa. Both Bogoria and Elmenteita have a spectacular scenery of escarpments and extinct craters. Elmenteita, whose former prehistoric larger lake shore was well above the present one, is adjacent to the important site of stone tools from *Homo erectus* about 0.75 Ma, Kariandusi.

The original inhabitants of the Rift Valley—the Maasai and their cousins—consider some of the EASL as important features in their cultural history, which they seek to maintain today. Maasai groups in northern Tanzania were a major contributor to the opposition to industrialisation proposals for Lake Natron involving soda ash extraction a decade ago, and a subgroup of the Tugen around Lake Bogoria—the Endorois—won a case in the African Commission on Human and People's Rights on 4 February 2010, ruling that the Kenyan government 'violated the Endorois' right as an indigenous people to property, health, culture, religion, and natural resources' when it created the Lake Bogoria National Reserve in 1973 HRW (2010). This represents the first recognition of the 'cultural ecosystem services of a soda lake', even though the facts of the case seem somewhat jumbled because there are 60,000 Endorois but only 24 families were moved in 1973 to outside the new reserve boundary. Moreover, access for the descendants of these family and their livestock herds to the reserve is not restricted.

Lake Magadi, in the south of the country, is in private ownership and extensively exploited on an industrial scale for its deposits of soda ash (called trona) left after evaporation. This represents a high economic value and, if biologically produced, this would be termed a provisioning service. The chemical—sodium sesquicarbonate or trona—is important in dyeing and glass manufacture. Lake Magadi is at low altitude (600 m). This makes it very hot, giving rise to evaporation, which over past millennia has created deposits tens of metres thick over about 100 km^2 of lake. Ironically they have recently been declared at risk due to mixing with silt brought in by streams draining Rift Valley escarpment uplands that have recently been deforested. The company extracting the trona claims that the purity of its product has dropped from 99 to 95 % as a consequence of silt covering some 14 km^2 of trona surface.

Just under a decade ago, the Tanzanian government announced plans to partner with the same company that manages Magadi Soda (Tata Chemicals) to extract trona from Lake Natron, but this proposal was, at least temporarily, shelved after an international campaign. The campaign was led by BirdLife International because of concern over the integrity of the lake. Natron is the only breeding site for the East African population of Lesser Flamingos and the main global breeding site, producing 75 % of the world's Lesser Flamingos (Childress et al. 2008). Their breeding success is dependent on the lake's water level (see also Chaps. 10 and 13). Flamingos breed only under certain lake conditions: they require an intermediate and receding lake level. Flooding of the lake triggers breeding as the lake levels fall, which previously occurred in 2007 and gave rise to a major breeding event (Tebbs et al. 2013a). Another one has most recently occurred (Birdlife International 2013), as the high lake levels in all East African lakes have been declining steadily since 2011–2012 (DM Harper, pers. obs.); Lake Bogoria was declining one vertical metre, September 2014–August 2015.

Provisioning ecosystem services are usually food products; ironically the phytoplanktonic food of Lesser Flamingos, *Arthrospira fusiformis*, is a highly valuable health food in industrial countries and is cultured widely (even at Kisumu in western Kenya, using water from Lake Victoria that is artificially made into saline alkaline for the culture vats (DM Harper, pers. obs.). It is known to have been used as food after drying and formation into cakes around Lake Chad, in Central Africa, but there is no known use of it from EASL.

14.1.6 Temporal and Spatial Variability

Several of the economic values of EASL are dependent upon their constancy, which is almost always assumed by non-scientists (such as most tourists and journalists), but rarely observed in reality. That assumption is based on a general lack of understanding of the ecohydrological processes within the lakes that lead to their most obvious visual character—large flocks of Lesser Flamingos. Several lakes, such as Lake Nakuru, are reported to have once held two million flamingos, but at other times have held almost none. The lake is at the bottom of its intensely farmed catchment with no outlet and also adjacent to the fourth largest city in Kenya, separated only by a wire fence. Unsurprisingly, the absence of flamingos has been seen as a sign of damage to the lake, and pollution is inevitably blamed, without evidence. In October 2006, the BBC news wrote 'Environmental experts are warning that the lake, which is home to millions of flamingos in their natural habitation, may dry up due to constant destruction of catchment areas and massive pollution' and '.... according to a government study, flamingos are now migrating elsewhere due to a scarcity of algae caused by the drop in water levels. The study also revealed that thousands of flamingos died after consuming toxic waste spilling into the lake from the nearby town of Nakuru' (BBC 2006).

Thirty years earlier, Ekkehard Vareschi (Vareschi and Jacobs 1985) had shown that the lake has two stable ecological states. The 'normal' one is when it has its usual water levels and its salinity favours *Arthrospira fusiformis*, which grows in high densities, and when Lesser Flamingos aggregate to feed upon it. In heavy rainfall years, lake level increases, salinity decreases, and *A. fusiformis* is replaced by smaller Cyanobacteria; 'they (Lesser Flamingos) may lack algae of suitable size and sufficient density for filtering and may then migrate to other feeding grounds' (Vareschi and Jacobs, *loc. cit.*, page 419; also Tuite 1979; Vareschi 1978). Much more recently it has become clear that even a hydrologically more stable lake, such as Bogoria (Harper et al. 2003; Schagerl et al. 2015), has periods of oscillating *A. fusiformis* density, causing flamingos to disperse in search of food elsewhere for completely natural reasons (Fig. 14.2; Kaggwa et al. 2012). Even though the causes of *A. fusiformis* decline

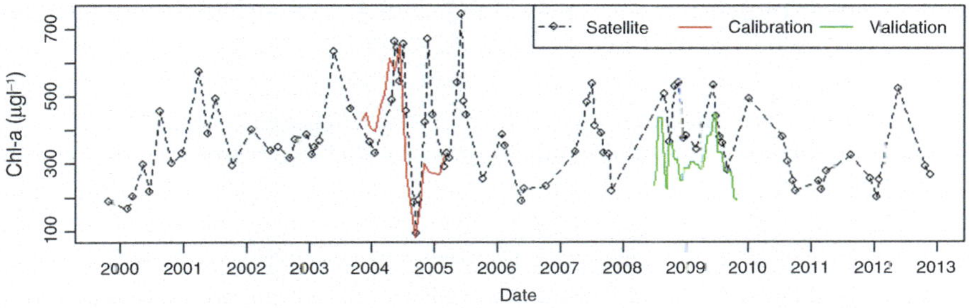

Fig. 14.2 Time series of chlorophyll based on interpretation of Landsat images (*black line*), calibrated using monthly measurements 2003–2005 taken by Odour (*red line*). The agreement between the in situ and satellite data was tested with another set of in situ chlorophyll data from 2008 to 2009 provided by Schagerl (*green line*). There is a good level of agreement between the satellite and in situ chlorophyll series, particularly given that the in situ datasets are based on three point measurements and the satellite data gives the lake wide mean. The *dashed line* between satellite observations is included for visualisation purposes and is not intended to imply a continuous series. For details of methods, see Tebbs (2014)

are not clearly understood, they are being sought within the ecosystem and not outside it (Schagerl and Oduor 2008; Tebbs et al. 2013b).

Needless to say, no evidence for 'thousands of flamingos died after consuming toxic waste spilling into the lake from the nearby town of Nakuru' (BBC 2006) has ever been produced by scientists or government officials. The soda lakes continue to have irregular fluctuations of high and low flamingo numbers, related to rainfall in the shallow lakes (Melack 1976; Tuite 1979; Vareschi 1978) but not in a correlative way because both high levels and low levels in the shallow lakes are unfavourable to *A. fusiformis* and hence to Lesser Flamingos.

14.2 Present Conservation Status and Management Protection of East African Saline Alkaline Lakes

The protected area management of EASL in the three East Africa countries where they usually occur includes most, but not all, of those important for Lesser Flamingo feeding but critically does not include the only breeding lake, Lake Natron, or important temporary feeding lakes such as Lake Eyasi in Tanzania or Lake Logipi in Kenya. We argue that effective conservation management needs to include the entire network of flamingo lakes in order to protect Lesser Flamingos at all stages of their life cycle and to provide the full natural range of feeding lakes, to take account of future climatic fluctuations. Moreover, such management will automatically protect the full range of saline alkaline ecosystem variations and the lower organisms within these types.

14.2.1 International and Regional

Two main international designations provide limited protection for these lakes, which is usually (but not always) enhanced by within-country protected area status. The first, and most recent, is UNESCO (United Nations Educational, Scientific and Cultural Organization) 'World Heritage Status' (WHS), which was given to the three central Kenyan soda lakes—Bogoria, Nakuru and Elmenteita—in 2013. WHS means the sites are recognised by UNESCO as of global importance to all citizens, added to the lists according to the criteria of the Convention on Protection of the World Cultural and Natural Heritage, adopted by UNESCO in 1972, which all three countries are signatories to. Three smaller Tanzanian lakes are already within the WHS—Serengeti National Park and Ngorongoro Conservation Area, together with smaller crater lakes on islands in Kenya's Lake Turkana. The reasons for those designations, however, are large mammal biodiversity in Tanzania's case and fossil remains of early hominids in Turkana's case. Ethiopia does not have any lake sites as or within WHS. Thus, only the three Kenyan lakes are protected because of their flamingos—a rather incomplete protection since it excludes breeding areas. There are several small, isolated soda lakes in volcanic craters in Kenya largely unprotected, but of limited value to Lesser Flamingos because of size; there are also eight in Uganda (Grant 2004). Several are in south-west Uganda, close to or within the Queen Elizabeth Conservation Area, but all have experienced degradation in the past few decades such that Lesser Flamingo densities, once in the tens of thousands, have only been in the hundreds this century.

UNESCO, as the only UN organisation with a remit to promote science as part of its task, has also established a 'Man and the Biosphere' (MAB) programme, which seeks to establish Biosphere Reserves in countries across the world. Tanzania has declared Manyara, Serengeti and Ngorongoro as MAB reserves, but those declared in both Kenya and Ethiopia exclude any EASL.

A few EASL have been designated under a convention initially ratified in the same year, the Ramsar Convention on Wetlands (in 1972). Ethiopia has not yet ratified this convention, although 25 'potential' Ramsar sites have been identified by BirdLife International, ready for such time as it does. Kenya has six sites declared as Ramsar sites, which include Bogoria, Nakuru and Elmenteita (the others are not lakes or are freshwater); Tanzania has four, of which only Natron is a saline alkaline lake. The Ramsar Convention therefore provides only a partial protection for the lake ecosystem type and the East African Lesser Flamingo population because it includes the breeding lake but an inadequate coverage of feeding lakes.

All three countries are, however, signatories to the 1992 Convention on Biological Diversity (CBD). This is important, not just for EASL and Lesser Flamingos but for all living organisms in the lake areas, because diversity is threatened by human actions, largely through habitat loss, over-exploitation and climate change, with an estimated 52 % of biodiversity lost between 1970 and 2010 (WWF 2014). Such losses are likely to decrease the stability and productivity of ecosystems, reducing the efficiency of ecological processes such as resource capture, nutrient cycling and decomposition (Cardinale et al. 2012). Recognition of the commitments to the World Heritage Convention places responsibility on signatories to 'recognize that the duty of ensuring the identification, protection, conservation, presentation and transmission to future generations of the cultural and natural heritage . . .belongs primarily to that State. It will do all it can to this end, to the utmost of its own resources and, where appropriate, with any international assistance and co-operation, in particular, financial, artistic, scientific and technical, which it may be able to obtain' (WHC 1972). The CBD similarly obliges them to establish 'a system of protected areas or areas where special measures need to be taken to conserve biological diversity' (CBD 2014).

The CBD obliges contracting parties (signatories) to identify and monitor the components of biodiversity and any processes that are likely to negatively impact conservation efforts. The Ramsar Convention has developed its advice about monitoring in a far more detailed fashion, such that it would be suitable for CBD as well as for Ramsar reporting [all international conventions have to be reported to by their signatories every 3 years in Conventions of the Parties (CoPs), held in different locations]. Ramsar distinguishes, in particular, between inventory, assessment and monitoring as separate processes. These three conventions, particularly CBD, thus provide international support for the national activities that underpin site protection for saline alkaline lakes.

14.2.2 National and Local

Each country's National Parks (all three countries), National Reserves (Kenya), conservation areas (Tanzania and Uganda) and Ramsar sites (Kenya and Tanzania) have to be reported on against management plans that have a 10- or more usually 5-year cycle. Guidance for writing management plans is provided by many organisations, but the most widely used is the IUCN (Springate-Baginski et al. 2009) and Ramsar (2015).

Under the Ramsar recommended structure, management plans should include a *Preamble*, giving a concise policy statement, and a *Description* consisting of existing data and an identification of any shortfall of relevant information. An *Evaluation* should then confirm the important features of management planning including ecological character, socio-economic, cultural and other factors. Measurable, achievable, non-prescriptive *Management objectives* should thereafter be shown, considering how a feature may change or be influenced. Thresholds for action should be defined, and what would happen if a limit might be exceeded should be shown. *Rationale* should identify and describe the management considered necessary to bring factors under control/maintain site features, and finally an *Action plan* should consider management projects in terms of who, when, where, priority and expenditure, including consideration for visitors and tourism. Annual and short-term reviews, along with major reviews, should be acknowledged for future management monitoring. Most management plans of EASL that are available for consultation do not follow this structure, and overall they are noticeably short of scientific data. An example (Mugo 2007) is reviewed in detail by Kaibos (2013) and Bell (2015). Both analyses show that there is a large gap between the scientific accumulation of data—which are brought together in books such as this one but otherwise are not made available in usable fashion by scientists or to conservation management—and inclusion in management plans for use by protected area management staff. Scientific data are analysed by research scientists for other research scientists to read but are rarely written for management scientists, particularly in the forms that they need to read.

14.2.3 Community Conservation

Involving members of local communities in the monitoring and management of areas of wildlife value (including but not exclusive to protected areas) has recently achieved prominence in 'western' countries as so-called citizen science. The recording of wildlife, especially birds, by amateurs has been common for over 100 years in some countries such as the UK and has recently been promoted, with particular reference to saline alkaline lakes, by the Agreement on the Conservation of African-Eurasian Migratory Waterbirds (AEWA). That intergovernmental treaty for the conservation of migratory waterbirds and their habitats across Africa, Europe, the Middle East, Central Asia, Greenland and the Canadian Archipelago has been signed by all three EASL countries.

This agreement has resulted in annual water bird counts—to a greater or lesser degree of success—in all three countries; many of the contributors to this are volunteers. It is these water bird counts (which include flamingos, even though they are not 'migratory' birds!) that have led to a quantification of the extent to which flamingos move from one lake to the other. The census has been continuous since 1990 on Kenyan lakes and so, when this has occasionally coincided with a census on a Tanzanian lake, it has demonstrated the importance of previously unknown lakes in the chain during the absence of substantial numbers of flamingos from all accessible lakes.

The importance of Lake Logipi, a temporary very shallow lake in Kenya just south of Lake Turkana, has been recognised in only the past few years, when almost no flamingos were censused at the three main lakes combined—Bogoria, Nakuru and Elmenteita—in January 2011 or 2012. This is shortly after nearly 1.5 million were censused at Lake Bogoria alone in 2006. The year 2011 was a year of high rainfall in Kenya, which saw unprecedented lake level rises in all the lakes, flooding roads and Kenya Wildlife Services administrative buildings in Nakuru, for example. This rise was maintained through to 2014, receding in 2015. In these years, although no censuses were taken in Logipi, several photographs were posted on the Internet of tens of thousands of Lesser Flamingos on or over that lake. It changed from a dry lake (e.g. in January 2006) to over 150 km^2 in January 2011 (Tebbs 2014), so its importance as the refuge of the 'missing' flamingos should not be underestimated, particularly since it was earlier identified as a resting site for one of seven satellite-tagged Lesser Flamingos in 2004 (Childress et al. 2007).

14.2.4 Synergy and Communication Between Nations and Between Lakes Management

The few times when censuses have been carried out simultaneously in Kenya and Tanzania (Mlingwa and Baker 2006 summarise dates of Tanzanian counts, which can be compared with Kenyan ones, Owino et al. 2001), interesting data have emerged. There were very few individuals counted on the accessible EASL in both Kenya (total about 300,000) and Tanzania (total about 430,000) in early 2004, supporting earlier circumstantial evidence that had suggested that East African Lesser Flamingos may fly to Botswana and Namibia to breed during periods when the Etosha and the Makgadikgadi salt pans are flooded (Borello et al. 1997; McCulloch and Borello 2000; Simmons 2000). Early 2004 did, however, have about 150 km^2 of water in Logipi, so it is possible that the missing flamingos might have been in shallow lakes like that one and also Natron, which has large lagoonal areas around the central trona (Tebbs et al. 2013a).

There seems to be little communication between the managers in EASL within countries or between them, other than the survey generated by the international waterbird census undertaken in many countries as part of the AEWA. This is a missed opportunity for advancing the conservation of this unique ecosystem type.

14.2.5 Overview of Conservation Gaps

There are three important gaps in the conservation of saline alkaline lakes and their biota, with Lesser Flamingo as the flagship species. The first is continual communication between every lake management in the three countries, particularly where they are protected areas (National Parks, National Reserves/conservation areas, or conservancies). The second is the dissemination of research outputs by individuals or groups of scientists, which presently are published in outlets to the benefit of the individual researcher, in a form that can be understood by all, but particularly by management agency staff. The third is the conservation status, which varies from National Parks with WHS, to nothing for industrial mining lakes. Can we fill these gaps?

14.3 How Might Management for Conservation Be Improved?

14.3.1 Regional Networking

The capacity of mobile phone technologies today and the concept of social networking mean that it is eminently possible for every EASL to be linked in a network which, at its simplest, could be a weekly sharing (and central storage) of photographs. Such an activity would immediately achieve three relatively crude but valuable data sets for every lake if photographs were taken at the same location(s) each time. The sets are water level fluctuations (even without a formal gauge structure), water colour (or even transparency with a stick Secchi disc) and relative abundance of Lesser Flamingos (prior agreement on a location 'representative' of overall lake density). The latter could be accompanied by overall weekly estimates of density. When numbers are high, density estimates are inaccurate, but knowing the difference between 1000, 10,000, 100,000 and 1,000,000 is easy to obtain and valuable to record.

14.3.2 National Communication and Data Sharing

The above suggestion is absolutely dependent upon the site managers and their government masters agreeing to the principle of data sharing and the consequences of that. One might say there are minimal consequences because tourists could do exactly the same thing and post on social networking sites such as Facebook, but such activities depend upon staff time and the (admittedly low) costs of data uploads, both of which require agreement. These activities, moreover, are far more likely to succeed if organised with maximum transparency and staff awareness.

14.3.3 International Partnerships

The second and third identified conservation gaps above can best be filled by international assistance. Ideally, one or a small group of scientists, such as the authors of this book, could all agree to provide a one-page synopsis, in plain English without jargon, to every paper that they have published or that others have published within their subject area. One or a few scientists could agree to establish and manage a website where such information is stored and accessible to all. The website could be a 'Facebook page', or it could be an individual site set up within an existing one, such as a university, or independently within an open-access website.

There are several existing lake database sets, with web pages, but none of them either include all lakes or have the ability to be updated regularly by an outsider to the organisation that runs them. Existing world lake websites appear to be little other than a location database—both the major examples are considerably out of date for EASL—the Global Lake Database (2008) or World Lakes Database (1999). The key value of such a new network would be its continual (minimal monthly) updating so that the persons dealing with each lake management could see how the others are changing. This requires either a very efficient and reliable organisation or a dedicated small team of conservationists, with links in each country to help maintain contacts with each lake's staff.

Research funded by countries outside Africa, mostly European, in partnership with national universities, research institutes and conservation agencies, has been responsible for most of the scientific advances underpinning conservation (where such advances have been applied). A notable exception in which scientific understanding has been not translated into layperson understanding is the 'two stable states' of the Nakuru ecosystem mentioned above. Another international effort, sadly uncoordinated between countries, has investigated the mortality events of Lesser Flamingos, so-called mass mortalities. Mortalities of a few hundred to a few hundred thousand birds have been reported since 1974 (Kock et al. 1999). Several papers in scientific journals in the early part of the twenty-first century have claimed that these mortalities are of 'increasing frequency', others 'frequent and unpredictable', which 'threaten the flamingo populations in the Rift Valley'. The evidence for these claims is somewhat thin. The largest mortality recorded occurred at Lake Bogoria between 2000 and 2001, when the total population at the lake was close to one million birds. A daily mortality, over 2 weeks of study, averaged 700 birds which, if consistent over the period that reserve staff reported the mortality as continuing, meant the death of 200,000 birds. This is approximately one-fifth of the maximum population at the lake over this time (Harper et al. 2003). This number of birds was replaced by breeding based on the numbers of immature and juvenile birds (Fig. 14.3) seen at Bogoria within 3 years (Harper et al. *loc. cit.*), indicating that mortalities are unlikely to cause population decline. There were small mortalities through the first 8 years of this century, but no mortalities have been reported since 2009, suggesting that they are infrequent.

Several causes have been forwarded for these earlier flamingo deaths. The first suggested cause was the pollution of water lakes by pesticides and heavy metals such as cobalt (Co), cadmium (Cd), copper (Cu), lead (Pb), nickel (Ni) and zinc (Zn) (Ochieng et al. 2007). These authors found concentrations exceeding the permissible level in drinking water (WHO) in all three soda lakes (note that comparing alkaline lake water and sediment samples with WHO standards from drinking water is a dubious exercise), but attributed most metals to geological processes,

Fig. 14.3 The three life history stages of Lesser Flamingo can easily be recognised and counted in the field. (**a**) Juvenile aged ca. 6 months with grey bill and legs, *brown* iris and *brown/white* plumage; (**b**) subadult aged 1–3 years with *grey/light red* bill, *grey* legs, *brown/orange* iris and *white/pink* plumage; (**c**) adult ca. 4 years + with *dark red* bill, *pink* legs, vibrant *orange* iris and *pale pink* plumage. This adult is fully sexually mature but its feathers will appear more vivid pink in breeding plumage, as more food has been ingested, enabling birds to prepare for the increased energetic demands of breeding

except for Nakuru. No published study has detected dangerous levels in flamingo tissues, merely suggesting they might occur based on modelling (Nelson et al. 1998).

The second possible cause of the Lesser Flamingo mass mortalities is toxicity of Cyanobacteria. This is still a contentious issue. Some species of Cyanobacteria have been detected as toxin producers, being able to produce toxins such as the neurotoxin detected in flamingo feathers (Metcalf et al. 2013) and the hepatotoxin microcystin-YR found in the stomach, intestine and faecal matter of Lesser Flamingo (Krienitz et al. 2003). Some cyanotoxins such as microcystins and anatoxin-a were isolated from *A. fusiformis* from Bogoria (Ballot et al. 2004). Their ability to cause deaths was never proven, only inferred, and the case has been recently weakened when other researchers showed that *A. fusiformis* and *Arthrospira maxima* were not toxic during a period when several small mortalities occurred (Straubinger-Gansberger et al. 2014).

The most likely direct cause, however, is changes in phytoplankton biomass (quantity and quality of food supply) (Krienitz and Kotut 2010; Ndetei and Muhandiki 2005). Krienitz and Kotut (2010) suggested that the crash of Cyanobacteria (*Arthrospira* spp.) and subsequent replacement with other Cyanobacteria (*Anabaenopsis*) might affect flamingos because the alternative cyanobacterial food has a larger diameter (200–300 μm) than the gaps of lamellae (50 μm); this would clog the filters in their beaks (Fig. 14.4).

Poor quantity and quality food sources leads to malnutrition, which itself leads to immunodeficiency, thus making birds susceptible to microbial infections such as *Mycobacterium avium* (Ballot et al. 2004; Kock et al. 1999). However, during the largest mortality ever recorded in 2000–2001, food levels were moderately high (Harper et al. 2003). Moreover, almost a decade later, Kaggwa et al. (2012) observed the dominance of *Arthrospira* exceeding 80 % when mortalities occurred. This shows that apart from food quantity and quality, other factors are involved in increased flamingo mortalities.

The most recently suggested scenario of food decline leading to flamingo deaths is infection of *Arthrospira* spp. by cyanophages. These can interrupt the food web by removing the primary producer (Peduzzi et al. 2014), by causing its sudden crash. Most flamingos can fly and therefore quickly disperse. The few, weaker/younger/injured birds stay and die of starvation. This latter situation has been observed several times at lakes as weight loss of weakened birds (e.g. Oaks et al. 2006). This may only be proximate: the ultimate cause may be the food available at the site from which the birds came from. The importance of this is indicated by what (little) is known of the 2006 mortalities. Four separate mortalities occurred in that year in three Kenyan lakes—in Nakuru twice, Elmenteita once and Oloidien once. In March 2006, at both Nakuru and Elmenteita, Lesser Flamingo numbers suddenly increased. Elmenteita's went from nearly nothing to 70,000 and Nakuru's from a few thousand to about 300,000. Later in the year (October–November), about 200,000 birds appeared at Oloidien (Harper et al. 2006; Donlan pers. comm.), and a low Nakuru population was once again swollen by at least 200,000 new arrivals (R Ngete, pers. comm. to DM Harper). The unexpected increase in numbers of healthy birds was accompanied by deaths—in each lake for about 2 weeks—of a small fraction, about 2–10 % of the numbers. It is not known where the birds came from but, in 2006, Bogoria was the only lake with good food supplies until about March, when they declined for the rest of the year; Nakuru had good food supplies only in the latter part of the year (Tebbs et al. 2015). It is thus likely that those birds that arrived at Nakuru, Elmenteita and Oloidien were already weakened by inadequate nutrition at their original lake(s), possibly Bogoria. According to this interpretation, they left as food was declining, then arrived, having expended energy and with weakened immune systems and were susceptible to the diseases that many individual birds carry but which a healthy bird can resist.

Abundant evidence from the early mortalities in the 1970s, which were investigated by veterinarians, and in some of the 2005–2006

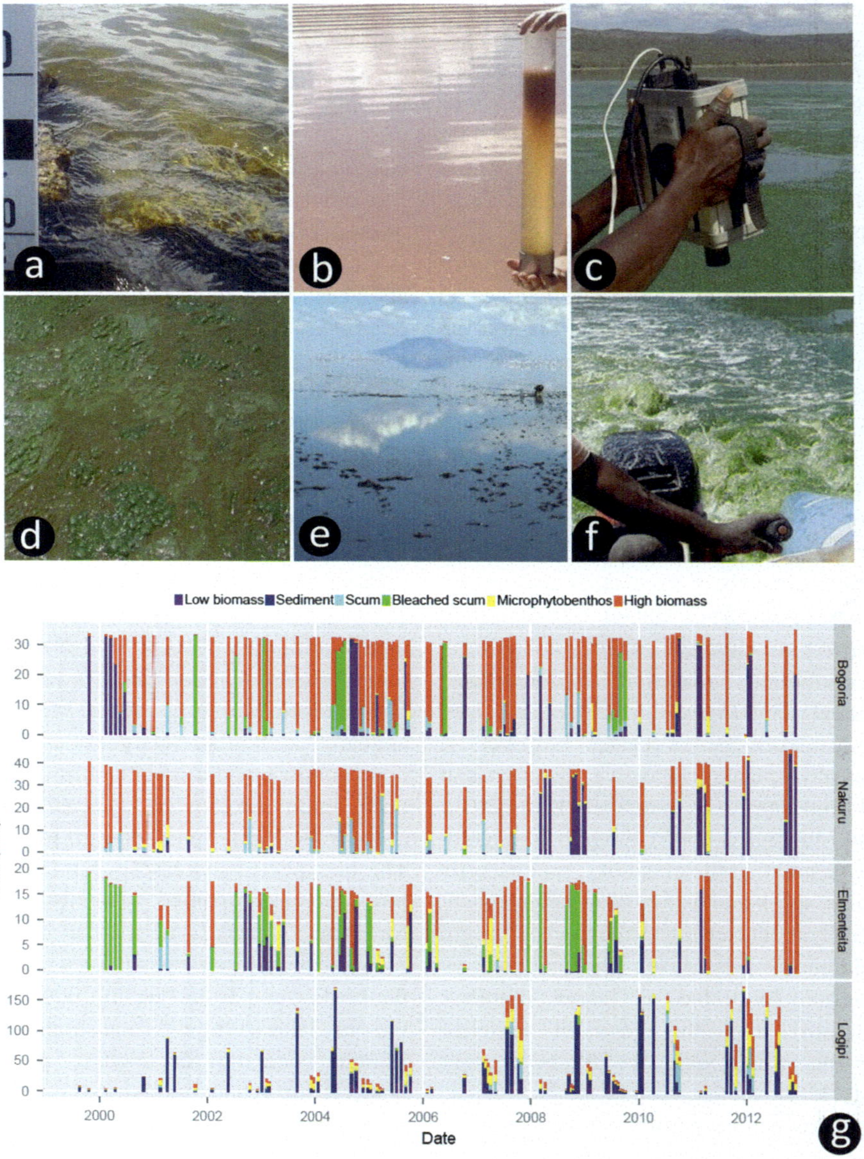

Fig. 14.4 Time series of lake ecological state (and Lesser Flamingo food value) for lakes Bogoria, Nakuru, Elmenteita and Logipi, showing, for each time step, the area of the lake surface belonging to each of six classes: (**a**) low biomass, Chl-a < 100 μgL^{-1} (low food value); (**b**) sediment (no food value); (**c**) scum, Chl-a > 800 μgL^{-1} (low food value); (**d**) bleached scum (no food value); (**e**) microphytobenthos (moderate food value); and (**f**) high biomass, $100 <$ Chl-a < 800 μgL^{-1} (high food value). The lake classes were assigned using the measured reflectance spectra from Landsat satellite imagery. (**g**) Lake ecological states for Bogoria, Nakuru, Elmenteita and Logipi. For details see Tebbs et al. (2015) and Tebbs (2014)

mortalities, investigated by the late Lindsay Oaks (Oaks et al. 2006), suggested that infectious microbial diseases could be endemic in the population and that they spread rapidly under any conditions where immune competence was weakened. Such conditions could be malnutrition and injuries caused by collisions with bushes and rocks (in 2005 such injuries accounted for 1 % of the population at Lake Bogoria, compromising their ability to fly, Oaks et al. *loc cit.*). Mortality was low at Lake Bogoria in 2005, not greater than the likely baseline figure for a species with a 50-year lifespan and sometimes large populations at single sites (Harper et al. 2003). Those birds that were examined, however, were thin and light; with an average body mass of 1202 g (range 1000–1280) compared to a normal average mass of 1780 for males and 1525 for females (Childress et al. 2006). They also had gross post-mortem lesions characteristic of disseminated infection by *Mycobacterium avium* (avian tuberculosis), subsequently confirmed by histopathology, demonstrating granulomatous inflammation with intralesional acid-fast bacteria. The bacteria were identified as *M. avium* by culture, by polymerase chain reaction specific for *M. avium* and by 16s rDNA sequencing (Oaks et al. *loc. cit.*). Pulsed-field gel electrophoresis showed that all five isolates were the same strain, strongly suggesting transmission between flamingos. Finding this disease at low levels in an otherwise healthy population suggests that *M. avium* infections may be endemic throughout the Lesser Flamingo population in East Africa.

The somewhat fragmented data available thus allow a working hypothesis to be formulated which can be tested by more careful, future recording of mortalities as and when they occur. It is that mortalities of Lesser Flamingos occur when there are environmental stressors, some observed by scientists, some not. Stressors are one or a combination of the following: (a) weakness due to lack of food in a previous lake, compounded by energy expended moving to a new lake, (b) injuries incurred when landing on rocks/bushes, (c) poor food quality if less than pure *A. fusiformis* is available (e.g. other Cyanobacteria causing clogging or containing toxins), (d) drought reducing available drinking water sources to semi-saline hot springs with potential toxins (Bogoria) or polluted sewage-treatment ponds (Nakuru), (e) overcrowding (flamingo density can be 2 m^{-2} (Robinson 2015) in shallow lake edges), where every individual obtains food from the same water that it defecates into, promoting disease spread and (f) other immunosuppressive factors such as viral disease or possibly toxins from *A fusiformis* strains or other Cyanobacteria.

Note that Lesser Flamingos carry the same genera of microbial diseases that humans do. These are endemic in healthy populations but rapidly spread to mortality epidemics under unfavourable conditions—crowding and where water supply (humans) or food (flamingos) is polluted, or which are lethal when individuals become immune-compromised (e.g. humans with AIDS; flamingos starved or injured). Typhoid, cholera, pneumonia, influenza, tuberculosis and aspergillus have all been found in post-mortems on birds, in most cases the probable cause of mortality (Oaks et al. 2006; Oaks unpublished).

14.4 Concluding Remarks

The entire book has celebrated the uniqueness of EASL; this chapter has used their conservation value and status to assess their vulnerability and look into the future. The lake network has existed for hundreds of thousands of years, expanding and contracting with climate shifts. The most recent severe decline in the network was about 125,000 years ago, when genetic evidence for Lesser Flamingos and by implication their habitat lakes shows severe reduction in numbers at a time of widespread aridity across Africa and Arabia (Parasharya et al. 2015), followed by rapid expansion from this bottleneck. We do not know the fluctuations since that time, although we can make speculations based on the evidence for climate shifts over the past few thousand years from the analysis of lake sediments (Verschuren et al. 2000) assuming that the periods of aridity shown resulted in a smaller lake network and a smaller population of Lesser Flamingos. The last arid spell, when Lake Naivasha, a 100–150 km^2 freshwater lake today, was dry, was in the mid-1800s (Verschuren, loc. cit.). Perhaps the lake network and numbers of Lesser Flamingos were both reduced at this time and have been growing since then: the high numbers (2–4 million individuals) speculated on by ornithologists in the middle of the last century and the current estimates (1–1.5 million) being natural fluctuations about a mean following fluctuations in lake area and food quantity, not dissimilar to those of wildebeest (*Connochaetes taurinus* Burchell) and as predicted by population theory for any R-selected species about its carrying capacity. The largest population of wildebeest in Africa migrates between the Serengeti National Park (Tanzania) and the Maasai Mara National Reserve (Kenya). Its population numbers have fluctuated between about 400,000 and 1.6 million since the mid-1960s, with major causes of mortality being starvation and diseases and a model based upon rainfall influencing grass availability accurately hindcasting census data over that period (Gereta and Wolanski 2008).

The evidence which we have of current dynamics in the lake network suggests important lakes like Nakuru drying once a decade approximately and over-filling with about the same frequency, both conditions unfavourable to flamingos. The much more limited evidence suggests Lesser Flamingo breeding at Lake Natron is very dependent upon hydrological dynamics. Thus, it is not perhaps unnatural for there to be population fluctuations, and our concern should be not to explain every single movement event but to evaluate continuously whether the fluctuations in lake area and flamingo numbers are exceeding natural fluctuation limits, once those have been agreed. Extensive but simple communication between lakes, along the lines suggested above, is the simplest way of achieving this and the first step in a truly integrated, multinational management of soda lakes and their unique inhabitants.

Acknowledgements This chapter has been made possible by funding from various donors over the past 15 years: the Earthwatch Institute, 2000–2007; the Darwin Initiative, 2003–2010; the British Council, 2010–2011; the University of Leicester graduate scholarship to EJT, 2008–2011, and the University of Leicester undergraduate and graduate students on field courses, 2004–2014. The Kenyan Government provided research permission through Permit NCST 5/002/R/020-D and its predecessors, supported by Professor Kenneth Mavuti who always provided endless intellectual support.

We dedicate this chapter to members of our team who died (of natural causes) too early: Dr. Brooks Childress, DMH's Ph.D. student after a successful business career, who initiated and led East African satellite tracking of Lesser Flamingos and Dr Lindsay Oaks, avian veterinarian at Washington State University, who joined our team in 2004 after he had successfully identified the cause of vulture mortalities in India and Pakistan. And Prof. Ekkehard Vareschi and his wife Angelika, who started research on soda lakes at Nakuru in the 1970s and were about to begin coordinated parallel work in Tanzania to ours in Kenya, were tragically killed by a falling tree in Arusha National Park. Lastly, we also remember fondly the logistical support provided to our research camps by the late Mrs. Jill Simpson and Mrs. Velia Carn, the latter always helped by the late John Kaba.

References

Ballot A, Krienitz L, Kotut K, Wiegand C, Metcalf JS, Codd GA, Pflugmacher S (2004) Cyanobacteria and cyanobacterial toxins in three alkaline Rift Valley lakes of Kenya – Lakes Bogoria, Nakuru and Elementeita. J Plankton Res 26:925–935

BBC (2006) Famous flamingos leave Kenya lake. http://news.bbc.co.uk/2/hi/africa/5405468.stm. Accessed 13 Nov 2015

Bell O (2015) Conservation management at Lake Bogoria National Reserve, Kenya. Bachelor thesis, University of Leicester

Bennun LA, Nasirwa O (2000) Trends in waterbird numbers in southern Rift Valley of Kenya. Ostrich 71:220–226

Bergner AGN, Strecker MR, Trauth MH, Deino A, Gasse F, Blisniuk P, Dühnforth M (2009) Tectonic and climatic control on evolution of rift lakes in the Central Kenya Rift, East Africa. Q Sci Rev 28 (25–26):2804–2816

Birdlife International (2013) Successful Flamingo breeding reported at Lake Natron! www.birdlife.org/africa/news/successful-flamingo-breeding-reported-lake-natron. Accessed 13 Nov 2015

Birdlife International (2015) Lake Manyara datasheet. http://www.birdlife.org/datazone/sitefactsheet.php?id=6967. Accessed 13 Nov 2015

Borello WD, Mundy PJ, Liversedge TN (1997) Movements of greater and lesser flamingos in southern Africa. In: Leshem E, Lachman E, Bertold P (eds) Migrating birds know no boundaries. Society for Protection of Nature in Israel, Tel Aviv, pp 201–218

Cardinale BJ, Duffy JE, Gonzalez A, Hooper DU, Perrings C, Venail P, Narwani A, Mace GM, Tilman D, Wardle DA, Kinzig AP, Daily GC, Loreau M, Grace JB, Larigauderie A, Srivastava DS,

Naeem DS (2012) Biodiversity loss and its impact on humanity. Nature 486:59–68

Childress B, Harper DM, Hughes B, Ferris C (2006) Adaptive benefits of differential post-fledging development patterns in the lesser flamingo (*Phoenicopterus minor*). Ostrich 77:84–89

Childress B, Hughes B, Harper D, van den Bossche W (2007) East African flyway and key site network of the lesser flamingo (*Phoenicopterus minor*) documented through satellite tracking. Ostrich 78:463–468

Childress B, Nagy S, Hughes BC (2008) International single species action plan for the conservation of the lesser flamingo, *Phoeniconaias minor*. CMS Technical Series No. 18, AEWA Technical Series No. 34, Bonn, Germany

Convention on Biological Diversity (CBD) (2014) Convention text article 2: use of terms. http://www.cbd.int/convention/articles/default.shtml?a=cbd-02. Accessed 19 Jan 2014

de Menocal PB (1995) Plio-Pleistocene African climate. Science 270:53–59

Gereta E, Wolanski E (2008) Ecohydrology driving a tropical Savannah system. In: Harper DM, Zalewski M, Pacini N (eds) Ecohydrology: processes, models and case studies. CABI International, Wallingford, pp 171–186

Global Lake Database (2008) LakeNet. http://www.worldlakes.org/. Accessed 13 Nov 2015

Grant WD (2004) Half a lifetime in soda lakes. In: Ventosa A (ed) Halophilic microorganisms. Springer, Berlin, p 350

Harper D, Childress B, Harper M, Boar R, Hickley P, Mills S, Otieno N, Drane T, Vareschi E, Nasirwa O, Mwatha W, Darlington J, Escute-Gasulla X (2003) Aquatic biodiversity and saline lakes: lake Bogoria National Reserve, Kenya. Hydrobiologia 500:259–279

Harper DM, Muchane M, Kimani DK, Mwinami T (2006) Thousands of lesser flamingos at lake Naivasha? Scopus 26:8–10

HRW (2010) Kenya: landmark ruling on indigenous land rights. https://www.hrw.org/news/2010/02/04/kenya-landmark-ruling-indigenous-land-rights. Accessed 13 Nov 2015

Kaggwa MN, Gruber M, Oduor SO, Schagerl M (2012) A detailed time series assessment of the diet of lesser flamingos: further explanation for their itinerant behaviour. Hydrobiologia 710:83–92

Kaibos DM (2013) Factors influencing the performance of national reserves: a case of lake Bogoria National Reserve, Kenya. Master thesis, University of Nairobi

Kock ND, Kock RA, Wambua J, Kamau GJ, Mohan J (1999) *Mycobacterium avium*-related epizootic in free-ranging lesser flamingos in Kenya. J Wildl Dis 35:297–300

Krienitz L, Kotut K (2010) Fluctuating algal food populations and the occurrence of lesser flamingos (*Phoeniconaias minor*) in three Kenyan Rift Valley lakes. J Phycol 46:1088–1096

Krienitz L, Ballot A, Kotut K, Wiegand C, Putz S, Metcalf JS, Codd GA, Pflugmacher S (2003) Contribution of hot spring cyanobacteria to the mysterious deaths of lesser flamingos at Lake Bogoria, Kenya. FEMS Microbiol Ecol 43:141–148

Mason IM, Guzkowska MAJ, Rapley CG (1994) The response of lake levels and areas to climatic change. Clim Change 27:161–197

McCulloch G, Borello W (2000) The importance of the Makgadikgadi salt pans in Botswana for flamingos in Africa. In: Baldassare GA, Arengo F, Bildstein KL (eds) Conservation biology of flamingos, Waterbirds, vol 23(Spec Publ 1), pp 64–68

Melack JM (1976) Limnology and dynamics of phytoplankton in equatorial African Lakes. Dissertation, Duke University North Carolina

Metcalf JS, Banack SA, Kotut K et al (2013) Amino acid neurotoxins in feathers of the lesser flamingo, *Phoeniconaias minor*. Chemosphere 90:835–839

Mlingwa C, Baker N (2006) Lesser flamingo *Phoenicopterus minor* counts in Tanzanian soda lakes: implications for conservation. In: Boere GC, Galbraith CA, Stroud DA (eds) Waterbirds around the world. The Stationery Office, Edinburgh, pp 230–233

Mugo K (2007) Lake Bogoria national reserve: world ramsar site (No. 1057): integrated management plan 2007–2012. WWF, pp 48. http://assets.panda.org/downloads/lake_bogoria_management_plan.pdf. Accessed 13 Nov 2015

Nasirwa O, Bennun LA (1994) Waterbirds in the southern Kenyan Rift Valley, July 1993 and January 1994. The National Museums of Kenya, Nairobi

National Geographic (2013) Unusual pictures: "Calcified" Birds, Bats Found at African Lake. http://news.nationalgeographic.com/news/2013/10/pictures/131003-calcified-birds-bats-africa-lake-natron-tanzania-animals-science/. Accessed 13 Nov 2015

Ndetei R, Muhandiki VS (2005) Mortalities of lesser flamingos in Kenyan Rift Valley saline lakes and the implications for sustainable management of the lakes. Lake Reserv Manag 10:51–58

Nelson Y, Thampy R, Motelin GK, Raini JA, DiSante C, Lion L (1998) Model for trace metal exposure in filter-feeding flamingos at alkaline Rift Valley lakes, Kenya. Environ Toxicol Chem 17:2302–2309

Oaks JL, Walsh T, Bradway D, Davis M, Harper DM (2006) Septic Arthritis and disseminated infections caused by *Mycobacterium avium* in lesser flamingos, Lake Bogoria, Kenya. In: Childress B, Béchet A, Arengo F, Jarrett N (eds) Flamingo, bulletin of the IUCN-SSC/Wetlands international flamingo specialist group, vol 14. Wildfowl and Wetlands Trust, Slimbridge, pp 30–31

Ochieng EZ, Lalah JO, Wandiga SO (2007) Analysis of heavy metals in water and surface sediment in five rift valley lakes in Kenya for assessment of recent increase in anthropogenic activities. Bull Environ Contam Toxicol 79:570–576

Owino A, Oyugi JO, Nasirwa OO, Bennun LA (2001) Patterns of variation in waterbird numbers on four rift valley lakes in Kenya. Hydrobiologia 458:45–53

Parasharya BM, Rank DN, Harper DM, Crosa G, Zaccara S, Patel N, Joshi CG (2015) Long-distance dispersal capability of lesser flamingo *Phoeniconaias minor* between India and Africa: genetic inferences for future conservation plans. Ostrich 86:221–229

Peduzzi P, Gruber M, Gruber M, Schagerl M (2014) The virus/'s tooth: cyanophages affect an African flamingo population in a bottom-up cascade. ISME J 8:1346–1351

Ramsar (2015) Ramsar handbooks. http://www.ramsar.org/resources/ramsar-handbooks. Accessed 13 Nov 2015

Remmert H (1980) Ecology: a textbook. Springer, Berlin

Richardson JL, Richardson AE (1972) History of an African rift lake and its climatic implications. Ecol Monogr 72:499–534

Robinson VJ (2015) The ecology of East African soda lakes: implications for lesser flamingo (Phoeniconaias minor) feeding behaviour. Dissertation, University of Leicester

Schagerl M, Oduor SO (2008) Phytoplankton community relationship to environmental variables in three Kenyan rift valley saline-alkaline lakes. Marine Freshw Res 59:125–136

Schagerl M, Burian A, Gruber-Dorninger M, Oduor SO, Kaggwa MN (2015) Phytoplankton succession of African soda lakes with a special focus on *Arthrospira fusiformis*. Fottea 15:245–257

Simmons RE (2000) Declines and movements of lesser flamingos in Africa. Waterbirds 23:40–46

Smol JP, Stoermer EF (2010) The diatoms. Applications for the environmental and earth sciences, 2nd edn. Cambridge University Press, Cambridge

Springate-Baginski O, Allen D, Darwell WRT (2009) An integrated wetland assessment toolkit: a guide to good practice. IUCN, Gland, xv + pp 144

Straubinger-Gansberger N, Gruber M, Kaggwa MN, Lawton L, Odour SO, Schagerl M (2014) Sudden flamingo deaths in Kenyan Rift Valley lakes. Wildlife Biol 20:185–189

Talling JF, Talling IB (1965) The chemical composition of African lake waters. Int Rev Ges Hydrobiol 50:421–463

Talling JF, Wood RB, Prosser MV, Baxter RM (1973) The upper limit of photosynthetic productivity by phytoplankton: evidence from Ethiopian soda lakes. Freshw Biol 3:53–76

Tebbs EJ (2014) Remote sensing for the study of ecohydrology in East African Rift lakes. Dissertation, University of Leicester

Tebbs E, Remedios J, Avery S, Harper D (2013a) Remote sensing the hydrological variability of Tanzania's Lake Natron, a vital lesser flamingo breeding site under threat. Ecohydrol Hydrobiol 13(2):148–158

Tebbs E, Remedios J, Harper D (2013b) Remote sensing of chlorophyll-a as a measure of cyanobacterial biomass in Lake Bogoria, a hypertrophic, saline-alkaline, flamingo lake, using Landsat ETM+. Remote Sens Environ 135:92–106

Tebbs EJ, Remedios JJ, Avery ST, Rowland CS, Harper DM (2015) Regional assessment of lake ecological states using Landsat: a classification scheme for alkaline–saline, flamingo lakes in the East African Rift Valley. Int J Appl Earth Obs Geoinf 40:100–108

Trauth MH, Maslin MA, Deino A, Strecker MR (2005) Late Cenozoic moisture history of East Africa. Science 309:2051–2053

Trauth MH, Maslin MA, Deino AL, Junginger A, Lesoloyia M, Odada EO, Olago DO, Olaka LA, Strecker MR, Tiedemann R (2010) Human evolution in a variable environment: the amplifier lakes of Eastern Africa. Quat Sci Rev 29:2981–2988

Tuite CH (1979) Population size, distribution and biomass density of the lesser flamingo in the Eastern Rift Valley, 1974–1976. J Appl Ecol 16:765–775

Vareschi E (1978) The ecology of lake Nakuru (Kenya) I Abundance and feeding of the lesser flamingo. Oecologia 32:11–35

Vareschi E, Jacobs J (1985) The ecology of Lake Nakuru VI Synopsis of production and energy flow. Oecologia 65:412–424

Verschuren D, Laird KR, Cumming BF (2000) Rainfall and drought in equatorial east Africa during the past 1,100 years. Nature 403:410–414

Verschuren D, Cumming BF, Laird KR (2004) Quantitative reconstruction of past salinity variations in African lakes: assessment of chironomid-based inference models (Insecta: Diptera) in space and time. Can J Fish Aquat Sci 61:986–998

WHC (1972) The world heritage convention. http://whc.unesco.org/en/convention/. Accessed 13 Nov 2015

World Lakes Database (1999) International lake environment committee foundation. http://wldb.ilec.or.jp/. Accessed 13 Nov 2015

WWF (2014) Living planet report 2014: species and spaces, people and places. In: McLellan R, Iyengar L, Jeffries B, Oerlemans N (eds). WWF, Gland

Soda Lakes of the East African Rift System: The Past, the Present and the Future

15

Steve Omondi Oduor and Kiplagat Kotut

Abstract

The soda lakes of the East African Rift System are unique water bodies whose formation, topography and locational settings have conferred upon them endorheic basin features with arid to semiarid climatic conditions. These features, together with the geological characteristics of their catchments, have favoured the development of saline alkaline properties that make them peculiar and important from a number of perspectives. Most of these lakes have extreme environmental characteristics shown by high ionic contents of their water, high temperature conditions and eutrophic states. This enables them to support growth of few tolerant species of extremophilic Cyanobacteria and other microbes. Some of these lakes such as the Lakes Nakuru and Bogoria are highly productive, yielding some of the highest primary production rates in the world. This high productivity supports hundreds of thousands to, sometimes, millions of Lesser Flamingos in these two lakes. These lakes, however, are highly vulnerable to climate changes and environmental degradation associated with human activities due to their unique environmental conditions. This paper explores some of these unique features, the threats to their existence, the challenges to their conservation and potential strategies to protect them.

15.1 Introduction

The East African Rift System (EARS) is a spectacular landmass that runs from the Red Sea and the Gulf of Aden through Ethiopia, Kenya and Tanzania to Mozambique, covering a distance of over 2200 km (Olaka et al. 2010). The system is one of the major lake regions of the world with a number of lakes whose formation is associated

S.O. Oduor (✉)
Department of Biological Sciences, Egerton University, P.O. Box 536-20115, Egerton, Kenya
e-mail: oduorsomondi@yahoo.co.uk

K. Kotut
Department of Biological Sciences, Embu University College, P.O. Box 6-60100, Embu, Kenya
e-mail: kkotut@gmail.com

M. Schagerl (ed.), *Soda Lakes of East Africa*, DOI 10.1007/978-3-319-28622-8_15

with tectonics and volcanism (Burgis and Morris 1987; Rosendahl 1987). The nature of their formation has played a key role in determining their morphological, hydrological and chemical properties (Schlüter 1997). The topography of the EARS and the surrounding regions, mainly the Ethiopian and the Kenyan Highlands, has created a unique climate characterized by semiarid to arid conditions within the valley. This is because the highlands deflect the north-east monsoonal wind flow southwards along the Somali Coast and the south-east flow northwards, moderating the climate here (Olaka et al. 2010). This interception of monsoonal rainfall by these highlands, together with the high temperatures, results in a strong moisture deficit in the Rift Valley floor, especially in the lakes' surroundings.

The lakes in the EARS vary widely in size and salinity, ranging from freshwater to hypersaline states (see Appendix). While many of them are relatively small and shallow, there are several large ones such as the Lakes Turkana (6400 km^2) in Kenya, Eyasi (about 450 km^2) and Natron (about 160 km^2) in Tanzania. The wide changes in area and volume exhibited by these lakes are associated with the rainfall and runoff patterns within their catchments. Seasonal changes in volume are usually accompanied by wide variations in alkalinity, conductivity, pH and other physico-chemical conditions, which in turn contribute to temporal variations in the biotic composition, abundance and productivity (Schagerl and Oduor 2008).

The East African soda lakes (EASL) are famous for their wealth of wildlife as well as the exceptional beauty of their natural scenery. This makes them important tourist areas that contribute significantly to the economy of the countries in which they are located. Globally, the EASL are recognized as major habitats for waterbirds, and a number of them have been classified as wetlands of international importance, the Ramsar sites, and UNESCO World Heritage sites. In Kenya, out of the six Ramsar sites identified, five are lakes within the EARS (Baringo, Bogoria, Elmentaita, Naivasha and Nakuru). The Lakes Nakuru, Elmentaita and Bogoria are also World Heritage Sites. Ramsar sites in Tanzania include Lake Natron. Amongst the waterbirds, the Lesser Flamingo *Phoeniconaias minor* Geoffroy Saint-Hilaire (Syn. *Phoenicopterus minor* Geoffroy Saint-Hilaire) is a characteristic bird of these lakes (see Chaps. 10 and 13). Its dense population has been rated as one of the most fascinating wildlife spectacles of the world (Mari and Collar 2000; Owino et al. 2001). The highest population densities occur in Kenya (about 1.5 million individuals) and Tanzania (about 600,000 individuals) (Childress et al. 2007). It is estimated that at times, over 50 % of the global flamingo population inhabits Lakes Bogoria and Nakuru (Owino et al. 2001; Vareschi 1978). Hence, these lakes hold the key to the conservation of Lesser Flamingos as well as other waterfowl (Childress et al. 2007; Owino et al. 2001).

The lakes, however, face a myriad of problems mainly associated with changes in their catchment's ecology, which alter their physico-chemical characteristics and biological community structure. Furthermore, changes in climate, resulting in major fluctuations in water level, are complicating their ecology as shown by the recent emigration of flamingos from Lake Nakuru in Kenya since 2012.

15.1.1 Soda Lake Formation

EASL originated following the splitting of the African Plate through tectonics coupled with volcanic eruptions that created a series of depressions (see Chap. 2). Most of these depressions have a small surface area (about 100–200 km^2), are shallow ($<$10 m) and exhibit a closed basin type (Ebinger et al. 1993); with the neighbouring highlands and escarpments deflecting the rain-bearing monsoonal winds away thereby causing low rainfall (Olaka et al. 2010).

The characteristics of soda lakes reflect a combination of features that include climate, geology and geography (Williams 2002). The high temperature and low rainfall, coupled with the endorheic nature of these lakes, make evaporation the dominant process of water loss. This promotes salt accumulation and salt precipitation to form crusts, especially near the lake edges. The geology of the catchments influences the types of ions that enter the lakes (mostly sodium, carbonate and bicarbonate), and their salinity and alkalinity therefore vary greatly (Beadle 1932; Baumgarte 2003; Holmes 1978; Millbrink 1977; Talling 2011; Williams 2002).

There are about 58 lakes identified in the EARS, most of which are located in Ethiopia (32 lakes), Kenya (12) and Tanzania (13). Some are freshwater systems and others are saline or alkaline, whereas others are not yet well studied and hence their limnological characteristics are poorly known. Those that are known to be saline or alkaline are listed in Table 15.1.

Over time, the East African soda lakes (EASL) have undergone substantial changes, with some having decreased in size, while others have fragmented into smaller bodies. For example, the area presently occupied by the Lakes Nakuru, Elmentaita and Naivasha was previously occupied by one large lake around 10,000 years ago that shrunk to form the three lakes (Harper et al. 1990; Richardson and Richardson 1972). The diatomite deposits mined around Elmentaita provide evidence for the remains of diatoms from the land area that was previously part of this large lake.

15.1.2 The Biota of Soda Lakes

The biology of some of the EASL has been extensively studied in the last few decades (Ballot et al. 2004, 2009; Dadheech et al. 2014; Krienitz et al. 2013; Melack 1988; Odada et al. 2003; Oduor and Schagerl 2007a; Schagerl and Oduor 2008; Vareschi 1978, 1982). Due to their high alkalinity, hypereutrophic state, and high temperatures amongst other peculiar characteristics, these lakes support a dense and

Table 15.1 The main saline alkaline lakes in the East African Rift Valley system

Country	Name of lake	Latitude	Longitude	Lake type
Djibouti	Assal	11.658961	42.406998	Saline
Ethiopia	Afrera (Giulietti)	13.255318	40.899925	Saline
Ethiopia	Gemeri	11.532507	41.664848	Probably saline
Ethiopia	Afambo	11.416091	41.682701	Saline
Ethiopia	Abbe	11.187832	41.784325	Saline
Ethiopia	Arenguade	8.695324	38.976388	Saline
Ethiopia	Abijatta	7.612998	38.597603	Saline
Ethiopia	Langano	7.594621	38.754845	Saline
Ethiopia	Shala	7.470730	38.522758	Saline
Ethiopia	Chitu	7.404516	38.420191	Saline
Ethiopia	Abaya	6.311204	37.847671	Saline
Ethiopia	Chamo	5.840081	37.560654	Saline
Kenya	Turkana	3.709412	36.031036	Saline
Kenya	Logipi	2.229319	36.554432	Saline
Kenya	Bogoria	0.252084	36.101317	Saline
Kenya	Solai	0.058794	36.148782	Alkaline
Kenya	Nakuru	−0.359113	36.092434	Saline
Kenya	Elementaita	−0.440135	36.242466	Saline
Kenya	Sonachi	−0.782666	36.261992	Saline
Kenya	Oloidien	−0.813648	36.276455	Saline
Kenya	Magadi	−1.923105	36.263266	Saline
Tanzania	Natron	−2.386090	36.011753	Saline
Tanzania	Magadi	−3.194250	35.538440	Alkaline
Tanzania	Eyasi	−3.599772	35.122490	Saline
Tanzania	Manyara	−3.601142	35.805588	Saline
Tanzania	Burunge	−3.884807	35.880661	Saline

diverse population of Bacteria and Archaea (Duckworth et al. 1996; Grant et al. 1990, 1999; Grant 2006; Jones et al. 1998; Willén 2011; see also Chap. 5). Earlier studies on the phytoplankton and zooplankton communities of these lakes revealed low diversity but high biomass, especially of phytoplankton, dominated by the cyanobacterium *Arthrospira. fusiformis* (Ballot et al. 2004; Harper et al. 2003; Oduor and Schagerl 2007a; Willén 2011; Schagerl et al. 2015). Recent investigations using molecular techniques, however, have revealed the existence of a high diversity of monadoid and coccoid green algae (Luo et al. 2012). Krienitz et al. (2012) report frequent changes in the phytoplankton community structure in which the dominant populations of *A. fusiformis* are usually replaced at irregular intervals partly by populations of the nostocalean *Anabaenopsis* sp. or the picoplanktonic chlorophyte *Picocystis salinarum* Lewin, amongst other Cyanobacteria or eukaryotic algae. These changes are driven by the changes in the physico-chemical variables in these lakes (Ballot et al. 2005, 2009; Harper et al. 2003; Krienitz and Kotut 2010; Oduor and Schagerl 2007a, b; Schagerl and Oduor 2008).

15.1.3 Primary Productivity of Soda Lakes

A combination of high solar radiation in the tropics, adequate nutrient supply and unlimited supply of inorganic carbon from their rich carbonate waters (Melack and Kilham 1974; Oduor and Schagerl 2007b; Talling 1965) makes the EASL highly productive with high rates of photosynthesis. They are amongst the most productive ecosystems in the world, with primary production rates that sometimes exceed 10 g C m $^{-2}$ day^{-1} as commonly recorded in the Lakes Nakuru, Bogoria in Kenya and Lake Abijatta in Ethiopia (Melack 1981; Melack and Kilham 1974; Oduor and Schagerl 2007b; Talling et al. 1973). This high productivity is the cornerstone of the food supply to the hundreds of thousands of Lesser Flamingos found in Lake Bogoria and previously in Lake Nakuru too. The recent heavy rains and flooding of Lake Nakuru, however, have caused a major ecological change in the lake that has altered water chemistry and reduced both phytoplankton biomass and primary productivity.

15.2 Anthropogenic Threats to the Future of Soda Lakes

15.2.1 Ecosystem Degradation

The EASL, like other lakes of the world, have in recent decades been subjected to ecosystem degradation that has been linked to human activities (Jellison et al. 2008) and, to some extent, natural forces. Some lakes such as Lake Nakuru are exposed to pollution from industries, factories, domestic effluents and farming activities (Gichuhi 2013; Odada et al. 2006). Major pollutants include nutrients washed from neighbouring farms into the lakes, solid wastes and heavy metals such as lead that has been identified in Lake Nakuru (Kairu 1996; Odada et al. 2006; Raini 2009). The pollution is more severe in lakes close to human settlements. Major indications that the environment of soda lakes is stressed can be traced back to the early 1990s following the widespread flamingo mortalities reported in a number of EASL (Ballot et al. 2004; Kairu 1996; Kock et al. 1999; Krienitz et al. 2003; Krienitz and Kotut 2010; Nelson et al. 1998). The closed hydrological nature of these lakes means that the impact of pollution is much more severe compared to exorrheic lakes which are mostly freshwater lakes: they accumulate and biomagnify many pollutants to a much greater degree, while the high salinity may also modify the toxicity of certain pollutants (Williams 1981).

15.2.2 Exploitation of Natural Resources

Soda ash is presently being mined in some of the EASL such as Lakes Magadi and Abijatta (see Chap. 13 for details). This mining involves infrastructure development and the use of machinery, which physically damages the lake habitats and releases various pollutants into the lake environment (Jellison et al. 2008). Plans are also underway to set up a soda ash industry in Lake Natron, whose soda ash reserve is estimated to be the single largest worldwide (Kadigi et al. 2014; Philip and Mosha 2012). Lake Natron, however, is a key feeding and breeding ground for the Lesser Flamingos in East Africa, an important tourist destination as well as a Ramsar site. As the lake is the only known breeding ground for the globally 'near-threatened' (according to the 2008 IUCN Red List Category of Threatened Species, Childress et al. 2007) Lesser Flamingos in East Africa, the proposal is facing a strong opposition from conservationists and local communities (Hughes 2008; RAM Team 2008). According to conservation experts, soda ash mining from Lake Natron would be the greatest ecological mistake Tanzania would ever commit. A study by experts from Sokoine University of Agriculture in Tanzania concluded that an investment in tourism, protection of the environment and promotion of alternative local livelihoods can earn the Tanzanian government between $1.28 and 1.57 billion in 50 years (Kadigi et al. 2014; Mwathe 2014). Mining of resources in bio-conservation areas has always been a cause of major disagreement between conservation and the investment groups on the likely environmental and economic benefits (Czech 2003, 2008). Owing to the large soda ash reserve in Lake Natron, it may only be a matter of time before the allure of 'a new era of meaningful resource prosperity' outweighs the environmental concerns and paves way for the soda ash exploitation. Small-scale mining of soda for domestic use also occurs in a number of these lakes. For example, in the Lakes Magadi and Elmentaita, members of the local community harvest soda ash near the lakeshore for sale along the highway as livestock salt licks.

The EARS is presently recognized as a potential source of geothermal energy and petroleum energy. The area has an estimated geothermal energy resource potential of more than 15,000 MW. Despite the high potential, however, only Kenya and Ethiopia have installed a capacity of about 217 MW. Efforts are currently underway to accelerate geothermal development in the countries of EARS (Zemedkun 2012). The western branch of the EARS and the broadly rifted zone of north Kenya and Ethiopia have experienced an unprecedented amount of hydrocarbon exploration in the last 12 years since Heritage Oil made the Turaco-1 discovery in the Albertine Graben in 2002. More than 20 oilfields have been discovered in these two basins. The commercial volume threshold has now been reached, but a pipeline to the Indian Ocean and/or an oil refinery is yet to be built to enable exploitation and export (Davison et al. 2014).

15.2.3 Effects of Catchment Activities on Soda Lakes

The location of saline lakes in closed basins makes them more sensitive to catchment activities than freshwater lakes (Jellison et al. 2008). For many years, the semiarid environment surrounding these lakes was exposed to limited human activities. The growing human population, however, is reflected in an increase in human settlements near these saline alkaline lakes and the associated human resource conflicts. For example, over the last 30 years, the Nakuru Basin, which was a sparsely populated and densely forested area, has been transformed into a heavily settled, highly populated and rapidly urbanizing region (Odada et al. 2003, 2006; Raini 2009). More recently, the area around Lake Elmentaita has witnessed an increase in human settlement and the construction of tourist-class hotels. These settlements practise rain-fed subsistence agriculture in an erosion-prone semiarid environment. Increased settlements, coupled with other catchment activities including overgrazing and excessive clearance of the natural vegetation, are likely to result in adverse hydrological changes with high sediment loads and associated pollutants being washed into the lakes. This is because, after rainfall, runoff volume from overgrazed and/or cleared catchments is usually larger and takes place over a shorter period than it would under natural conditions. Furthermore, the vegetation clearance creates unfiltered runoff: the entire load is deposited in the lakes, resulting into rapid sedimentation (Jellison et al. 2008; Odada et al. 2006; Raini 2009).

15.2.4 Tourism Development Effects on Soda Lakes

The EASL are spectacular, with the most fabulous bird spectacle in the world (Williams 1967). The Lakes Nakuru, Naivasha (Oloidien section), Eyasi, Abijatta and Natron are amongst the Rift Valley lakes that are highly visited by tourists. They are major bird sanctuaries, with the Lesser Flamingo being one of the most important in attracting tourists and therefore of great economic importance to the region. Although the tourism potential and the associated economic gains provide compelling reasons for their conservation, the anticipated increase in visitor number and the associated services present an emerging challenge. For example, the rising demand for infrastructure to support the tourist industry has translated into a scramble for land, especially within the vicinity of these lakes. Most of the land around Lake Elmentaita, for example, is now in private hands. Apart from altering the natural beauty of the landscape and therefore diminishing its tourist potential, the new infrastructure is restricting access to various sections of the lakes.

15.2.5 Challenges to Soda Lakes Associated with Climate Change

In addition to the periodic annual lake level changes, another important characteristic of many EASL is a longer-term (5–10 years) periodic to non-periodic fluctuation in lake levels brought about by extreme changes in rainfall patterns (Olaka et al. 2010). Although the long-term periodicity is a natural phenomenon, these fluctuations appear in the recent past to have been exacerbated by climate change. The predictions suggest extreme rainfall intensities in some regions and prolonged droughts in other areas (IPCC 2007). The resulting major lake level fluctuations cause dramatic changes in the immediate environmental conditions and thus adversely affect the lakes' biological characteristics (Melack 1996; Odada et al. 2006). For example, intense studies carried out in the 1970s in the lakes Elmentaita and Nakuru demonstrated that, during periods of low rainfall, an abrupt increase in salinity resulted in a decrease in phytoplankton abundance and altered zooplankton composition (Melack 1979, 1988; Vareschi 1982; Vareschi and Vareschi 1984). Whereas a number of the EASL here were showing signs of drying up a few years ago, over the last 2 years (2013–2014), many of them have flooded extensively. Lakes Baringo, Nakuru and Elmentaita reclaimed new land areas and displaced human settlements along the flooded shores. Although a number of possible reasons have been advanced to explain this phenomenon (e.g. increased precipitation, underground water flow between lakes, seismic pressure, siltation, etc.), no convincing and acceptable explanation has been provided (Olago et al. 2009). The flooding of these lakes causes ecological problems, whereby the large-scale lake water dilution results in changes in physical, chemical and biological characteristics. In Lake Nakuru, for example, the water chemistry changed significantly: conductivity fell from a high of 96 mS cm^{-1} and salinity from about 63.5 ‰ (Jirsa et al. 2013) to a low of only 7279 μS cm^{-1} and 4.1 ‰, respectively, in September 2013. This resulted in the disappearance of *A. fusiformis* the main food for the Lesser Flamingos from the phytoplankton. This food scarcity caused the Lesser Flamingos to emigrate from this lake.

15.3 Possible Approaches to the Conservation of Soda Lakes

Effectively conserving the soda lakes requires properly identifying the anthropogenic activities that contribute to the degradation of each lake. A number of such factors have been identified and possible remedial measures to help conserve these important resources are suggested (Czech 2008; Harper et al. 1990; Jellison et al. 2008; Kadigi et al. 2014). This identification should be followed by an aggressive campaign to raise awareness about the ecosystem goods and services as well as the value of conserving EASL to the communities using them. Conservation efforts can only be successful if inclusive community-based approaches that recognize efforts by individuals and organizations are employed. Although natural phenomena such as siltation and the evaporative concentration of solutes in the lakes may not be avoidable, human activities known to exacerbate these natural events need to be controlled. Such activities include vegetation clearing in the catchment areas, which enhances siltation, uncontrolled mining of resources from these water bodies and excessive abstraction of water from the rivers flowing into these lakes.

In summary, the following measures can go a long way to conserve these ecosystems:

- Maintaining catchment integrity through controlled land use practices and afforestation of areas vulnerable to soil erosion. This will also help restore the water balance and promote sustainable water management.
- Entrenching conservation ethics amongst catchment residents and promoting sustainable conservation efforts through capacity building and linking conservation with the achievement of development aspirations.
- Monitoring trends in aquatic biodiversity of the soda lakes and environmental conservation efforts in order to evaluate progress and identify new threats.

15.4 The Future of the Soda Lakes

The EASL face a grim future. Though they have faced major changes in the past, mainly associated with climate dynamics, the increase in destructive human activities that is characteristic of most of their catchments will lower their resilience to these perturbations. Land use, effluent discharges and climate change are critical challenges that will shape the future of saline alkaline lakes. Amongst the notable adverse environmental changes associated with increased pollution is a rise in potentially toxic species of algae and Cyanobacteria as well as other aquatic pathogens. One common feature within the catchments is human resource conflicts. The root causes of these conflicts are the ever-increasing human populations, poor enforcement of environmental regulations and the unsustainable exploitation of natural resources (Odada et al. 2006). Unless sustainable methods of natural resource use are put in place, these conflicts are set to increase as more people settle here. Climate change is also expected to increase the complexity of problems associated with altered catchment ecology.

The rising demand for freshwater brought about by human population growth and economic development is expected to lead to an increased diversion of river flow throughout Africa. Reduced inflows into saline lakes will shrink them and alter water quality. For example, Lake Abbe on the border of Ethiopia and Djibouti has shrunk by 67 % since the 1930s as a result of irrigated agriculture in Ethiopia (UNEP 2000). The ongoing construction of the irrigation and hydroelectric power generation dam on the Omo River in Ethiopia, the major river feeding Lake Turkana, will no doubt affect the water availability in this lake. Although most of the key saline alkaline lakes such as Bogoria or Nakuru are not currently affected by diversions, this condition is likely to occur as irrigated agriculture takes root in the catchments of the endorheic basins. In conjunction with surface diversions, groundwater pumping for agricultural purposes, though not presently a major issue in the EARS, will probably lead to desiccation of saline lakes as more boreholes are sunk to provide irrigation water. This situation has been reported in central Mexico, where most of the shallow permanent and temporary salt lakes have disappeared, while the deeper ones have shrunk rapidly because of overpumping of groundwater for irrigation (Alcocer and Escobar 1990).

Other social challenges with the potential to escalate include the political competition for supremacy as well as turf wars between conservation agencies and the various development agencies, which have also contributed in slowing down conservation efforts. For example, periodic politically instigated ethnic clashes have rocked the Lake Nakuru catchment. Such periods are characterized by massive destruction of natural resources such as forests. Conservation efforts come to a standstill. Conflicts between the demands for the expansion of urban settlements and the need to protect the fragile ecosystems that make up the catchments of these lakes have resulted in a complex situation. Unless well managed, this can reverse the conservation gains made in the past. Other conflicts are associated with an increasingly enlightened population demanding a share of the levies collected in conservation areas such as National Parks and reserves.

References

Alcocer I, Escobar E (1990) The drying up of the Mexican Plateau axalapazcos. Salinet 4:34–36

Ballot A, Krienitz L, Kotut K, Wiegand C, Metcalf SJ, Codd AG, Pflugmacher S (2004) Cyanobacteria and cyanobacterial toxins in three alkaline rift valley lakes of Kenya-Lakes Bogoria, Nakuru and Elmenteita. J Plankton Res 26:925–935

Ballot A, Krienitz L, Kotut K, Wiegand C, Pflugmacher S (2005) Cyanobacteria and cyanobacterial toxins in the alkaline crater Lakes Sonachi and Simbi, Kenya. Harmful Algae 4:139–150

Ballot A, Kotut K, Novelo E, Krienitz L (2009) Changes of phytoplankton communities in Lakes Naivasha and Oloidien, examples of degradation and salinization of lakes in the Kenyan rift valley. Hydrobiologia 632:359–363

Baumgarte S (2003) Microbial diversity of soda lake habitats. Dissertation, Gemeinsame Naturwissenschaftlichen Fakultät der Technischen Universität Carolo-Wilhelmina zu Braunschweig

Beadle LC (1932) Scientific results of the Cambridge expedition to the East African Lakes 1930–1. 4. The waters of some East African Lakes in relation to their fauna and flora. J Linn Soc Lond Zool 38:157–211

Burgis M, Morris P (1987) The natural history of Lakes. Cambridge University Press, Cambridge

Childress B, Hughes B, Harper D, Van den Bosche W (2007) The East African flyway and key site network of the lesser flamingo (*Phoenicopterus minor*) documented through satellite tracking. Ostrich 78:463–468

Czech B (2003) Technological progress and biodiversity conservation: a dollar spent, a dollar burned. Conserv Biol 17:1455–1457

Czech B (2008) Prospects for reconciling the conflict between economic growth and biodiversity conservation with technological progress. Conserv Biol 22:1389–1398

Dadheech PK, Mahmoud H, Kotut K, Krienitz L (2014) *Desertifilum fontinale* sp. nov. (Oscillatoriales, Cyanobacteria) from a warm spring in East Africa, based on conventional and molecular studies. Fottea Olomouc 14:129–140

Davison I, Steel I, Taylor M, Beirne EO, Faull T (2014) Exploration: the East African rift system. Geol Explor 11:26–31

Duckworth AW, Grant WD, Jones BE, Van Steenbergen R (1996) Phylogenetic diversity of soda lake alkaliphiles. FEMS Microbiol Ecol 19:181–191

Ebinger CJ, Yamane T, Kelley S (1993) Volcanism and extension between the main Ethiopian and Gregory Rifts. In: Thorweihe U, Schdelmeier H (eds) Geoscientific research in Northeast Africa. Balkema, Rotterdam, pp 301–304

Gichuhi M (2013) Ecological management of the Mau catchment area and its impact on Lake Nakuru National Park. J Agric Sci Technol 15:81–101

Grant WD (2006) Alkaline environments and biodiversity. In: Gerday C, Glansdorff N Extremophiles. Encyclopedia of life support systems (EOLSS), Developed under the Auspices of the UNESCO. Eolss Publishers, Oxford. http://www.eolss.net. Accessed Nov 2015

Grant WD, Mwatha WE, Jones BE (1990) Alkaliphiles: ecology, diversity and applications. FEMS Microbiol Rev 75:255–270

Grant S, Grant WD, Jones BE, Kato C, Li L (1999) Novel archaeal phylotypes from an East African alkaline saltern. Extremophiles 3:139–145

Harper DM, Mavuti KM, Muchiri SM (1990) Ecology and management of Lake Naivasha, Kenya, in relation to climatic change, alien species' introductions and agricultural development. Environ Conserv 17:328–335

Harper DM, Childress RB, Harper MM, Boar RR, Mills SC, Otieno N, Drane T, Vareschi E, Nasirwa O, Mwatha WE, Darlington JPEC, Escuté-Gasulla X (2003) Aquatic biodiversity and saline lakes: Lake Bogoria National Reserve, Kenya. Hydrobiologia 500:259–276

Holmes A (1978) Holmes principles of physical geology, 3rd edn. Van Nostrand Reinhold, New York

Hughes L (2008) Mining the Maasai reserve: the story of Magadi. J East Afr Stud 2:134–164

IPCC (Inter-governmental Panel on Climate Change) (2007) Contribution of working group I to the 4th assessment report of the inter-governmental panel on climate change. In: Solomon S, Qin D, Manning M, Chen Z, Marquis M, Averyt KB, Tignor M, Miller HL (eds) Cambridge University Press, Cambridge, p 996

Jellison R, Williams DW, Timms B, Alcocer J, Aladin NV (2008) Salt lakes: values, threats and future. In: Polunin NVC (ed) Aquatic ecosystems. Cambridge University Press, Cambridge, pp 94–110

Jirsa F, Gruber M, Stojanovic A, Oducr SO, Mader D, Körner W, Schagerl M (2013) Major and trace element geochemistry of Lake Bogoria and Lake Nakuru, Kenya, during extreme drought. Chem Erde-Geochem 73:275–283

Jones BE, Grant WD, Duckworth AW, Owenson GG (1998) Microbial diversity of soda lakes. Extremophiles 2:191–200

Kadigi RMJ, Mwathe K, Dutton A, Kashaigili J, Kilima F (2014) Soda ash mining in lake Natron: a reap or ruin for Tanzania? J Environ Conserv Res 2:37–49. doi:10.12966/jecr.05.01.2014

Kairu JK (1996) Heavy metal residues in birds of Lake Nakuru, Kenya. Afr J Ecol 34:397–400

Kock ND, Kock RA, Wambua J, Kamau GJ, Mohan K (1999) Mycobacterium avium-related epizootic in free ranging lesser flamingos in Kenya. J Wildl Dis 35:29–300

Krienitz L, Kotut K (2010) Fluctuating algal food populations and the occurrence of lesser flamingos (*Phoeniconaias minor*) in three Kenyan rift valley lakes. J Phycol 46:1088–1096

Krienitz L, Ballot A, Kotut K, Wiegand C, Pütz S, Metcalf JS, Codd GA, Pflugmacher S (2003) Contribution of hot spring cyanobacteria to the mysterious deaths of lesser flamingos at Lake Bogoria, Kenya. FEMS Microbiol Ecol 43:141–148

Krienitz L, Bock C, Kotut K, Luo W (2012) *Picocystis salinarum* (Chlorophyta) in saline lakes and hot springs of East Africa. Phycologia 51:22–32

Krienitz L, Dadheech PK, Kotut K (2013) Mass developments of a small sized ecotype of *Arthrospira fusiformis* in Lake Oloidien, Kenya, a new feeding ground for lesser flamingos in East Africa. Fottea Olomouc 13:215–225

Luo W, Kotut K, Krienitz L (2012) Hidden diversity of plankton in the soda lake Nakuru, Kenya, during a phase of low salinity revealed by SSU rRNA gene clone library. Hydrobiologia. doi:10.1007/s10750-012-1310-y

Mari C, Collar C (2000) Pink Africa. The Harvill Press, London

Melack JM (1979) Photosynthesis and growth of *Spirulina platensis* (Cyanophyta) in an equatorial lake (Lake Simbi, Kenya). Limnol Oceanogr 24:753–760

Melack JM (1981) Photosynthetic activity of phytoplankton in tropical African soda lakes. Hydrobiologia 81:71–85

Melack JM (1988) Primary producer dynamics associated with evaporative concentration in a shallow, equatorial soda lake (Lake Elmenteita, Kenya). Hydrobiologia 158:1–14

Melack JM (1996) Recent development in tropical limnology. Verh Internat Verein Theor Angew Limnol 26:211–217

Melack JM, Kilham P (1974) Photosynthetic rates of phytoplankton in East African alkaline saline lakes. Limnol Oceanogr 19:743–755

Millbrink G (1977) On the limnology of two alkaline lakes (Nakuru and Naivasha) in the East African rift valley system in Kenya. Int Rev Ges Hydrobiol 62:1–17

Mwathe K (2014) Fresh lake Natron soda ash push threatens the future of East Africa's Flamingos. Birdlife International, Africa. http://www.birdlife.org/africa/news. Accessed 11 2015

Nelson YM, Thampy RJ, Motelin GK, Raini JA, DiSante CJ, Lion LW (1998) Model for trace metal exposure in filter-feeding flamingos at alkaline rift valley lakes, Kenya. Environ Toxicol Chem 17:2302–2309

Odada EO, Olago DO, Bugenyi F, Kulindwa K, Karimumuryango J, West K, Ntiba M, Wandiga S, Aloo-Obudho P, Achola P (2003) Environmental assessment of the East African rift valley lakes. Aquat Sci 65:254–271

Odada E, Raini J, Ndetei R (2006) Lake Nakuru: experiences and lessons learned brief. www.worldlakes.org/uploads/Nakuru. Accessed 10 Jan 2015

Oduor SO, Schagerl M (2007a) Temporal trends of ion contents and nutrients in three Kenyan rift valley saline–alkaline lakes and their influence on phytoplankton biomass. Hydrobiologia 584:59–68

Oduor SO, Schagerl M (2007b) Phytoplankton primary productivity characteristics in response to photosynthetically active radiation in three Kenyan rift valley saline–alkaline lakes. J Plankton Res 19:1041–1050

Olago D, Opere A, Barongo J (2009) Holocene palaeohydrology, groundwater and climate change in the lake basins of the Central Kenya Rift. J Hydrol Sci 54:765–780. Special issue: Groundwater and Climate in Africa

Olaka LA, Odada EO, Trauth MH, Olago DO (2010) Sensitivity of East African rift lakes to climate fluctuations. J Palaeolimnol 44:629–644

Owino AO, Oyugi OJ, Nasirwa OO, Bennun AL (2001) Patterns of variation in Water bird numbers on four rift valley lakes in Kenya, 1991–1999. Hydrobiologia 458:45–53

Philip JYN, Mosha DMS (2012) Salt lakes of the African rift system: a valuable research opportunity for insight into nature's concerted multi-electrolyte science. Tanzan J Sci 38:1–12

Raini AK (2009) Impact of land use changes on water resources and biodiversity of Lake Nakuru catchment Basin, Kenya. Afr J Ecol 47:39–45

Richardson JL, Richardson AE (1972) History of an African rift lake and its climatic implications. Ecol Monogr 42:499–534

Rosendahl BR (1987) Architecture of continental rifts with special reference to East African. Annu Rev Earth Planet Sci 15:445–503

Schagerl M, Oduor SO (2008) Phytoplankton community relationship to environmental variables in three Kenyan rift valley saline-alkaline lakes. Mar Freshw Res 59:125–136

Schagerl M, Burian A, Gruber-Dorninger M, Oduor SO, Kaggwa MN (2015) Algal communities of Kenyan soda lakes with a special focus on *Arthrospira fusiformis*. Fottea 15:245–257

Schlüter T (1997) Geology of East Africa. Beiträge zur regionalen Geologie der Erde, 27. Gebrüder Bornträger, Berlin, Stuttgart
Talling JF (1965) The photosynthetic activity of phytoplankton in East African lakes. Int Rev Ges Hydrobiol 50:1–32
Talling JF (2011) Some distinctive subject contributions from tropical Africa to fundamental science of inland waters. Inland Waters 1:61–73
Talling JF, Wood BR, Prosser VM, Baxter MR (1973) The upper limit of photosynthetic productivity by phytoplankton: evidence from Ethiopian soda lakes. Freshw Biol 3:53–76
RAM Team (2008) Ramsar Advisory Mission No. 59: Tanzania (2008). Mission Report
UNEP (2000) Global environmental outlook. Earthscan, London
Vareschi E (1978) Ecology of Lake Nakuru (Kenya). I Abundance and feeding of lesser flamingo. Oecologia 32:11–35
Vareschi E (1982) The ecology of Lake Nakuru (Kenya). III Abiotic factors and primary production. Oecologia 55:81–101
Vareschi E, Vareschi A (1984) The ecology of Lake Nakuru (Kenya). IV Biomass and distribution of consumer organisms. Oecologia 61:70–82
Willén E (2011) Cyanotoxin production in seven Ethiopian rift valley lakes. Inland Waters 1:81–91
Williams WD (1967) The chemical characteristics of lentic surface waters: a review. In: Weatherley AH (ed) A review of the Australian inland waters and their fauna. Australian National University Press, Canberra, p 287
Williams WD (1981) The limnology of Saline lakes of Western Victoria. Hydrobiologia 81–82:233–259
Williams WD (2002) Environmental threats to Salt lakes and the likely status of inland saline ecosystems in 2025. Environ Conserv 29:154–167
Zemedkun MT (2012) Overview of geothermal resource exploration and development in the East African rift system. Presented at short course VII on exploration for geothermal resources, organized by UNU-GTP, GDC and KenGen, at Lake Bogoria and Lake Naivasha, Kenya, Oct 27–Nov 18 2012

Appendix: Lakes of the East African Rift System

Country	Lake	Latitude	Longitude	Type	Salinity [‰]	pH	Alkalinity [meq L^{-1}]
Djibouti	Assal	11.658961	42.406998	Hypersaline	158–277[a,d]	NA	NA
Ethiopia	Abaja/Abaya	6.311204	37.847671	Fresh-subsaline	<0.1–0.9[j,t]	8.65–9.01[j,t]	8.8–9.4[j,t]
Ethiopia	Ābay Hāyk'	7.940916	38.357706	NA	NA	NA	NA
Ethiopia	Abbe/Abhe Bad	11.187832	41.784325	Hypersaline	160[a]	8[f]	NA
Ethiopia	Abijata/Abiyata	7.612998	38.597603	Hypo-mesosaline	4.41–43.84[f,j,t]	9.3–10.0[f,t]	102.3–33.0[f,j]
Ethiopia	Afambo	11.416091	41.682701	NA	NA	NA	NA
Ethiopia	Afrera (Giulietti)/ Afdera	13.255318	40.899925	Hypersaline	169.49[g]	6.55[g]	NA
Ethiopia	Ara Shatan	8.044060	38.351119	NA	NA	NA	NA
Ethiopia	Arenguade/Arenguadi	8.695324	38.976388	Sub-hyposaline	1.58–5.81[r,t]	9.75–10.30[r,t]	34.8–46.0[t]
Ethiopia	Awassa/Awasa	7.049786	38.437614	Fresh-subsaline	<0.1–0.8[f,j,t]	8.70–8.92[f,t]	3.8–8.4[f,t]
Ethiopia	Basaka/Beseka/ Metahara	8.876931	39.869957	Sub-hyposaline	0.89–5.3[c,j,t]	9.40–9.45[j,t]	27.5–46.5[j,t]
Ethiopia	Bischoftu/Bishoftu	8.741815	38.982053	Fresh-subsaline	<0.1–0.6[t]	9.49–9.53[t]	14.6[t]
Ethiopia	Budamada	7.096464	38.090773	NA	NA	NA	NA
Ethiopia	Caddabassa	10.208165	40.477524	NA	NA	NA	NA
Ethiopia	Chamo	5.840081	37.560654	Fresh-subsaline	<0.1–1.0[j,t]	8.90–9.20[j,t]	12.0–13.2[j,t]
Ethiopia	Chelekleka	8.769979	38.972268	NA	NA	NA	NA
Ethiopia	Chitu/Chiltu/Chittu	7.404516	38.420191	Meso-hypersaline	19.01–64.16[f,j,t]	10.10–10.50[f,j,t]	22.6–715.0[f,t]
Ethiopia	Gemeri/Gummare/ Gamari	11.532507	41.664848	Fresh-subsaline	<0.10–0.70[a,f]	8[f]	NA
Ethiopia	Guda/Babogaya/ Babogay/Pawlo	8.787114	38.993297	Freshwater	<0.1–0.2[t]	9.08–9.36[t]	8.6[t]
Ethiopia	Hora/Biete Mengist/ Mengest	8.762769	38.991666	Fresh-subsaline	0.06–1.00[f,t]	8.80–9 02[f,t]	21.0–26.8[f,t]
Ethiopia	Kilotes/Kilole	8.802975	39.083319	Fresh-hyposaline	<0.1–6.9[f,r,t]	8.75–9.60[f,t]	2.8–63.4[f,t]
Ethiopia	Koftu/Kuftu	8.834188	39.048815	Fresh	<0.1[t]	7.93[t]	1.8[t]
Ethiopia	Koka	8.410564	39.091301	Fresh	<0.1–0.2[j,t]	7.73–9.00[t]	2.3–2.6[j,t]
Ethiopia	Kuriftu	8.779183	39.000428	Fresh	<0.1[t]	7.76–8.40[t]	2.6[t]
Ethiopia	Langano/Langeno	7.594621	38.754845	Fresh-subsaline	<0.1–2.4[f,j,t]	8.95–9.46[f,j,t]	12.0–14.6[f,t]
Ethiopia	Mechelera	7.040416	38.086138	NA	NA	NA	NA
Ethiopia	No name 1	7.887698	38.386545	NA	NA	NA	NA
Ethiopia	No name 2	7.785068	38.360982	NA	NA	NA	NA
Ethiopia	No name 3	7.779455	38.394885	NA	NA	NA	NA
Ethiopia	Shalla/Shala	7.470730	38.522758	Hyposaline	9.96–18.10[f,j,t]	9.65–10.10[f,j]	162.2–218.0[f,j]

(continued)

M. Schagerl (ed.), *Soda Lakes of East Africa*, DOI 10.1007/978-3-319-28622-8

Country	Lake	Latitude	Longitude	Type	Salinity [‰]	pH	Alkalinity [meq L^{-1}]
Ethiopia	Tilo	7.063586	38.094892	Hyposaline	8.83[w]	NA	178.9–186.5[l]
Ethiopia	Ziqualla	8.542149	38.855073	Fresh	0.24[e]	7.91[e]	1.5[e]
Ethiopia	Ziway/Zway/Zwai	7.980358	38.826942	Fresh	<0.1–0.4[f,j,t]	8.23–8.70[f,j,t]	3.8–4.0[j,t]
Kenya	Baringo	0.652385	36.065111	Fresh-subsaline	<0.1–0.7[f,o,t]	8.14–9.10[f,o,t]	4.4–5.5[f,t]
Kenya	Bogoria/Hannington	0.252084	36.101317	Mesosaline	29.96–35.99[b,f,s,t]	10.3–10.5[f,o,s,t]	54.6–992.8[s,t]
Kenya	Elmentaita	−0.440135	36.242466	Hypo-mesosaline	3.11–28.42[f,n,p,t]	9.4–10.9[f,h,n,t]	4.8–352.1[n,t]
Kenya	Lake 92	0.479144	36.095624	NA	NA	NA	NA
Kenya	Logipi	2.229319	36.554432	NA	NA	NA	NA
Kenya	Magadi	−1.923105	36.263266	Hypersaline	69.63–140.10[i,t]	9.8[t]	NA
Kenya	Naivasha	−0.770393	36.353703	Fresh	<0.1[f,t]	9.51–7.80[f,t]	3.2[t]
Kenya	Nakuru	−0.359113	36.092434	Hypo-hypersaline	5.34–62.10[f,n,q,t]	9.8–10.9[f,h,n,t]	16.5–540.3[n,s,t]
Kenya	Oloidien	−0.813648	36.276455	Fresh-hyposaline	<0.10–4.11[f,t]	9.10–10.01[f,t]	6.7–73.0[h,t]
Kenya	Solai	0.058794	36.148782	Fresh-subsaline	<0.10–2.29[f,u]	8.6[u]	885.7[u]
Kenya	Sonachi	−0.782666	36.261992	Sub-mesosaline	1.98–21.75[f,m,t]	9.90–10.17[f,t]	27.5–437.4[t]
Kenya	Turkana	3.709412	36.031036	Fresh-subsaline	0.45–2.89[a,f,t]	9.3–9.7[f,t]	19.5–24.5[f,v]
Tansania	Babati	−4.294375	35.721245	NA	NA	NA	NA
Tansania	Balangida	−4.351204	35.349770	Hypersaline	63.74[h]	9.7[h]	925.6[h]
Tansania	Balangida Lehu	−4.683192	35.230293	NA	NA	NA	NA
Tansania	Basodesh	−4.297004	35.128136	NA	NA	NA	4.2[k]
Tansania	Basotu	−4.385083	35.079490	Fresh	<0.1[f]	6.60–8.58[f,k]	4.5[k]
Tansania	Big Momela	−3.223280	36.909454	Hyposaline	8.75–10.52[h,n]	10.4[h]	168.2–239.0[h,n]
Tansania	Burunge	−3.884807	35.880661	Fresh	0.26[t]	9.37[t]	1.1[t]
Tansania	El Kekhotoito	−3.233605	36.877993	Subsaline	1.52[h]	9.7[h]	42.1[h]
Tansania	Empakaai/Embagai	−2.913039	35.840807	Hyposaline	7.73[h]	10.1[h]	183.3[h]
Tansania	Eyasi	−3.599772	35.122490	Subsaline	2.39[t]	9.5[h]	116.4[h]
Tansania	Gawali	−4.490318	35.058711	NA	NA	NA	NA
Tansania	Ghama	−4.434263	35.085287	Hyposaline	6.56[f]	9.37–10.50[f,k]	163.0–163.3[h,k]
Tansania	Gidabuid	−4.303892	35.125905	Hyposaline	4.01[f]	8.48–8.50[f,k]	27.4[k]
Tansania	Gidaburk	−4.394862	35.090193	Hyposaline	6.10[f]	9.00–10.35[f,k]	87.4[h,k]
Tansania	Gidamuniud	−4.412473	35.067441	Hyposaline	NA	9.0–9.5[k]	42.8–323.0[k]
Tansania	Gidamur	−4.322480	35.104181	Saline	3.91[f]	9.36–10.30[f,h]	70.0[h,k]
Tansania	Kindai	−4.843996	34.733849	Subsaline	1.77[h]	8.3[h]	2.6[h]
Tansania	Kitangiri	−4.048522	34.347954	NA	NA	NA	NA
Tansania	Kusare	−3.225866	36.880826	Subsaline	0.78[h]	9.3[h]	38.7[h]
Tansania	Laja	−4.336721	35.090968	Fresh	<0.1[k]	9.4[k]	85.2[k]
Tansania	Lekandiri/Lekandiro	−3.210823	36.895440	Subsaline	1.80[h]	10.1[h]	50.6[h]
Tansania	Lgarya	−2.990502	35.033441	Mesosaline	28.79[h]	9.8[h]	554.0[h]
Tansania	Magad	−3.194250	35.538440	Hyposaline	5.02[h,n]	10.2[h,n]	84.1[h,n]
Tansania	Manyara	−3.601142	35.805588	Fresh-hyposaline	0.17–4.38[n,t]	9.20–10.13[n,t]	3.5–78.0[n,t]
Tansania	Mikuyu	−4.562222	34.944906	Hyposaline	5.33[f]	8.54[f]	NA
Tansania	Nadnakid	−4.309418	35.113286	Hyposaline	3.23[h]	8.9–10.5[h,k]	30.2[h,k]
Tansania	Natron	−2.386090	36.011753	Hyposaline	12.27[t]	9.46[t]	13.1[t]
Tansania	Ndobot	−4.317556	35.107573	NA	NA	9.5[k]	144[k]
Tansania	Reshitani/Rishateni	−3.231811	36.907589	Hyposaline	7.73–10.05[h,n]	10.1[h,n]	164–233[h,n]

(continued)

Country	Lake	Latitude	Longitude	Type	Salinity [‰]	pH	Alkalinity [meq L^{-1}]
Tansania	Silver Sea	−3.230232	36.867149	Fresh	<0.1[h]	9.1[h]	14[h]
Tansania	Singida	−4.785837	34.756508	Subsaline	2.42[h]	8.6[h]	5.7[h]
Tansania	Small Momela	−3.226931	36.896157	Subsaline	1.72[h]	9.8[h]	46.7[h]
Tansania	Tulusia	−3.210952	36.906877	Hyposaline	8.33[h]	10.4[h]	188.8[h]

Data compiled from various sources (see below). Classification according to: <0.5 freshwater; 0.5 < 3.0 subsaline (transition zone between freshwater and saline); 3 < 20 hyposaline; 20 < =50 mesosaline; and >50 hypersaline. For some lakes, only conductivity data were available. These were converted via: Salinity [‰] = Conductivity [mS cm^{-1}]–1.509 ($n = 695$, $r = 0.975$, unpublished data from various soda lakes). Fresh = freshwater, NA = no data available

Sources

[a]Baxter RM (2002) Lake morphology and chemistry. In: Tudorancea C, Taylor WD (eds) Ethiopian Rift Valley lakes. Backhuys Publishers, Leiden: 45–60

[b]Beadle LC (1932) Scientific results of the Cambridge Expedition to the East African lakes, 1930–31. 4. The waters of some East African lakes in relation to their fauna and flora. J Linn Soc Zool 38:157–211

[c]Belay EA (2009) Growing lake with growing problems: integrated hydrogeological investigation on Lake Beseka, Ethiopia. Dissertation, Friedrich-Wilhelms-University Bonn

[d]Boutaiba S, Hacene H, Bidle KA, Maupin-Furlow JA (2011) Microbial Diversity of the Hypersaline Sidi Ameur and Himalatt Salt Lakes of the Algerian Sahara. J Arid Environ 75(10):909–916. doi:10.1016/j.jaridenv.2011.04.010

[e]Degefu F, Herzig A, Jirsa F, Schagerl M (2014) First limnological records of highly threatened tropical high-mountain crater lakes in Ethiopia. Trp Conserv Sci 7(3):365–381

[f]Gasse F, Talling JF, KIlham P (1983) Diatom assemblages in East Africa: classification, distribution and ecology. Rev Hydrobiol Trop 16(1):3–34

[g]Getahun A (2001) Lake Afdera: A threatened saline lake in Ethiopia. SINET: Ethiop J Sci 24(1):127–131

[h]Hecky RE, Kilham P (1973) Diatoms in alkaline, saline lakes: Ecology and geochemical implications. Limnol Oceanogr 18(1): 53–71

[i]Jones BF, Eugster HP, Rettig SL (1977) Hydrochemistry of the Lake Magadi basin, Kenya. Geochim Cosmochim Acta 41:53–72

[j]Kebede E, Zinabu G-M, Ahlgren I (1994) The Ethiopian Rift Valley Lakes: chemical characteristics of a salinity-alkalinity series. Hydrobiologia 288:1–12

[k]Kilham P, Cloke PL (1990) The evolution of saline lake waters: gradual and rapid biogeochemical pathways in the Basotu Lake District, Tanzania. Hydrobiologia 197:35–50

[l]Lamb LL, Leng MJ, Lamb HF, Mohammed MU (2000) A 9000-year oxygen and carbon isotope record of hydrological change in a small Ethiopian crater lake. The Holocene 10:167. doi: 10.1191/095968300677444611

[m]McIntyre S, Melack JM (1992) Meromixis in an equatorial African soda lake. Limnol Oceanogr 27:595–609

[n]Melack JM, Kilham P (1974) Photosynthetic rates of phytoplankton in East African alkaline saline lakes. Limnol Oceanogr 19:743–755

[o]Odada EO, Onyando J, Obhudho PA (2006) Lake Baringo. Experience and lessons learned Brief. http://iwlearn.net/iw-projects/1665/experience-notes-and-lessons-learned/lakebaringo_2005.pdf/view. Accessed November 2015

[p]Oduor SO, Schagerl M (2007) Temporal trends of ion contents and nutrients in three Kenyan Rift Valley saline alkaline lakes and their influence on phytoplankton biomass. Hydrobiologia 584:59–68. doi:10.1007/s10750-007-0605-x

[q]Peduzzi P, Gruber M, Gruber M, Schagerl M (2014) The virus's tooth: cyanophages affect an African flamingo population in a bottom-up cascade. The ISME Journal, advance online. doi:10.1038/ismej.2013.241

[r]Prosser MV, Wood RB, Baxter RM (1968) The Bishoftu Crater Lakes: A bathymetric and chemical study. Arch Hydrobiol 65(3):309–324

[s]Gruber-Dorninger M (2015) East African Soda Lakes—Biogeochemistry and interactions between organismic compartments. PhD, University of Vienna

[t]Schagerl M (unpublished results)

[u]Yasindi AW (2001) The ecology of planktonic ciliates in tropical lakes of East Africa. Dissertation, University of Waterloo, Canada

[v]Yuretich RF, Cerling TE (1983) Hydrochemistry of Lake Turkana, Kenya: mass balance and mineral reactions in an alkaline lake. Geochim Cosmochim Acta 47:1099–1109

[w]Zinabu GM, Taylor WD (1997) Bacteria-chlorophyll relationships in Ethiopian lakes of varying salinity: are soda lakes different? J Plankton Res 19(5):647–654

Taxonomy Index

M. Schagerl (ed.), *Soda Lakes of East Africa*, DOI 10.1007/978-3-319-28622-8

Subject Index

M. Schagerl (ed.), *Soda Lakes of East Africa*, DOI 10.1007/978-3-319-28622-8

D

G

The manufacturer's authorised representative in the EU is Springer Nature Customer Service Centre GmbH, Europaplatz 3, 69115 Heidelberg, Germany. If you have any concerns regarding our products, please contact ProductSafety@springernature.com

Printed and bound by CPI Group (UK) Ltd, Croydon CR0 4YY
05/10/2026
02228127-0020